AF815822

Organic and Inorganic Low-Dimensional Crystalline Materials

NATO ASI Series

Advanced Science Institutes Series

A series presenting the results of activities sponsored by the NATO Science Committee, which aims at the dissemination of advanced scientific and technological knowledge, with a view to strengthening links between scientific communities.

The series is published by an international board of publishers in conjunction with the NATO Scientific Affairs Division

A	**Life Sciences**	Plenum Publishing Corporation
B	**Physics**	New York and London
C	**Mathematical and Physical Sciences**	D. Reidel Publishing Company Dordrecht, Boston, and Lancaster
D	**Behavioral and Social Sciences**	Martinus Nijhoff Publishers
E	**Engineering and Materials Sciences**	The Hague, Boston, Dordrecht, and Lancaster
F	**Computer and Systems Sciences**	Springer-Verlag
G	**Ecological Sciences**	Berlin, Heidelberg, New York, London,
H	**Cell Biology**	Paris, and Tokyo

Recent Volumes in this Series

Volume 162—Quantum Uncertainties: Recent and Future Experiments and Interpretations
edited by William M. Honig, David W. Kraft, and Emilio Panarella

Volume 163—Thin Film Growth Techniques for Low-Dimensional Structures
edited by R. F. C. Farrow, S. S. P. Parkin, P. J. Dobson, J. H. Neave, and A. S. Arrott

Volume 164—Techniques and Concepts of High-Energy Physics IV
edited by Thomas Ferbel

Volume 165—Relativistic Channeling
edited by Richard A. Carrigan, Jr., and James A. Ellison

Volume 166—Incommensurate Crystals, Liquid Crystals, and Quasi-Crystals
edited by J. F. Scott and N. A. Clark

Volume 167—Time-Dependent Effects in Disordered Materials
edited by Roger Pynn and Tormod Riste

Volume 168—Organic and Inorganic Low-Dimensional Crystalline Materials
edited by Pierre Delhaes and Marc Drillon

Series B: Physics

Organic and Inorganic Low-Dimensional Crystalline Materials

Edited by

Pierre Delhaes

CNRS, Paul Pascal Center
University of Bordeaux I
Talence, France

and

Marc Drillon

EHICS
Strasbourg, France

Springer Science+Business Media, LLC

Proceedings of a NATO Advanced Research Workshop on
Organic and Inorganic Low-Dimensional Crystalline Materials,
held May 3–8, 1987,
in Minorca, Baleares, Spain

Library of Congress Cataloging in Publication Data

NATO Advanced Research Workshop on Organic and Inorganic Low-Dimensional Crystalline Materials (1987: Minorca, Spain)
Organic and inorganic low-dimensional crystalline materials / edited by Pierre Delhaes and Marc Drillon.
p. cm.—(NATO ASI series. Series B, Physics; v. 168)
"Proceedings of a NATO Advanced Research Workshop on Organic and Inorganic Low-Dimensional Crystalline Materials, held May 3–8, 1987, in Minorca, Baleares, Spain"—T.p. verso.
"Published in cooperation with NATO Scientific Affairs Division."
Bibliography: p.
Includes index.
ISBN 978-1-4899-2093-5 ISBN 978-1-4899-2091-1 (eBook)
DOI 10.1007/978-1-4899-2091-1

1. Solid state physics-Congresses. 2. Solid state chemistry—Congresses. 3. One-dimensional conductors—Congresses. 4. Superconductors—Congresses. 5. Magnetic materials—Congresses. I. Delhaes, Pierre. II. Drillon, Marc. III. North Atlantic Treaty Organization. Scientific Affairs Division. IV. Title. V. Series.
QC176.A1N28 1987
530.4′1—dc19 87-28561
CIP

Originally published by Plenum Press, New York in 1987
Softcover reprint of the hardcover 1st edition 1987

SPECIAL PROGRAM ON CONDENSED SYSTEMS OF LOW DIMENSIONALITY

This book contains the proceedings of a NATO Advanced Reserach Workshop held within the program of activities of the NATO Special Program on Condensed Systems of Low Dimensionality, running from 1983 to 1988 as part of the activities of the NATO Science Committee.

Other books previously published as a result of the activities of the Special Program are:

Volume 148 INTERCALATION IN LAYERED MATERIALS
edited by M. S. Dresselhaus

Volume 152 OPTICAL PROPERTIES OF NARROW-GAP LOW-DIMENSIONAL STRUCTURES
edited by C. M. Sotomayor Torres, J. C. Portal, J. C. Maan, and R. A. Stradling

Volume 163 THIN FILM GROWTH TECHNIQUES FOR LOW-DIMENSIONAL STRUCTURES
edited by R. F. C. Farrow, S. S. P. Parkin, P. J. Dobson, J. H. Neave, and A. S. Arrott

PREFACE

The research of unitary concepts in solid state and molecular chemistry is of current interest for both chemist and physicist communities. It is clear that due to their relative simplicity, low-dimensional materials have attracted most of the attention. Thus, many non-trivial problems were solved in chain systems, giving some insight into the behavior of real systems which would otherwise be untractable.

The NATO Advanced Research Workshop on "Organic and Inorganic Low-Dimensional Crystalline Materials" was organized to review the most striking electronic properties exhibited by organic and inorganic sytems whose space dimensionality ranges from zero (0d) to one (1d), and to discuss related scientific and technological potentials. The initial objectives of this Workshop were, respectively:

i) To research unitary concepts in solid state physics, in particular for one dimensional compounds,

ii) To reinforce, through a close coupling between theory and experiment, the interplay between organic and inorganic chemistry, on the one hand, and solid state physics on the other,

iii) To get a salient understanding of new low-dimensional materials showing "exotic" physical properties, in conjunction with structural features.

To reach these goals, the meeting was organized in three different steps as reported in the table of contents. The first sessions were devoted to plenary lectures on the different classes of low-dimensional materials presenting either a mixed-valence or a magnetic character. Topics relative to organic as well as inorganic materials were rewieved; this is rather unusual in standard conferences, since researches on conducting and magnetic compounds are traditionally segregated. These presentations were followed by a series of review papers on the physical properties, mainly on magnetic aspects (phase transitions, linear and nonlinear excitations, Haldane conjecture,..etc). One special session was organized on the new superconducting materials, the organic ones as well as the new ternary oxides which present superconductivity above the boiling point of liquid nitrogen. In addition to the first series of review papers, several lectures were given on special topics in connection with the main subject. These lectures were devoted to the problem of charge transfer in biological systems, studies on large metallic clusters, which are regarded as "zero-dimensional systems", and potential applications of mixed valence compounds, including equally organized monomolecular layer systems (Langmuir-Blodgett films). The purpose of these special sessions was both to increase the convergence between researches whose goals are different, and to get further information before the round table discussions.

For a complementary level of information, each participant presented one poster relevant to these topics; the new results on magnetic chains as well as on conducting materials are included as short contributions in this book.

The last part was devoted to the round tables on prospects and developments. The proposed subjects (new conducting systems, new magnetic compounds, ferromagnets, competition between magnetism - superconductivity and charge density waves, potential technology directions) were presented by the group leaders, then discussed in small groups of about ten persons. Schemes for designing new types of materials were reported. Summaries of these discussions are given in the last part of this volume.

The concluding remarks were given by Prof. V.J. Emery, who emphasized the interest in using large equipments, such as different neutron reactors or synchrotron sources, for analyzing crystal structures (long-range order but also local organization, structural phase transitions,..etc). An outstanding example of new unexpected and very stimulating findings is the discovery of the new "high T_c" superconductors.

We wish to mention that this NATO workshop was prepared and organized thanks to the positive contribution of the organizing committee (D. Beltran (Spain), R. Carlin (USA), L.J. de Jongh (Netherlands), and J. Rouxel (France)), and the enthusiasm during the preparation of study-group reports of the leaders (A. Barraud (France), D. Brom (Netherlands), R. Carlin (USA), H.J. Keller (Germany) and F. Palacio (Spain)).

Finally, we wish to thank the panel on "Condensed Systems of Low Dimensionality" of NATO's Scientific Affairs Division for the financial and administrative support which made this workshop possible.

P. Delhaes and M. Drillon

CONTENTS

PLENARY LECTURES

COMMUNICATIONS ON MOLECULAR CONDUCTORS

COMMUNICATIONS ON MAGNETIC CHAINS

REPORTS ON ROUND TABLES

SOLID STATE CRYSTAL CHEMISTRY OF 1D INORGANIC CHAINS

Jean Rouxel

Laboratoire de Chimie des Solides, U.A. CNRS n° 279
Universite de Nantes - 2, Rue de la Houssiniere
44072 Nantes Cedex 03

INTRODUCTION

Low dimensional solids can be regarded as the result of a stacking of slabs or a juxtaposition of fibers. This particular situation directly reflects a very high anisotropy in the chemical bonding. Strong iono-covalent or metallic bonds are responsible of the cohesion of slabs or fibers, whereas these units are separated by weak interactions mostly of the van der Waals type.

The van der Waals gap, between slabs or fibers, is bonded on each side by atomic layers of the same nature, anionic layers in the case of 2D or 1D conductors. This has two consequences :

i) this surrounding of metal atoms layers or rows by the non metal elements implies a chemical formula MX_n, with high non metal content, specialy for 1D phases, and going from 2D to 1D dimensionality will indeed correspond to the increase of n

ii) for a structural type to be stable, it is necessary that the bond through the van der Waals gap, always weak, be however sufficiently strong to maintain the atomic arrangement against the coulombian repulsions between similar anionic layers. The occurrence of a low dimensional solid corresponds to a compromising between these terms. The more ionic the materials, the stronger the tendancy to exhibit low dimensionality through repulsion and hence to obtain a separation of the sheets and fibers, but also the less stable they become. The limit of low dimensionality is the intrinsic instability that it implies.

Now, if we come back to chemistry, oxides are two much ionic and thus the rutile structure is observed in most of the MO_2 transition metal oxides and not the layered forms of the parent MS_2 chalcogenides. Layered oxides can only exist for the highest oxidation states of transition metals with a strong polarization of the electrons of the anions towards the inside of the slabs (MoO_3 case), if $(OH)^-$ groups are present allowing hydrogen bonds to stabilize the structures (layered silicates) and also if extra cations are located between the slabs separating negative layers and introducing attractive coulombic forces and this is typically the case of Na_xCoO_2 and related phases.[1] Between the two much ionic oxides and the quasi metallic mostly three dimensional tellurides, sulfides and selenides represent one of the most favorable domains in chemistry where 1D and 2D derivatives may be found.

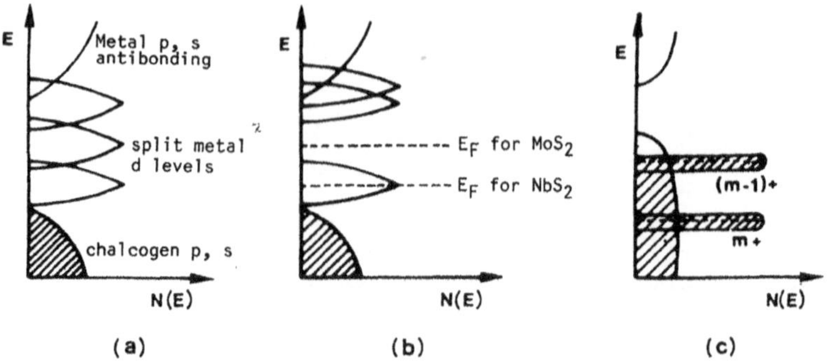

Fig. 1 – Simplified band scheme for transition metal chalcogenides : (a) general, (b) half-filled dz^2 band in NbS_2 and filled in MoS_2, (c) situation at the end of a period (FeS_2).

But now, this evolution in the direction of sulfides or selenides brings new problems which arise from a redox competition as can be understood when looking at the usual and general band scheme for transition metal chalcogenides (figure 1). Between a valence band essentially sp anionic in character and antibonding levels mostly formed from the corresponding cationic levels, the d orbitals of the metal, split by the crystal field, play an essential role concerning the physical properties and also the behavior towards intercalation reactions. But the position of these d levels also governs redox processes inside the structure itself, which may induce drastic structural modifications. In front of sulfur a maximum cationic oxidation state of four is generally observed leading to d^0 configuration and no electronic conductivity in the case of IVA elements with octahedral symmetry and a broad t_{2g} band. A d^1 configuration in the case of V, Nb, Ta is responsible of a lowering of the symmetry to a trigonal prismatic situation with a half-filled $dz^2(a'_1)$ band and a metallic behavior. Immediately after, with a d^2 configuration and same symmetry, MoS_2 and WS_2 are diamagnetic semi-conductors. Going further to the right we should come back to octahedral symmetry with half-filled t_{2g} band. But the expansion of the d orbitals decreases, reducing the chance of overlapping and leading to localized levels. The d levels are progressively lowered and at a certain point will enter the sp anionic valence band. If a filled d level enters the valence band there is no drastic change in the electrical properties. But if an empty d level is in this situation, it will fill up at the expense of the anionic sp band, at the higher part of which holes appear.[2] It means that anionic pairs may form from this oxidation of anions. This is the case upon reduction of Fe^{3+} to Fe^{2+} in pyrite or marcasite with the occurrence of $(S_2)^{2-}$ pairs. This redox competition between cationic and anionic species will still become more important on going from sulfides to selenides, the top of the valence band being at a higher level.

TRICHALCOGENIDES CHEMISTRY AND C.D.W. PHENOMENON

All of the transition metal trichalcogenides have a common structural framework, i.e. $[MX_3]$ trigonal prismatic chains made up of $[MX_6]$ units.[3] Irregular $[MX_6]$ trigonal prisms are stacked on top of each other to form the $[MX_3]$ chains, which run parallel to the b axis of a usually monoclinic unit cell. This represents a direct transposition, through chalcogen content increase from the NbX_2 or TaX_2 bidimensional structures to a one dimensional one. A very important point is that the trigonal prism is irregular. A chalcogen pair has been formed which allows to assign the general formulation $M^{4+}(S_2)^{2-}S^{2-}$ to these compounds. But things may be much more complicated due to the possible length change of the pair. This is once more the result of the already mentioned redox competition between cationic levels and the top of the valence band.

$NbSe_3$ represents the most important compound as well as the most complicated structure.[4] Three different chains are found (Figure 2) according to three different values taken by the Se–Se bond length in their triangular basis (respectively 2.37, 2.48, 2.91 Å). The first one corresponds to a true $(Se_2)^{2-}$ pair and reflects the situation $Nb^{4+}(Se_2)^{2-}Se^{2-}$. The longest distances in the two others mean that we have started to populate the antibonding level of the Se_2 pair. Less than one electron per metal atom is left. The Se–Se pair behaves as an electron reservoir which takes or gives less or more electrons to the adjacent metallic chain and governs its properties. This situation illustrates the proximity of the Nb^{4+} level and of the antibonding Se–Se level in the energy diagram. The two charge density waves[5] in $NbSe_3$ (T_c = 145 and 59 K as shown in Figure 3) occur in the two first types of chains, the third chain being probably an insulating one. However the transverse distances between one metal atom and chalcogen atoms belonging to neighboring chains (translated from b/2) are not long enough to exclude any bonding. Vibrational spectroscopy studies show that the strength of these lateral Nb–Se bonds is a substantial fraction (0.15 to 0.28) of that of the intrachain (M–X) bonds. $NbSe_3$ is not a perfect 1D material, but rather a 2D one with a high anisotropy within its slabs. Only a portion of the Fermi surface nests and $NbSe_3$ remains semi-metallic below the transition.

In order to get a more one dimensional compound it is necessary to increase the ionicity of the chemical bonds. One has to change niobium for the more electropositive tantalum and selenium for the more electronegative sulfur.[6] TaS_3, which was already known with an orthorhombic unit cell, testifies effectively of the increased 1D character in two ways :

i) a transition at 210 K which goes as far as a metal to insulator transition,[7]

ii) important pretransitional effects manifested by diffuse lines which condense at the transition into spots between the main Bragg rows of spots. These pretransitional effects mean that the C.D.W. exists dynamically above the transition but without phase coherence from one chain to the other.

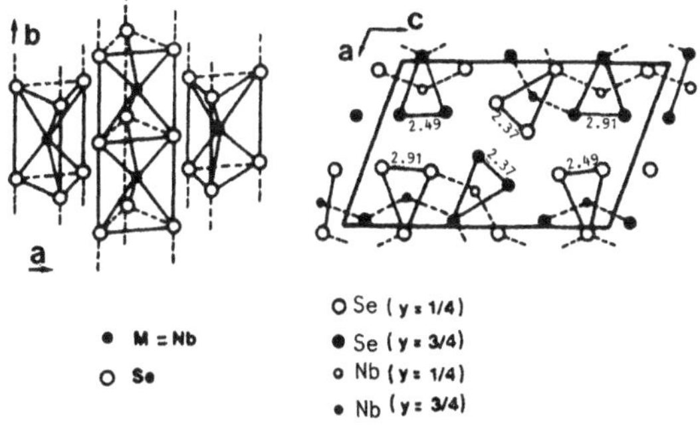

Fig. 2. $NbSe_3$ structural type.

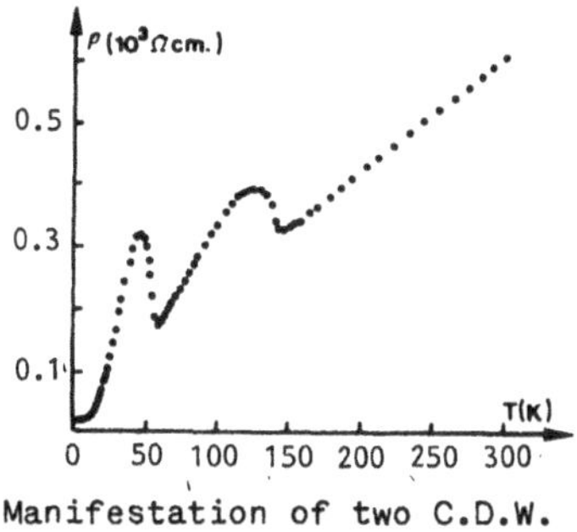

Fig. 3. Manifestation of two C.D.W. in $NbSe_3$.

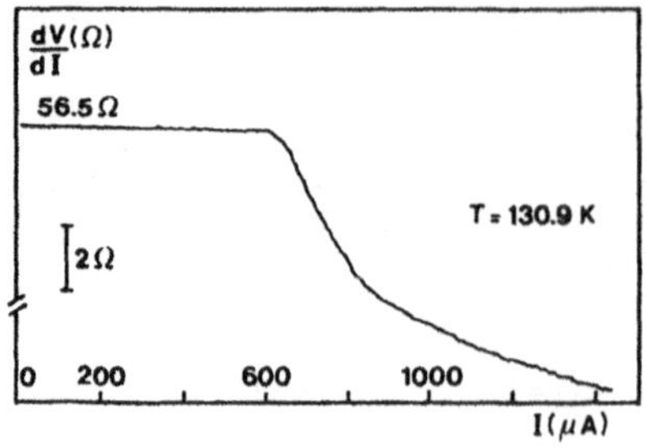

Fig. 4. Differential resistance dV/dI as a function of applied current for different temperatures.

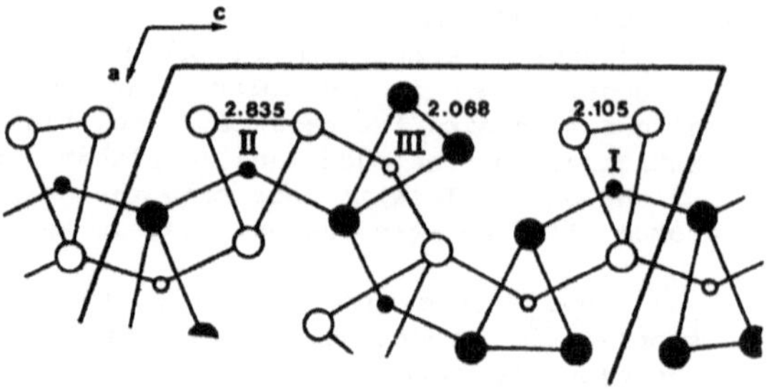

Fig. 5. The structure of TaS_3 projected along [010]

A very new thing that was brought by $NbSe_3$ is the possible depinning of the C.D.W. Above a threshold field, non linear effects[8] are observed which are attributed to a collective motion of the C.D.W.. This is manifested by a break (Figure 4) in the voltage-intensity characteristics of the crystal.[9] dV/dI is constant at the beginning, the Ohm law is followed, then appears an additional current brought by the sliding motion of the C.D.W.. Noise is also generated in the sample, i.e. a broad band noise and a periodic one which can be decomposed into its fundamental and harmonics through a Fourier analysis. A first problem was to try to find some indications about the role of factors possibly influencing the threshold value of the field. Starting from one material one can either change the association of the various chains or modify the electronic properties through chemical alterations.

The first point concerns the existence of polytypes. Like in the case of most low dimensional compounds polytypism is expected in trichalcogenides. It should associate the same chains, or different ones in new arrangements. In spite of carefull studies involving different temperatures and pressures of selenium, no polytype was observed in the case of $NbSe_3$. However trials were successfull for TaS_3. This is once more in agreement with the increased low dimensional character of TaS_3 as compared to $NbSe_3$. In addition to the orthorhombic form a monoclinic TaS_3 was found.[10] It has chains with 2.835, 2.068 and 2.105Å S-S distances (Figure 5). The c parameter of the orthorhombic structure (which was not known in details) is equal to b monoclinic. This corresponds to one [TaS_6] prism height or equivalently to a Ta-Ta distance along the chain axis. The orthorhombic unit cell[11] includes four zig-zag arrangements of TaS_3 trigonal prisms (Figure 6). Resistivity measurement on the monoclinic form shows also a metal to non metal transition, at 240 K.[12] But the superstructure is now incommensurate with the initial sublattice. In both cases (orthorhombic and monoclinic form) a depinning of the C.D.W. is possible, but the threshold field is clearly dependant on the gap to commensurability (Table I). In o.TaS_3 with a commensurate distortion, depinning is much easier than in m.TaS_3 with an incommensurate one[13].

A way to modify the electronic properties of low dimensional solids is to change their chemical composition by performing either intercalation between the slabs or fibers or substitution inside these

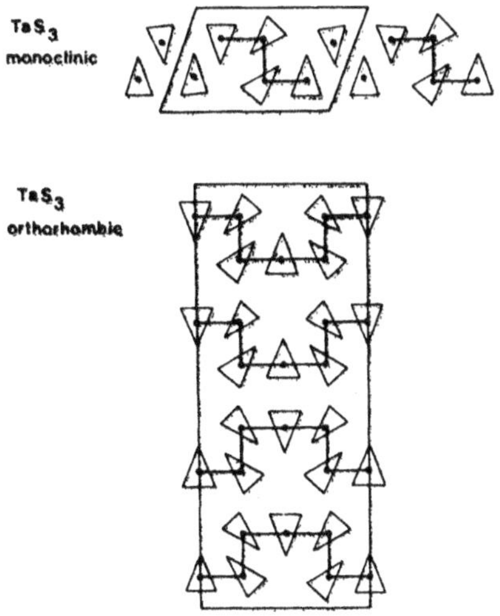

Fig. 6. Tantalum position for both types of TaS_3.

Table I. Distortion wave vectors of $NbSe_3$ and TaS_3.

$NbSe_3$	T_2 = 59 K 0, 0.241(3), 0 1/2, 0.259(3),1/2	T_1 = 145 K 0, 0.243(3), 0	T > 145 K No visible diffuse scattering
TaS_3 (monoclinic)	T_2 = 160 K 0, 0.254(3), 0	T_1 = 240 K 0, 0.254(3), 0	T > 240 K Diffuse scattering
TaS_3 (orthorhombic)	T ⩽ 130 K Commensurate	T ⩽ 210 K Incommensurate 1/2, 1/8, 0.250(3) (roughly)	T > 240 K Diffuse scattering

units. Intercalation seems to be a mild and convenient way to change the position of the Fermi level, and thus the q vector of the distortion, through the guest-host electronic exchange. An adaptation of the q vector to a commensurate situation could be, in particular, envisioned. Indeed intercalation is easy in trichalcogenides. But is leads to badly organized samples, not suitable for electrical measurements. Now, it is possible to introduce chemical impurities in the chains. By doing so, the C.D.W. will experience a disordered distribution of non equivalent potentials. The Nb atoms were substituted with either isoelectronic (Ta) or non isoelectronic (Ti, Zr) impurities. It was noted[14] that the threshold field depends upon the impurity concentration c as follows : $E_T \propto C^2$ for a weak pinning with isoelectronic impurities, and $E_T \propto C$ for a strong pinning (non isoelectronic impurity). The role of disorder created by irradiation has also been studied.[15] A great advantage of this technique is that at low irradiation doses, the number of defects is proportional to the irradiation dose. But the radiation damage was found to produce a different behavior than substitutional impurities in $NbSe_3$: the threshold field increases linearly with defect concentration and, even at large doses, the radiation does not destroy the C.D.W. unlike in the case of chemical impurities.

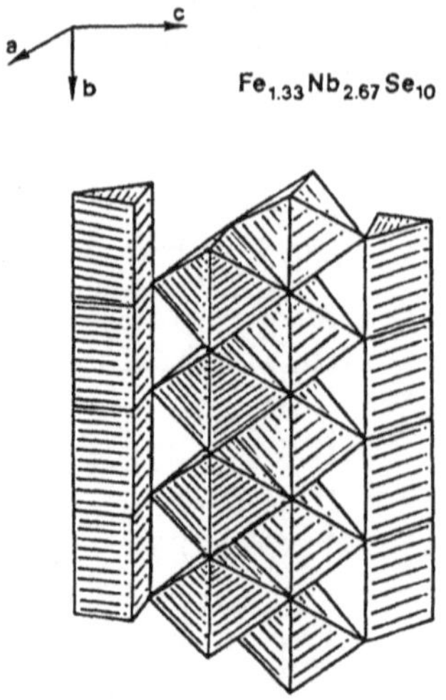

Fig. 7. Condensation of two types of chains in the $FeNb_3Se_{10}$ structural type.

1D CONDUCTORS FROM CONDENSATION OF CHAINS

In the course of the studies about the role of chemical impurities, new compounds were isolated. $FeNb_3Se_{10}$ is the prototype of this new class of 1D conducting chains.[16] It has a very particular structural type (Figure 7) with two types of chains running along the b axis of a monoclinic unit cell, each chain being repeated twice. One chain corresponds to a trigonal prismatic frame of selenium atoms around a first type of niobium atom, the other develops an octahedral arrangement of selenium atoms around both iron atoms and a second type of niobium atoms. Iron is octahedrally coordinated whereas niobium presents both types of coordination. This compound has two interests :
(i) the trigonal prismatic chain is exactly similar to the chain with the shortest Se-Se bond length in $NbSe_3$ (d(Se-Se) = 2.34 Å). Because $FeNb_3Se_{10}$ exhibits a metal to non-metal transition at 140 K with a distortion having the same distortion vector as the first C.D.W. in $NbSe_3$,[17] this can be taken as a proof that this C.D.W. was really seated in the corresponding chain in $NbSe_3$.
(ii) The $FeNb_3Se_{10}$ structural type can be regarded as the result of condensations in a given structure of various chains, each of them corresponding to the usual situation of a given element. Isotypic compounds have been prepared with other transition metal elements such as V, Cr, Ta which allows to modulate the properties increasing the pinning of the C.D.W. and the influence of an additional Anderson type localization effect upon the transition. But the concept has been nicely extended to condensations of the trigonal prismatic chains with chains having a square planar section (Pd) or chains built up from tetrahedra (Ni).[18]

1D CONDUCTING CHAINS STABILIZED BY COUNTER IONS

The mineral patronite, VS_4, presents a 1D organization with S_4 rectangles, built up from two true $(S_2)^{2-}$ pairs, rotating around -V-V- chains and giving a rectangular antiprismatic coordination to the metal atoms. Vanadium is at the 4+ oxidation state, but there is a d^1-d^1

pairing which gives a diamagnetic semi-conducting situation. Attempts to change the properties by applying a pressure have been unsucessfull. Heating the samples does not allow to remove this Peierls-type distortion before the compound decomposes.

It was of interest to prepare the Nb and Ta homologues of VS_4. More expanded d orbitals could help to get an electronic delocalization along the chains. In fact, even with the help of pressure it was not possible to get NbS_4 and TaS_4. This is probably due to an increased ionicity which destabilizes the structure. The way to manage with that is to prepare selenides instead of sulfides, or to separate the chains by counter ions.

1.1. <u>Halogenotetrachalcogenides</u> : $NbSe_4$ and $TaSe_4$ cannot be made. But adding some iodine one gets $(MSe_4)_nI$ phases with M = Nb, Ta, n = 1/3, 1/2, 10/3 etc.[19] The needle shaped crystals of $(NbSe_4)_3I$ crystallize with the tetragonal symmetry (space group P4/mnc). The structure (Figure 8) is built up from $NbSe_4$ chains aligned along the c axis and separated by iodine columns. The niobium coordination is a rectangular antiprism made up of true Se-Se pairs (d(Se-Se) = 2.34 Å). Along the metallic chains, long and short Nb-Nb distances alternate in the sequence of two long (3.25 Å) for one short (3.06 Å) distance. From one compound to the other this sequence changes according to the population of the counter ion, and also to the way the Se_4 rectangles rotate around the metallic chain. Two successive Se_4 rectangles are rotated by an angle of 45° which corresponds to a minimization of the energy of a MSe_8 "molecular" group (minimization of the repulsion between their occupied π and π^* orbitals, along with the obtention of the best set of symmetry adapted orbitals to interact with the metal orbitals). But the relative displacement of such successive rectangles can refer to a right- or left-handed rotation, which can be described with the help of arrows pointing towards the midle of a Se-Se pair. Thus $(NbSe_4)_3I$ is a 123432 tetrachalcogenide, $(TaSe_4)_2I$ is 1234 and $(NbSe_4)_{10/3}I$ is a 1234565432 chalcogenide. In the M-Se combination all the metal orbitals are concerned, except dz^2 which remains isolated and built a dz^2 band along the chain. This band governs the electronic properties of the system.[20] Assuming that each iodine takes one electron, the average number of d electrons per metal is (n-1)/n with a corresponding filling of the dz^2

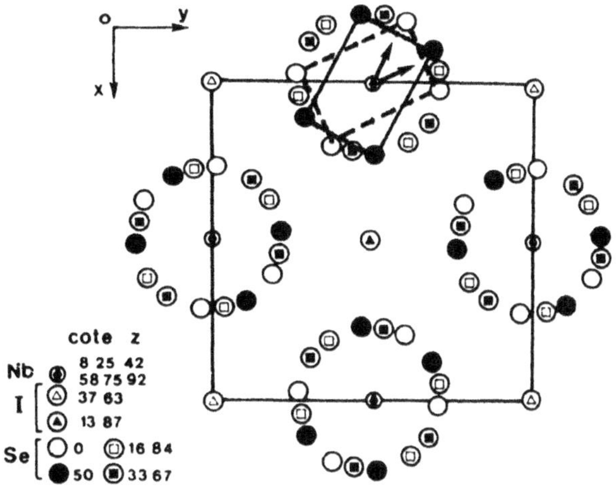

Fig. 8. $(NbSe_4)_3I$ structural type.

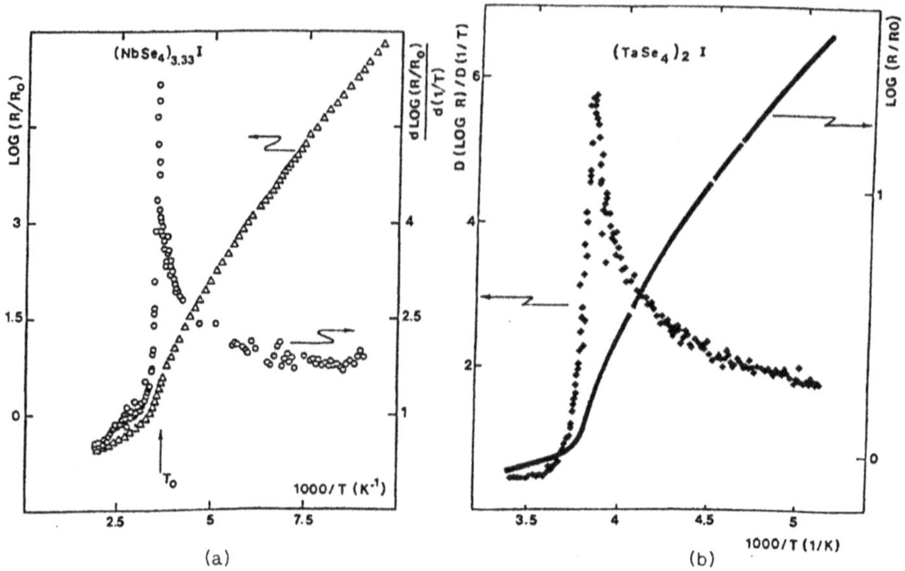

Fig. 9. C.D.W. transitions in (a) $(NbSe_4)_{10/3}I$ and (b) $(TaSe_4)_2I$.

band of f = (n−1)/2n, possibly leading to a distortion increasing the repeat distance by a factor 1/f. As n increases the dz^2 band of $(MSe_4)_nI$ becomes closer to half filled. Vanadium tetrasulfide constitutes finally an example of the $n \to \infty$ limit. It has consequently the 2c insulating Peierls distortion. More precisely, tight-binding calculations have been performed which modify slightly the rough approach mentioned just above. They are in rather good agreement with the observed distortion. For example the calculations on $(TaSe_4)_2I$ lead to an estimation of $2k_F$ = 0.88c* and the distortion is observed at 0.94 c* in ref[21] and 0.91 c* in ref[22].

Two materials, $(TaSe_4)_2I$ and $(NbSe_4)_{10/3}I$, have been deeply investigated due to the nice C.D.W. type transitions they show.[23] Figure 9 represents the logarithmic derivative curves $d\log(R/R_0)/d^{1}(T)$ with sharp peaks at the transitions (respectively 285 and 263 K). The very good single crystals of $(TaSe_4)_2I$ allowed precise experiments. In the non linear regime, i.e. above the threshold field associated to the depinning of the C.D.W., the current is the summ of an ohmic contribution and a contribution due to the moving of the C.D.W. with velocity v :

$$J = J_{ohm} + n\ e\ v,$$

where n e, is the electronic concentration in the bands affected by the C.D.W. gap. The velocity v can be written as $v = \lambda\nu$ with $\lambda = 2\pi/q$. ν is manifested through the periodic noise which corresponds to a modulation of the C.D.W. velocity when passing the pinning centers. Measuring J gives $J_{CDW} = n\ e\ v$. A linear relationship between J_{CDW} and ν should be checked :

$$J_{CDW} = (n\ e\ \lambda)\ \nu,$$

which is beautifully demonstrated in the case of $(TaSe_4)_2I$ at two different temperatures on figure 10.

$(NbSe_4)_{10/3}I$ shows similar effects below 285 K. But, in addition, an hysteresis phenomenon takes place in the 285–220 K temperature range.

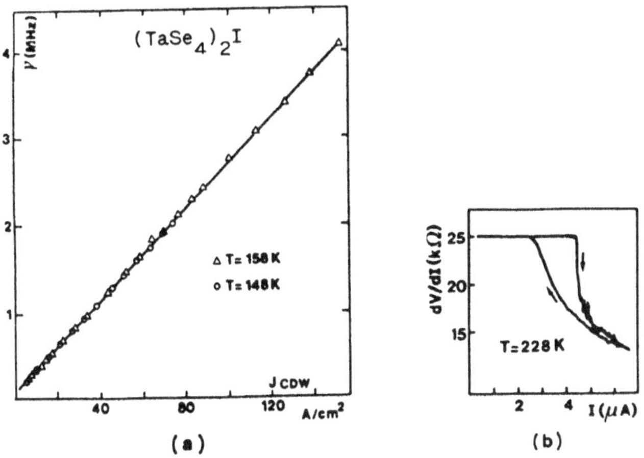

Fig. 10. (a) linear relationship between J_{CDW} and in $(TaSe_4)_2I$,
(b) memory effect in the monohmic behavior of $(NbSe_4)_{10/3}I$.

There is a threshold field when the field is increased and a different one when it is decreased. This type of memory effect has also been observed in o.TaS_3 [24] and in the blue bronze $K_{0.30}MoO_3$.[25]

All these $(MSe_4)_nI$ crystals are obtained in similar conditions from a mixture of the elements in evacuated and sealed silica tubes heated during a few days at temperatures ranging from 450 to 600°C. Due to their closely related structures intergrowth phenomena are to be expected, associating small domains of one phase to a matrix corresponding to an other one. This would simply result in local rearrangements of the iodine positions and slight variations in the sequence of metal-metal bonds along the chains. Such defects may have been depicted through electron diffraction studies, however the nice fringe figures that were obtained could also correspond to faults in the rotational orientation of the Se_4 rectangles. More likely it is revealed by the electrical behavior of some $(NbSe_4)_3I$ crystals. Two types of crystals are found, which cannot be distinguished by X-rays techniques. Both types of crystals have a room temperature resistivity of about 1 Ω.cm. Type I behavior is characterized by a very small semi-conducting gap below 275 K. In type II crystals, after the same decrease of the gap at 275 K, the resistivity increases again at lower temperatures showing a behavior very similar to that of $(NbSe_4)_{10/3}I$ at these temperatures. This suggests that type II crystals could contain micro-domains of the $(NbSe_4)_{10/3}I$ composition, these domains acting as local defects perpendicular to the chain axis. Such a situation may shed some light on the real nature of defects pinning the C.D.W. state. As well as in trichalcogenides there is a high probability that such defects are not represented by chemical impurities but also by chemical defects such as a local variation of the Se-Se bond lengths or structural defects due to intergrowth mechanisms. The transition at 275 K is a structural transition which changes the P4/mnc space group to its $P42_1c$ subgroup. Now the inversion center is missing. It is a non displacive second order transition which changes the Nb-Nb bonds sequence and hence the electrical behavior.

Finally the tetrachalcogenides remind us of some organic one dimensional conductors[26] such as $(TSeT)I_x$ with a stacking of TSeT molecules and iodine columns.

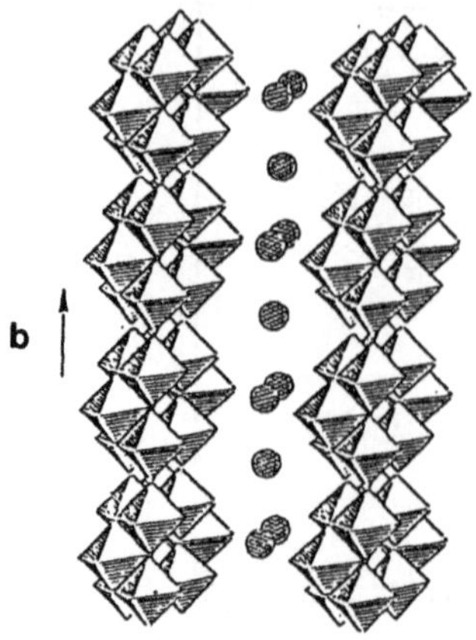

Fig. 11. Structural type of blue bronzes $A_{0.30}MoO_3$ with chains of Mo(2) and Mo(3) MoO_6 octahedra separated by A^+ ions.

The so-called blue bronzes and related phases represent a symmetrical situation in respect to the $(MX_4)_nI$ derivatives. With a general formulation A_xTO_n (A = alkali metal, T = transition metal) they are particular transition metal oxygen bronzes in which the TO_n framework, instead of being a three dimensional tunnel arrangement, is made of chains separated by the A^+ ions (Figure 11). The potassium blue bronze $K_{0.30}MoO_{3}$ was known since the works of Wold,[27] Graham and Wadsley[28] and Bouchard[29] to present a metal to non metal transition around 180 K. It is only recently that the conductivity anisotropy of the material and the C.D.W. origin of the transition were recognized.[30,31,32] Non linear effects were soon observed.[32,33] Single crystals of $K_{0.30}MoO_3$ are usually grown by the electrolytic reduction of a K_2MoO_4-MoO_3 melt at 550°C. Monoclinic crystals elongated along the b-axis are obtained. The structure is built up with groups of ten distorted MoO_6 octahedra that are linked together via corner sharing along the b axis and the 102 direction in order to form infinite slabs. Three different Mo sites allow to reduce the two dimensional "geometrical" features to a one dimensional electronic situation. Indeed the 4d electrons are mainly associated to the so called Mo(2) and Mo(3) sites which are involved in infinite chains of MoO_6 octahedra, parallel to the b axis.

Diffuse streaks in shape of flattened cigars, situated at $\pm q_b\vec{b}^*$ from planes with odd k indices are observed with q_b = 0.28 at 292°K. Below 180 K they condense in well defined superstructure reflections with wave vector $q = 0\vec{a}^* + q_b\vec{b}^* + 0.5\vec{c}^*$.[31] The q wave vector is strongly temperature dependant and $K_{0.30}MoO_3$ is found to undergo a lock-in incommensurate to commensurate transition at 110 K.[34,35,36]

The non linear transport properties found in the Peierls semi-conducting state are attributed to C.D.W. transport. Besides the incommensurate to commensurate lock- in at 110 K, $K_{0.30}MoO_3$ shows particular features such as low frequency voltage fluctuations with a slow time scale, hysteresis, time dependent and memory effects. A predominant mechanism of nucleation and motion of C.D.W. discommensurations and/or dislocation was proposed as a possible explanation of these phenomena.[33]

This field of investigation has been extended to $Rb_{0.30}MoO_3$ as well as other molybdenum bronzes such as the violet bronzes (AMo_6O_{17}) or the red bronzes ($A_{0.33}MoO_3$). New titanium bronzes[37] are also being considered (they have a different structural type).

TETRATELLURIDES OR PENTACHALCOGENIDES

Going down to tellurium, a $NbTe_4$ chain becomes stable. $NbTe_4$ crystallizes with the tetragonal symmetry (a = b = 6.499 ; c = 6.837 Å). The average structure is quite simple with $NbTe_4$ chains running along the c axis. Niobium is surrounded by a square antiprism of tellurium. Additional spots in the diffraction pattern where attributed to a 2a ×2b × 3c superstructure.[38] It was shown later[39,40] that the superstructure along c is an incommensurate one. In fact the reflection pattern of $NbTe_4$ presents two classes of reflections : (i) a group of strong reflections at the nodes of the reciprocal lattice spanned by the vectors $\vec{a}^*$, $\vec{b}^*$, $\vec{c}^*$, (ii) a group of weaker reflections which can be described with the above reciprocal lattice and a fourth vector :

$$\vec{q}' = 1/2\ \vec{a}^* + 1/2\ \vec{b}^* + 0.309\ \vec{c}^*,$$

according to an integral linear combination $\vec{S} = h\vec{a}^* + k\vec{b}^* + l\vec{c}^* + m\vec{q}'$. This modulated structure recently solved[41,42] shows that the modulation wave resides mainly on the Nb atoms which have an amplitude of 0.33 Å. Between each pair of Nb atoms there are four Te atoms with the same z coordinate, the average position of which are related to each other by the fourfold axis (equal displacements along c, equal magnitudes but different orientations in the a b plane). As a whole the square of Te atoms can have a displacement along c. In addition there are the shrinking and expansion of the square and its rotation around c. This form of displacement established by van Schmalen was already suggested by Böhn and von Schnering. Whangbo[43] gives an attractive explanation. The experimental value close to c*/3 means that dz^2 bands along the Nb chains are 1/3 filled. Now, an ideal $NbTe_4$ structure made of undistorted Te square planes implies the formal oxidation state $Nb^{4+}(Te_2^{2-})_2$ and a half-filled dz^2 band. The discrepancy between the two values should be resolved through Te^{2-} species, which are produced by lengthening the Te–Te bonds. As the formula unit is represented by $Nb_{24}(Te_2)_{48}$ for the real unit cell (2a, 2a, 3c), we may assume an increase of 4Te–Te bonds out of 48, corresponding to a change from $4(Te_2)^{2-}$ to $8Te^{2-}$. in this way 16d electrons, instead of 24, are available for 24Nb. This would lead to a 1/3 filled dz^2 band. The fundamental question which remains is to know the origin of the distortion. A C.D.W. transition is probalby not to be considered due to the importance of the distortion.

NEW TRENDS : CHAIN–CHAIN STABILIZATION, WAVED STRUCTURES, CLUSTER CONDENSATION

A first evolution, starting from $(MX_4)_nI$ phases would be to replace I^- ions by other types of counter ions. Replacement of iodine by other halogens may lead to two types of predictable :

(i) If the structural type is maintained it will be less stable (increased overall charge on the chains due to a more electronegative halogen and, at the same time smaller separation between chains). This may allow to solubilize the $(NbX_4)_\infty$ chains in solvants. Considering the work done by J.M. Tarascon and F. Di Salvo on the $M_2Mo_6Se_6$ phases to get Mo_6Se_6 chains dispersion in organic solvant,[44] a recombination could be attempted to get a mixed chain structure associating (MX_4) and Mo_6Se_6

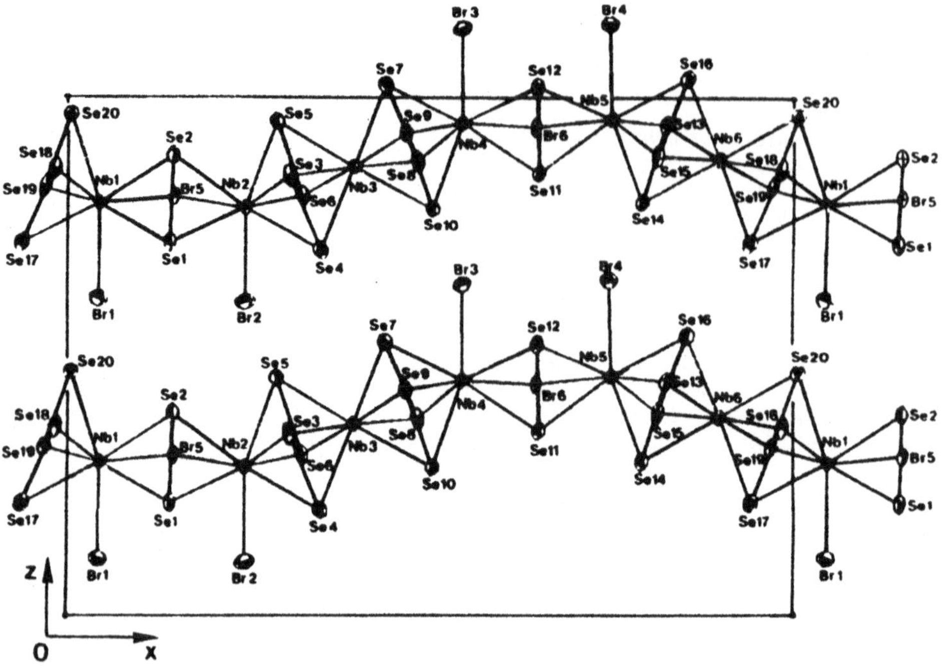

Fig. 12. Structural type of $Nb_6Se_{20}Br_6$.

chains. All the trials have been, up to now, unsuccessfull. Soft chemistry process such as the action of chlorine diluted in argon at room temperature allows to remove iodine (ICl and ICl_3 are formed) but the MX_4 framework is also attacked. Direct reaction between the elements do not lead to $(MX_4)_nCl$ phases, only a few bromine derivatives, badly crystallized, have been obtained. This is clearly related to the loss of stability of the structure.

(ii) Working at higher temperature one can expect new structural arrangements in which the halogen becoming more electronegative will tend to participate directly in the coordination polyhedra of the metal by pushing apart the selenium framework. This has been realized in a series of materials in which niobium atoms directly bonded to the halogen are found along with niobium atoms that are still in $(Se_4)_2$ rectangular antiprismatic coordination. In fact a more correct approach is to consider M_2X_4 bipyramids. $Nb_6Se_{20}Br_6$[45] with a nice waved structure (Figure 12), $Nb_4Se_{16}Br_2$[46], appear then to derive from a partial condensation of these bipyramids in Nb_3Se_9 and Nb_4Se_{12} groups separated by the halogen-bonded niobium atoms. The same bipyramid are found isolated in $Nb_2S_2Cl_2$ or Nb_2Se_8 where they are separated by chlorine atoms or by a Se_5 group. On the other hand they give rise to an infinite condensation in VS_4 where the MX_4 chains are isolated or in $(MSe_4)_nI$ where they are separated by halogen atoms (Figure 13).

A way to stabilize chains is to separate them by counter ions. An other solution would be to play with two types of chains. This is illustrated by the very new compound $Nb_3Se_{10}Br_2$, the structure of which[47] is built from an alternation of $NbSe_4$ chains centered along a 2-fold axis and $NbSe_3Br_2$ chains running along a 2_1 axis parallel to the former one (Se-Br exchanges on certain sites explains the rather complicated formulation) studies on these materials are in progress.

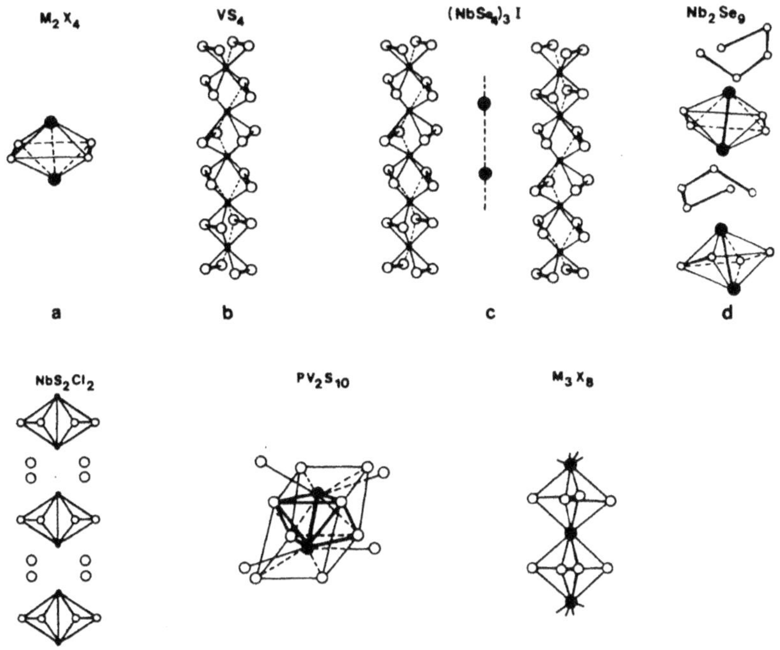

Fig. 13 - 1D structures from infinite or finite condensation of Nb_2X_4 bipyramids.

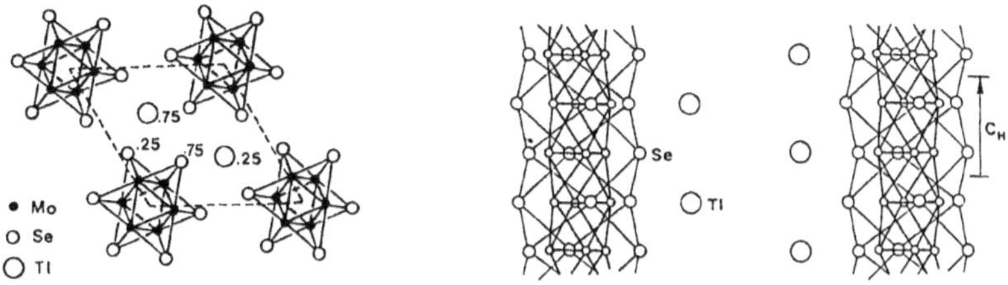

Fig. 14 - $Tl_2Mo_6Se_6$ structural type.

The Chevrel phases[48] are built up from Mo_6 octahedral clusters enclosed in a pseudo cube of chalcogen atoms. These Mo_6X_8 units, usually isolated, can accept between them extra cations and particularly rare earths cations. But it has been shown also that the clusters can condensate to form infinite chains. This is the case of $M_2Mo_6Se_6$ phases (Figure 14), the chains being separated by M^+ cations (M = Tl, K, Rb, Cs...). Among these compounds, $Tl_2Mo_6Se_6$[49] which is metallic, is still being extensively investigated, largely in connection with its superconductive transition around 3 K. The $M_2Mo_6X_6$ compounds, although no sharp Peierls transition has been seen, are remarkable 1D materials because of the very high anisotropy of their transport properties[50] and also because of the anisotropy of their polarized reflectance spectra.[51] The architecture of chain-like $M_2Mo_6X_6$ compounds is very attractive because of its extreme sensitivity to the interchain coupling which suggests many chemical developments.

CONCLUSIONS

This paper was an attempt to discuss some pathways in the search of quasi 1D inorganic conductors. Isolated chains are to be found rather in sulfides or selenides chemistry. An increased ionic character will increase the low dimensionality but at the same time the compound may become unstable. In that case counter ions are necessary to stabilize the structure. This is represented by the $(MX_4)_nI$ phases as well as by the so-called blue bronzes $K_{0.30}MoO_3$. The stabilization can also be reached by a juxtaposition of different chains or by their condensation in a 2D arrangement ($FeNb_3Se_{10}$). Finally chains can be formed by an infinite condensation of clusters.

Chalcogen pairs are often time present in these structures. With a variable length, which means a variable population of their antibonding state, they can play the role of electron reservoirs able to give or take more or less electrons to the adjacent metallic chains. They often govern the electronic properties of the adjacent metallic chain. Let us come back, as an illustration of the importance of that point, to the $NbSe_3$ situation. Using a tight binding scheme based on the extended Hückel method, calculations have been performed[52] on mononuclear [$NbSe_6$] and binuclear [Nb_2Se_9] building blocks. Each is examined in a D_{3h} symmetry corresponding to a regular frame with an equilateral triangular base and also a C_{2v} symmetry which is the actual symmetry of the chains. The d block orbitals of these building blocks are represented in Figure 15. With the D_{3h} symmetry we have the classical splitting with a'_1 (z^2) and $e'(x^2-y^2, xy)$ below $e''(xz, yz)$. The highest orbital set derived from the six selenium atoms is above a block of filled Se levels. In this scheme an $NbSe_6$ complex corresponds to Nb^{5+} with empty d levels and a hole in the Se e'' levels. When a distortion occurs a Se-Se bond is formed with a breaking of the degeneracy of the $(e'')^3$ configuration. Two electrons remain on the Se-Se bonding level, one goes on the a'_1 level reducing the metal from +5 to +4 oxidation state. The latter situation corresponds to type III chain in $NbSe_3$ (Se-Se = 2.37 Å). The former is close to chain I (Se-Se = 2.91 Å). Chain II with Se-Se = 2.48 Å represents an intermediate situation. The question arises to know why the

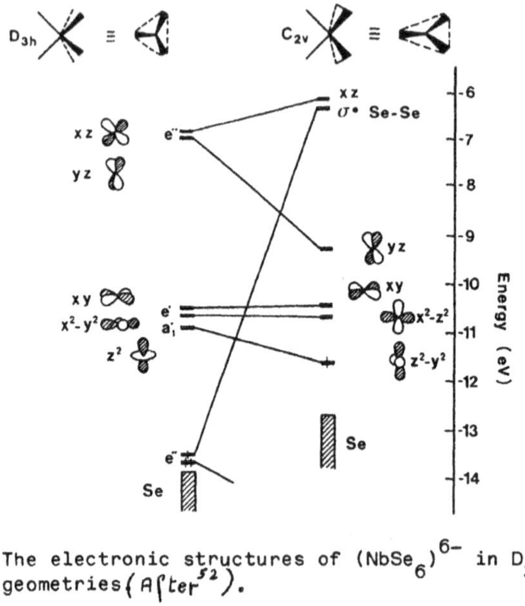

Fig. 15 – The electronic structures of $(NbSe_6)^{6-}$ in D_{3h} and C_{2v} geometries (After[52]).

situation corresponds to a segregation of various chains in $NbSe_3$ and to a segregation of different polyhedra along the same chain in MoS_3. The easy and quasi elastic electronic exchange between metals and chalcogen pair is certainly also concerned with the catalytic activity of these materials.

REFERENCES

1. C. Fouassier, C. Delmas and P. Hagenmuller, Mat. Sci. Eng. 31:97 (1977).
2. F. Jellinek, Inorganic Sulfur Chemistry, G. Nickless Ed., Elsevier, Amsterdam (1968).
3. A. Meerschaut and J. Rouxel, in "Crystal Chemistry and Properties of Materials with quasi 1D structures, J. Rouxel Ed., D. Reidel (1986).
4. A. Meerschaut and J. Rouxel, J. Less Common Met. 39:197 (1975) ; J.L. Hodeau, M. Marrezio, C. Roucau, R. Ayroles, A. Meerschaut, J. Rouxel and P. Monceau, J. Phys. C : Solid State Phys. 11:4117 (1978).
5. P. Haen, P. Monceau, B. Tissier, G. Wayssand, A. Meerschaut, P. Molinié and J. Rouxel, Proc. 14th Int. conf. on Low Temp. Phys., Otaniemi, Finland, 5:445 (1975).
6. E. Bjerkelund and A. Kjekshus, Z. anorg. allg. Chem. B, 328:235 (1964).
7. T. Sambongi, K. Tsutsumi, Y. Shiozaki, K. Yamaya and Y. Abe, Sol. State Comm. 22:729 (1977).
8. P. Monceau, N.P. Ong, A.M. Portis, A. Meerschaut and J. Rouxel, Phys. Rev. Lett. 37:602 (1976).
9. R.M. Fleming, Phys. Rev. B 22(12):5606 (1980) ; P. Monceau, J. Richard and M. Renard, Phys. Rev. B 25(2):931 (1982).
10. A. Meerschaut, L. Guémas and J. Rouxel, J. Solid State Chem. 36:118 (1981).
11. A. Meerschaut and J. Rouxel, in "Crystal Chemistry and Properties of Materials with quasi 1D structures, J. Rouxel Ed., D. Reidel (1986).
12. C. Roucau, R. Ayroles, P. Monceau, L. Guémas, A. Meerschaut and J. Rouxel, Phys. Stat. Sol. 62:483 (1980).
13. M. Nunez Regueiro, Thesis, Grenoble (1979) ; K. Hasegawa, A. Maeda, S. Uchida and S. Tanaka, Physica, 117-118B:599 (1983) ; T. Takoshima, M. Ido, K. Tsutsumi, T. Sambongi, S. Honna, K. Yamaya and Y. Abe, Solid State Comm. 35:911 (1980) ; A. Zettl, G. Grüner and A.H. Thompson, Phys. Rev. Lett. 47(1):64 (1980).
14. J.W. Brill, N.P. Ong, J.C. Eckert, J.W. Savage, S.K. Khanna and R.B. Somsano, Phys. Rev. B 23(4):1517 (1981).
15. W.W. Fuller, P.M. Chaikin and N.P. Ong, Solid State Comm. 39:547 (1981) ; P. Monceau, J. Richard and R. Lagnier, J. Phys. C, 14:2995 (1981) ; L. Zuppiroli, Radiation Effects 54 (1982).
16. A. Meerschaut, P. Gressier, L. Guémas and J. Rouxel, Mat. Res. Bull. 16:1035 (1981) ; R. Cava, V.L. Hines, A.D. Mighell and R.S. Roth Phys. Rev. B 24(6):3634 (1981).
17. A. Ben Salem, A. Meerschaut, H. Salva, Z.Z. Wang and T. Sambongi, J. Physique 45:771 (1984) ; R.J. Cava, F.J. Di Salvo, M. Eibschutz and J.V. Waszczak, Phys. Rev. B 27(12):7412 (1983).
18. D.A. Keszler and J.A. Ibers, J. Solid State Chem. 52:73 (1984).
19. P. Gressier, A. Meerschaut, L. Guémas, J. Rouxel and P. Monceau, J. Solid State Chem. 51:141 (1984).
20. P. Gressier, M.H. Whangbo, A. Meerschaut and J. Rouxel, Inorg. Chem. 23:1221 (1984).
21. C. Roucau, R. Ayroles, P. Gressier and A. Meerschaut, J. Physique C, Solid State Phys. 17:2993 (1984).

22. H. Fujishita, M. Sato and S. Hoshino, Sol. State Comm. 49(4):313 (1984).
23. Z.Z. Wang, M.C. Saint-Lager, P. Monceau, M. Renard, P. Gressier, A. Meerschaut and J. Rouxel, Sol. State Comm. 46(4):325 (1983) ; M. Maki, M. Kaiser, A. Zettl and G. Grüner, Sol. State Comm. 46:497 (1983) ; Z.Z. Wang, P. Monceau, M. Renard, P. Gressier, L. Guémas and A. Meerschaut, Sol. State Comm. 47(6):439 (1983).
24. G. Mihaly and L. Mihaly, Solid State Comm. 48:449 (1983).
25. J. Dumas, C. Escribe-Filippini, J. Marcus and C. Schlenker, "Nato-Davy advanded Institute" Cambridge (1983) ; R.M. Fleming and L.F. Schneemeyer, Phys. Rev. B 28:6996 (1983).
26. P. Delhaes, C. Coulon, S. Flandrois, B. Hilti, C.W. Mayer, G. Riho and J. Rivory, J. Chem. Phys. 73:1452 (1980).
27. A. Wold, W. Kunnmann, R.J. Arnott and A. Ferretti, Inorg. Chem. 3:545 (1964).
28. J. Graham and A.D. Wadsley, Acta Cryst. 20:93 (1966).
29. G.H. Bouchard, J.H. Perlstein and M.J. Sienko, Inorg. Chem. 6:1682 (1967).
30. R. Brusetti, B.K. Chakraverty, J. Devenyi, J. Dumas, J. Marcus and C. Schlenker in "Development in Condensed Matter Physics, Plenum, New York, Vol. 2:181 (1981).
31. J.P. Pouget, S. Kagoshima, C. Schlenker and J. Marcus, J. Phys. Lett. 44:L113 (1973).
32. R.J. Cava, R.M. Fleming, P. Littlewood, E.A. Rietman, L.F. Schneemeyer and R.F. Dunn, Phys. Rev. B 30:3228 (1984).
33. C. Schlenker and J. Dumas in Crystal Chemistry and Properties of Quasi 1D structures, J. Rouxel Ed., D. Reidel (1986), and references there in.
34. R.M. Fleming and L. F. Schneemeyer, Bull. Am. . Soc. 29:470 (1984).
35. J.P. Pouget, A.H. Moudden, R. Moret, C. Escribe-Filippini, B. Hennion, J. Marcus and C. Schlenker, in Proc. Int. Conf. on the Physics and Chemistry of Low Dimensional Synthetis Metals (ICSM84) (1984) Vol. 121, 1985 Mol. Cryst. and Liquid Cryst.).
36. P. Butaud, P. Segransan, C. Berthier, J. Dumas and C. Schlenker, Phys. Rev. Lett. 55:253 (1985).
37. L. Brohan, R. Marchand and M. Tournoux, personal communication.
38. K. Selte and A. Kjekshus, Acta Chem. Scand. 18:690 (1964).
39. J. Malvy, G.A. Wiegers, J. Van Landuyt and S. Amelinck, Mat. Res. Soc. Sym. Proc. 21:181 (1984).
40. F. Boswell, A. Prodan, Mat. Res. Bull. 19:93 (1984).
41. H. Böhn and H.G. Von Schnering, Z. Kritallogr. 162:26 (1983).
42. S. Van Smaalen, K.D. Bronsema, Acta Cryst. B42:43 (1986).
43. M.H. Whangbo, in Crystal Chemistry and Properties of Materials with Quasi 1D Structures, J. Rouxel Ed., D. Reidel (1986).
44. J.M. Tarascon and F.J. Di Salvo, C.H. Chen, P.J. Carroll, M. Walsh and L. Rupp, J. Sol. State Chem. 58:290 (1985).
45. A. Meerschaut, P. Grenouilleau and J. Rouxel, J. Sol. State Chem. 61:90 (1986).
46. P. Grenouilleau, A. Meerschaut, L. Guémas and J. Rouxel, J. Sol. State Chem., 66:293 (1987).
47. P. Grenouilleau, L. Guémas, A. Meerschaut and J. Rouxel, J. Sol. State Chem., in press.
48. R. Chevrel, M. Sergent and J. Prigent, J. Sol. State Chem. 3:515 (1971).
49. M. Potel, R. Chevrel and M. Sergent, Acta Cryst. B36:1545 (1980).
50. R. Lepetit, P. Monceau, M. Potel, P. Gougeon and M. Sergent, J. Low Temp. Phys. 56:219 (1984).
51. H.P. Geserich, to be published.
52. R. Hoffmann, S. Shaik, J.C. Scott, M.H. Whangbo and M.J. Foshee, J. Sol. State Chem. 34:263 (1980).

STRUCTURES AND PROPERTIES OF LOW-DIMENSIONAL METAL COMPLEXES

Allan E. Underhill

Department of Chemistry and
Institute of Molecular and Biomolecular Electronics
University College of North Wales
Bangor, Gwynedd LL57 2UW, UK

INTRODUCTION

Interest in the structures and properties of low-dimensional metal complexes was stimulated by the work of Krogmann in the late 1960s.[1] However the compounds which aroused so much interest then have a much longer history. Indeed many of the compounds that were studied intently in the 1970s were first prepared in the middle of the 19th century although the true nature of the materials was to remain a mystery for over a hundred years.

Low dimensional complexes must possess structures which facilitate highly anisotropic inter-molecular interactions in the solid state. In practice this is found to be most readily achieved by basing the materials on planar metal complexes. The formation of columns of these planar molecules enables strong interactions in the direction of the column through either the overlap of metal or of ligand orbitals. Orthogonal to the columns the interactions are much weaker due to a combination of longer inter-atomic distances and incorrect spatial orientation of the complexes to give good orbital overlap. Planar complexes are most commonly found for ions possessing a d^8 configuration. The most important complexes are those of Ni(II), Pd(II) and Pt(II) although Ir(I) compounds have also been investigated. Other metals are also important when used in conjunction with ligands which superimpose a planar stereochemistry on the resulting complex.

The interactions between the planar complexes in the stack can be of three basic types.

Direct metal orbital overlap This is predominantly found for complexes which possess small ligands and is favoured by the presence of third row transition metals possessing large 5d orbitals. Examples are the partially oxidised $[Pt(CN)_4]$ complexes.

Direct ligand orbital overlap This is found for complexes containing large macrocyclic ligands or second row donor atoms such as sulphur. Since the interaction is ligand based the size of the central metal atom is not so important although the overlap of the metal orbitals with those of the ligand can affect the extent of the interaction.

Indirect orbital overlap between the complexes Over the past 10 years new materials have been synthesised in which large macrocyclic metal complexes are linked together by bridging groups such as oxygen or -CH=CH-.

The importance of these new materials that have been sythesised over the past 20 years lies in their remarkable and unusual properties. Their inherent structural anisotropy results in highly anisotropic solid state properties. The extent of anisotropy in these physical properties is often much higher for metal complexes of this type than it is for compounds based purely on organic materials. The compounds exhibit remarkable optical anisotropy,for example, $K_2[Pt(CN)_4]Br_{0.3}.3.2H_2O$ behaves optically as a metal with a plasma edge in the metal atom chain direction but as a transparent insulator orthoganal to this crystal axis. Similarly electrical conduction anisotropy as high as 10^5 has been observed.

KROGMANN SALTS

K. Krogmann in his review article in 1969 opened up the new area of one-dimensional metals by his studies of partially oxidised platinum and iridium complexes.[1] The complexes are restricted to those of Pt or Ir with small ligands capable of π back-bonding with the central metal. The most commonly studied compounds are based on $[Pt(CN)_4]^{2-}$ or $[Pt(C_2O_4)_2]^{2-}$ but some studies have also been made on $[Ir(CO)_2Cl_2]^-$. The simple integral oxidation state compounds behave as either semiconductors or insulators. The $[Pt(CN)_4]^{2-}$ of Group 1 cations form anionic stacks with anisotropic optical properties and with electrical conductivities in the range 10^{-4} to 10^{-9} S cm^{-1}.

The Krogmann salts are prepared by partial oxidation of the integral oxidation state compounds either by use of a chemical oxident such as Cl_2 or by electrocrystallisation. There are basically two different types of Krogmann salts.

(a) The anion-deficient (AD) salts. In these salts the additional positive charge on the Pt atoms due to partial oxidation is compensated by the incorporation of a non-stoichiometric number of anions compared with the parent complex. For example, in the tetracyanoplatinate series the salts have the general formula $M_2[Pt(CN)_4]\ X_n.pH_2O$(where M is a monovalent cation, X an anion, and n non-integer (0.19 to 0.49)).

(b) Cation-deficient (CD) Salts - In these salts a non-stoicheiometric deficiency of cations compared with the parent Pt(II) complex compensates for the partial oxidation of the platinum atoms. The general formulae is, therefore $M_m[PtL_4].pH_2O$(where m is a non-integer and L is CN or 1/2 (C_2O_4)).

Tetracyanoplatinates (tcp)[2]
(i) Anion-deficient Salts

The ADtcp complexes are the most extensively studied series of Krogmann salts. In particular the solid state properties of $K_2[Pt(CN)_4]Br_{0.30}3.2H_2O$, KCP(Br) have been studied in great detail because of the availability of large high quality single crystals.

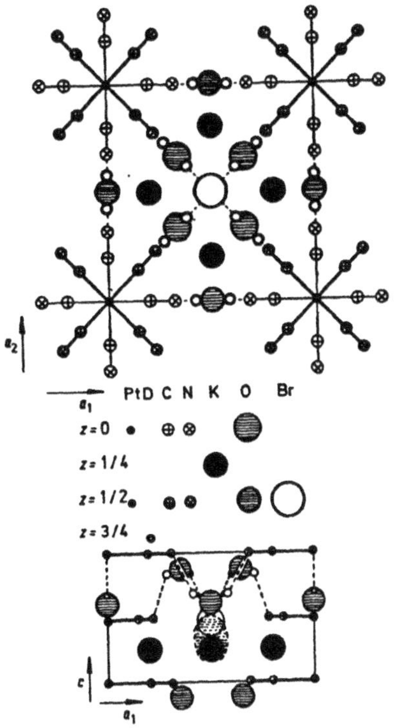

Figure 1. Crystal structure of KCP(Br). Upper-projection along the conducting c-axis; lower-projection along one of the a-axes. (Reproduced by permission from *Phys. Rep.*, 1978, **40**, 204).

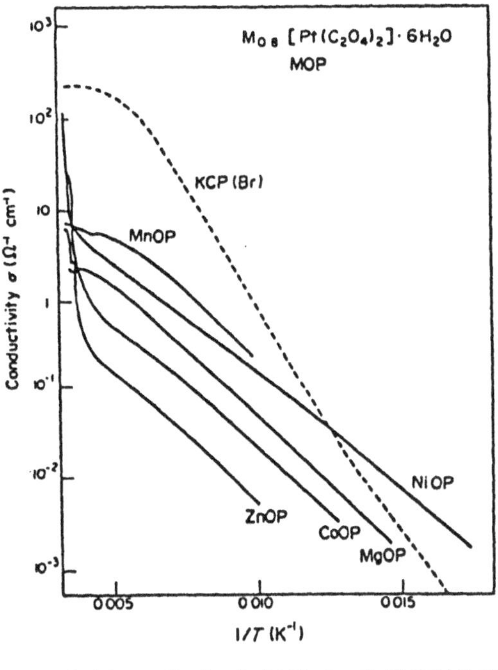

Figure 2. Variation of electrical conductivity with inverse temperature for a series of bis(oxalato)platinate salts

In KCP(Br) the degree of partial oxidation, DPO, indicates that 0.30 electrons per platinum atom have been removed from the highest filled band (predominantly $5d_{z^2}$) to leave 1.70 electrons per platinum atom in the band. The structure of KCP(Br) contains columnar stacks of $[Pt(CN)_4]^{1.7-}$ ions with a linear Pt-atom chain aligned along the c axis of the unit cell (Figure 1). The d_{Pt-Pt} is 2.89A, which is only slightly longer than that observed in Pt metal (2.774 A). The inter-chain Pt-Pt distance is 9.89 A. The Br^- ions are located at the centre of the unit cell but only 60% of the available crystallographic sites are occupied by Br^- and the remainder contain an additional water molecule. The K^+ ions are located in one half of the unit cell whilst the remaining water molecules occupy the other half. The water molecules form a hydrogen-bonded network between the cyanide ligands and either the bromide ion at the centre of the unit cell or the cyanide group of a $[Pt(CN)_4]^{1.7-}$ ion is an adjacent chain.

At room temperature the electrical conductivity in the metal atom chain direction, $\sigma_{\|}$, is about 300 Scm^{-1} and the dependence of conductivity with temperature is slightly negative with the degree of anisotropy $\sigma_{\|}/\sigma_{\perp}$ of about 10^5.[4] On lowering the temperature the conductivity passes through a maximum and then follows a complicated temperature dependence. The first partial derivative of $\ln\sigma_{\|}$, with respect to 1/T has a maximum at 100K. At low temperature the behaviour is that of a semiconductor with a well defined band gap.

The high temperature metallic behaviour of KCP(Br) can be understood in terms of the overlap of the $5d_{z^2}$ orbitals of the Pt to form a delocalised band. Because of the non-integral charge in the anion the band is only approximately 7/8 filled resulting in one-dimensional metallic properties at room temperature. However R.E. Pierls predicted that a one-dimensional system would be inherently unstable with respect to a lattice distortion in the metal-atom chain direction.[5] This Peierls distortion (PD) would split the partially filled band into filled and empty bands of lower and higher energy respectively, thus lowering the total electronic energy of the system. The period of the PD will depend on the degree of band filling and, therefore, on the degree of partial oxidation (DPO) of the complex. Thus the Peierls distortion converts a 1-D metal into a semiconductor by opening up a band gap at the Fermi surface. The band gap will be twice the activation energy for conduction in the semiconductor region. Peierl's argument is for the zero-temperature case and at some finite temperature well above absolute zero the band gap becomes temperature dependent and at some higher scale temperature (T_P) the band gap will disappear and the system will display metallic transport properties.

In the PD state the electron density of the conduction band is perturbed by the displacement of the positive metal ions from their undistorted positions. This is known as a charge density wave (CDW). In a perfect 1-D system a phase change can only occur at 0K and it is through the presence of weak 3-D effects that real compounds exhibit metal-semiconductor transitions at finite temperatures. In metal complexes the chains of metal atoms are surrounded by ligands, cations, water molecules and sometimes anions and these chemical species allow weak interchain interactions. These interactions are important in allowing coulombic interactions to become strong enough at low temperatures to cause 3-D ordering of the CDW's on adjacent chains. The temperature at which this occurs is known as the three-dimensional ordering temperature T_{3D}. Studies at microwave frequencies on KCP(Br) have shown that the conductivity is frequency dependent below T_{3D}.[6]

Studies under pressure indicate that T_{3D} increases and T_P decreases with increasing pressure suggesting that the PD might be suppressed at pressure greater than 70K bar.[7] An alternative explanation of the metal-non-metal transition in KCP(Br) has been proposed in which a mixed valence state of 0.85 Pt(II) and 0.15(Pt(IV) is the ground state and a Pt(III) state an excited state.[8]

Following the discovery of the 1-D metallic properties of KCP(BR) a whole series of ADtcp's have been prepared and their structures and electrical conduction properties studied.[2] A number of these compounds are isostructural with KCP(Br) whilst others have a somewhat similar structure but are anhydrous. The variation in electrical conduction and solid state properties with changes in the structure and chemical composition of the lattice for this series of ADtcp's is well understood. The conductivity at room temperature is very dependent on the intrachain Pt-Pt distance varying from over $10^3\Omega^{-1}cm^{-1}$, for the FHF salt with a d_{Pt-Pt} of 2.798 A, and dropping to $0.4\Omega^{-1}cm^{-1}$ for the NH_4^+ salt with an average d_{Pt-Pt} of about 2.92 A. This variation is clearly related to the extent of orbital overlap and hence to the width of the conduction band. As first pointed out by Williams, there is a close relationship between d_{Pt-Pt} and DPO[9]. Hence, there is also a relationship between the d_{Pt-Pt} and the extent of band filling and the value of the Fermi wave-vector k_F.

Bis(oxalato)platinates[10]

1-D metals based on the $[Pt(C_2O_4)_2]$ anions are all of the cation deficient type. Unlike the tcp's they are found with either mono or divalent cations.

A series of compounds of general formulae $M_x[Pt(C_2O_4)_2]\cdot 6H_2O$ (M-OP, where $0.8 < x < 0.84$ and M = Mg, Fe, Co, Ni, Cu, Zn) has been prepared and studied in single crystal form.[11] All the salts apart from the Cu(II) salt exhibit the same structure over a characteristic temperature range. This structure (space group Cccm) was first reported for the Mg^{2+} salt and consists of a columnar stack of $[Pt(C_2O_4)_2]^{n-}$ anions with an intra-stack separation of about 2.84 A. The divalent metal ions are located in planes in between the anions and are surrounded by an octahedral arrangement of water molecules. The water molecules form a hydrogen-bonded network involving the oxalate ligands on the adjacent anion columns. Only about 40% of the cation sites are occupied.

Optical reflectivity measurements of many of these salts have revealed a plasma edge for light polarized parallel to the metal atom chain direction and the absence of a plasma edge orthogonal to this direction. The band-width deduced from these measurements is about 3.9 eV, slightly smaller than that observed for KCP(Br), although the intra-chain Pt-Pt separation is less. Although the inter- and intra-stack separations in these compounds are comparable to KCP(Br), the electrical conduction properties at room temperature show the compounds to be considerably less anisotropic ($\sigma_{||}/\sigma_{\perp} \approx 3 \times 10^2$) than KCP(Br). This is probably in part a reflection of the greater amount of inter-stack hydrogen bonding present in these compounds. These salts can be divided into two categories dependent upon the nature of a phase transition near room temperature. The Co^{2+}, Ni^{2+} and Zn^{2+} salts all undergo a Cccm → Pccn transition. This corresponds to a small slippage of the anion stack at the centre of the unit cell relative to the plane formed by the anions at the corners of the unit cell. For the Ni^{2+} salt the movement is about 0.1 A. In the Mg^{2+} and Fe^{2+} compounds the change of space group is from orthorhombic Cccm to twinned monoclinic with an

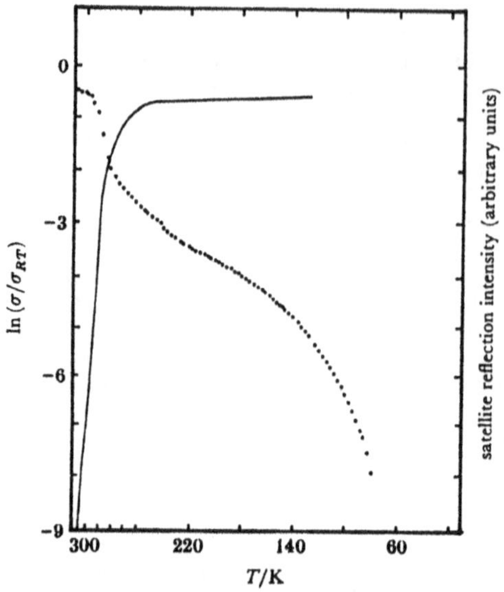

Figure 3. Comparison of the variation of (···) electrical conductivity and of (——) the intensity of the satellite reflection with temperature for $Co_{0.82}[Pt(C_2O_4)_2]\cdot 6H_2O$.

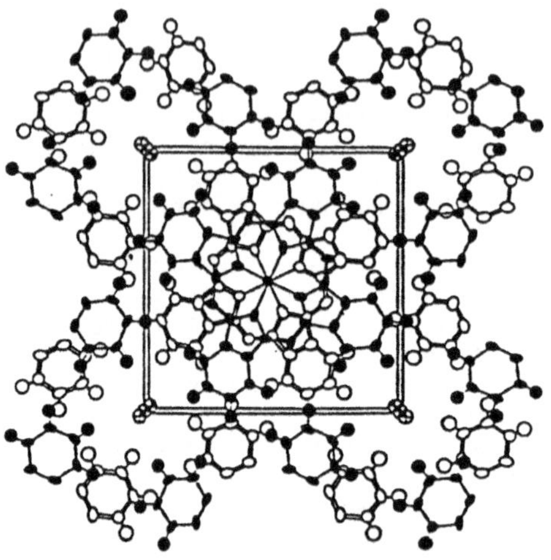

Figure 4. View down the *c* axis of the unit cell of Ni(PC)I (reproduced by permission from *Struct. Bonding (Berlin)*, 1982).

angle of about 1°. Although the distortions involved in these phase transitions are small the compounds belonging to the two classes exhibit quite different temperature-dependent conduction properties.

The three salts that undergo the Cccm → Pccm transition all exhibit a similar type of temperature-dependent conductivity. At some temperature above 300 K they exhibit a metallic-type temperature dependence whereas below this temperature the conductivity decreases rapidly with decreasing temperature and below about 250 K they exhibit semiconductor-type temperature dependence. Figure 2 compares the temperature dependence of the conductivity of M-OP's with that of KCP(Br).[11] It can be seen that the metal to semiconductor transition in Co-OP occurs over a much smaller temperature range than in KCP(Br), partly as a result of the decreased anisotropy of this compound.

These bis(oxalato)platinate salts also differ from the partially oxidised tcp's in the nature of the three-dimensional ordering that accompanies the metal to semiconductor transitions. In KCP(Br) extensive studies showed that only limited three-dimensional ordering accompanied the metal to semiconductor transition and that inter-chain correlations extended over only about five platinum atom chains. The bis(oxalato) platinate salts are characterized by the appearance of sharp satellite reflections indicative of a highly correlated superstructure. The appearance of these satellite reflections above 300 K corresponds to the onset of the transition as revealed by the conductivity studies. As the temperature decreases, the intensity of the satellite spots increases. At the temperature of the Cccm → Pccn phase transition the pattern of satellite reflection changes and they continue to increase the intensity until saturation is observed at approximately 250 K. Figure 3 shows the increase of intensity of the satellite reflections for the cobalt salt, Co-OP and the variation of electrical conductivity over this temperature range.[12] A small positive temperature-independent thermopower characteristic of a metal is observed for Co-OP above 300 K and this changes to a temperature-dependent thermopower below 300 K. The completion of the superstructure formation to 250 K corresponds to a change in the slope of the thermopower with temperature. There is no evidence for the Cccm → Pccn transition in the electrical conduction or thermopower measurements.

Of the compounds that exhibit the Cccm to twinned monoclinic transitions only the magnesium salt (Mg-OP) has been studied in detail. The phase transition is again accompanied by the formation of a superstructure, as shown by the appearance of satellite reflections. However, in this case, the superstructure is not commensurate with the Peierls distortion in the platinum atom chain direction. The change in electrical conductivity with temperature in the region of the transition is much less marked than those observed for the first series of compounds.

Partially oxidised bis(squarato)platinum and bis(croconato)platinum salts have been reported but detailed studies have been restricted by the lack of suitable single crystals.[13]

NEUTRAL COMPLEXES AND DERIVATIVES

Neutral square coplanar complexes of divalent transition metal ions and mono-anionic chelate or dianionic tetrachelate ligands have been widely studied. Columnar stack structures are common but electrical conductivities in the metal atom chain direction are very low and the temperature dependence is that of a semiconductor or insulator. However, many of these compounds have been shown to undergo partial oxidation when heated with iodine or sometimes bromine and the resulting crystals exhibit high conductivities with occasionally a metallic-type temperature dependence.[14,15] The electron transport mechanism may be located either on predominantly metal orbitals, predominantly ligand π-orbitals and occasionally on both metal and ligand orbitals.

The chemical flexibility of the porphyrin-like metallomacrocycles has been used to vary the electronic properties of the molecular units and hence the resulting solid. This has been brought about by the choice of the basic ligand, the peripheral substituents, and the central metal. A large number of compounds has been obtained as conducting polycrystalline powders by oxidation of the parent metallomacrocycle with iodine.

Ni(L)I (where L = PC, TBP or TATBP)

Iodine oxidation of Ni(PC), Ni(TBP) and Ni(TATBP) in each case yields crystals of composition Ni(L)L. Full X-ray diffraction, resonance Raman and ^{129}I Mossbauer studies show that the first two compounds are isostructural and isomeric. They contain columnar stacks of planar metallomacrocycles with parallel chains of I_3^- counter ions (Figure 4). In Ni(PC)I and Ni(TBP)I the iodine superlattice spacing is a multiple of the Bragg spacing so that the superlattice is commensurate with the Bragg lattice.

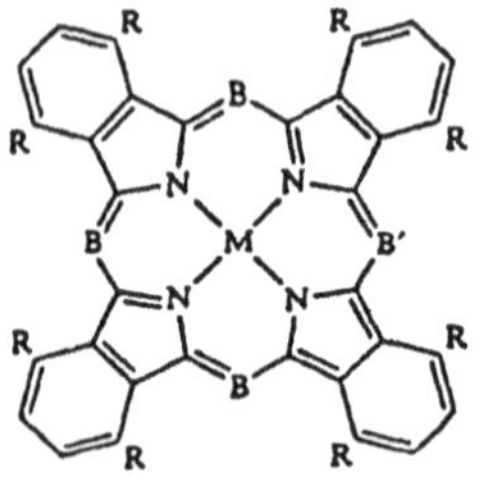

(a)

(a) Metallomacrocycle: R = H, B = B^1 = N, M(PC)
R = H, B = N, B^1 = CH, M(TATBP)
R = H, B = B^1 = CH, M(TBP)

Single crystals of all these compounds behave as molecular metals at room temperature but the temperature dependence of the conductivities varies significantly from one another. The room temperature conductivities and mean free paths for carrier motion along the stack are comparable with other 1-D metals including organic charge-transfer salts such as $(TTT)_2I_3$. The most striking feature is the abrupt metal to semiconductor transition observed for Ni(PC)I at approx. 60 K. There is no change in the resonance Raman spectra at the transition and therefore the extent of charge transfer does not change. ESR studies indicate the presence of a ligand centred π-cation radical suggesting that the 'hole-type carriers' are located on the ligand orbitals.[16] In Ni(TBP)I and Ni(TATBP)I the carrier spin g-value and line width are anomalous and these results have been interpreted as indicating that these are the first partially oxidised complexes in which charge carriers exhibit both metal and ligand properties. The electron 'hole' created by iodine oxidation is associated with both π and d 'bands' and can jump between the metal and the macrocyclic ligand as well as between one Ni(L) unit and its neighbours in the stack.[17]

Metallomacrocyclics do not often form stacked structures unless they are in the 'partially oxidised' form. One successful approach which allows the formation of stacked metallophthalocyanines and tetra benzoporphyrins with variable distances between the planar macrocycles is to form polymers by linking together the planar macrocycles by bisaxially metal bonded bridging ligands. A large variety of central atoms has been used in the macrocycles. For main group elements Si, Ge, Sn, Al etc bridging ligands such as O^{2-} S^{2-} and F^- have been used. In this type of polymer the electronic pathway in the subsequently doped polymer occurs exclusively by π-π overlap of the macrocycle and the bridging ligand is not involved in the conduction mechanisms.

Both $(PcSiO)_n$ and $(PcGeO)_n$ ($n \approx 70$-140) have high thermal and chemical stability. The M-O-M distances vary from 333pm (M = SI) to 382pm (M = Sn).[16]

The electrical conductivity of the unoxidised polymer varies from 10^{-6} (M = Si) to 10^{-9} Scm^{-1} (M = Sn). Oxidation with iodine results in partially oxidised products of the type $[(PcMO)I_y]_n$ (where y has a maximum value of about 1.1). The iodine is present as I_3^- or I_5^- and the maximum partial oxidation is 0.33. The conductivity of the polymer in the partially oxidised form is 10^{-1} Scm^{-1} for the Si complex.[19]

When transition metals, which prefer octahedral coordination (eg. Fe, Ru, Co), are at the centre of the macrocycle, linear organic molecules containing delocalised π-electrons (eg. pyrazine, 1,4-diisocyanobenzene cyanide) may be used as the bridging ligand in the polymer. In these cases an additional electronic pathway along the central axis of this polymer can be established. These polymers may exhibit comparatively high semi-conducting properties without 'partial oxidation'. For example: $[PcRu(tz)]_n$ exhibits a conductivity of 1 x 10^{-2} Scm^{-1}.[20] Partial oxidation by iodine increases the conductivity of many of the polymers by several orders of magnitude but that of $[PcRu(tz)]_n$ is only slightly increased.

Recent studies of [PcCo(CN)] prepared by electrocrystallisation of K[PcCo(CN)] have shown it to have an overlapping 3D structure resulting in a room temperature conductivity of 5 x 10^1 Scm^{-1} without doping.[21]

METAL COMPLEXES OF SULPHUR DONOR LIGANDS

Metal complexes of sulphur donor ligands have received increased study over the past few years as a consequence of their molecular similarity to many of the organic donors such as TTF and BEDT.TTF.

Transition metal complexes of 1,2-dicyanoethylene-1,2-dithiol (maleonitrile, mnt), have received special study since this ligand can stabilise transition metals in a variety of oxidation states and the high electron affinity of the terminal cyanide groups aids the delocalisation of charge within the complex, thus reducing potential coulombic repulsions.

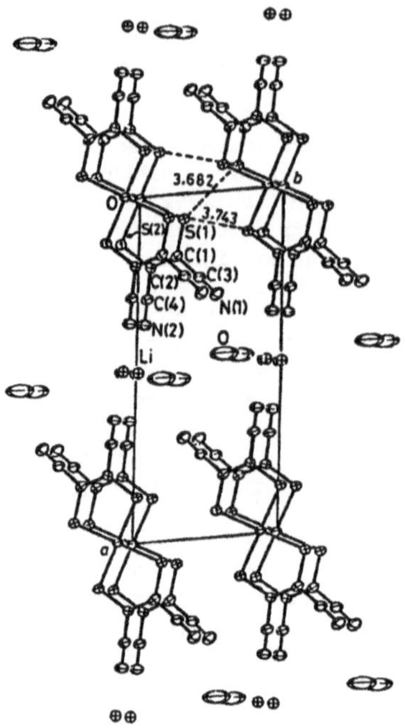

Figure 5. The c^* axis projection of the crystal structure of LiPt(mnt).

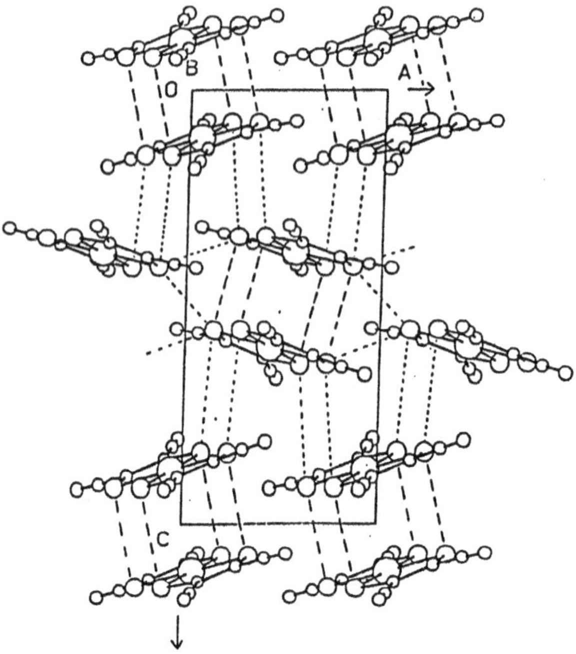

Figure 6. View down the *b* axis of the crystal structure of $Cs[Pd(S_2C_2(CN)_2)_2]\cdot 0.5H_2O$.
The short S-S contacts are shown by dashed lines.

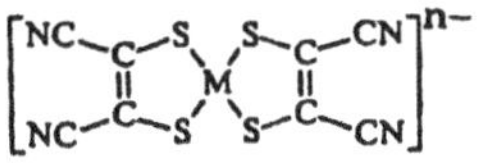

(b)

(b) $[M(mnt)_2]^{n-}$

A wide range of metal dithiolene complexes has been prepared and their electrical conduction properties reported. They include neutral, mono-anion, and di-anion complexes with a variety of substituents on the ligand (R = Ph, Me, CN, H, CF_3) and a variety of cations. The choice of cation has often been determined by the desire to obtain easily crystallised products and has resulted in the use of rather bulky counter ions such as $[N(C_2H_5)_4]^+$.

In an attempt to maximise interactions between the metal dithiolene anions Underhill and Ahmad studied the lithium salt of the $[Pt(S_2C_2(CN)_2)_2]$ cation. Slow aerial oxidation of a 50% aqueous acetone solution of $H_2[Pt(S_2C_2(CN)_2)_2]$ and LiCl yielded a black microcrystalline product of small shining black needles and black platelets. 4-Probe dc conduction studies on the needle shaped crystals showed the room temperature conductivity along the needle axis to be $\approx 100\ \Omega^{-1}\ cm^{-1}$.[22] The needle shaped crystals were a cation deficient compound $(H_3O)_{0.33}Li_{0.8}[Pt(S_2C_2C(CN)_2)_2]\cdot 1.67H_2O$.(LiPt(mnt)). X-Ray studies show that the room temperature structure (Figure 5) of LiPt(mnt) consists of equidistant stacks of nearly eclipsed $[\{Pt(mnt)_2]$ anions along the c-axis, with c = 3.639 A.[23] The unit cell is triclinic and the stacks form sheets along b separated along a by Li^+ and H_2O (Figure 5). There are short sulphur contacts between the chains within the sheets as well as within the chain, suggesting a relatively two dimensional network. Although the compound is cation deficient, the long Pt-Pt interchain separation shows that the bonding within the chain cannot be of the simple metal orbital overlap type found for the Krogmann salts. The high conductivity must therefore result from interaction of the whole anion in the chain direction. The temperature dependence of $\sigma_{||}$ for freshly prepared crystals slowly increases with decreasing temperature with $\sigma_{||}$ passing through a maximum at about 250 K. The conductivity falls to its room temperature value at about 200 K and then falls rapidly with decreasing temperature. Below 100 K the behaviour is that of a semiconductor with activation energy of about 34 meV.

Crystallographic studies have shown a superstructure below Tc = 215 K which is preceeded by one-dimensional diffuse scattering typical of a Peierls transition. The position of the diffuse line along c indicates that the extent of band filling is 0.41 or 0.59. Studies have shown that the thermopower above T_c is approximately constant and positive implying hole characteristics for the carriers in the metallic region and therefore, the extent of band filling is 0.59. Below T_c the thermopower reflects the intrinsic properties of the charge density wave semiconductor. Magnetic susceptibility studies have revealed the Pauli paramagnetism of the conduction electrons above Tc and a sharp drop in χ_p at 200 K corresponding to the metal-semiconductor transition. Optical reflectivity measurements in the metal atom chain direction show a plasma edge characteristic of a metal with a plasma frequency of $\approx$ 6 - 7000 cm^{-1}. Assuming a one-dimensional tight binding approximation to model the electronic structure of LiPt(mnt) leads to a band width of ≈ 0.4eV somewhat lower than that derived from the Pauli susceptibility results.[24]

Recent studies have centred on the mono-anion salts $[M(S_2C_2(CN)_2)_2]^-$ (where M = Ni, Pd, or Pt) of Group 1 cations and ammonium.

The crystal structure of $NH_4[Ni(S_2C_2(CN)_2)_2].H_2O$ has been determined and shown to be the first equidistant stack d^7 monoanion dithiolene complex.[25] The structure comprises equidistant anionic stacks in which the coordination plane of the complex is tilted with respect to the stacking axes. Within the stack the S-S and Ni-Ni distances are 3.918A, which is the cell repeat distance. The stacks are arranged in sheets parallel to the ab face of the cell. Within the sheets of anions the neighbouring stacks are arranged so that the metal centres are stepped by b/2 and are associated with S..S contacts of 3.79 and 3.91 A. Preliminary X-ray studies indicate that the Na^+, K^+, and Rb^+ salts also possess an equidistant stack structure. $NH_4[Ni(S_2C_2(CN)_2)_2].H_2O$ exhibits the properties of a Heisenberg antiferromagnetic chain at room temperature.[26] However at low temperatures the compound becomes diamagnetic with a smeared out transition in the region 200-130K. The transition is associated with dimerisation of the structure resulting in a pairing of the single electrons on pairs of adjacent anions. A similar temperature dependency of magnetic susceptibility has now been found for the Na^+ salt. The NH_4^+ salt exhibits a conductivity of around 10^{-2} Ω^{-1} cm^{-1} at room temperature and although it behaves as a semiconductor throughout the temperature range studied, the temperature dependence of the conductivity shows evidence of the phase transition found from magnetic and structural studies.[25]

The Na^+, K^+, NH_4^+ and Rb- salts of $[Pt(S_2C_2(CN)_2)_2]^-$ have been prepared and studied. The structure of $Rb[Pt(S_2C_2(CN)_2)_2].H_2O$ contains planar anions arranged in a dimeric eclipsed configuration to form a columnar stack structure.[27] The interplanar spacing within the dimers is 3.356 A and between dimers 3.512 A. The stacks of anions are arranged in sheets with short S...S contacts (3.655 A) between anions in adjacent stacks. All the compounds have conductivities in the region of 10^{-5}-10^{-6} Ω^{-1} cm^{-1} at room temperatures and behave as semiconductors with activation energies in the range 100-400 meV. All the compounds are diamagnetic at room temperature.[28]

The crystal structure of $Cs[Pd(S_2C_2(CN)_2)_2].0.5H_2O$ has recently been determined (see Figure 6) and is quite different from that of any previously reported structure of a $[M(S_2C_2(CN)_2)_2]^-$ salt.[29] The structure contains eclipsed dimers of $[Pd(S_2C_2(CN)_2)_2]$ with S - S contacts of approx 3.36 A, somewhat shorter than those found in $Rb[Pt(S_2C_2(CN)_2)_2].H_2O$ (≈ 3.45 A). The dimers are arranged in a staggered column arrangement along the c axis of the unit cell. Orthogonal to the c axis the staggered columns are arranged to form a sheet structure in the ac plane with short inter-dimer contacts of about 3.6 A. Within the staggered columns there are inter-dimer S - S contacts of ≈ 3.8 A, somewhat shorter than the inter-dimer contacts of 3.9 A in the $Rb[Pt(S_2C_2(CN)_2)_2].H_2O$. From preliminary studies it appears likely that NH_4^+ and Rb^+ salts probably possess similar structures.[28]

All the $[Pd(S_2C_2(CN)_2)_2]^-$ salts apart from the NH_4^+ salt exhibit much higher room temperature conductivities than the analogous Ni or Pt complexes. These vary from ≈ 5 S^{-1} cm^{-1} for the Cs salt to 7 x 10^{-3} S^{-1}

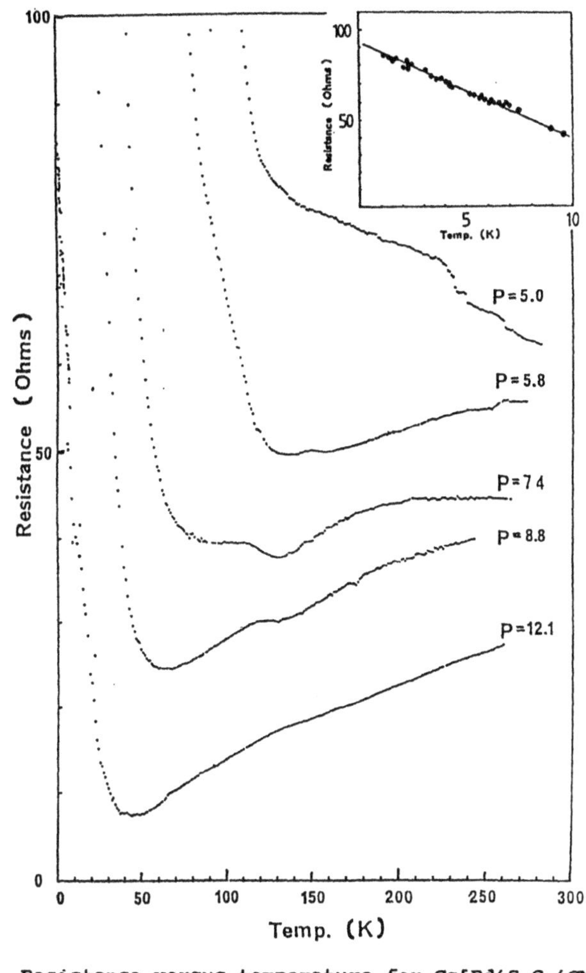

Figure 7. Resistance versus temperature for $Cs[Pd(S_2C_2(CN)_2)_2]0.5H_2O$

cm^{-1} for the NH_4^+ salt. The high conductivity of the Cs salt is surprising in view of the recently determined crystal structure. The salt is unusual in many of its other properties.[30] It has a small temperature independent paramagnetism at room temperature and the ESR signal clearly indicates the presence of mobile electrons at room temperature. When the compound is cooled below room temperatures, the static susceptibility, conductivity and ESR results show evidence of a transition and the compound becomes a well behaved semiconductor at low temperatures.[29] However the conductivity at room temperature rises rapidly with increasing pressure whilst at lower temperatures the changes are more dramatic, with a transition from semiconducting to metallic behaviour. At the highest pressure investigated, 12.1 kbar, the compound remains metallic down to 40K whilst the resistance rise below 40K is very weak, and the conductivity at 1.4K is still above 10 Scm^{-1}.[31] (Figure 7). This is the first example of a π-electron molecule of this class to be an integral oxidation state compound. The metallic properties must be related to the two-dimensional studies.

A new class of conductive and superconductive materials derived from transition metal complexes containing the extended sulphur containing ligand dmit, $[M(dmit)_2]^{n-}$ has recently been discovered. These compounds were expected to exhibit enhanced interstack interaction due to the large number of S atoms on the ligand. Salts of the type $(NBu_4)_{0.29}[Ni(dmit)_2]$ exhibited low anisotropy in conductivity but were semiconductors at room temperature.[32]

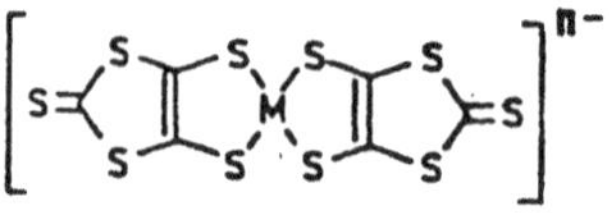

(c)'

(c) $[M(Dmit)_2]^{n-1}$

The combination of the $[Ni(dmit)_2]$ anion with the organic donor TTF produced a new type of compound $TTF[Ni(dmit)_2]_2$ in which the 2D $Ni(dmit)_2$ network resulting from S...S interactions between the $Ni(dmit)_2$ units are linked through the TTF molecules by additional S...S interactions resulting in a 3D network of interactions. $TTF[Ni(dmit)_2]_2$ retains metal-like conductivity down to very low temperatures. Recently under a pressure of 7kbar the compound has been shown to become superconducting with Tc = 1.62K.[33]

CONCLUSIONS

It is clear from the work described in the article that the past 20 years have seen the development of a whole new range of materials based on metal complexes and exhibiting novel properties due to their low-dimensional structures. This work has culminated in the development of compounds with anisotropic optical properties and unusual magnetic phase transitions. Perhaps the most spectacular successes have been the synthesis of molecular metals and recently a molecular superconductor.

I would like to thank the SERC for continued support and I acknowledge the contribution of my many coworkers and collaborators.

References

1. K. Krogmann, Angew. Chem. Int. Ed. Eng., 8, 35(1969).
2. J.M. Williams, A.J. Schultz, A.E. Underhill and K. Carneiro, in 'Extended Linear Chain Compounds', ed. J.S. Miller, Plenum, New York, 1981, vol. 1, p73.
3. G. Heger, H.J. Deiseroth, and H. Schultz, Acta Crystallogr., B34, 725(1978).
4. H.R. Zeller and A.J. Beck, J. Phy. Chem. Solids, 35, 77(1974).
5. R.E. Peierls, 'Quantum Theory of Solids' Oxford University Press, London 1955 p108.
6. A.S. Berenblyum, L.I. Buravov, M.D. Khidekel, I.F. Schegolev and E.B. Yakimov, J.E.T.P., LeH., 13, 440(1971).
7. M. Theilemans, R. Deltour, D. Jerome and J.R. Cooper, Solid State Commun., 19, 21(1976).
8. H. Nagasawa, Phys. Status Solidi B, 109, 749(1982).
9. J.M. Williams, Inorg. Nucl. Chem. Lett., 12, 651(1976).
10. A.E. Underhill, D.M. Watkins, J.M. Williams and K. Carneiro, in 'Extended Linear Chain Compounds' ed. J.S. Miller, Plenum, New York, 1981, vol 1, p119.
11. A. Braude, K. Carneiro, C.S. Jacobsen, K. Mortensen, D.J. Turner and A.E. Underhill, Phys. Rev. B, in press.
12. A. Braude, A. Lindegaard-Andersen, K. Carneiro and A.E. Underhill, Synth-Metals 1, 35(1979).
13. H. Toftlund, J. Chem. Soc. Chem. Commun., 837(1979).
14. J.A. Ibers, L.J. Pace, J. Martinsen and B.M. Hoffman, Struct. Bonding (Berlin), 50, 1(1982).
15. T.J. Marks and D.W. Kalina in 'Extended Linear Chain Compounds' ed. J.S. Miller, Plenum, New York, 1981, vol 1 p157.
16. C.J. Schramm, R.P. Scaringe, D.R. Stojakovic, B.M. Hoffman, J.A. Ibers and T.J. Marks, J. Am. Chem. Soc., 102, 6702(1980).
17. J. Martinsen, L. J. Pace, T.E. Phillips, B.M. Hoffmann and J.A. Ibers, J. Am. Chem. Soc., 104, 83(1982).
18. C.A. Dirk, T. Inabe, K.F. Schoch jr. and T.J. Marks, J. Am. Chem. Soc., 105, 1539(1983).
19. B.N. Diel, T. Inabe, J.W. Lyding, K.F. Schochk jr, K.R. Kannerwurt and T.J. Marks, J. Amer. Chem. Soc., 105, 1551(1983).
20. M. Hanack, S. Deger, U. Keppeler, A. Lange, A. Liverenz and M. Rein, Synth. Metals, 19, 739(1987).
21. Y. Orihashi, N. Kobayashi, H. Ohno, E. Tsuchida, H. Matsuda, H. Nakanishi and M. Kato, Synth. Metals 19, 751(1987).
22. M.M. Ahmad and A.E. Underhill, J. Chem. Soc., Dalton Trans., 1065(1982).
23. A. Kobayashi, Y. Sasaki, H. Kobayashi, A.E. Underhill and M. M. Ahmad, J. Chem. Soc., Chem. Commun., 390(1982).
24. M.M. Ahmad, D.J. Turner, A.E. Underhill, C.S. Jacobsen, K. Mortensen and K. Carneiro, Phys. Rev. B, 29, 4796(1984).
25. P.I. Clemenson, A.E. Underhill, M.B. Hursthouse and R.L. Short. J. Chem. Soc., Dalton Trans, submitted for publication.
26. L.C. Isett, D.M. Rosso and G.L. Bottger, Phys. Rev. B 22, 4739(1980).
27. M.M. Ahmad, D.J. Turner, A.E. Underhill, A. Kobayashi, Y. Sasaki and H. Kobayashi, J. Chem. Soc. Dalton Trans., 1759(1984).
28. P.I. Clemenson and A.E. Underhill unpublished results.
29. M.B. Hursthouse, R.L. Short, P.I. Clemenson and A.E. Underhill, J. Chem. Soc. Dalton Trans., in press.
30. P.I. Clemenson, A.E. Underhill, M.B. Hursthouse, R.L. Short, G.J. Ashwell, I.M. Sandy, M. T. Jones, C. Jacobsen and K. Carneiro, Synth. Metals 19, 579(1987).

31. I.D. Parker, R.H. Friend, P.I. Clemenson and A.E. Underhill, Nature, 324, 547(1986).
32. L. Valade, J - P Legros, M. Bousseau, P. Cassoux, M. Garbauskas and L.V. Interrante, J. Chem. Soc. Dalton Trans. 783(1985).
33. P. Cassoux, L. Valade, J - P Legros, L. Interrante, and C. Roueau, Physica, 143B, 313(1986).

LIGAND-BRIDGED MIXED-VALENCY METAL CHAIN COMPLEXES: PROTOTYPE N-SITE, N-ELECTRON SYSTEMS WITH ELECTRON-PHONON COUPLING

Peter Day

Oxford University
Inorganic Chemistry Laboratory
South Parks Road
Oxford OX1 3QR, England

1. INTRODUCTION

The fundamental distinction between metal chain compounds in which neighbouring metal atoms are in direct contact with one another and those where they are linked together by anions or neutral molecules has been established for many years (Day 1974, 1978, 1980). Summarizing the relation between structures and physical properties of metal chain compounds in one sentence, we can state that when there is direct overlap between cation orbitals, mixed valency is enough to produce a partly filled conduction band and one-dimensional metallic behaviour, while interposing ligand orbitals between those of the cation attenuates the cation-cation overlap sufficiently to give an insulating ground state, even when the metal ions have mixed oxidation states. Of course, there is astonishing richness in the physics of the one-dimensional Fermi surface, and chemistry has played an essential part in its exploration. Systematic fine tuning of dimensionality, electronic bandwidth and band filling is achieved by subtle chemical variation. Recent progress and present understanding of mixed valency metal chain compounds with metallic conductivity is described in the chapter of Underhill.

Two further distinctions must be made between the mixed valency metal chain compounds with bridging ligands and those that lack this feature. The first is that in principle the degree of partial oxidation in the latter can take any value, since the counterbalancing anions occur in channels outside the columns of co-facially organized square planar complexes that form the metal chains. Therefore the Peierls instability occurs at a wavevector bearing no simple relation to the metal-metal repeat distance. The resulting charge density wave (CDW) is then necessarily incommensurate with the underlying lattice. In contrast, all the examples of mixed valency chains containing bridging groups have average metal oxidation states that are integers or rational fractions, so the CDW is commensurate along the chain. The second difference between the two kinds of chain concerns the role of phonons. In the chains consisting of metal ions alone, the longitudinal acoustic modes, of very low frequency, are the ones whose softening leads to the CDW. Contrastingly, when the ligands form part of the chain, it is the much higher frequency metal-ligand vibrations that are the important ones. This is the reason why the CDW is commensurate in the ligand-bridged systems, and the oxidation states of the metal ions are much more deeply trapped. All known examples of unbridged metal chains are class III (Robin & Day 1967) mixed valency compounds while all ligand-bridged systems are class II insulators.

Given these facts, it might be thought that the ligand-bridged metal chains would be inherently less interesting than the others. Paradoxically, this is not the case, because the very strength of the electron-phonon coupling that yields the insulating ground state enables us to use these compounds as a test bed to explore the physics of the Mott-Hubbard-Peierls model. Wolfram's Red Salt (WRS), the prototype of ligand-bridged chains, was examined quite early in the history of one-dimensional studies (Thomas & Underhill 1971). Structural, vibrational and optical data on a wide range of WRS-type compounds is very well summarized and correlated in the reviews by Clark (1984a,b) and will not be repeated here. A typical crystal structure of a WRS compound is shown in Fig.1 (Clark et al. 1986). However, much more

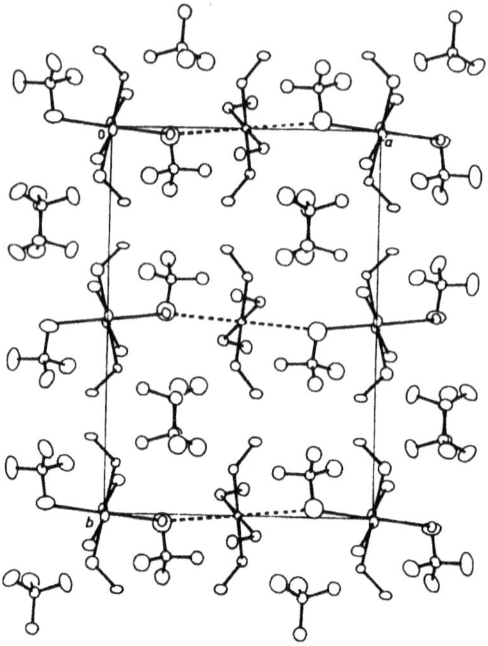

Fig. 1. A typical halogen-bridged mixed-valence metal chain compound. Crystal structure of $[Pt(CH_3NH_2)_4]$-$[Pt(CH_3WH_2)_4Br_2][ClO_4]_4$ viewed perpendicular to the chains (Clark et al. 1986).

recent interest in the Mott-Hubbard model with electron-phonon coupling (e.g. Nasu 1983, 1984) has led to a spate of new information, which it is the aim of this chapter to summarize. On the theoretical front, too, recognition of the importance of the concepts of 'negative U' and biopolarons have given new impetus to the field. Finally, by way of introduction, one should note the relevance of this physics to the new high temperature ceramic superconductors. It seems to me very probable that the square planar CuO_4 units characteristic of the latter are two-dimensional analogues of the linear bridged metal systems, albeit occupying a different part of the phase diagram derived from electron bandwidth, one-centre electron repulsion and electron-phonon interaction. In the following sections, the recent theory in this area will be summarized and spectroscopic and structural evidence adduced.

2. ELECTRON CONFIGURATION AND CHAIN SYMMETRY

The operational part of a ligand(X)-bridged chain of metal ions M can be written as in Figure 2(a), where X is usually a mono-atomic ligand such as a halide ion, but could also be trans-bidentate aromatic heterocycle like pyrazine. Best known among M are Pt, Pd, Ni in Group 10 or Au, Ag (though not apparently Cu) in Group 11. A crucial feature in describing the electronic structures, though, is that the M-X distances are not equal. Among possible patterns of dimerization (or higher multiplication of the primitive unit cell that can be imagined, see Figure 2), only that of Figure 2(c) is found in practice. It

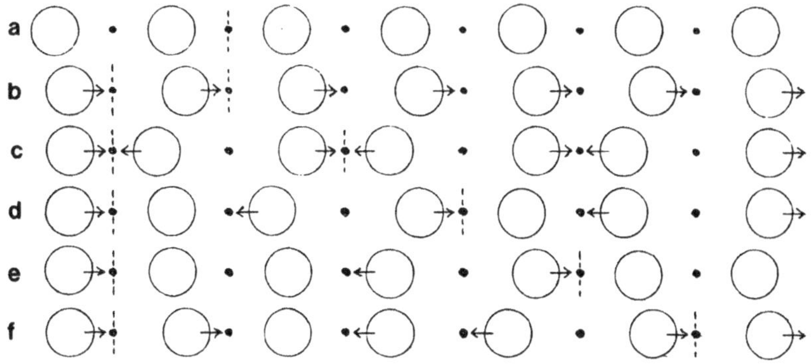

Fig. 2. Ligand bridged (large circles) metal atom (small dots) chain. (a) Uniform chain. (b) Asymmetrically dimerized. (c) Symmetrically dimerized. (d) Trimerized. (e) Tetramerized. (f) Pentamerized.

corresponds to a symmetric dimerization which one could envisage as resulting from condensation of a longitudinal acoustic phonon at the Brillouin zone boundary of the dimerized unit cell. Another, more chemical, way of looking at it is to say that there is an alternation of metal oxidation states along the chain, M(II,IV) in the WRS series and M(I,III) in the Group 11 compounds. Taking into account the ligand field splitting of d-orbitals in octahedral and square planar complexes, one can state that the d-orbital occupancy is $(z^2)^2(x^2-y^2)^0$, $(z^2)^0(x^2-y^2)^0$ in the M(II,IV) and $(z^2)^2(x^2-y^2)^2$, $(z^2)^2\ (x^2-y^2)^0$ in the M(I,III) series, with all other d-orbitals being occupied.

As first pointed out by Day (1985), the symmetry of the dimerized diatomic metal halide chain, with a centre of inversion at each metal centre, is the same as that of trans-polyacetylene $(CH)_x$ in which there is an inversion centre in the middle of each double and single bond. In both cases, therefore, there is a two-fold degeneracy, as shown in Figure 3. This leads to the suggestion that, like trans-$(CH)_x$, the WRS-like chain might support the existence of domain walls or solitons. Some hypothetical formulae are given in Figure 4.

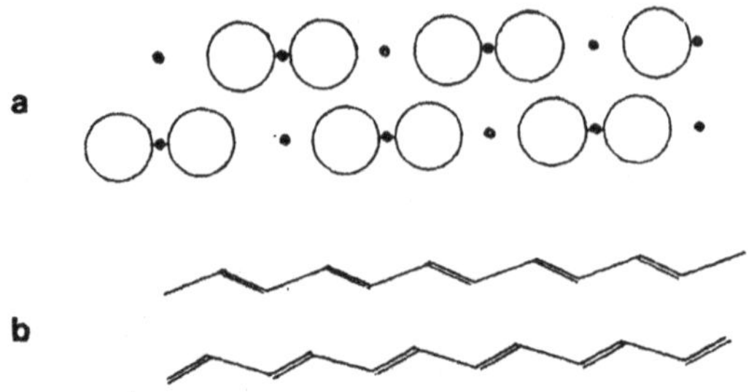

Fig. 3. Comparison between the dimerization patterns of (a) ligand bridged mixed valency metal chains and (b) trans-polyacetylene.

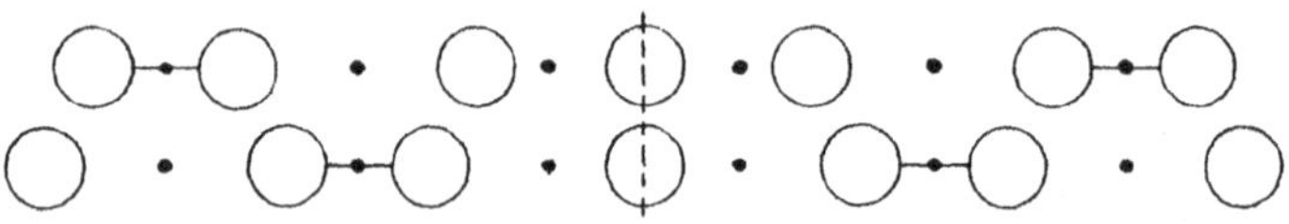

Fig. 4. Hypothetical solitons or domain walls in a ligand-bridged mixed valency metal chain.

On the other hand, though the symmetries are similar, the microscopic Hamiltonians are certainly different because in trans-$(CH)_x$, displacing the C atoms modifies the transfer integral for the π-electrons while in WRS, the on-site energies of the d-electrons are modified by displacing the halide ions. Thus in a sense the electron-lattice interaction in the former is site off-diagonal while in the latter it is site-diagonal. Onadera (1987) calculates that the effective mass of a soliton in WRS could be as high as 600 electron masses. No compelling physical evidence has yet been described for solitons in WRS, though occasional reports of broad weak e.p.r. signals indicate charged defects that could repay close study.

3. THE N-SITE, N-ELECTRON PROBLEM; NEGATIVE U

The dimerized WRS chain with electron configuration $(z^2)^2_A(z^2)^0_B(z^2)^2_C$.......can be schematically represented as in Figure 5, where the vertical scale represents electron binding energy. This makes it clear that the system is a simple example of 'negative U' (Anderson 1975), i.e. that the electron-lattice interaction has produced

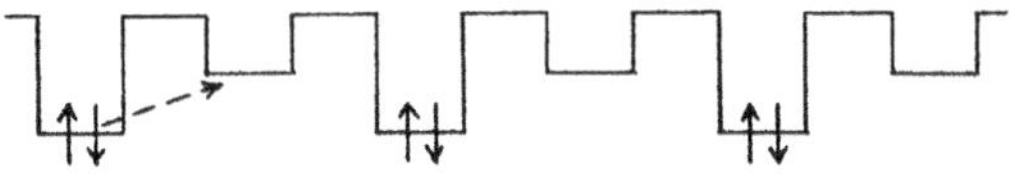

Fig. 5. Schematic potential energy variation along the metal chains in WRS. The dotted arrow indicates the visible intervalence transition.

a ground state in which there is an apparent attraction between electrons, causing them to pair on alternate sites rather than distributing themselves with one per site. Inorganic chemists will recognize this as the very common situation of a mixed valency compound with a difference in oxidation state of two, that occurs throughout the transition and post-transition series (Robin & Day 1967, Prassides and Day, 1985). The physics of the problem is conveniently developed (Prassides et al. 1986) by considering only two sites A, B and two electrons, enclosed in a deformable lattice (Toyazawa 1981). Considering only singlets, zero-order wavefunctions are

$$\psi_a = |\psi^B\alpha\psi^B\beta\rangle$$

$$\psi_b = 2^{-\frac{1}{2}}[\,|\psi^A\alpha\psi^B\beta\rangle + |\psi^B\alpha\psi^A\beta\rangle\,] \qquad (1)$$

$$\psi_c = |\psi^A\alpha\psi^A\beta\rangle$$

The electronic Hamiltonian is

$$H_{el} = H^A_{el} + H^B_{el} + V_{AB} + e^2/r_{12} \qquad (2)$$

where H^A_{el}, H^B_{el} are the electronic Hamiltonians associated with the two sites, V_{AB} represents the interaction between them and e^2/r_{12} is the electron repulsion operator. The vibronic matrix is constructed by evaluating $\langle\psi_i|H_{el}|\psi_j\rangle$ (i,j = a,b,c), assuming zero overlap between sites, using the harmonic approximation. The vibrational potential energy is

$$W^j = W^{oj} + l^jQ_j + (1/2)k^jQ_j^2 \qquad (3)$$

The zero of energy is defined by $W^{OA} + W^{OB} = 0$. Following Wong and Schatz (1981) we introduce the coordinates

$$Q_\pm = 2^{-\frac{1}{2}}(Q_A \pm Q_B) \qquad (4)$$

The problem is enormously simplified (though at the cost of some realism) by assuming that the force constants $k^{A,B}$ and $l^{A,B}$ are independent of the number of electrons on each site. With this approximation $W_{a,b,c}$ have identical dependence on Q_+, which can be eliminated. Dimensionless variables in units of $\hbar\omega_-$ are a convenience at this point, as follows

$$q = 2\pi(\omega_-/h)^{\frac{1}{2}}Q_-$$

$$\lambda = (8\pi^2 h\omega_-{}^3)^{-\frac{1}{2}}l \qquad (5)$$

where $\omega_- = (2\pi)^{-1}k^{\frac{1}{2}}$ is the frequency associated with the coordinate Q_-. With this notation the matrix elements are

$$\begin{aligned}
\langle\psi_a|H_{el}|\psi_a\rangle &= (\tfrac{1}{2})(q+\lambda)^2 - \lambda^2/2 + V^0_{aa}/h\omega_- + U \\
\langle\psi_b|H_{el}|\psi_b\rangle &= (\tfrac{1}{2})q^2 + V^0_{bb}/h\omega_- + U' \\
\langle\psi_c|H_{el}|\psi_c\rangle &= (\tfrac{1}{2})(q-\lambda)^2 - \lambda^2/2 + V^0_{cc}\ h\omega_- + U \\
\langle\psi_a|H_{el}|\psi_b\rangle &= \langle\psi_a|H_{el}|\psi_c\rangle = \varepsilon \\
\langle\psi_a|H_{el}|\psi_c\rangle &= \varepsilon'
\end{aligned} \qquad (6)$$

where

$$\begin{aligned}
V^0_{aa} &= V^0_{cc} = \langle\psi_a|V^{AB}|\psi_a\rangle \quad ; \quad V^0_{bb} = \langle\psi_b|V^{AB}|\psi_b\rangle \\
Uh\omega_- &= \langle\psi_a|e^2/r_{12}|\psi_a\rangle = \langle\psi_c|e^2/r_{12}|\psi_c\rangle \\
U'h\omega_- &= \langle\psi_b|e^2/r_{12}|\psi_b\rangle
\end{aligned} \qquad (7)$$

If the centres A and B are far enough apart U'= 0 and similarly, if $\langle\psi^A|\psi^B\rangle = 0$ then ε' is also zero. The competition between the on-site Coulomb repulsion U and the electron-phonon interaction is clearly evident in the diagonal terms of eq.(6), where the stabilization energy due to the latter is the $\lambda^2/2$ contribution. To obtain the matrix (eq.(8) below) first derived by Toyazawa (1981), it is only necessary to add $(\lambda^2/2 - V^0_{aa}/h\omega_- - U)$ to the diagonal elements in eq.(6), and put $\lambda^2/2 - U = 2W$:

$$\begin{bmatrix}
(\tfrac{1}{2})(q+\lambda)^2 & \varepsilon & \varepsilon' \\
\varepsilon & (\tfrac{1}{2})q^2 + 2W & \varepsilon \\
\varepsilon' & \varepsilon & (\tfrac{1}{2})(q-\lambda)^2
\end{bmatrix} \qquad (8)$$

A similar matrix to eq.(8) was also derived independently by Prassides and Day (1984) while examining the $Sb^{V}Cl_6{}^{-}$ - $Sb^{III}Cl_6{}^{3-}$ mixed valency system. In the latter instance, sufficient structural and dynamic information is available from elastic and inelastic neutron scattering (Prassides and Day 1985a,b) to enable quite precise estimates of λ and U to be made. The complete phonon density of states has not been measured in any of the WRS compounds. Because they are so strongly one-dimensional though it turns out that Raman and electronic spectra alone contain enough data to estimate λ, U and ε (see section 4.1)

A few special cases of eq.(8) have been worked out by Prassides et al. (1986), as shown in Figure 6. If $\varepsilon = \varepsilon' = 0$ (Figure 6(a)) the ground state is either single of mixed valency according as $\lambda^2/2 \lessgtr U$. With W>0 and ε small, the ground state potential energy surface has three minima, a shallow one at q = 0 and deeper ones at $q = \pm q_1$ (Figure 6(b)). As ε increases the q = 0 minimum disappears and the two $\pm q_1$ ones become shallower (Figure 6(c)) until, at large ε, only a single minimum remains

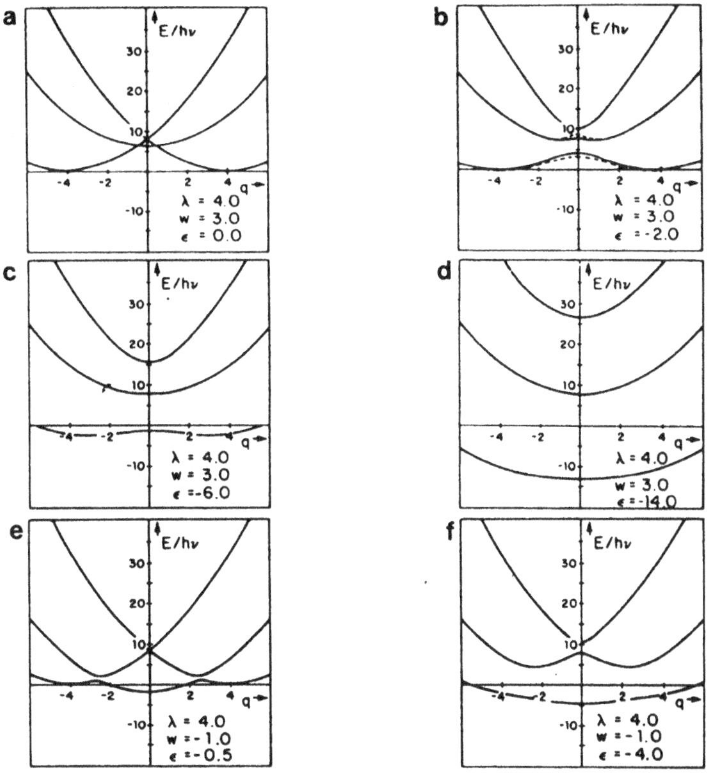

Fig. 6. Potential energy surfaces for the 2-site, 2-electron case with various values of λ, W and ε (eq.(8)).

at q = 0 (Figure 6(d)). A similarly evolving pattern with increasing ε was also found for $W<0$ (Figures 6(e,f)).

Extending these arguments from two-sites and two electrons to the one-dimensional N-site, N-electron case that directly models the compounds of interest in this Chapter, Nasu and Toyazawa (1982) derived a phase diagram as a function of U, λ and ε. They showed that only two ground states, CDW (mixed valency in our terminology) or spin density wave, SDW (single valency with a magnetic interaction between sites) could exist, though on both sides of the dividing line $U = \lambda^2/2$ lay regions where metastable minima of the other kind of ground state were to be found (Figure 7(a)). In Figure 7(a) the schematic potential energy surfaces shown next to each category of ground state correspond to those of Figure 6. In particular the region between the dashed lines corresponds to ground state potential energy surfaces exhibiting three minima, as in Figure 5(e). Despite the fact that the dividing line between SDW and CDW is given by $\lambda^2/2 = U$, two interactions are unsymmetrical in the sense that in the CDW phase the energy gain due to electron-phonon interaction is always reduced by interelectron repulsion because the two electrons with opposite spins occupy the same lattice site while in contrast, the SDW phase is unaffected by electron-phonon coupling because the charge density is spatially uniform. Thus in the former the energy gain is $\lambda^2 - U$ while in the latter it is merely U.

Equation (8) and its generalized form for N sites and electrons, is valid in the so-called adiabatic approximation, or static limit of zero nuclear kinetic energy. If the nuclear kinetic energy operator $T_n(q)$

is included we have

$$(H_{el} + T_n(q))\Phi_\nu = E_\nu \Phi_\nu \tag{9}$$

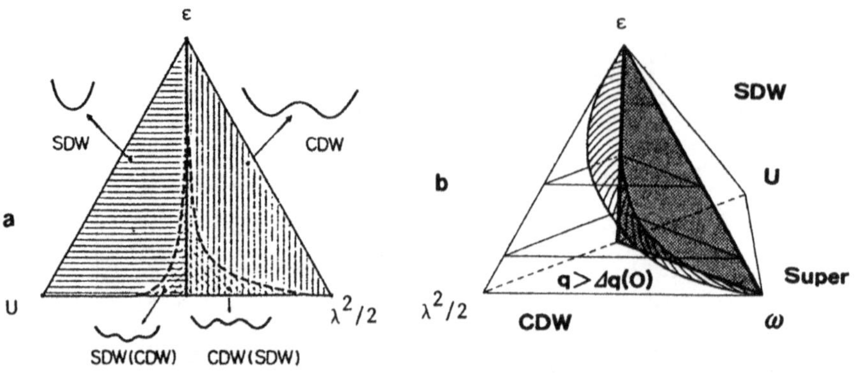

Fig. 7. Phase diagrams of the one-dimensional N-site, N-electron system (Nasu and Toyazawa 1982, Nasu 1985). (a) The adiabatic case. (b) The non-adiabatic case.

for which the solutions are vibronic functions in which the electronic and vibrational wavefunctions ψ and χ can no longer be separated. Prassides et al. (1986) worked out the eigenvalues and wavefunctions of the discrete 2-site, 2-electron case of eq. (9) and used them to calculate absorption and resonance Raman (RR) profiles for WRS. In addition to solving the full dynamical problem, they showed that in the strong electron-phonon coupling region $\varepsilon,\varepsilon' \ll \lambda^2/2$ almost identical results could be obtained by using the exact vibronic wavefunctions of W^{0A} etc. (eq.(3)) as the unperturbed bases with V^{AB} as the perturbation. Results of these calculations are described in sections 4.1 and 4.2. The static limit, determined by the ratio ω/ε, where ω is the charateristic vibrational frequency, neither permits a superconducting ground state nor fluctuations of the CDW. When ω/ε is snall one can say that the electron 'follows' the lattice but that the lattice is too heavy to 'follow' the electron. However, as ω becomes important relative to ε, the electrons become polarons and the metallic ground state that would arise from $\varepsilon \gg \lambda$,U becomes superconducting. Nasu (1985) has expressed these extra possibilities in the tetrahedral ($\varepsilon,\lambda T,\omega$) phase diagram shown in Figure 7(b).

It should be emphasised that no experimental evidence has ever been found for fluctuation of the CDW in WRS compounds which are in any case insulators (Interrante et al. 1974). On the other hand at least one mixed valency oxide with an oxidation state difference of two ($BaBi_{0.8}Pb_{0.2}O_3$) is a superconductor, and it seems possible that the key to an understanding of the exceptionally high temperature superconductors based on square planar CuO_3 networks may lie in their closeness to the valence disproportionation threshold Cu(II) → Cu(I),Cu(III), for which CsAuCl , with its trapped d^8,d^{10} sites, constitutes the deeply localized analogue (Day et al. 1978). Another way of looking at the d^{n-2},d^n superlattice in the WRS and Group 11 compounds is as static analogues of bipolarons, which are commonly defined as pairs of electrons stabilized by a local lattice distortion (Alexandrov et al. 1986).

4. PHYSICAL PROPERTIES OF WRS COMPOUNDS

In this section we concentrate on experiments performed in the last few years because although WRS compounds have been studied since the 1950's (Yamada & Tsuchida 1956) the situation has become much clearer with low temperature and single crystal measurements. Earlier work is summarized by Miller and Epstein (1976). We also place the optical and vibrational properties in the context of the theoretical treatment of the previous section.

4.1 Vibrational Spectra

Because they are so nearly one-dimensional, the phonons that dominate the spectra and electron-phonon coupling mechanisms in the WRS compounds are the longitudinal metal-ligand stretching modes. Most dramatic among the manifestations of the coupling between electron transfer and the vibrational modes are the remarkable resonance Raman (RR) spectra first observed, and extensively documented by Clark and his colleagues (Clark 1984a,b). A recent example measured on a single crystal (Tanaka et al. 1985) is shown in Figure 8, together with the onset of the intense absorption. The existence of such long progressions in the single symmetric Cl-Pt-Cl stretching mode (17 members in Figure 8(a)) indicate very clearly that only this mode is substantially coupled to the electronic transition which has its onset near 2.3eV. The ratio between the intensities of successive members in the progression I(n+1)/I(n) is a measure of the displacement of the minimum in the upper state potential energy surface. Applying the theory of Mingardi and Siebrand (1975), Clark (1978) has estimated that this displacement is of the same magnitude as the difference between the symmetrical and unsymmetrical structures in Figures 2(a) and 2(c). This is to be expected if the excited state in question is the intervalence transition shown as the dotted arrow in Figure 5.

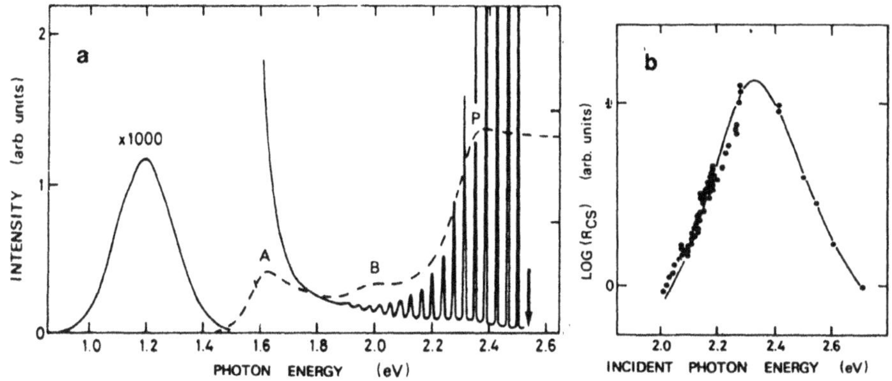

Fig. 8. (a) Resonance Raman, luminescence (solid lines) and absorption spectrum (dashed line) of $[Pt(en)_2][Pt(en)_2Cl_2](ClO_4)_4$ at 2K excited at 2.54 eV. (b) Observed Raman cross-section and that calculated from eq.(10) with E_0=2.310, γ=0.14 eV; $\hbar\omega_0$=38.4 meV (Tanaka et al. 1985).

It has been observed in many WRS compounds (see, e.g. Clark 1984b) that the excitation profile, which measures the resonance enhancement of the Raman scattering, peaks at lower frequency than the maximum of the intense intervalence absorption. Where the crystals are very small, absorption measurements have had to be made on mulls, but in favourable cases (Figure 8) it was possible to work with the same single crystal, both for RR and absorption. Thus the Raman cross-section as a function of incident photon energy $R(E_i)$ can be quantitatively evaluated by taking account of the absolute absorption and reflectivity at the incident and scattered photon energy. It is then found that $R(E_i)$ has a resonance peak precisely at the absorption edge. Given the extreme one-dimensionality of the system and the fact that the absorption in this energy region arises solely from the $(z^2)^2{}_A \rightarrow (z^2)^0{}_B$ excitation, Tanaka et al. (1985) have shown that $R(E_i)$ agrees very well with the expression (Martin and Falicov 1975):

$$R(E_i) \propto E_i^4 \int[\rho(E)/(E_i-E-i\gamma)(E-\hbar\omega-E-i\gamma)]dE \qquad (10)$$

if the interband density of states $\rho(E)$ is assumed to be zero for $E < E_0$ and proportional to $(E-E_0)^{-\frac{1}{2}}$ for $E > E_0$. The E^{-1} divergence at E_0 is removed by the damping factor, and the excellent agreement between theory and experiment for $[Pt(en)_2][Pt(en)_2Cl](ClO_4)_4$ is shown in Figure 8(b).

An intriguing feature of the Raman spectra of WRS compounds that has recently come to light with the use of single crystals and low temperatures is the broad emission band (1.2 eV in Figure 8(a)) that occurs far from the absorption edge. This has variously been assigned as emission from the self-trapped state to which the system has relaxed after excitation into the intervalence band (Tanino and Kobayashi 1983) or as an electronic Raman transition (Wong and Schatz 1984). In the vibronic model of section 3 the latter could arise from transitions between the two lowest vibronically coupled surfaces like those in Figure 6(b). Prassides et al. (1986) recalculated the RR profile for WRS using the 2-site, 2-electron model, both in the exact vibronic treatment and its perturbational approximation. They found that within these models, no choice of λ, W and ε could be found that reproduced the separation between the first component of the RR progression and the broad peak, or its width, although the energy of the intervalence band and the relative intensities of the two RR features were successfully modelled. On the other hand if one assumes that on excitation into the intervalence band the system relaxes to the ground vibrational level of the ψ_b potential energy surface of eq.(1) the envelope of the emission back to $\psi_{a,c}$ can be synthesised. The level of agreement with experiment can be gauged by comparing Figure 8(a) with Figure 9, which was calculated for $\varepsilon = -1.0$, $\varepsilon'=0$, $\lambda=5.8$, $W=23.4$, in units of $\omega=38.4$ meV. These values contrast with the suggestion by Nasu (1984) that in WRS the electron-phonon coupling energy $\lambda^2/2$ is of the same magnitude as ε and U, all being about 1 eV. Our estimate puts $\varepsilon \ll \lambda^2/2$, i.e. at the limit of weak electronic interaction. We believe this to be more realistic because the ω progression in the RR spectrum of WRS is extremely harmonic, at least up to 16 ω (Figure 8(a)) showing that the lower potential energy surface remains harmonic up to 0.6 eV. Furthermore, the very low electrical conductivity of WRS (Interrante et al. 1974) could hardly be compatible with a large ε. On the absolute values of U and λ, one must remain agnostic because the 2-site, 2-electron treatment does not take into account other vibrational degrees of freedom that would be present in an extended lattice or the long range Coulomb energy. The latter have been included in the more complete calculations on the three-dimensional mixed valency salt Cs_2SbCl_6 (Prassides and Day 1985a). Clearly the WRS compounds lie near the ε vertex of the phase diagrams in Figure 7, on the CDW side.

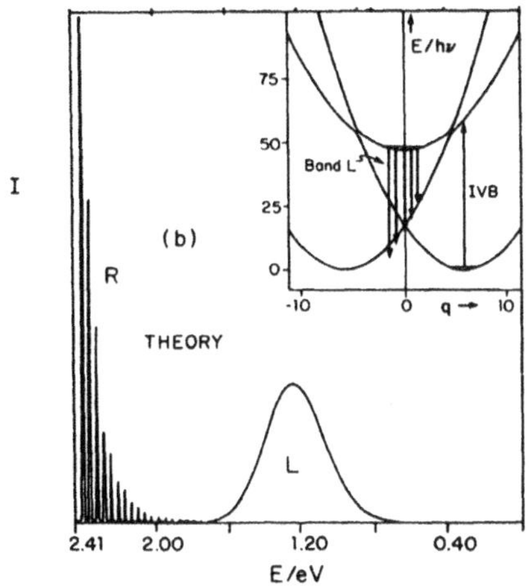

Fig. 9. Calculated RR (R) and emission (L) profiles of WRS. Inset shows the assumed emission process (Prassides et al. 1986).

4.2 Electronic Excited States

The WRS compounds are strongly pleochroic since the major absorption in the visible due to the intervalence transition $(z^2)^2_A \to (z^2)^0_B$ or $\psi_a \to \psi_b$ (eq.(1)) is strongly polarized parallel to the chain axes. It is also sufficiently intense to give high reflectivity, though only when the electric vector of the incident light is parallel to the chains.

The early transmission spectra of Yamada and Tsuchida (1956) were truncated by stray light but the microscopic measurement by Day (1974) was more satisfactory in this respect (Figure 10(a)). Comparable transmission data, in agreement with this older work, have been reported much more recently (Tanino and Kobayashi 1983; Tanaka et al. 1984). In all these spectra the intervalence absorption has rather an unusual shape, in that it rises sharply near 2.1 eV to a peak at 2.4 eV, then remains almost constant up to about 3 eV, when it rises sharply again.

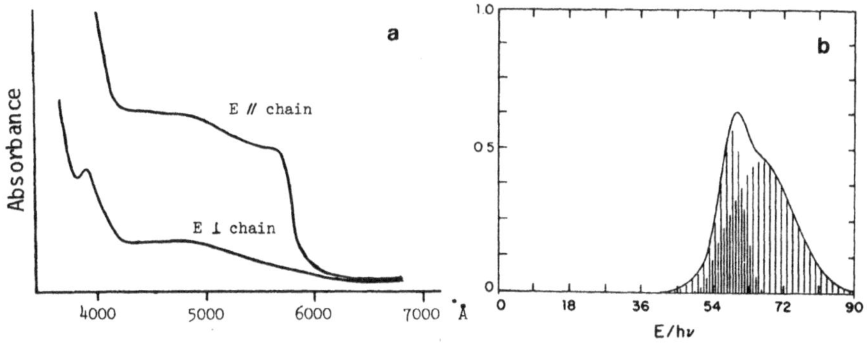

Fig. 10. (a) Polarised transmission spectra of WRS (Day 1974).
(b) Simulated composite intervalence absorption (Prassides et al. 1986).

Such behaviour suggests that the absorption may be composite, and Prassides et al. (1986) considered the possibility that there might be two intervalence transitions, arising from excitations $\phi_{a,c} \rightarrow \psi_b$ and $\psi_{a,c} \rightarrow \psi_{c,a}$. The second transition could gain intensity if ε' in eq.(8) was not zero, and a single asymmetric peak can indeed by simulated for reasonable values of the parameters ε, ε', W, λ (Figure 10(b)). It should be noted that ε'/ε should be of the magnitude $\langle\psi^A|\psi^B\rangle$, and therefore much less than unity.

Another complicating feature in the transmission spectra of relatively thick crystals is the presence of further weak peaks at lower energy than the main absorption edge, e.g. those labelled A and B in Figure 8(a). They do not contribute to any RR intensification nor to luminescence at higher energy. Neither does absorption in this region give rise to any photoconductivity (Haraki et al. 1987). Nevertheless they seem to be characteristic features of the chain structure, and only appear when the electric vector is parallel to the chain axis. Their energies are too low for them to be ligand field transitions of the Pt(II) or Pt(IV) units, one of which accounts for the sharp peak at 3950Å in the E// chain spectra of Figure 10(a). Thus for the time being their origin remains obscure.

For crystals as strongly absorbing as the WRS ones in the intervalence region, specular reflectivity is a convenient way of measuring the optical constants. Quite a lot of such polarised reflectivity data are available (e.g. Breer et al. 1978, Papavassiliou and Adetsis 1980, Wada et al. 1985) but the most complete are those of Wada et al. (1985) since they studied Pd and Pt with an equatorial ligand and bridging Cl,Br,I at room temperature and low temperature from 0.5 to 6.5 eV. In every case the reflectivity peak only occurs when the E// chain axis, as expected for a $(z^2)^2_A \rightarrow (z^2)^0_B$ transition. Kramers-Kronig transformation, yields the imaginary part of the complex dielectric constant, from which the energy, E_{abs}, and oscillator strength f can be derived (Table 1).

Table 1. Room temperature absorption energies E_{abs} and oscillator strengths f derived from polarised reflectivity spectra of WRS compounds (Wada et al. 1985).

	E_{abs}(eV)	f
Pt-(en)-Cl	2.72	3.0
Pt-(en)-Br	1.95	4.0
Pt-(en)-I	1.37	6.7
Pd-(en)-Cl	2.05	
Pt-(en)-Br	1.13	5.2

Several features of Table 1 require comment. For a given M(Pd,Pt), E_{abs} decreases in the order Cl > Br > I while f increases as Cl < Br < I. The values of f are all extremely high, but they agree with comparable data from the earlier authors. One reason could be that, viewed as localized (class II) intervalence charge transfer transitions between neighbouring M(II), M(IV), the electric dipole transition moment μ is proportional to the distance R the charge moves, and to the 'valence delocalization coefficient' (Robin and Day 1967) that measures the extent of mixing between ψ_a and ψ_b (eq.(1)). (see eq.(7) of Atkinson and Day 1969). Thus $f \propto (\alpha^2 R)^2$. In the WRS compounds R is quite large (5.5–5.7Å), though this cannot be the whole story since the observed f are so large that it would violate the sum rule if only the filled and empty $M(z^2)$ were involved in the transition. However, given that the M-M distance is so large, direct overlap between M orbitals is negligible, and the transition is mediated by the intervening X. In other words, ψ^A, ψ^B in eq.(1), which contribute to ε in eq.(6), are not pure M orbitals, but are molecular orbitals with a large contribution from $M(z^2)$ but also a smaller though crucial one from X(p). Given that such covalency normally increases from Cl to I, the order of the f's is explained.

An alternative way of taking the orbitals of the bridging ligand into account explicitly, especially if the degree of admixture is relatively small, is to use a perturbation approach involving configuration interaction between $\psi_{a,b,c}$ of eq.(1) and 'local' charge transfer configurations like $(\psi x)^{-1}(\psi^A)^{+1}$. (Mayoh and Day 1973, Day 1981a,b). We have used this approach to account successfully for the intensities of intervalence transition in discrete ligand-bridged mixed-valency dimers. The absorption corresponding to excitations of this type has been located in the reflectivity of WRS compounds with Cl and I bridges at 5.9 and 4.2 eV respectively (Tanaka et al. 1985). Table 1 also reveals that $f_{a,b,c}$ is larger for Pd than Pt compounds with the same bridging halogen. This is also to be expected since X→Pd(IV) charge transfer states are lower in energy than X→Pt(IV), and should therefore mix more with the M(II)→M(IV) state.

5. CONCLUSION

Although ligand-bridged mixed valency compounds of the WRS type have known for 100 years, and were first studied 30 years ago, it is only in the last few years that structural, optical and vibrational experiments have been carried out on them with enough precision to give a coherent picture of the statics and dynamics of the ground and excited states. At the same time theory has progressed towards a deeper understanding of vibronic coupling in intervalence electron transfer and the extension of such models from the chemists' prototypes of discrete dimeric molecules to infinite lattices. Of the latter, chains are the easiest to handle and the WRS compounds are valuable models. An important point of contact with wider theoretical implications is established by considering WRS materials in the context of N-site, N-electron problem, where the competing roles of electron repulsion, electron-phonon coupling and electronic bandwidth are starkly exemplified. WRS lies at the limit of small electronic bandwidth, with sufficiently strong electron-phonon coupling to stabilize an insulating CDW ground state or, put another way, a state with negative U. In other parts of the ε, U, λ, ω phase diagram lie superconductors, to seek which we may find it useful to argue from simple limiting cases like WRS.

REFERENCES

Alexandrov, A.S., Ranninger, J. and Robaszkiewicz, S., 1986, Phys.Rev. B33, 4526.

Anderson, P.W., 1975, Phys.Rev.Lett. 34, 953.

Atkinson, L. and Day, P., 1969, J.Chem.Soc. A, 2423.

Breer, H., Endres, H., Keller, N.J. and Martin, R., 1978, Acta Cryst. B34, 2295.

Clark, R.J.H., 1978, Ann.N.Y.Acad.Sci. 313

Clark, R.J.H., 1984a, Chem.Soc.Rev. 13, 219.

Clark, R.J.H., 1984b, in 'Adv. in Infrared & Raman Spectroscopy' eds. R.J.H. Clark and R.E. Nester, Wiley-Heyden, vol.11, p.95.
Clark, R.J.H., Croud, V.B., Davies, H.M. and Hursthouse, M.N., 1986, J.Chem.Soc., Dalton Trans. 403.
Day, P., 1974, in 'Extended Interactions between Transition Metal Ions', ed. L. Interrante, Amer.Chem.Soc.Symp.Ser. 5, 234.
Day, P., 1978, Ann.N.Y.Acad.Sci., 313, 9.
Day, P., 1980, ch. in 'Physics and Chemistry of Low-dimensional Solids', ed. L. Alcacer, Plenum, p.305.
Day, P., 1981a, Int.Rev.Phys.Chem. 1, 149.
Day, P., 1981b, Comments on Inorg. Chem. 1, 155.
Day, P., 1985, Phil.Trans.Roy.Soc. A314, 144.
Day, P., Vettier, C. and Parisot G., 1978, Inorg.Chem. 17, 2319.
Haruki, M., Tanaka, M. and Kurita, S., 1987, Synthetic Metals, 19, 901.
Interrante, L.V., Browall, K.W. and Bundy, F.B., 1974, Inorg.Chem. 13, 1158.
Martin, R.M. and Falicov, L.M., 1975, in 'Light Scattering in Solids' , ed. M. Cardona, Springer, p.79.
Mayoh, B. and Day, P., 1973, J.Chem.Soc. Dalton Trans., 846
Miller, J.S. and Epstein, A.J., 1976, Prog.Inorg.Chem., 20, 2.
Mingaroli, M. and Siebrand, W., 1975, J.Chem.Phys., 62, 1074.
Nasu, K., 1983, J.Phys.Soc.Jap., 52, 3865,.
Nasu, K., 1984, ibid., 53, 427.
Nasu, K., 1985, ibid., 54, 1933.
Nasu, K. and Toyazawa, Y., 1982, J.Phys.Soc.Jap. 51, 2098.
Onadera, Y. 1987, J.Phys.Soc.Jap. 56, 250.
Papavassiliou, G. and Adetsis, A.D., 1980, J.Chem.Soc., Faraday Trans. II, 76, 104.
Prassides, K. and Day, P., 1984, J.Chem.Soc. Faraday Trans. II, 80, 85.
Prassides, K. and Day, P., 1985a, Inorg.Chem., 24, 1109.
Prassides, K. and Day, P., 1985b, Inorg.Chem., 24, 3035.
Prassides, K., Schatz, P.N., Wong, K.Y. and Day, P., 1986, J.Phys.Chem., 90, 5588.
Robin, M.B. and Day, P., 1967, Adv.Inorg.Chem. & Radiochem., 10, 248.
Tanaka, M., Kurita, S., Koyima, T. and Yamada, Y., 1984, Chem.Phys., 91, 257.
Tanaka, M., Kurita, S., Kada, Y., Kojima, T. and Yamada, Y., 1985, Chem.Phys., 96, 343.
Tanaka, M., Kurita, S., Fujisawa, M. and Matsumoto, S., 1985, J.Phys.Soc.Jap., 54, 3632.
Tanino, H. and Kobayashi, K., 1983, J.Phys.Soc.Jap., 52, 1446.
Toyazawa, Y., 1981, J.Phys.Soc.Jap., 50, 1861.
Wada, Y., Mitani, T., Yamashita, M. and Koda, T., 1985, J.Phys.Soc.Jap., 54, 3143.
Wong, K.Y. and Toyazawa, Y., 1982, J.Phys.Soc.Jap., 51, 2098.
Wong, K.Y. and Schatz, P.N., 1984, Chem.Phys.Lett., 108, 484.

ELECTRONIC STRUCTURE OF LINEAR CHARGE TRANSFER SOLIDS

Z. G. Soos

Department of Chemistry
Princeton University
Princeton, New Jersey 08544

INTRODUCTION

Until recently organic conductors were so scarce that recipes for their synthesis were deemed necessary. Now these recipes seem quaint as organic and polymeric conductors multiply.[1] Segregated, rather than mixed, stacking of π-molecular donors (D) and acceptors (A) is indeed required. But conducting polymers show crystallinity not to be essential, while excellent D_2X conductors and superconductors have neither crystallographically equivalent sites nor strictly equal transfer integrals t. The physical idea of facile charge fluctuations among nearly equivalent, partly-filled sites supercedes any mathematical statement of equality. Polarizable counter ions, disorder, dimensionality, overlap patterns, etc. have all provided useful but limited insights.

The unusual physical properties[2,3] of π-radical solids are associated with nonintegral oxidation states, or mixed valence, in narrow bands (4t~1ev). This central idea for high conductivity is a goal rather than a recipe and has been anticipated theoretically by Mott.[4] A hypothetical regular chain of H atoms is metallic for small lattice spacing R, when 4t(R) is comparable to the ionization potential I or electron affinity A of H. When R exceeds a critical value R_c, however, a paramagnetic semiconductor is expected. Nonintegral oxidation states are represented by extra electrons or holes at H^- or H^+ sites, respectively, and in a rigid lattice lead to metallic conductivity even for $R > R_c$ for finite t(R). Crystallographically equivalent D^+ or A^- stacks of radical ions are in fact semiconductors, while isostructural mixed-valence systems are conductors.[5] Thus electron-electron interactions $U \sim \Delta E_{CT} \sim 1$ ev are comparable to 4t in π-radical solids. Conjugated polymers are also in the intermediate regime of $U \sim 4t$, for roughly tenfold larger microscopic parameters, and nonintegral oxidation states are achieved by doping, ionization, or heteroatoms. Mixed valence in Cu(II)/Cu(III) layers is apparently central for high-T_c superconductors. The sharp distinction between half and other fillings is fundamentally associated with electron-electron correlations. An optical consequence treated below is the different length dependence of $\pi \rightarrow \pi^*$ absorption in linear polyenes and cyanine dyes.

The magnetic, optical, and transport properties of π-radical organic solids may be understood[2] in terms of Wannier orbitals ϕ_p forming partly-filled narrow bands. The frontier orbitals are the highest occupied MO of

D, the lowest unoccupied MO of A, or the $2p_z$ AO of atoms in a conjugated molecule. Face-to-face stacking of planar D and/or A leads to one-dimensional quantum cell models defined by the transfer integrals t along the stack. The backbone of conjugated polymers also gives a one-dimensional problem. Segregated A^- or D^+ stacks of ion radicals correspond to one electron per ϕ_p and are formally related to an H-atom chain. Mixed DA stacks are solid-state generalizations of Mulliken's treatment[6] of DA dimers. Many qualitative features follow[2] immediately on distinguishing between mixed and segregated stacking, between one electron per ϕ_p and other filling, and between regular and alternating t along the stack.

Quantum cell models give precise but approximate descriptions of mixed-valence systems and are particularly important for excited states. Accurate solutions that include both electron-phonon and electron-electron interactions remain of great current interest. Models based on one ϕ_p per site, linear electron-phonon coupling, and on-site (Hubbard U) or simplified (Pariser-Parr-Pople, PPP) electron-electron interactions have the additional advantage of providing a common framework to conjugated molecules and polymers, to ion-radical and charge-transfer (CT) solids, and to magnetic insulators described by Heisenberg exchange models. Noninteracting electrons reduce to familiar Hückel models. Exact Hückel, Hubbard, PPP, and Heisenberg results can be compared rationally to see which accounts most naturally for the electronic structure of widely different systems. Such Hückel and PPP models for small conjugated molecules have now been superceded by more extensive, all-electron calculations that will not soon be practical for solids or polymers. A common theoretical framework invites common, or at least related, parametrization. Since polyacetylene (PA) or polydiacetylenes (PDAs) retain the bond lengths and angles of small polyenes, the same microscopic parameters are indicated originally. In my opinion, wider comparisons with a common parameter set offer more insights than exhaustively fitting PA or some other system of special interest. The following examples deliberately mix molecules, π-molecular solids, and polymers.

The relevant one-dimensional quantum cell models[7] have nearest-neighbor transfer $t_p = \langle\phi_{p+1}|H|\phi_p\rangle$,

$$H = \sum_{p\sigma} t_p(a^+_{p\sigma}a_{p+1\sigma} + a^+_{p+1\sigma}a_{p\sigma}) + \sum_p \epsilon_p n_p + \sum_{p,p'} V(p,p')(n_p - z_p)(n_{p'} - z_{p'}) \quad (1)$$

The fermion operators $a^+_{p\sigma}(a_{p\sigma})$ create (annihilate) an electron with spin σ in ϕ_p; the transfers in (1) thus conserve spin and S commutes with H. The number operator $n_p = a^+_{p\alpha}a_{p\alpha} + a^+_{p\beta}a_{p\beta}$ is restricted to 0,1, or 2 for nondegenerate ϕ_p. The site energies $\epsilon_p = \langle\phi_p|H|\phi_p\rangle$ may be taken as a common reference energy for chemically or crystallographically equivalent sites. The spin-independent potential V(p-p') is restricted to on-site (p=p') terms in Hubbard models and involves Coulomb interactions in PPP models with site charges z_p for empty ϕ_p. For $\epsilon_p = \epsilon$ and arbitrary V(p,p'), the cell model (1) has electron-hole, or alternancy, symmetry and the Lieb-Mattis theorem[8] insures a singlet ground state for an even number of valence electrons, N_e. For $N_e = N$ and strong repulsive on-site interaction $U > 0$, the ground and low-lying states of (1) have $n_p = 1$. The spin degrees of freedom are described by the $s = ½$ Heisenberg hamiltonian

$$H_{spin} = \sum_p 2J_p(\vec{s}_p \cdot \vec{s}_{p+1} - 1/4) \quad (2)$$

with antiferromagentic exchange $J_p = 2t_p^2/U_e$ due to virtual transfers through ionic states at $U_e = U$ in Hubbard models, $U_e = U - V_1$ for adjacent electron-hole pairs in PPP models, or $U_e = \Delta E_{CT}$ in ion-radical arrays. Heisenberg models are widely applied to magnetic insulators and provided early insights[9] for triplet spin excitons in organic D^+ or A^- crystals. The Taylor expansion of t_p or J_p about the equilibrium value of $\vec{r}_{p+1} - \vec{r}_p$ couples the electronic system to the lattice. Current work[1] is generally based on linear electron-phonon or spin-phonon coupling within the adiabatic approximation. The resulting possibilities for (1) and (2) encompass very many cases indeed.

OPTICAL EXCITATIONS AND GAP STATES

Correlations in half-filled narrow bands lead to a finite optical gap E_g even for regular t_p, when band or one-electron theory unequivocally predicts metallic behavior and vanishing E_g. The excitation $E_g \sim U$ in Hubbard models amounts to forming a AA^{-2} of DD^{+2} pair in the ion-radical lattice. It may consequently be interpreted as a CT process ΔE_{CT} or as I-A - e^2/R in an in H-atom chain or as $\pi \rightarrow \pi^*$ excitations of carbon chains in polymers. The infinite Hubbard chain with uniform t is exactly soluble[10] for arbitrary filling $\rho = N_e/N$. The optical gap $E_g(t/U)$ is finite for $U > 0$ and $\rho = 1/2$; it vanishes for $\rho \neq 1/2$, when the oxidation state is not integral.

Coulomb interactions among C^+ and C^- sites are also included in PPP models[11] for conjugated molecules or polymers. In the Ohno formula,[12] the interaction between charges separated by R is interpolated from U at R = 0 to e^2/R as $R \rightarrow \infty$,

$$V(R) = U(1 + U^2R^2/e^4)^{-\frac{1}{2}} \tag{3}$$

Thus U is again the only adjustable parameter, which in practice is taken from atomic ionization data rather than adjusted. Such a procedure is not practical for the smaller $U \sim 1$ ev in D^+ or A^- solids, where U and a nearest neighbor V(a) are fit to optical, magnetic, and other data. The potential V(R) above has no free parameters, although dielectric constants or shielding could be incorporated as needed for solid-state applications.[13] While a single property like E_g may be fit with either on-site or long-range interactions, the simultaneous modeling of several observables discriminates among potentials.[14] The PPP choice is clearly indicated for conjugated molecules.

The lowest-energy $\pi \rightarrow \pi^*$ absorption of gas-phase polyenes are compared in Table 1 with exact PPP results based on standard molecular parameters. Linear polyenes are half-filled, with one π-electron per site, and have alternating transfer integrals $t_\pm = t(1 \pm \delta)$ with $-t = 2.40$ ev and $\delta \sim 0.07$ for nominally double and single bonds. The short polyene ions in Table 1 have equal bond lengths, or $\delta = 0$. The PPP model with nearest-neighbor t has electron-hole symmetry; this leads to identical spectra[7] for carbanions ($N_e = N + 1$) and carbocations ($N_e = N - 1$). The E_g values in Table 1 are again satisfactory without any adjustable parameter and decrease as expected with increasing length. Neutral polyene radicals with uniform bond lengths are also included in Table 1, although experimental E_g are not available. Their negative spin densities at even sites is correctly given by exact PPP results[19] and describe properly this correlation effect in PA. The linear cyanine dyes in Table 1 also have $N_e = N + 1$, since each terminal N contributes two π-electrons, and have essentially uniform bond length.

Peierls' theorem indicates that all these systems become alternating, with $\delta \neq 0$, at sufficiently large N. Such dimerization transitions are found on cooling many crystals with regular D^+ or A^- stacks, while conjugated polymers have alternating t at all accessible temperature. The sharp distinction between integral and nonintegral oxidation states is best

Table 1. Lowest dipole-allowed $\pi \rightarrow \pi^*$ excitations of all-trans conjugated molecules with alternation δ. The exact PPP results (ref. 7) are based on standard molecular parameters.

Length, N	Alternation, δ	PPP(ev)	E_g(ev) expt.	reference
Polyene, C_NH_{N+2}	0.07			
8		4.561	4.40	15[a]
10		4.234	4.02	
12		4.001	3.65	
Ions, $(C_NH_{N+2})^{\pm}$	0.0			
5		3.456	3.43/3.13	16[a]/17[b]
7		2.799	2.88/2.64	
9		2.343	2.25	
11		2.009	1.98	
Radicals, C_NH_{N+2}	0.0			
7		4.389		
9		3.969		
11		3.669		
Cyanine Dyes	0.0			
$(R_2N(CH)_{N-2}NR_2)^+$				
5		3.075[c]	3.97	18[d]
7		2.128	2.98	
9		1.547	2.39	
11		1.221	1.98	

a - gas-phase data

b - $(CH_3)_2C(CH)_{N-2}C(CH_3)_2^+$ in concentrated H_2SO_4; slightly higher energies are reported for N = 7, 9, and 11 in heptafluorobutyric acid

c - without correcting for different C-N bond lengths.

d - $[(CH_3)_2N(CH)_{N-2}N(CH_3)_2]^+ClO_4^-$ in methylene chloride

shown for the regular array, however, when for $N \to \infty$ correlations lead to $E_g > 0$ for $N_e = N$ and to $E_g = 0$ for $N_e \neq N$. The polyene E_g in Fig. 1 have been recalculated for $\delta = 0$ and still extrapolate to $E_g \sim 2.3$ ev at $N \to \infty$, slightly above the PA or PDA value. Comparison with the $\delta = 0.07$ line in Fig. 1 shows that correlation rather than alternation dominates for standard PPP parameters. These gas phase values are clearly too high, but red shifts of ~ 0.5 ev are found in polar solvents[15] and the polymer also provides a polarizable medium. The polyene ions and cyanine dyes in Fig. 1 extrapolate to $E_g = 0$ within the accuracy of the results in Table 1. The different length dependence of the $\pi \to \pi^*$ excitations of polyenes and cyanine dyes has long been emphasized,[11] but could not be resolved within one-electron theory.

The additional low-energy excitations of complex TCNQ salts, with $N_e = 2N/3$ or $N/2$, were first studied by Iida,[20] fit naturally in the classification of π-radical solids,[2] and have been extensively discussed by Torrance.[3] Similar low-energy excitations[21] occur in nonstoichiometry TTF X_γ halides, with $N_e = \gamma N$ and $\gamma \sim 0.8$ and may be modeled[22] qualitatively via an extended Hubbard model. The higher-energy excitation at E_g, which we have identified as characteristic of the half-filled array, still occurs around 1.0 to 1.5 ev. As shown by Tanaka and coworkers[23] and Kuroda and coworkers,[24] however, the onset of molecular ($\pi \to \pi^*$) excitations is in the same region. Polarized light and single crystals lead to convincing assignment of the transition dipole for CT processes normal to the D or A planes and $\pi \to \pi^*$ excitations within the plane, even though the molecular planes are not usually perpendicular to the stacking axis. Such excitations are beyond models with one ϕ_p per site, but may reasonably be estimated for isolated molecules or ions. The optical excitations of π-radical solids offer striking confirmation of the difference between integral and nonintegral oxidation states, even though quantitative comparisons are difficult.

Half-filled chains of identical sites and nearest-neighbor but arbitrary t_p have, via electron-hole symmetry, precisely one electron per site in the ground state. For odd $N_e = N$, the HOMO is a nonbonding orbital with nodes at even sites.[11] The lowest optical excitation of a polyene radical, or neutral soliton[25] S, is consequently at $E_g/2$. Topological solitons[25] in the infinite polyene are alternation crossovers between the two Kekulé structures and have a width associated with reduced alternation in the region over which the spin is delocalized. A single S or S^+, S^- occurs in the ground state of an odd-length polyene or ion in Table 1. Midgap excitations are also expected for S^- and S^+, which correspond to $N_e = N \pm 1$ and have singlet ground states for odd N. Correlations strongly affect[26] the doubly degenerate excitation of S. Their optical gaps in Fig. 1 fall on the same $\delta = 0$ curve as even polyenes. The degeneracy of the second one-electron excitation is responsible for the two-photon 2^1A_g state being <u>below</u> E_g in even polyenes with four or more double bonds[27], as found in exact solutions of PPP models with standard prameters. The 2^1A_g bond orders indicate that this gap state consists of a pair of neutral solitons. By contrast, the nondegenerate excitation of S^- or S^+ falls considerably below E_g, as shown in Fig. 1, and rationalizes the low-energy PA absorption at 0.43 ev associated with charged solitons.[28]

NEUTRAL-IONIC INTERFACE

The mixed regular array in TTF-chloranil consists of D^{+q} and A^{-q} sites with equal Mulliken transfer integrals $t = \langle DA|H|D^+A^-\rangle$ along the stack at 300 K. Optical and vibrational data indicate[29,30] that the partial ionicity q changes discontinuously from $q \sim 0.3$ to $q \sim 0.7$ on cooling through $T_c = 82K$. The neutral-ionic (n-i) transition is accompanied by dimerization.[31] It may also be induced by hydrostatic pressure for $T > T_c$. The resulting phase diagram is rich and complicated.[32] As in the case of metal-insulator

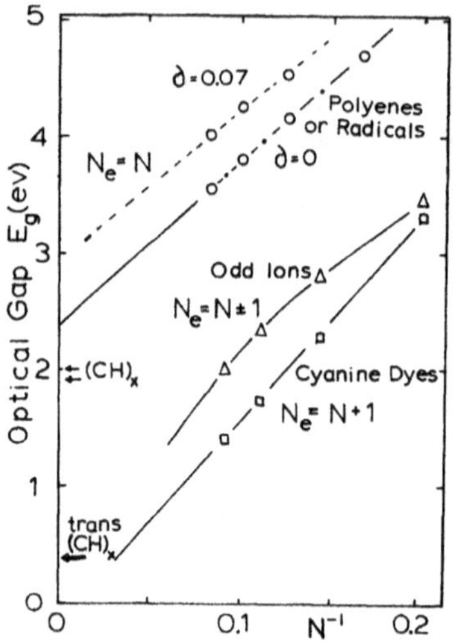

Fig. 1. Lowest $\pi \to \pi^*$ absorption of all-trans polyenes, radicals, ions and cyanine dyes based on exact PPP solutions and molecular parameters from ref. 7. The alternation δ is discussed in the text. Related experimental values are collected in Table 1.

transitions, the concomitant occurrence of electronic and structural changes leaves open the driving force. Crystal perfection, purity, and the accurate determination of partial ionicity and small structural changes are some of the experimental problems. The role of interchain interactions and of excitonic (molecular) states of D, D^+ and A, A^-, are theoretical concerns. The general features of n-i transitions[1] must be separated from those restricted to TTF-chloranil. Almost inevitably, however, it must be rare to match lattice energies for completely different ionicities q within kT per molecule. Early investigations[2] of TMPD-TCNQ, TMPD-chloranil, and PD-chloranil have direct bearing on the ionic side, while M_2P-TCNQ and M_2P-$TCNQF_4$ illustrate[33] other possibilities for q and alternation.

DA dimers illustrate polar covalent bonding, with both electrons in ϕ_D in $|n\rangle = |DA\rangle$ and equal sharing in $|i\rangle = |D^+A^-\rangle$. The admixture of $|n\rangle$ and $|i\rangle$ depends on their relative energies and on t. There is always precisely one electron per ϕ_p whose unequal sharing between ϕ_D and ϕ_A leads to partial ionicity $0 \le q \le 1$. The relevant energies are t, the ionization potential I of the donor, the electron affinity A of the acceptor, the nearest-neighbor interaction V ($\sim e^2/R$) of a D^+A^- pair, and the Madelung constant α of the lattice. The solid is neutral (q = 0) and diamagnetic for $I - A - \alpha V > 0$ and $t = 0$, when the nondegenerate singlet $|n\rangle = |\cdots DADA \cdots\rangle$ is the ground state and the lowest CT exciton occurs at $\Delta E_{CT} \simeq I - A - V$. The lattice becomes ionic (q = 1) for $I - A - \alpha V < 0$ and $t = 0$, with 2^N degenerate spin states in $|i\rangle = |\cdots D^+A^-D^+A^- \cdots\rangle$ leading to a paramagnetic (Curie-law) solid and $\Delta E_{CT} = | - I + A + (2\alpha - 1)V |$ now describing the back transfer $D^+A^- \rightarrow DA$. Both $|n\rangle$ and $|i\rangle$ are stabilized by long-range three-dimensional Coulomb interactions in lattices with $\alpha > 1$, since $\Delta E_{CT} = V(\alpha - 1) > 0$ even at the $t = 0$ interface at $I - A = \alpha V$. Perturbation expansions[34] in $x = |t|/V(\alpha - 1)$ then suffice for the n-i phase boundary and ionicity jump. Finite t lifts the spin degeneracy of $|i\rangle$ and leads to (2) with uniform $J = t^2/\Delta E_{CT}$ since there is only one pathway for virtual transfers when D^{+2} or A^{-2} sites are excluded. In regular arrays, the lattice remains paramagnetic and the n-i interface at 0 K is a paramagnetic-diamagnetic transition.

The electronic structure of TTF-chloranil thus illustrates an extreme sensitivity of q to T or P due to the near degeneracy of $|i\rangle$ and $|n\rangle$ and the low-lying $\Delta E_{CT} \sim 0.55$ ev at 82 K. Torrance and coworkers[29,35] have combined t = 0 results for the solid with $|t| \sim 0.2$ ev results for dimers to obtain the principal features of the n-i transition in TTF-chloranil. Charge resonance among ionic and neutral domains at the n-i interface leads[34] to novel ΔE_{CT} behavior in the solid, however, which is shown below to be opposite to dimer spectra.

The exclusion of high-energy D^{+2} and A^{-2} sites gives a quantum cell model with three states per site: D, $D^+\alpha$, or $D^+\beta$ and A, $A^-\alpha$, or $A^-\beta$ for nondegenerate ϕ_p. Quite aside from computational convenience, this restriction avoids poorly understood parameters for D^{+2} and A^{-2}. The model is highly idealized, however, and may be more satisfactory for contrasting theories than for detailed fits. Finite $\Delta E_{CT} = V(\alpha - 1)$ for solids with $\alpha > 1$ leads to[34]

$$q_n = 4x^2 \qquad\qquad q_i = 1-4x^2 \ln 2 \tag{4}$$

for the ground-state ionicity at the n-i interface, with $x = |t|/V(\alpha - 1)$. The perturbation result is supported by exact solutions[34] up to nine DA pairs. The ionicity jump $\Delta q(x) = q_i - q_n$ is unity at t = 0, as expected, decreases for x > 0, and vanishes at $x_c = 0.33 \pm 0.02$, as estimated from extrapolations of finite fragments rather than from (4). The TTF-chloranil jump of $\Delta q \sim 0.4$ leads to[34] $x \sim 0.25$ and $|t| = 0.23$ ev on evaluating V and α for the actual structure. The position and intensity of the CT absorption are then completely determined and may be compared with experiment.

The t = 0 lines in Fig. 2 shows a V-shaped minimum for ΔE_{CT} in the vicinity of the n-i interface. While the dimer has vanishing CT energy when $|n\rangle$ and $|i\rangle$ cross, the minimum is $V(\alpha - 1)$ in either the neutral or ionic lattice and is about twice the observed[29] CT peak at 0.55 ev. The CT intensity goes at t^2 for small t and increases the minimum dimer gap to $8^{1/2}|t|$. In the neutral solid, however, the energy for creating adjacent ionic $\cdots(D^+A^-)_r\cdots$ sequences is almost independent of r at the interface. Conversely, adjacent neutral $\cdots(DA)_r\cdots$ sequences have comparable energy for any r in ionic lattices. The resulting charge resonance[34]

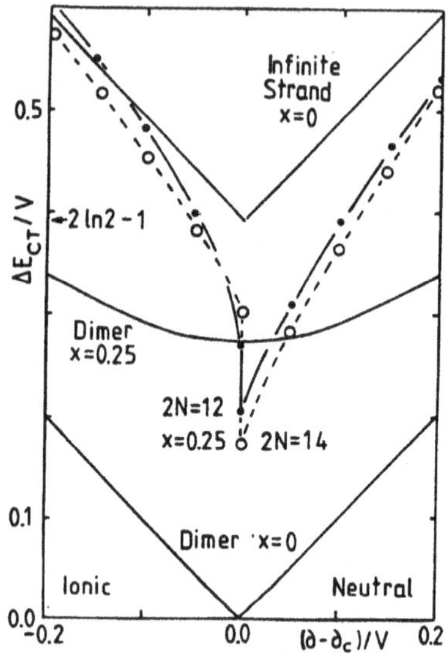

Fig. 2. Charge-transfer absorption, in units of the nearest-neighbor Coulomb interaction V, of a crystal and dimer on the neutral and ionic side of the n-i interface. Finite $x = |t|/V(\alpha-1)$ increases ΔE_{CT} for the dimer but decreases ΔE_{CT} due to charge resonance in the crystal. The indicated ΔE_{CT} are for intensity-weighted absorptions in finite rings, as given in ref. 34.

now lowers ΔE_{CT}, as shown in Fig. 2 for direct solutions to rings of 6 and 7 DA pairs. The lowering is strongest at the interface and $|t| = 0.23$ ev now underestimates the observed ΔE_{CT} by 10-20%. The oscillator strength is correctly given within the combined experimental and theoretical uncertainties. The numerical results in Fig. 2 indicate higher ΔE_{CT} on the ionic side, as observed, even in the regular chain. The evaluation of additional increases due to dimerization on the ionic side requires accurate displacements and electron-phonon coupling constants. As already noted, the parametrization of π-radical solids is considerably more difficult.

Charge resonance at the n-i interface is expected on physical grounds. A conductivity peak[36] at the n-i interface is also natural and occurs in TTF-chloranil. The importance of charge resonance decreases rapidly on moving away from the interface, as seen in Fig. 2 and understood in terms of an energy cost proportionation to the lengths of $\cdots(D^+A^-)_r\cdots$ or $\cdots(DA)_r\cdots$ fragments. Thus CT complexes with quasi-neutral or quasi-ionic ground states may reasonably be approximated by dimer or perturbative models that miss charge-resonance contributions.

Dimerized ionic lattices lead naturally to alternating Heisenberg spin chains and to triplet spin excitons, as observed[37] in M_2P-TCNQ ($q \sim 0.40$) and in many segregated-stack and inorganic chains. Triplet excitons are not seen in TTF-chloranil at $T < 82$ K, presumably because the singlet-triplet gap is too high. Rather, resolved D^+ and A^- epr signals are identified[36] through the known g-tensors of TTF^+ and chloranil$^-$. They are interpreted as "misfits" in the dimerized chains and may be viewed as topological solitons frozen in the lattice. Such partnerless spins are also seen in alternating segregated stacks. The nominally regular mixed stacks in TMPD-TCNQ, TMPD-chloranil, and other ionic complexes, by contrast, show[2] a single activated $g = 2$ resonance, approximately at the average g value. The simultaneous modeling of optical and magnetic excitations naturally provides more complete information about electronic structure.

The exact separation of spin and CT excitations requires $t \to 0$, $\Delta E_{CT} \to \infty$ such that $J = 2t^2/\Delta E_{CT}$ remains constant. This is sensible but not exact for $t \sim 0.2$ ev and $\Delta E_{CT} \sim 1.0$ ev in either mixed or segregated stacks. Finite t inevitably leads to small admixtures of real rather than virtual excited states and thus to weak dipole-allowed transitions among "spin" states that are readily found numerically.[34] Formally, then, the minimum CT gap on the ionic side becomes arbitrarily small in a regular chain.[38] The 2^N spin states form a continuum on the ionic side, while the nondegenerate ground state on the neutral side is separated by a finite gap from the lowest excited state. This is the essence of the paramagnetic-diamagnetic transition[39] of the regular array. The finite ΔE_{CT} in Fig. 2 is based on the intensity-weighted sum of CT transitions. The point is that typical values of t and ΔE_{CT} give very weak transitions among the spin states. Distinguishing between spin and CT excitations thus resembles such useful fictions as exact spin states in systems with finite spin-orbit coupling.

CHARGE FLUCTUATIONS AND VIBRATIONS

CT complexes, π-radical salts or linear polyenes all feature low-energy charge fluctuations at ΔE_{CT} or E_g in which electrons move along the backbone. Still lower-energy charge fluctuations occur in mixed-valence conductors or in complex salts, where electron motions such as $A\ A^- \to A^-\ A$ or $D\ D^+ \to D^+\ D$ are possible without creating A^{-2} or D^{+2} sites. In the context of quantum cell models with one orbital ϕ_p per site, all other electrons merely contribute to some force field, typically harmonic, for vibrations about the observed geometry. Cell models are inappropriate for determining the ground-state geometry, which for crystals can be taken as an input, and deal instead with low-energy magnetic or optical excitations. The partial double and single bonds in PA directly reflect π-electron contributions that modify the force constants associated with the σ-framework.[40] Electronic ($\pi \rightarrow \pi^*$) excitations or ionization induced by chemical doping disrupts[25] the π-bonding locally for suitable electron-phonon coupling. The resulting modulation of C-C and C-H force constants affects both the IR and Raman modes.[41].

As discussed by Rice and co-workers,[42] charge fluctuations in π-radical solids are strongly coupled to totally symmetric a_{1g} vibrations that are forbidden in the isolated molecule. Typical intermolecular separations in face-to-face stacks are less than the Van der Waals contact of ~ 3.5Å for π-systems. The resulting narrow bands ($4|t|$ ~ 1 ev) are far weaker than typical chemical bonds, but so are the energies for charge fluctuations at or below $E_g \sim U \sim 1$ ev. The distinction between half and other fillings now appears as far stronger vibronic coupling[42] in 1:2 TCNQ salts with tetramerized unit cells, or two electrons for four ϕ_p, than in 1:1 systems. Still weaker but analogous spin-phonon coupling[9] occurs in dimerized spin chains, where thermally excited triplets correspond to "antibonding" dimers.

The complete assignment of normal modes in conjugated molecules like benzene[43] or short polyenes[27] is a separate active field. One goal is to obtain transferable force constants in a reliable force field for entire classes of molecules. Such a program is not yet practical for conjugated polymers or π-molecular solids. Rather, charge fluctuations contributions to vibrational spectra are treated[41,42,44] phenomenologically by taking the relevant normal frequencies, ω_n, and dimensionless electron-phonon coupling constants, λ_n, from experiment. Pecile and coworkers[44] take ω_n for either segregated or mixed π-molecular solids from reference compounds in which the D or A are known chemically to be neutral or singly charged. Since crystal shifts of intromolecular vibrations are often small compared to difference in $\omega_n(D) - \omega_n(D^+)$ or $\omega_n(A) - \omega_n(A^-)$, comparison of several modes gives some of the best determination of partial ionicity and its variations at phase transitions.

Horowitz and coworkers[45] have developed an analogous approach to vibronic contributions in PA. Resonance Raman enhancement is seen for three modes. Although the unit cells in polydiacetylenes are far larger, only modes associated with the backbone show resonance enhancement.[46] In either case, the relevant modes are found experimentally. Three frequencies ω_n° and coupling constants λ_n are chosen for $(CH)_x$ and for $(CD)_x$ to discuss the shifts and intensities of both ir and Raman modes, the latter varying with the photon energy $\hbar\omega_L > E_g$. The polymer's inhomogeneity is explicitly recognized. Excellent fits are simultaneously obtained[45] for Raman and ir modes in related PA systems excited in various ways. Alternative models based on finite segments have also been proposed.[47]

As in the case of magnetic data fit to spin hamiltonians, however, additional questions arise in relating the model to the actual electronic structure. We consider here Horowitz's observation[45] that, in the adiabatic limit, the curvature of the ground-state energy per site, $\epsilon''(\delta_0)$, fixes the coupling to the enhanced modes. The π-electronic contribution to the assumed harmonic σ-electron force field leads to a dimerized lattice with $\delta_0 > 0$. Lattice vibrations modulating $\delta_0(t)$ in phase are sensitive to electronic excitation at $E > E_g$ or to ionization. The resonance Raman data for several $E > E_g$ then probes different PA regions with different optical gaps. The energy dependence of $\epsilon''(\delta_0)$ is consistent with uncorrelated electrons in a Peierls-Hückel model. Yet $\epsilon''(\delta_0)$ includes all electron-electron interactions in the adiabatic approximation.

Both PA and PDAs have small alternation, $\delta_0 \sim 0.07$, that modulate the transfer integrals t_p. The potential V(p,p') in (3) also reflects alternation and contributes to the stronger dimerization of PPP systems with standard molecular parameters. In the following, we retain V(p,p') for a regular, N = 4n ring (equal bond lengths, 180° bond angles) and compare[48] $\epsilon_0''(\delta,\text{PPP})/2t$ with Hückel results for the same $t_{\pm} = t(1 \pm \delta)$. The $N \rightarrow \infty$ Hückel curve in Fig. 3 shows the logarithmic divergence associated with the Peierls instability. The optical gap $E_g = 4|t_0|\delta$ is the same for all N = 4n Hückel rings and directly reflects correlations in the corresponding Hubbard or PPP models. The N = 8 and 12 PPP curves in Fig. 3 are close to each other and to the Hückel result. The area under the $\epsilon''(\delta)$ curves can readily be shown to be conserved.[48] Without extrapolation, we conclude that neither the magnitude nor the functional dependence of $\epsilon_0''(\delta)$ is particularly sensitive to PPP correlations around $\delta \sim 0.07$.

It has long been known that ground-state properties (bond lengths, rotational barriers, ir frequencies, etc.) may accurately be calculated by semi-empirical or one-electron schemes that contain large absolute errors. There is almost complete cancellation of correlation contributions associated with a given number of electron pairs in closed shells. The strong correlation effects for standard PPP parameters or for $U = 4|t|$ in Hubbard models largely cancels in $\epsilon_0(\delta) - \epsilon_0(0)$. The $\epsilon_0''(\delta)$ curves in Fig. 3 show the cancellation to persist for the curvature. Conversely, open electronic shells and excitation energies are frequently poorly described by semi-empirical methods and, as illustrated above for optical gaps and charge resonance, correlation contributions are then essential.

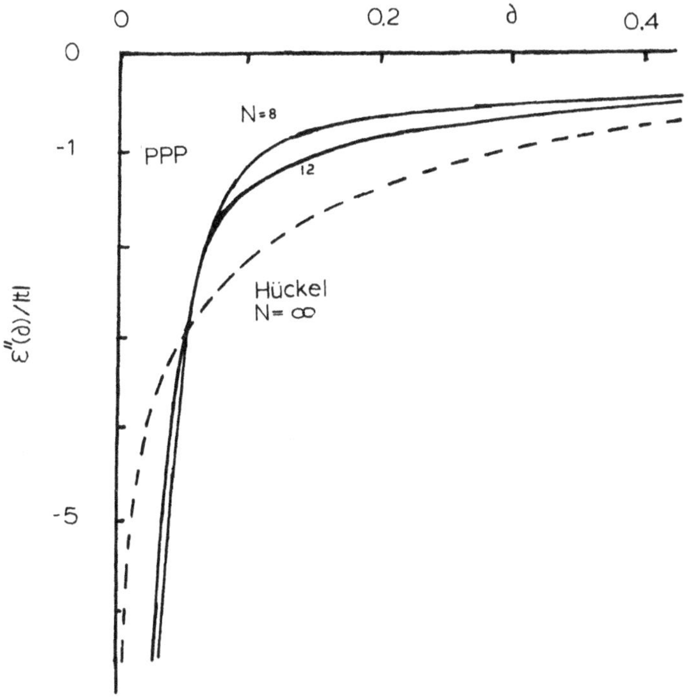

Fig. 3. π-electron contribution $\epsilon''(\delta)/|t|$ to the dimerization of infinite Hückel rings and finite PPP rings with alternation $0 \leq \delta \leq 0.40$. The divergence of the Hückel curve at $\delta = 0$ reflects the Peierls instability. PPP results with standard molecular parameters lead to similar $\epsilon''(\delta)/|t|$ around typical polymeric values of $\delta \sim 0.10$.

DISCUSSION

The electronic structure of organic ion-radical and CT solids has many aspects. Individual D or A molecules and their ions pose typical molecular problems and provide the starting point for solid-state perturbations in molecular solids. The collective magnetic, transport, and optical properties associated with partly-filled narrow bands are generally of central interest, however, and lead to quantum cell models (1) based on one frontier orbital ϕ_p per site. Both electron-phonon and electron-electron interactions are important and can readily be incorporated in (1).

Similar quantum cell models are found in chemically very disparate systems. Both inorganic and organic magnetic insulators are generally described by Heisenberg exchange models (2), although with potentially quite different sign and magnitudes of $J(R_0)$, different anisotropic or antisymmetric corrections, different linear spin-phonon coupling $(\partial J/\partial R)_0$, and different internal structure in the linear array. All these systems nevertheless have sufficiently strong on-site correlations U to separate the spin and charge degrees of freedom and, for one ϕ_p per site, to localize a spin-½ at each site. Conversely, noninteracting electrons in ϕ_p lead to simple Hückel models or s-bands and now t_p and $\alpha = (\partial t/\partial R)_0$ are of interest. The intermediate regime of $U \sim 4|t|$ is appropriate for π-radical solids and, for larger parameters, for conjugated polymers. Long-range V(R) in (3) are required for conjugated molecules, while extended Hubbard models with U and a nearest-neighbor V(a) suffice currently for π-radical solids.

Quantum cell models are inherently phenomenological. They would not be fully quantitative even in the unlikely event that exact solutions became routine. The models capture such essential features as the electrical and optical distinction between half-filled and other arrays or the magnetic distinctions between regular and alternating semiconductors. Heisenberg, Hubbard, PPP, and Hückel have become intrinsically interesting, especially when lattice deformations are included, in several areas of solid-state physics. A very large number of possibilities are encompassed. The requisite approximations for including both electron-electron and electron-phonon coupling remain areas of current interest, with occasional exact results in various limits serving as benchmarks.

Quantum cell models provide a consistent and so far adequate framework for collective properties. We have emphasized here a related experimental advantage associated with choosing favorable cases for detailed comparisons. Optical excitations, dipole moments, or spin densities are more accessible in conjugated molecules whose bond lengths and bond angles, and thus zeroth-order microscopic parameters, are retained in conjugated polymers. Magnetic properties are best explored in insulating single crystals with regular or alternating Heisenberg chains. Only qualitative comparisons may be possible for limited data in intermediate cases requiring parameters for exchange, electron transfers, correlations and electron-phonon coupling. The parameters from related but more favorable systems are then natural within a common theoretical model for the valence electrons.

The choice of ϕ_p underscores the phenomenological nature of quantum cell models. The partition into valence and core electrons is neither unique nor exact. The $2p_z$ carbon orbitals suffice for low-lying excited states, while all 2s/2p orbitals are important for bonding and geometry. Conduction and localized orbitals in metals present similar choices. As already noted, the onset of molecular excitations in π-radical solids is only slightly above E_g. Possible roles for other orbitals consequently depend on the system and the topic of interest.

In polydiacetylenes, ϕ_p is usually associated with the extended π-system normal to the PDA plane. The in-plane π-electrons of triple bonds now formally belong to the σ-framework, but their identity is well preserved[49] in small ene-yne molecules. Such π'-electrons are localized in triple bonds and blend into the self-consistent background in one-electron theory. Low-energy charge fluctuations of the π'-dimers are strongly coupled[50] to charge fluctuations of the delocalized π-electrons, however, even on neglecting small π-π' transfer integrals. Without readjusting molecular PPP parameters for triple bonds, the expanded π, π' system has[50] dipole-allowed 1^1B_n and 2^1B_u excitations shifted to the red, 2^1A_g and 1^3B_u states shifted to the blue. Coulomb correlations between π and π' electrons thus alters the splittings of the extended π system and explains the comparable E_g in PA and PDAs in spite of the stronger alternation of the latter. Correlations are again more important than alternation for the optical gap, but E_g increases with alternation even in PPP theory. Thus $\pi \rightarrow \pi^*$ transitions in ene-ynes or in PDAs point either to different microscopic parameters or to new possibilities such as Coulomb correlations between π and π'-electrons. Physical considerations motivate the inclusion of additional ϕ_p for PDA and illustrate possible evolution of quantum cell models.

ACKNOWLEDGEMENTS

It is a pleasure to thank S. Mazumdar, S. Ramasesha and S. Kuwajima for their contributions to diagrammatic valence-bond methods and to acknowledge gratefully the National Science Foundation for financial support under DMR-8403819

REFERENCES

1. Proceedings of the International Conference on Science and Technology of Synthetic Metals, Kyoto, Japan (1986), Synth. Metals, 17-19 (1987).
2. Z. G. Soos, Ann. Rev. Phys. Chem. 25:121 (1974); Israel J. Chem. 23:37 (1983); Z. G. Soos and D. J. Klein, Charge-Transfer in Solid State Complexes, in Molecular Association, Vol. 1, R. Foster, ed., Academic Press, New York (1975) and references therein.
3. J. B. Torrance, Acc. Chem. Res. 12: 79 (1979).
4. N. F. Mott, Proc. R. Soc. 62:416 (1949); Rev. Mod. Phys. 40:677 (1968); Metal-Insulator Transitions, Taylor and Francis, London (1974).
5. J. B. Torrance, J. J. Mayerle, K. Bechgaard, B. D. Silverman and Y. Tomkiewicz, Phys. Rev. B23:4960 (1980).
6. R. S. Mulliken, J. Amer. Chem. Soc. 74:811 (1952); J. Phys. Chem. 56:801 (1954); R. S. Mulliken and W. B. Person, Molecular Complexes: A Lecture and Reprint Volume, Wiley, New York (1969).
7. Z. G. Soos and S. Ramasesha, Phys. Rev. B29:5410 (1984).
8. E. H. Lieb and D. C. Mattis, Phys. Rev. 125:164 (1962). Appendix.
9. P. L. Nordio, Z. G. Soos, and H. M. McConnell, Ann. Rev. Phys. Chem. 17:237 (1966).
10. E. H. Lieb and F. Y. Wu, Phys. Rev. Lett. 20:1445 (1968); G. V. Uimin and S. V. Fomichev, Soviet Phys. JETP 36:1001 (1973); A. A. Ovchinnikov, I. I. Ukranskii, and G. V. Kventsel, Soviet Phys. Uspekhi 15:575 (1973).
11. L. Salem, The Molecular Orbital Theory of Conjugated Systems, Benjamin, New York (1964) and references therein.
12. K. Ohno, Theor. Chim. Acta. 2:219 (1964)
13. M. Sasai and H. Fukutome, Prog. Theor. Phys. 70:1471 (1983); 69:373 (1983).
14. S. Ramasesha and Z. G. Soos, Phys. Rev. B32:5368 (1985).
15. K. L. D'Amico, C. Manos, and R. L. Christensen, J. Amer. Chem. Soc. 102:1777 (1980).
16. A. H. Zimmerman, R. Gygax, and J. I. Brauman, J. Amer. Chem. Soc. 100:5595 (1978).
17. T. S. Sorenson, J. Amer. Chem. Soc. 87:5075 (1965).
18. S. Malhorta and M. Whiting, J. Chem. Soc. 1960:3812 (1960).
19. Z. G. Soos and S. Ramasesha, Phys. Rev. Lett. 51:2374 (1983).

20. Y. Iida, Bull. Chem. Soc. Japan 42:637 (1969).
21. J. B. Torrance, B. A. Scott, B. Welber, F. B. Kaufman, and P. E. Seiden, Phys. Rev. B19:730 (1979).
22. S. Mazumdar and Z. G. Soos, Phys. Rev. B23:2810 (1981).
23. J. Tanaka, M. Tanaka, T. Kawai, T. Takabe, and O. Maki, Bull. Chem. Soc. Japan 49:2358 (1976); C. Tanaka, J. Tanaka, K. Dietz, C. Katayama, and M. Tanaka, ibid 56:405 (1983).
24. K. Yakushi, M. Iguchi, G. Katagiri, T. Kusaka, T. Ohta, and H. Kuroda, Bull. Chem. Soc. Japan 54:348 (1981); T. Sugano, K. Yakushi, and H. Kuroda, ibid 51:1041 (1978).
25. W. P. Su, J. R. Schrieffer, and A. J. Heeger, Phys. Rev. Lett. 44:1698 (1979); Phys. Rev. B22:2099 (1980).
26. Z. G. Soos and L. R. Ducasse, J. Chem. Phys. 78:4092 (1983).
27. B. S. Hudson, B. E. Kohler, and K. Schulten, Excited States 6:1 (1982) and references therein.
28. J. Orenstein and G. L. Baker, Phys. Rev. Lett. 49:1042 (1982); C. V. Shank, R. Yen, R. L. Fork, J. Orenstein, and G. L. Baker, ibid 49:2660 (1982); see also ref. 1.
29. C. S. Jacobsen and J. B. Torrance, J. Chem. Phys. 78:112 (1983); A. Girlando, R. Bozio, C. Pecile, and J. B. Torrance, Phys Rev. B26:2306 (1982).
30. Y. Tokura, Y. Kaneko, H. Okamoto, S. Tanuma, T. Koda, T. Mitani, and G. Saito, Mol. Cryst. Liq. Cryst. 125:71 (1985) and references therein; Y. Tokura, T. Koda, G. Saito, and T. Mitani, J. Phys. Soc. Japan 53:4445 (1984).
31. S. Kagoshima, Y. Kanai, M. Tani, Y. Tokura, and T. Koda, Mol. Cryst. Liq. Cryst. 120:9 (1985).
32. Y. Kaneko, T. Mitani, S. Tanuma, Y. Tokura, T. Koda, and G. Saito, Phys. Rev. B (in press). N. Nagaosa and J. Takimoto, Synth. Met. 19:497 (1987).
33. Z. G. Soos, H. J. Keller, K. Ludolf, J. Queckbörner, D. Wehe, and S. Flandrois, J. Chem. Phys. 74:5287 (1981)
34. Z. G. Soos, S. Kuwajima, and R. H. Harding, J. Chem. Phys. 85:601 (1986).
35. R. M. Metzger and J. B. Torrance, J. Am. Chem. Soc. 107:117 (1985).
36. T. Mitani, G. Saito, Y. Tokura, and T. Koda, Phys. Rev. Lett 53:842 (1984).
37. D. Nöthe, W. Moroni, H. J. Keller, and Z. G. Soos, Solid State Commun. 26:713 (1978).
38. B. Horowitz and J. Solyom, Phys. Rev. B35:7081 (1987); A. Girlando and A. Painelli, ibid B34:2131 (1986) and private communication.
39. Z. G. Soos, S. R. Bondeson, and S. Mazumdar, Chem. Phys. Letters 65:331 (1979).
40. H. C. Longuet-Higgins and L. Salem, Proc. Roy. Soc. London, Ser. A 251:172 (1959).
41. E. J. Mele, Phonons and the Peierls Instability in Polyacetylene, in Handbook of Conducting Polymers, Vol. 2, T. A. Skotheim, ed., Marcel Dekker, New York (1986) and references therein.
42. M. J. Rice, N. O. Lipari, and S. Strässler, Phys. Rev. Lett. 21:1359 (1977); M. J. Rice, L. Pietronero, and P. Brüesch, Solid State Commun. 21:757 (1977).
43. A. Ozkabak, L. Goodman, S. Thakur, and K. Krogh-Jespersen, J. Chem. Phys. 83:6047 (1985).
44. R. Bozio and C. Pecile, Molecular Vibration Studies of Quasi-One-Dimensional Organic Charge-Transfer Compounds, in the Physics and Chemistry of Low-Dimensional Solids, NATO ASI Series C56, L. Alcacer, ed., D. Reidel, Dordrecht, The Netherlands (1980); A. Girlando, A Painelli, and C. Pecile, Mol. Cryst. Liq. Cryst. 120:17 (1985); M. Meneghetti, R. Bozio, and C. Pecile, Synth. Met. 19:451 (1987); R. Bozio, M. Meneghetti, C. Pecile, and F. Moran, ibid 19:309 (1987).

45. B. Horowitz, Z. Vardeny, and O. Brafman, Synth. Met. 9:215 (1984); E. Ehrenfreund, Z. Vardeny, O. Brafman, B. Horowitz, H. Fujimoto, J. Tanaka, and M. Tanaka, ibid 17:263 (1987); E. Ehrenfreund, Z. Vardeny, O. Brafman, and B. Horowitz, Mol. Cryst. Liq. Cryst. 117:367 (1985).
46. D. N. Batchelder, The Study of Electronic Excitations of Polydiacetylenes by Optical and Resonance Raman Spectroscopy, in Polydiatetylenes, D. Bloor and R. R. Chance, eds. NATO ASI Series E102, Martinus Nijhoff, Dordrecht, The Netherlands (1985) and references therein.
47. H. Kuzmany, Pure and App. Chem. 57:235 (1985); G. Brivio and E. Mulazzi, Phys. Rev. B30:876 (1984).
48. Z. G. Soos, N. A. Fisher, and G. W. Hayden, unpublished.
49. U. Dinur and M. Karplus, Chem. Phys. Lett. 88:171(1982).
50. Z. G. Soos, S. Mazumdar, and S. Kuwajima, Extended PPP Models for Polydiacetylene Excitations, in Crystallographically Ordered Polymers, D. J. Sandman, ed., ACS Symp. Ser. 337, (1987); Physica 143B:538 (1986).

INSULATING MAGNETIC CHAINS: IONOCOVALENT COMPOUNDS

M. Steiner

Hahn-Meitner-Institut
D-1000 Berlin 39

INTRODUCTION

The determination of the ground state of a magnetic system is of fundamental importance for the understanding of the basic physics leading to the particular ground state. In 3D-systems the ab initio calculation is usually impossible, since the interactions involved cannot be predicted quantitatively. However, phase-diagrams have been investigated in great detail theoretically as well as experimentally [1]. Such studies can be accomplished, since one deals with a system with true long range order (LRO). In this situation a thorough theoretical description is possible, and usually no particular care must be taken to consider quantum effects. Most of the effort in the studies has been devoted to the phase transitions between the different regions in the phase-diagram. The points in the phase-diagram, where phase-lines meet, are of particular interest, since special types of phase-transitions may occur: bicritrical, tricritical or tetracritical points are found. At these special points, the critical behaviour is different from a usual continuous phase transition. In particular the critical exponents have special values at these particular points [2]. In 1D-systems the situation is fundamentally different, since LRO exists at $T = 0$ only, if at all. Thus, phase-diagrams in external parameters, as magnetic field and temperature, can not be realized in one dimension. Thus at $T = 0K$, only phase-diagrams can be considered for a variation of the parameters in the Hamiltonian: exchange energy J and/or anisotropy energy A. Since these considerations apply to a system at $T = 0K$, whereas at $T > 0K$ fluctuations destroy LRO, it is understandable that quantum effects have to be taken very seriously. Because of this complication, detailed theoretical descriptions of the ground state of 1D-magnets have become available only recently. The situation on the experimental side is characterized by first approaches to this problem. In the first part of this review we will describe the present status of the theory in this field and the second part will deal with the experimental situation.

I. THEORY

In our discussion of the ground state of 1D-magnets, we will discuss systems which are described by the following Hamiltonian:

$$H = -2J \sum_i \vec{S}_i \vec{S}_{i+1} + A \sum_i (S_i^z)^2 \tag{1}$$

or

$$H = -2 \sum_i (J^x S_i^x S_{i+1}^x + J^y S_i^y S_{i+1}^y + J^z S_i^z S_{i+1}^z) \tag{2}$$

while (1) can be used for $S > 1/2$, (2) has to be used for systems with $S = 1/2$. One can easely classify the systems according to their respective parameters in (1) or (2): if $J > 0$, then one will find ferromagnetic correlations while for $J < 0$ antiferromagnetic correlations are observed. If in (1) $A > 0$ the spins will show easy plane anisotropy, while for $A < 0$ Ising-behaviour should occur. Up to now, it was general and well accepted knowledge that in systems with spin-dimensionality $n = 1$ ($A < 0$) the excitation-spectrum has a gap, while for $n > 1$ ($A > 0$) a low energy mode exists with no gap [3], the dispersion relation being the following for $S > 1/2$ for FM-chains:

$$\hbar\omega = 2S\,\{(2J - 2J\cos\pi q_c)\,(2J - 2J\cos\pi q_c + A)\}^{1/2} \tag{3}$$

It is well known as well that ferromagnetic and antiferromagnetic chains are quite different in their ground state behaviour: the ferromagnetic ground state, $M = \Sigma S_i$, is a "good" or a true ground state, while the Neél-state with a perfect up-down arrangement of the spins is not a "good" ground state [4]. A consequence of this is seen even in the 3D-ordered state of real quasi-1D-magnets: the observable magnetic moment of the spin, $\langle\mu\rangle$, is close to the expected moment according to the electronic level scheme for systems with ferromagnetic chains, while for antiferromagnetic chains $\langle\mu\rangle$ is much smaller than expected [3]. Such "zero-spinreductions" have been found experimentally to reduce the observable moment by as much as a factor of two [5]. Since such effects have been explained on the basis of rather conventional approaches like spin wave calculations, not much emphasis was put into the detailed study of the exact ground state of 1D-antiferromagnets. Only the effect of a magnetic field on the ground state of such a system has been studied [6]. It was Haldane, who triggered a new wave of interest in the problem of the ground state of a 1D-antiferromagnet by his now famous conjecture [7]:

The excitation spectrum of a 1D-antiferromagnet
is gapless, if $S = n/2$, n odd integer;
has a gap, if $S = n$, n even integer.

This conjecture was based upon a very complicated field theoretical approach, which will not be presented here, since it does not yield much insight into the physics of this gap. It is impossible to review all papers following Haldane's conjecture, and therefore only a few, which mark important steps in the progress of clarification and understanding of the phenomenon, can be cited. This selection clearly represents an experimentalist's viewpoint rather than the one of a theoretician.

Strong support for Haldane was the result of a computer study on corresponding finite systems [8]. On the other hand, there were others who challanged Haldane's conjecture very strongly [9]. This did not come as a great surprise, since non of the results were exact analytical results. A rather general treatment was published 1983 for Heisenberg-Ising chains leading to a very interesting phase diagram [10]. In the last years the research in this field resulted in a rather general agreement, that Haldane was right. A very recent review describes the present theoretical situation in detail [11]. The results are summarized in fig 1 and fig. 2. They represent the behaviour of a system with integer spin described by (1) at T = OK and zero magnetic field. Let us first consider the general features of the phase diagramm shown in fig. 1. For sake of clarity this is a simplified phase diagramm; some details, which are still unclear, have been omitted. For $\Delta < 0$ the ordering is ferromagnetic; interesting features appear in the region enclosed by the solid line on the left, the ordinate and the broken line. Systems in this area have xy-symmetry and order. Beyond the broken line A - C, no ordering occurs: the strong xy-anisotropy prevents ordering and leads to a singlet ground state. This means for the elementary excitations of these systems that below the line A - C in the ordered phase there is no gap, while in the singlet ground state regime a gap develops. The phase boundary between the two regions is the line A - C. Interesting enough, the phase transition along this line is of the Kosterlitz-Thouless-type [12]. Before turning to the right hand part of the phase diagram it is essential to realize that the pysical process producing the singlet ground state is a single ion property: each individual ion becomes nonmagnetic, if the the parameters Δ and A are beyond the line A - C.

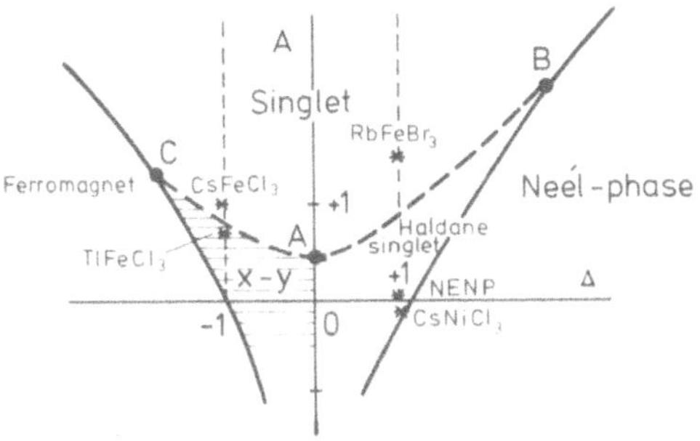

Fig. 1. Phase diagram for S = 1-chains (according to [10], [11]). Δ exchange anisotropy; A single ion anisotropy.

Turning to the $\Delta > 0$-regime, we consider antiferromagnetically coupled systems. For parameters to the right of the solid line, one finds the conventional antiferromagnetic ground state, the Neél-state. Again as in the ferromagnetic case, we find a region enclosed by the solid line on the right, the ordinate at the bottom and the broken line at the top. The properties of the phase within this region are completely different from the ferromagnetic case, where ferromagnetic ordering with xy-symmetry occured. In the antiferromagnet this is a phase with a singlet ground state as well! Since this region includes the Heisenberg AFM as well, this phase is called the Heisenberg AFM or Haldane phase. It is just here, where the Haldane prediction applies. However, this phase diagram shows that the Haldane prediction does not only apply to the Heisenberg AFM itself but as well to systems with additional anisotropy. This is very important for the possible experimental study of the Haldane prediction. The broken line A - B is a Kosterlitz-Thouless-line again as in the ferromagnetic case. But now it separates two phases with singlet ground state: the single-ion singlet state for large A and the Haldane-type singlet state. A comparative study of the properties of these different singlet states has not been performed yet. Some of the characteristics of this new phase can be seen rather clearly when considering the excitation gap along a special path in the phase diagram: one goes parallel to the A-axis with $\Delta = 1$, starting at the Heisenberg-point, and runs up to the Kosterlitz-Thouless-line. In fig. 2 the gap energy along this path is shown as predicted by the Haldane approach. The spin wave calculation would yield a gapless xy-behaviour for $0 < A < 1$ and an increasing gap for $A < 0$ in the Ising-regime. Thus no gap at the Heisenberg-point. Completely different behaviour is seen for the Haldane-approach: a gapless phase is observed only for $A = 1$ and $-1 < A < A_{critical}$. For A around zero a finite gap is observed being maximal at the Heisenberg-point.

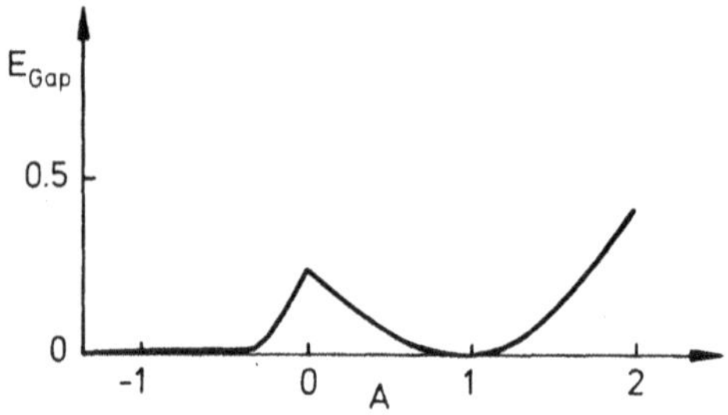

Fig. 2. Energy of the Haldane-gap versus single ion anisotropy (from [10]).

In concluding this description of the phase diagram of S = 1-chains, we should recall those results which are new and therefore interesting for possible experimental investigations. Obviously, the S = 1-AFM chain close to the Heisenberg-point is expected to show new and complex features in connection with the Haldane gap. Another interesting feature of the phase diagram of fig. 1, which does apply to both ferromagnetic and antiferromagnetic chains, is the Kosterlitz-Thouless-line, which terminates the singlet phase with decreasing A. The Kosterlitz-Thouless transition is a very special type of phase transition, which was first discussed in the context of the phase transition in the 2-dimensional xy-system [13]. It is characterized by a very particular critical behaviour [13] and could not be studied in great detail experimentally up to now, due to a lack of suitable 2D-systems.

II. EXPERIMENT

In this section, possible experimental studies of the features discussed above will be considered. This includes realizations of the appropriate systems and suitable experimental methods.

Let us start with real crystals, which can be described with Hamiltonian (1) and have the right parameter values to be in the interesting regions of the phase diagram. In order to select the proper systems, we recall what is wanted:

a) S = 1, always
b) A along the Kosterlitz-Thouless-line ($J \gtrless 0$)
c) $A \approx 0$ and $J < 0$

As far as a) is concerned, in ionic transitionmetal compounds, which will be discussed here, S = 1 would be found for Ti^{2+}, V^{3+}, Cr^{4+}, Ni^{2+} and Cu^{3+}, if a pure spin magnetism was assumed to occur in the crystal of interest. This, however, is usually not the case, and orbital contributions to the magnetism lead to anisotropy energies A of the single ion type. As A is due to crystal field effects, it is very sensitive to details and changes of the crystal structure. It is therefore very difficult to predict the size and sign of A. In table I real systems are collected, which are close to the conditions a) - c). Isotropic exchange has been assumed troughout.

Table I. Systems of Interest for Studies of the Haldane-Gap and the KT-Line

	S	J/k [K]	A/k [A]	T_N[K]	References
$CsNiCl_3$	1	16.6	-0.63	4.85	[14]
NENP	1	25	1	?	[15]
$CsFeCl_3$	1	7.40	15.4	-	[16]
$TlFeCl_3$	1	3.14	4.84	2.0	[17]

All of the systems listed in tab. I have the ABX_3-structure [5], except NENP, whose structure is orthorhombic [15]. The above systems have been put in the phase diagram, fig. 1, according to the respective microscopic parameters. It is obvious that some of them come very close to the interesting regions of the phase diagram. The Ni^{2+}-systems are all very close to the HAFM-point. Interestingly enough, there are only ferrommagnetic systems close to the KT-line, although AFM-chains exist with large A. These are far away from the KT-line, however. A few words are necessary about the rather unusual S-value of the Fe^{2+}-systems in table I: pure spin-magnetism would yield S = 2 for Fe^{2+}. Since Fe^{2+} is notorius for its orbital contribution, it is not surprising to find an effective S-value different from 2. In these particular systems the large anisotropy A leads to a singlet ground state with m = 0 and a first excited state m = ± 1. Thus at temperatures comparable or lower than the splitting, the system can be considered to have an effective S = 1.

All of the systems listed in table I have been studied experimentally, but with respect to the phase diagram studies have been performed on the Ni^{2+}-compounds only. We will discuss experiments relevant to the Haldane-conjecture first, followed by a discussion about possible experiments on KT-line.

a) <u>The HAFM-chain with S = 1</u> The main aim of such experiments must be the identification of a Haldane-gap in such systems. Although such a gap has consequences for a number of experimentally accessible quantities, like susceptibility or magnetic heat capacity, the most direct way is a measurement of the gap by inelastic neutron scattering. But even in this case, the mere detection of a gap in the excitation spectrum of a HAFM-chain does not necessarily prove that the Haldane-gap was found. This is because real "quasi-1D-near-Heisenberg-AFM's" with S = 1 do always have a finite anisotropy, which can produce a gap in the spectrum as well. As can be seen from table I, there is always some anisotropy A present in real systems available for experimental studies. Thus, more is needed than simply a gap energy, which might even be close to the expected value. The discussion of possible contributions to the gap can be performed on a very sound basis, if the microscopic parameters of the Hamiltonian (1) are precisely known. This is by no means a trivial demand, in particular if different theoretical approaches yielding different parameters can be used. A further problem concerns the isotropy of the exchange in (1): for systems with $S > 1/2$ it has not been possible up to now to unambiguously identify an anisotropic contribution to the exchange in competition with single ion anisotropy contributions [18]. Thus, we will assume the exchange to be isotropic, i.e. $|\Delta| \equiv 1$, and consider a line parallel to the A-axis through the HAFM-point in fig. 1.

The first system, which was studied with respect to the Haldane problem, was $CsNiCl_3$ [14, 19]. The basic properties of this compound were known for some time [5], but they were not sufficient to draw any conclusions about the existence of the Haldane-gap. The main obstacle was the existence of a weak interchain coupling J' and the small easy axis anisotropy A, whose values were not precisely known. In order to overcome this difficulty, the following approach was applied: below T_N in the 3D-ordered phase caused by J' there is no such effect as the Haldane-gap, and conventional spin wave theory is applicable and yields the microscopic parameters very precisely. The basis of such a calculation is the exact knowledge of the magnetic structure. A least squares fit of the so derived dispersion relations of the spin wave excitations to the experimentally derived ones yields the microscopic parameters of the Hamiltonian. Such a study by Buyers et al. [14] yielded excellent agreement between calculation and experiment for the following set of parameters:

$J/k = 16.6$ K
$J'/k = 0.29$ K
$A/k = -0.63$ K

In order to take care of the interchain interaction, (1) has to be extended to

$$H = -2J \Sigma \vec{S}_i \vec{S}_{i+1} + A \Sigma (S_i^z)^2 + 4J' \Sigma \vec{S}_i \vec{S}_k$$

The value of A so derived for $T < T_N$ does not agree with the value of A derived from a spin wave calculation at $T > T_N$ in the 1D-regime. They found, taking into account the weak interchain interaction, J', as the best 1D-value of A:

$A/k_{1D} = -1.78$ K as compared to $A/k_{3D} = -0.63$K

From this inconsistency it was concluded that at $T > T_N$ the gap was due to the Haldane-mechanism and the corresponding gap-energy was deduced

$E_g = 0.32$ $THz = 0.46 * 2 * J$

The good agreement of this value with the theoretically predicted .4(2J) and the inconsistency found for A at $T \gtrless T_N$ lead Buyers et al. to conclude that their results were the first experimental evidence for the existence of the Haldane-gap. Three weak points in their argument should be mentioned, which could not easily be clarified by their experimental procedure of using unpolarized neutrons: the 3D-magnetic structure they used was only one of two possible, by then not distinguishable, possibilities [20], and second they did not determine the eigenvectors of the different modes, which is a more stringent test of the calculation than checking the spin wave energies alone. The third point is the lack of other experimental evidence at $T > T_N$ that the observed gap is indeed inconsistent with an explanation different from the one based on the Haldane conjecture.

These points were adressed in addition to the mere measurement of spin wave dispersions in the 3D- and the 1D-regime by a detailed study of the excitations in $CsNiCl_3$ using unpolarized and polarized neutrons [21]. The dispersions observed by unpolarized neutrons agreed completely with the results of Buyers et al. Using polarized neutrons the ambiguity about the magnetic structure was removed first: it was found that the field dependence of the spin-flip- (SF) and the (NSF) non-spin-flip-scattering did fully agree with the structure considered as the most probable one and used by Buyers et al. By using the possibility of distinguishing the different spin components through their different selection rules for SF- and NSF-scattering, a determination of the eigenvectors of the different modes was possible. In order to do this properly, a magnetic field of 8 kG was applied perpendicular to the scattering plane. This field was sufficient to produce a single-domain system as observed in the Bragg scattering but was still very small with respect to the energies of the excitations. It is therefore expected that this field does not obscure the interpretation of the results.

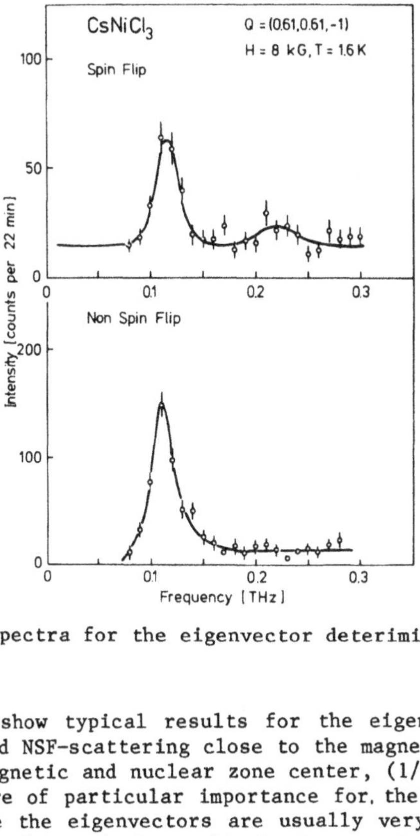

Fig. 3. Typical spectra for the eigenvector deterimination in $CsNiCl_3$.

In fig. 3 we show typical results for the eigenvector determination obtained for SF- and NSF-scattering close to the magnetic zone center. Measurements at the magnetic and nuclear zone center, (1/3, 1/3, 1) and (0, 0, 1) respectively, are of particular importance for, the eigenvector determination, since there the eigenvectors are usually very simple. This can be clearly seen in the theoretical results by Buyers et al [22] shown in fig. 4a. The corresponding experimental results are shown in fig. 4b. Inspection of fig. 4 shows that the experimentally determined eigenvectors do agree with the calculation except for the high energy mode at (1/3, 1/3, 1): this mode was assumed to be due to the fluctuation in the y-spin components by Buyers et al. [14], while the polarized neutron experiments unambigously lead to the conclusion that this mode is due to fluctuations in the x- and z-components of the spin and not the y-components. The identification of the y-components is particularly clear, since it is purely NSF-scattering in the given configuration as can be seen nicely in fig. 3. This discrepancy in the assignement of this mode is of special importance, since it is the gap energy of this mode at (1/3, 1/3, 1), which yields directely A, the gap energy being $\sqrt{J * A}$. No other mode has such a simple connection to the value of A. Since there is such a good agreement between the other measured and calculated energies, one is tempted to conclude that the y-mode has not been observed because of its low energy. This, however, needs further theoretical efforts in two directions: first, one needs the intensities of the different modes across the zone; second, the sensibility of the observed mode energies to changes in A has to be investigated. Such studies are in progress. The result of this eigenvector determination, though not fully explained yet, does not contradict Buyers's conclusion about the size of A for $T < T_N$ but rather leads to an even lower value of A. A low value of A is certainly in agreement with obserevd low spin-flop-field of 1.9 T [23].

Thus, the inconsistency in the value of A for $T \gtrless T_N$ remains and needs further investigations: assume the value of A for $T > T_N$ was due to the Ising-type single ion anisotropy; let us not discuss at the moment, why that anisotropy should change at T_N: we would then consider the case of a near Heisenberg antiferromagnetic chain with weak Ising anisotropy. This system would have a characteristic energy for the fluctuatiuons of the size of the gap energy $Eg/k \approx 5$ K. Consequentely, at $T < 5$ K the fluctuation would show Ising character: the propagating spin waves would be in the transvers spin-components only, i.e. in the x- and y-components. At $T > 5$ K the fluctuations would gradually change over to Heisenberg behaviour: spin waves show up in all three components, i.e. the cross-section would be completely isotropic. At 5 K a significant anisotropy in the cross-section is expected. Similar behaviour has been observed in $CsNiF_3$ [18]. Fig. 5 shows the spin waves at 5 K and different $\vec{Q}$ for SF- and NSF-scattering. The corresponding background has been determined under exactely the same conditions using the possibilties of the polarized neutron set up [24]. These results clearly demonstrate the complete isotropy of the cross section for spin wave scattering in the 1D-regime. This is a further evidence that the energy gap can not be explained by anisotropy but must have another origin. Since Heisenberg-behaviour is observed, this result can be considered as a further support for the assumption that the gap is a manifestation of the Haldane conjecture. In conclusion the polarized neutron experiments confirm and strengthen the interpretation of the unpolarized experiments that there is strong experimental evidence for the existence of the Haldane-gap in $CsNiCl_3$ in the 1D-regime.

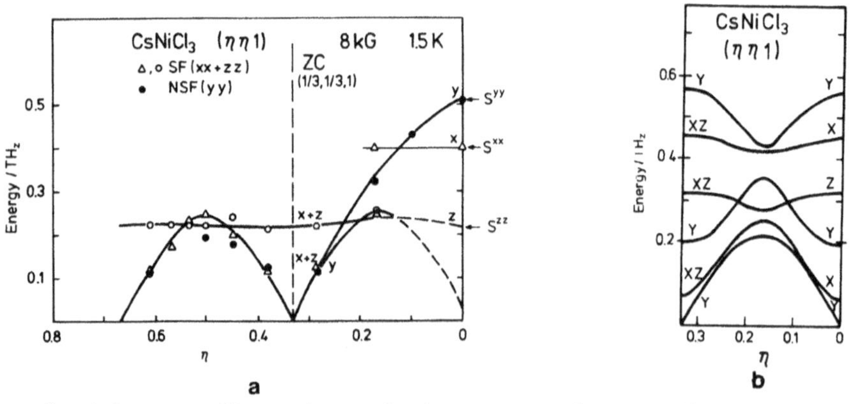

Fig. 4. Spin wave dispersion and eigenvectors for $CsNiCl_3$ at low T: a) Theory; b) Experiment.

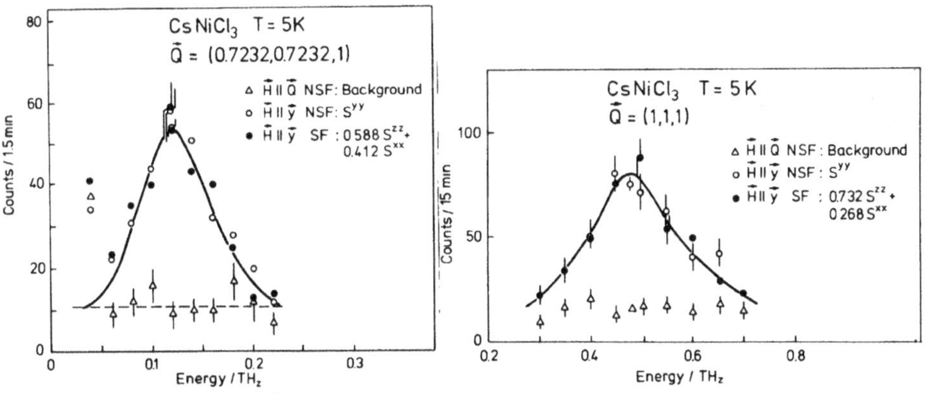

Fig. 5. Spin waves in $CsNiCl_3$ at $T > T_N$: SF and NSF at two different wave vectors Q.

Very recently a study on NENP has led to similar conclusions [15]. NENP $(Ni\ (C_2H_8N_2)\ 2NO_2ClO_4)$ has orthorhombic symmetry with a large spacing between the chains running along the b-axis [15] thus leading to a very weak interchain interaction. Single crystal susceptibility data show a very strong decrease of χ with decreasing T below 60 K. These results [25] are in agreement with $J'/J = 4 \cdot 10^{-4}$ and $J/k = 50$ K. From this drop in χ, at low T a gap energy of about 14 K was deduced [26]. χ is clearly different for all three directions, in agreement with crystal symmetry. No evidence for 3D-ordering was found down to 1.2 K. This together with the strong drop in χ with T is argued to be evidence for a singlet ground state of this system. This in turn was argued to be experimental evidence for the gap to be due to the Haldane mechanism. This line of arguments has to be supported by other experimental results, however: the interpretation of $\chi(T)$ is not known theoretically, and the orthorhombic symmetry complicates the interpretation further; the absence of long range order, LRO, for $T > 1.2$ K isn't a proof that LRO will not occur at all, since the estimation of T_N on the basis of J and J' as given in [15] is an upper limit only. First neutron scattering results for the excitations in this system show two energy gaps in agreement with the symmetry [25]: one easy-axis-anisotropy along the chain and one perpendicular to it. The gap values of .26 J and .58 J are argued to be inconsistent with the anisotropy deduced from high T-susceptibilities and thus to constitute a proof for them beeing due to the Haldane mechanism. This conclusion needs further experimental support, but NENP is a very promising system due to its very small interchain interaction J' as compared to $CsNiCl_3$. In particular, the anisotropy energy needs to be studied in more detail in order to compare their values with the observed gap energies. A low T study for $T < 1.2$ K to search for a possible 3D-ordering appears necessary as well.

Thus, up to now available experimental results on S = 1 systems do indeed yield evidence for the observability of the Haldane-gap in real quasi-1D-systems. It ist clear on the other hand that further investigations are necessary for a full understanding of this interesting phenomenon. It might be helpful for the design of future experiments to try to develop a physical picture of a system with Haldane-gap. The following picture might not contain all details, but it shows the main features, hopefully. In contrast to systems where each <u>individual</u> ion is in a singlet ground state in the Haldane case the ground state of the <u>total system</u> is a singlet as a consequence of quantum effects in the system with strong correlations. This is similar to the case of antiferromagnetically coupled pairs of Heisenberg-spins [27] studied recently: although the spin system shows complete Heisenberg-symmetry, the spectrum shows gaps due to transitions from the singlet ground state to the first excited magnetic state. Thus, one main aim of future experiments must be to find and explore the difference between the "local" and the "correlation" singlet ground state. This task would be easier, if more theoretical results were available for the behaviour of the Haldane-gap with T and H. One other possibility would be to study S = 3/2-systems in the same spirit as the S = 1-systems to find out whether they behave differently. One such study on $CsVCl_3$ did in fact yield the behaviour expected from spin wave theory in contrast to $CsNiCl_3$ [21]. Since all these studies essentially rely on the detection of inconsistencies in the standard description, a combination of different experimental techniques is of great importance.

b) The Kosterlitz-Thouless-line Since there are no experimental results available along this line or aimed at the study of the phase transition at this line, the importance of such studies will be emphasized in this section. Some possible candidates for such studies will be proposed as well.

A phase transition at the KT-line is characterized by very unusual critical behaviour and special critical indices. In particular [13] the correlation length ξ goes like

$$\xi \sim \exp[\pi/2\ (A/A_{critical} - 1)^{1/2}]$$

and the critical susceptibility χ as

$$\chi \sim \xi^{2-\eta}$$

with $\eta \approx 1/4$. Therefore

$$S(q) \sim \frac{1}{(\xi^{-2} + q^2)^{1-\eta/2}}$$

with $\eta \approx 1/4$ is expected. Up to now, the KT-transition has been adressed for 2D-systems with xy-symmetry only, the situation for which the KT-transition was discussed first. It appears, however, very difficult to find good realizations for such systems in which the statics and dynamics of the critical fluctuations can be studied in detail [13]. In 2D-systems, the peculiar critical behaviour has its physical origin in the unbinding of vortex- antivortex-pairs, characteristic for $T < T_{KT}$, into free vortices at $T > T_{KT}$. Whether such topological structures are typical for the 1D-case considered here as well, is not clear yet, but these systems are strongly nonlinear and soliton-like structures can be expected. In fact they have been observed in systems far below the KT-line [13]. Thus, experiments exploring the behaviour of the spin dynamics are badly needed for a better understanding of this interesting phase transition.

From fig. 2 it is clear that $CsFeCl_3$ [16] and $TlFeCl_3$ [17] are close to the KT-line beeing just above and below of it, respectively: $CsFeCl_3$ has a singlet ground state, while $TlFeCl_3$ is magnetic. The experimental difficulty is obviously the continous variation of A necessary to study the KT-transition. Since A is very sensitive to small lattice changes, a possible way to vary A seems to be the application of external pressure. The order of magnitude of the necessary changes to be introduced can be seen from the lattice parameters of the two systems as shown in table II.

Table II.

	a[Å]	c[Å]
$CsFeCl_3$	7.237	6.045
$TlFeCl_3$	5.989	5.960

Similar pressure experiments have been performed on $RbFeCl_3$ and showed an increase of T_N upon application of pressure [28]. Further evidence for the delicate sensitivity of A on the crystal structure is the wide scattering of $AFeX_3$-compounds across the phase diagram of fig. 1. Another possibility of going from the singlet to magnetic ground state is the application of an external magnetic field on the ferromagnetic chain systems. That this is indeed possible has been demonstrated for $CsFeCl_3$ [29]. This question should be treated theoretically in order to find out whether in this experimentally rather easy case the KT-line can be approached as well. A further complication for interpretation of the results for the above systems is the appearance of 3D-LRO in real quasi 1D-systems.

In conclusion, the conjecture made by Haldane has stimulated a new wave of investigations concerning the ground state of S = 1-chains. The theoretical results obtained up to now have revealed a phase diagram with a rich variety of very interesting phases. Two topics appear of very special interest for experiments: the Haldane-gap itself, for which some experimental evidence has been found, and the Kosterlitz-Thouless-line, at which the singlet ground state changes into a magnetic ground state. For experiments in this direction, new materials with S = 1, which are either close to the HAFM of to the KT-line, are needed. Both topic represent a challenge to experimentallists, if a clear identification of the phases and gaps is to be obtained.

REFERENCES

[1] Lindgard, P.A. in Neutron Diffraction (Dachs, H. ed.), Springer Verlag Berlin 1978, p. 197.

[2] Fisher, M.E. in Multicritical Phenomena (Pynn, R. and Skjeltorp, A. eds.), Plenum Press New York 1984, p.1;
Wolf, W.P., ibid, p. 13.

[3] Villain, J., J. Physique 35, 27, 1974.

[4] Montano, P.H., Cohen, E.H., Shechter, H., Phys. Rev. B 6, 1053, 1972.

[5] Steiner, M., Villain, J., Windsor, C.G., Adv. Phys. 25, 87,1976.

[6] Griffiths, R.B., Phys. Rev. 133, A 768, 1964.

[7] Haldane, D.M., Bull. Am. Phys. Soc. 27, 181, 1982.

[8] Botet, R., Jullien, R., Kolb, M., Phys. Rev. B 27, 613, 1983.

[9] Bonner, J.C., Müller, G., Phys. Rev. B 29, 5216, 1984.

[10] Botet, R., Jullien, R., Kolb, M., Phys. Rev. B 28, 3914, 1983.

[11] Bonner, J.C., Proc. Magn. Magn. Materials Conf. Baltimore, 1987, in print.

[12] Kosterlitz, J.M., Thouless, D., J. Phys. C 6, 1181, 1973.

[13] Steiner, M., Bishop, A.R. in "Solitons" (Trullinger, S.E., Zakharov, V.E., Pokrovski, V.L. eds.), Elsevier Publishers 1986.

[14] Buyers, W.J.L., Morra, R.M., Armstrong, R.L., Hogan , M.J., Gerlach, P., Hirakawa, K., Phys. Rev. Lett. 56, 371, 1986.

[15] Renard, J.P., Clement, S., Verdaguer, M., Proc. Indian Academy SCI. (Chem. Sci.) 98, 1987, in print.

[16] Steiner, M., Kakurai, K., Knop, W., Dorner, B., Pynn, R., Happek, U., Day, P., McLeen, G., Sol. State Communication 38, 1179, 1981.

[17] Visser, D., Knop, W., Pynn, R., Annex to ILL-report 1981.

[18] Kakurai, K., Pynn, R., Dorner, B., Steiner, M., J. Phys. C 17, L 123, 1984.

[19] Petitgrand, D., Kjems, J.K., Steiner, M., Risø Nat. Lab. (Denmark) Report No. R 517, 1985, unpublished.
[20] Cox, D.E., Minkiewicz, V.J., Phys. Rev. B 4, 2209, 1971.
[21] Steiner, M., Kakurai, K., Kjems, J.K., Proc. of Magnetism and Magn. Mat. Conf. Baltimore 1987, in print.
[22] Morra, R.M., Buyers, W.J.L., Armstrong, R.L., Briat, B., to be publisched.
[23] Johnson, P.B., Rayne, J.A., Friedberg, S.A., J. appl. Phys. 50, 1853, 1979.
[24] Shirane, G., Physica 137 B, 43, 1986.
[25] Renard, J.P., Verdaguer, M., Regnault, L.P. Erkelens, W.A.C., Rossat-Mignod, J., Stirling, W.G., Europhys. Lett., submitted.
[26] Meyer, A., Gleizes, A., Girerd, J.J., Verdaguer, M., Kahn, O., Inorg. Chemistry 21, 1729, 1982.
[27] Feile, R., Kjems, J.K., Hauser, A., Güdel, H.U., Falk, U., Furrer, A., Sol. State Communications 50, 435, 1984.
[28] Harrison, A., Visser, D., Vettier, C., Annex to the ILL-report 1984.
[29] Knop, W., Steiner, M., Day, P., J. MMM 31 - 34, 1023, 1984.

INSULATING MAGNETIC CHAINS: RECENT MODELS AND MATERIALS

Christopher P. Landee

Department of Physics
Clark University
Worcester, Massachusetts

1. INTRODUCTION

"Studies of one-dimensional systems are characterized by a rich coupling between theory and experiment because certain idealized models are exactly soluble, and because Nature offers us real physical systems that correspond remarkably closely to these idealized models." This quote, from the article by Birgeneau and Shirane in 1978 [1], is as true today as it was then and accounts for the steady interest and growth within the field of one-dimensional magnetism. Even though the first experimental studies on 1-d magnets on TMMC and CMC were reported nearly twenty years ago [2,3], and even though the seminal review article by de Jongh and Miedema appeared 13 years ago [4], the pace of work in this field is, if anything, quickening. In the four years since the Castiglione Conference on "Magneto-Structural Correlations in Exchange Coupled Systems" [5], major developments have occurred in both the theoretical and magneto-chemical areas of this discipline, developments which we shall scrutinize both in this review and in the course of the conference. It is one of the strengths and joys of low-dimensional magnetism that the breakthroughs in theory stimulate further experimental efforts and that the development of new classes of magnets push the theorists into uncharted waters.

Indeed, the quote which opens this paper is, if anything, too modest; the theorists now have many arrows in their quivers with which to attack challenging problems; it is no longer necessary to work only with models for which exact analytic solutions can be found. These techniques, such as finite chain methods, finite size scaling and quantum Monte Carlo, enable them to make detailed, quantitative statements about static and dynamic properties as well as phase diagrams. A good example of the results which can be obtained from such a multiple-front attack is found in the recent review by Bonner [6] in which the present state of knowledge of the Haldane question is outlined. On the experimental side, we no longer need to restrict ourselves to the materials Nature offers us. The past decade of work by magnetochemists has uncovered many of the rules which determine the magnitude, sign and nature (Ising, planar, Heisenberg) of superexchange interactions [7]; this rapid growth of knowledge has allowed the systematic development of many of the new materials discussed in this article and at this Conference.

In the following sections, I present a review of significant recent advances in the field of one-dimensional coordination compounds. Uniform chains are discussed briefly, followed by more lengthy reviews of homogeneous alternating chains and bimetallic, ferrimagnetic chains.

2. UNIFORM CHAINS

Uniform magnetic chains (one exchange constant, homometallic) are in a mature state of development; examples are known for most of the members of the "magnetic zoo" [4,8] and further work is more evolutionary than revolutionary in nature. However, the past several years have seen the appearance of copper pyrazine nitrate, $Cu(py)(NO_3)_2$ [9], the most well-isolated of all the Heisenberg S=1/2 antiferromagnetic chains, as well as several new members of the well-studied $[(CH_3)_3NH]MCl_3 \cdot 2H_2O$ series [10]. Both the nickel analog, NiTAC [11], and the iron analog, FeTAC [12], consist of ferromagnetic chains coupled ferromagnetically to adjacent chains in one direction; long-range antiferromagnetic order occurs in the helium region brought about by the much weaker interaction in the third dimension. Metamagnetic transitions occur in both materials at low fields. The nickel chain shows evidence of orthorhombic anisotropy while FeTAC is an Ising system for which the easy axis coincides with the chain axis; no evidence of spin-canting is seen.

The search for solitons [13] has maintained interest in S=1 systems in which the single-ion anisotropy places the singlet state low. This class of magnetic chains has received an additional stimulus from the prediction of Haldane [14] that integer spin chains are intrinsically different from those containing half-integer spins, one important distinction being the existence of a gap in the energy level spectrum of the S=1 antiferromagnetic chain. Such a gap is known <u>not</u> to exist for the S=1/2 chain but recent theoretical work [6] seems to indicate the presence of such a gap for a restricted range of exchange and single-ion anisotropies. Experimental work on the Haldane question is in its early stages and the results of the experimental tests of this conjecture are only now appearing.

3. HOMOGENEOUS ALTERNATING CHAINS

The field of alternating exchange in linear chain systems has been particularly active in the past several years. This field has been driven by its close connection to the related problem of the Spin-Peierls transition, in which a uniform Heisenberg or xy magnetic chain is coupled to a 3-d system of phonons; depending on the relative energies of lattice distortions and dimerization, such a chain can undergo a lattice distortion, breaking the chain into a series of coupled dimers [15]. The interdimer exchange is weaker by definition than the intradimer exchange by the ratio α. While several spin-Peierls systems have been actively investigated, many more alternating systems are now known which do not undergo the dimerization. De Jongh has explained this failure in terms of the relative stiffness of most of these compounds, which prevents the dimerization from occurring before long-range magnetic order is induced by the three-dimensional magnetic interactions, at which point the spin-Peierls transition becomes impossible [16].

$$H = -2J\Sigma[\vec{S}_{2i}\cdot\vec{S}_{2i-1} + \alpha\vec{S}_{2i}\cdot\vec{S}_{2i+1}]$$

Properties of many of the known alternating chain systems are listed in Table 1. The exchange strengths quoted have been derived by comparison to the predictions for magnetic susceptibility of magnetization based upon the Hamiltonian above. Note that the exchange constant is written as -2J,

TABLE 1. Properties of alternating magnetic chains. The exchange strengths J are based on the 2J Hamiltonian described in the text.

COMPOUND	EXCHANGE J/k	α	COMMENTS	REF.
		Compounds with Alternating Spacing		
$CuCl_2\cdot$(4-Mepy)	−13.4 K	0.6		23, 24
$IPACuCl_3$	−19.7 K −14 K	0.35 0.6		21, 22 29
$LiCuCl_3\cdot 2H_2O$	−5.33 K	0.89	Transition at 4.4 K	21
$PiperaziniumCuCl_3$	−12.5 K	≈ 0.6		39
$CuBr_2\cdot(N\text{-}MeIm)_2$	−10.4 K	0.4 0.1		23, 40
Cu-OTS	−18.6 K	0.90	Spin-Peierls (?)	23, 26
$(Paraquat)Cu_2Cl_6$	−19 K	0.1		29
$CuCl_3\cdot$(4-Bzpip)	+ 30 K	−0.10	Ferromagnetic dimer	24, 29
$[(CH_3)_4N]CuCl_3$	+ 29 K	?	Five alternating J's	31, 32
		Compounds with Alternating Ligands		
$Cu(3,6\text{-}dto)Cl_2$	−3.9 K	0.69	Alternating at 293 K	33
$Cu(2,5\text{-}dth)Cl_2$	−14.4 K	0.87	Transition at 26 K	34
$Cu(4,7\text{-}dtd)Cl_2$	−29.7 K	0.35		34
$Cu_2L(CH_3COO)_2 2CH_3OH$	−11.4 K	0.19		41
$Cu_2(amoxa)(CH_3COO)_2$	−310 K	> 0.0002		35
Cu(hfac)•TEMPOL	+9.5 K	−0.004		36, 37
$Cu(hfac)_2$NIT-Me	+18.5 K	?	Alternating ferromagnetic chain	38

introducing a factor of two which is frequently left out. The alternation parameter, α, lies between 0 and 1. In the case where the sign of the exchange varies as well as the magnitude, α takes on a negative sign. For convenience, the following discussion breaks the systems into those for which the alternation is due to alternating spacings within the chains and those with alternating ligands. While this distinction is useful in terms of the magneto-structural correlations, there is no difference found between the magnetic behavior of the two classes.

3.1 Alternating Spacing The chemistry literature is rich with examples of copper halide oligomers which stack together to form alternating chains [17]. When the oligomer is a dimer, as seen in Fig. 1, or a monomer, Fig. 2, the possibility exists of simple alternating exchange as illustrated in the examples to follow. However, magnetic ions can be coupled to more than just their two neighboring magnetic ions, giving rise to so-called ladder chains shown in Fig. 2-c such as $Cu(NO_3)\cdot 2.5H_2O$ [18] and $(N_2H_5)CuCl_3$ [19], or the stacked dimers found in $KCuCl_3$, Fig. 1-a. These magnets with multiple exchange pathways have an unrecognized potential as a new class of spin-Peierls systems since Sutherland [20] has predicted that spontaneous dimerization will be found in the S=1/2 Heisenberg antiferromagnetic chain with competing nearest and next-nearest-neighbor exchange. However, since the magnetic analysis of such complex systems is still in its infancy, for the remainder of the discussion I will concentrate on magnetic chains whose magnetic pathways require no more than one alternation parameter to be understood.

A clear example of alternating spacing in an alternating chain is is found in $(IPA)CuCl_3$, (IPA = isopropylammonium) [21] whose chain structure is of the type represented by Fig 1-c. The copper-chloride bonds within the dimeric units are short, 2.30 and 2.31 Å, while the bond between a copper and the chloride in the axial position is considerably longer, at 2.70 Å. [22] The magnetic data is well described by the predictions of the alternating chain with an intradimer exchange of J = -19.7 K and α = 0.35, Fig 3. The chains within lithium copper(II) trichloride dihydrate, $LiCuCl_3\cdot 2H_2O$ are structurally similar but a smaller alternation parameter has been found [21]. A phase transition is observed to occur in this compound at 4.4 K so parameters obtained by an analysis in terms of a one-dimensional model will be subject to some uncertainty.

Even though there are many systems known which display alternation in their superexchange pathways, the absence of such an alternation in a room-temperature x-ray study does not guarantee uniform chain behavior. One of the best-studied alternating chains is $Cu(4\text{-Mepy})_2Cl_2$, where 4-Mepy is 4-methylpyridine [23, 24] which is structurally uniform at room-temperature, Fig 2-a. However, a structural transition, attributed to a collective freezing of the methyl group rotations, occurs near 75 K and leads to a dimerization of the chain [23, 25]. Attempts to produce a dimerized chain through use of the asymmetric ligand 3-methylpyridine, surprisingly failed, provided the product was produced by slow crystallization. For samples prepared at greater rates, the low-temperature susceptibilities are characterized by divergent contributions superimposed on the behavior of the uniform chain; the greater the rate of crystallization, the larger the excess susceptibility [25]. This behavior has been attributed to the presence of stacking faults which break the chains into weakly interacting segments. The resulting spins on the segments with an odd number of copper ions behave as a random exchange Heisenberg antiferromagnetic spin system (REHAC). Such behavior was previously seen in the uniform chain $Cu(py)(NO_3)_2$ [9].

Alternating chains have most frequently been characterized by powder susceptibility measurements but these alone are not adequate to distinguish

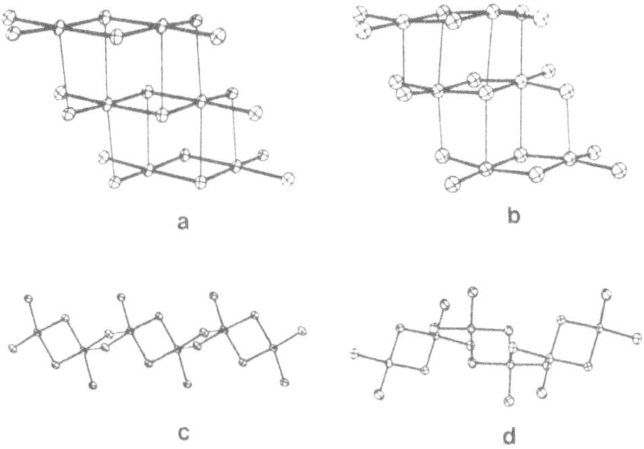

Figure 1. Stacking of dimeric units into alternating chains. Ref. 17.

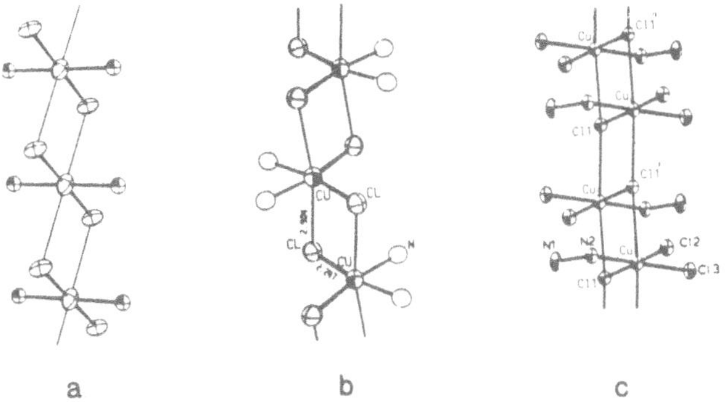

Figure 2. Stacking of monomeric units to form uniform chains, zig-zag chains and ladder chains. Ref. 17.

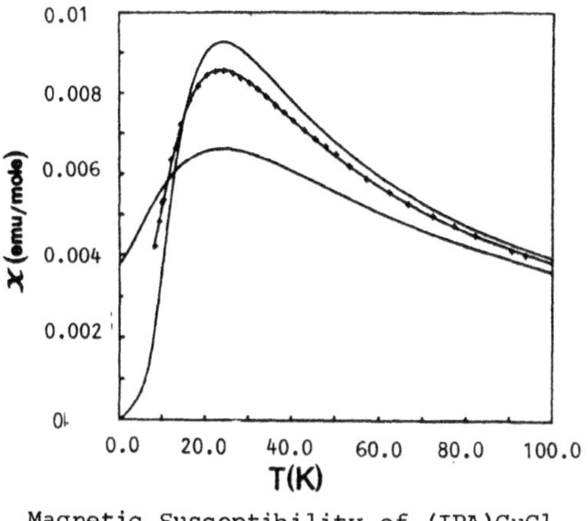

Figure 3. Magnetic Susceptibility of (IPA)$CuCl_3$. Ref. 21.

between a simple alternating chain and a system which undergoes a spin-Peierls transition. The compound Cu-HTS (catena[hexanedione bis(thiosemicarbanonato)]copper(II)) was originally described [23] as an alternating chain, based an an analysis of the powder susceptibility data. However, a more complete study involving high-field magnetization as well as susceptibility has led to an reinterpretation [26] of a system which undergoes a spin-Peierls transition near 12 K. More detailed studies of Cu-HTS and its isomorphs Cu-KTS and Cu-OTS [23] are called for.

Alternating chains are not restricted to materials which contain purely antiferromagnetic interactions. Ferromagnetic interactions are frequently found to occur in copper dimers [27] and linear chains [8]. When ferromagnetic dimers are stacked, the conditions are therefore right for an alternating system in which both the sign as well as the magnitude of the exchange may vary. (Note that the standard predictions for the susceptibility of an alternating chain [23, 28] cannot be used in such a case since they are based upon a Hamiltonian in which all the interactions are antiferromagnetic.) Such a material has been found in $CuCl_3$(4-Bzpip), where 4-Bzpip is 4-benzylpiperidinium,. The high temperature ($T > 80$ K) susceptibility can be understood as arising from ferromagnetic dimers with an exchange constant of $J/k = 30$ K; the decrease of susceptibility seen at lower temperatures is attributed to the presence of weak antiferromagnetic interactions between the dimers [29]; a mean-field approach yields a value of $J'/k = -4$ K, $\alpha = -0.13$. The high-field magnetization data [24] has been interpreted in terms of an antiferromagnetic $S = 1$ chain with an effective exchange strength of $J'(S=1)/k = -1.2$ K, yielding equivalent results.

An extreme case of alternating ferromagnetic exchange is found in TMCuC, $[(CH_3)_4N]CuCl_3$. This compound consists of infinite chains of tribridged copper ions with <u>five</u> inequivalent copper sites [30]. The symmetric Cu-Cl-Cu bridging angles are in the range 86°-90°, which lead to ferromagnetic coupling. The variations in the angles therefore lead to five somewhat different exchange parameters. There of course does not exist any theoretical model for such a system so the interpretation of the data has been in terms of a uniform ferromagnetic Heisenberg chain with an exchange constant of $J/k = 29$ K [31, 32].

3.2 <u>Alternating Ligands</u> Alternating chains with the same metal ion can also be formed by alternating the ligands bridging the ions in a ... -L-M-L'-M-L-... pattern. The respective exchange strengths mediated by the two different ligands can be much different so the alternation parameter can become quite small. An illustration of this behavior is seen in the alternatingly bridged compound catena-dichloro(3,6-dithiaoctane)copper(II), Cu(3,6-dto)Cl_2 [33]. Copper ions are bibridged along the a direction by alternating sulfur and chlorine bridges, Fig. 4; an important feature of this packing is the relative rotation of the Cu_2Cl_2 and Cu_2S_2 planes. The magnetic data, Fig. 5, was fit to the alternating chain model with an exchange parameter of $J/k = -3.9$ K and an alternation parameter of 0.69. Replacing the substituted octane by either 2,5-dithiahexane or 4,7-dithiadecane yields structures in which the bibridged copper dimers are maintained but in which the room temperature packing in the solid state has been modified; the Cu_2S_2 bridging units are no longer formed [34]. The room temperature structure is such that a given copper atom is nearly equidistant from both neighboring sulfur atoms. Evidence of a phase transition has been seen in the powder susceptibility data of the Cu(2,5-dth)Cl_2 near 26 K so the packing of the dimers in this compound at low temperature is unknown. The exchange strength of the low-temperature phase of the (2,5-dth) compound was found to have increased significantly to $J/k = -14.4$ K with an alternation parameter of 0.87. The phase transition has not appeared to have altered the packing within the dimers, since the data above the transition could be fit to a model of a simple Heisen-

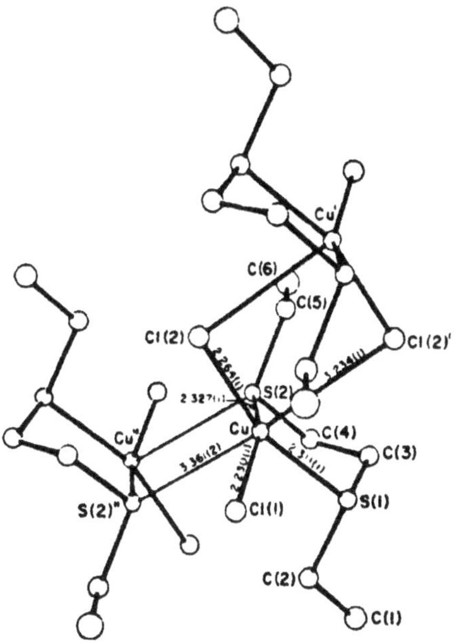

Figure 4. Structure of the dimers in (Cu(3,6-DTO)Cl_2. From Ref. 33.

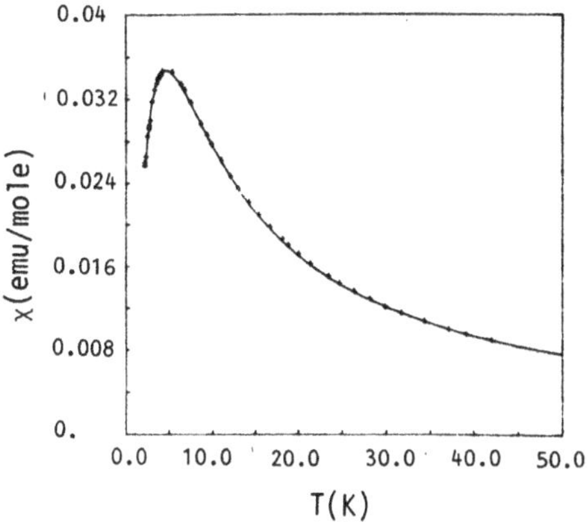

Figure 5. Magnetic Susceptibility of Cu(3,6-DTO)Cl_2. Ref. 33.

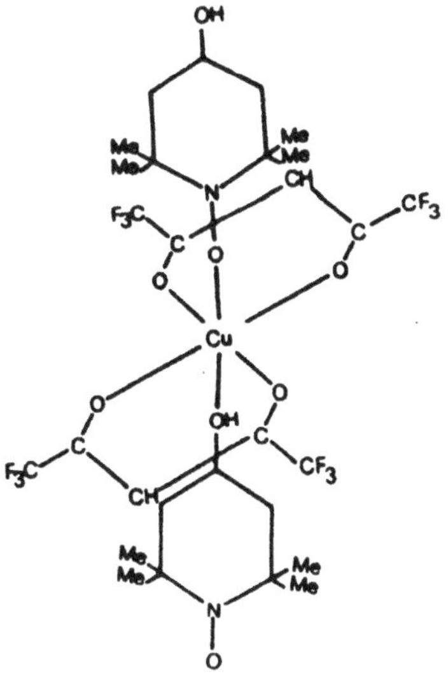

Figure 6a. The structure of Cu(hfac)•TEMPOL. Ref. 37.

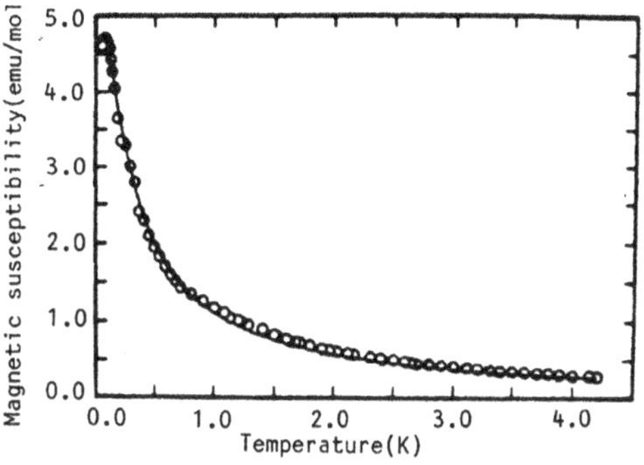

Figure 6b. Low temperature magnetic susceptibility of Cu(hfac)•TEMPOL. Ref. 37.

berg dimer with the same exchange parameter. The J/k, α parameters for the (4,7-dtd) compound were found to be -29.7 K and 0.35, respectively.

An extreme case of alternation is found in the behavior of $Cu_2(amoxa)(CH_3COO)_2$, where amoxa = N,N'-bis-(2(aminomethyl)-2-methylpropyl)oximidate [35]. The crystal structure consists of zigzag chains of copper atoms alternately bridged by amoxa and acetate groups. The acetate group acts as a monodentate ligand, contributing an oxygen atom to bridge between two adjacent coppers while the amoxa group links copper pairs by contributing one oxygen and two nitrogens to each, completing the square planar configuration of the coppers. The magnetic susceptibility shows evidence of very strong antiferromagnetic interactions and could be modeled with the Bleaney-Bowers equation with an exchange strength of J/k = -310 K; this strong interaction was attributed to the oxamidate ligand. No alternation parameter was necessary to fit the magnetic data but the absence of any hyperfine splitting or half-field transition in the EPR spectra indicated that the interdimer coupling αJ/k had to be of the order 0.1 K, yielding a lower bound on $\alpha > 0.0002$.

Alternating ferromagnetic/antiferromagnetic exchange has also been found in the linear chain $Cu(hfac)_2$•TEMPOL, bis[(hexafluoroacetyl)acetonato][(4-hydroxy-2,2,6,6-tetramethylpiperidinyl] -N-oxy]copper(II) [36, 37],. Susceptibility data above 4 K and EPR data were originally interpreted [36] as arising from weak ferromagnetic coupling between the copper ions and the unpaired electron on the nitroxide radical, Fig. 6-a. The data were fit to an isolated dimer model with a ferromagnetic exchange constant of J/k = 9.5 K. Measurements at lower temperatures revealed a peak in the susceptibility near 80 mK, indicating the existence of antiferromagnetic interactions between the triplet states, Fig 6-b. The data could be fit to the model of a S=1 Heisenberg antiferromagnetic chain with an antiferromagnetic exchange strength J'/k = -39 mK; consequently the alternation parameter was determined to be $\alpha = -0.004$.

A similar ferromagnetic interaction between copper ion and a nitronyl nitroxide radical has been observed in the analogous compound Cu(hfac)NIT-Me, where NIT-Me = 2,4,4,5,5-pentamethylimidazoline-3-1-oxy [38]. Unlike the TEMPOL compound, however, the alternating exchange constant is also ferromagnetic, giving the first example of a ferromagnetic chain involving alternating spins. The susceptibility data in the range 6-300 K could be analyzed with a Heisenberg hamiltonian for a uniform S=1/2 ferromagnetic chain with an exchange strength of +18 K; it is not possible to form an estimate for the amount of exchange alternation in such a chain.

4. FERRIMAGNETIC CHAINS

The major advance in magneto-structural chemistry in the past four years has been the systematic development of structurally ordered bimetallic chains. An entire new class of magnetic materials are now available in which metal ions can be substituted with ease, with the consequent control of the size of the magnetic moments and the size and nature of the exchange interaction. Three different families of these compounds are now known to exist, all of which consist of chains in which two crystallographically distinct metal sites alternate. Ordered bimetallic compounds arise when different metal ions show distinct preferences for one site over the other. Since the different metal ions have different moments, the chains consist of alternating moments coupled, in every known case, by an antiferromagnetic exchange. The Hamiltonian used to characterize the magnetic behavior of these chains is listed below where, once again, the exchange constant is written as 2J. Properties of the known ferrimagnetic chains are displayed in Table 2.

TABLE 2. Properties of ordered bimetallic chains. The exchange strengths J are based on the 2J Hamiltonian described in the text. The letters inside the parentheses refer to the model employed: (FC) refers to a finite chain calculation based on a Heisenberg interaction between alternating S=5/2, S=1; (H) = Heisenberg; (I) = Ising; (Cl) = classical spin model.

COMPOUND	EXCHANGE J/k	COMMENTS	REF.
(Dithio)oxalate Family			
$MnCu(DTO)\cdot 7.5H_2O$	−20.4 K	min. in χT at 130 K	43–45
$MnA(DTO)\cdot 7.5H_2O$	none	diamagnetic A = Ni, Pd, Pt.	43, 44
$CuNi(oxalato)\cdot 4H_2O$	−38 K	less well isolated chains, no minimum observed in χT	46
EDTA Family, $MM'EDTA\cdot 6H_2O$ (M on hydrated site, M′ on chelated site)			
$MnNiEDTA\cdot 6H_2O$	−0.75 K (FC) −0.93 K (∞)	minimum in χT at 2.8 K T_c = 0.66 K	48,50
$MnCoEDTA\cdot 6H_2O$	−0.95 K (I)	T_c = 1.06 K	51
$CoCoEDTA\cdot 6H_2O$	—	weak exchange, dimeric	47,49
$CoCuEDTA\cdot 6H_2O$	—	AFM, dimeric	47,49
$CoNiEDTA\cdot 6H_2O$	—	AFM, no maximum in χ	47
$NiNiEDTA\cdot 6H_2O$	−4.17 K (H) −4.12 K (I)	No divergence seen in χT	47, 52
$CuX_2\cdot 4/3TMSO$ Family (TMSO = tetramethylenesulfoxide)			
$CuBr_2\cdot 4/3TMSO$	J_d/k = 18 K J_m/k = −1.5 K	Ferromagnetic dimer Antiferromagnetic to monomer	53
$CuCl_2\cdot 4/3TMSO$	J_d/k = 16 K J_d/k = 1.3 K	Ferromagnetic dimer Ferromagnetic to monomer	53
1,3-Propylenebis(oxamato) Family, $MM'(pba)(H_2O)_3\cdot 2H_2O$			
$MnCu(pba)(H_2O)_3\cdot 2H_2O$	−16.9 K (FC)	Minimum in χT at 115 K Maximum in χ at 2.2 K	54, 55
$NiCu(pba)(H_2O)_3\cdot 2H_2O$	−59.9 K	Minimum in χT at 83 K Maximum in χT at 7.3 K	55
$MnCu(pbaOH)(H_2O)_3$	≈ −17 K (?)	Orders at 4.6 K	56

$$H = -2J\Sigma[\vec{S}_{2i} \cdot (\vec{s}_{2i-1} + \vec{s}_{2i+1})]$$

The distinctive magnetic behavior of these chains arises from the inability of the antiferromagnetic interaction to cancel the alternating, unequal moments. As the temperature is lowered, the antiferromagnetic coupling first reduces the effective moment as the adjacent spins are rendered antiparallel. At still lower temperatures, the net moment within each unit cell begin to align parallel as the correlation length grows, causing a ferromagnetic-like divergence of the product χT; in between the divergence and the decline is the minimum in χT, which is taken as the characteristic of the ferrimagnetic chain. A lucid discussion of compensation within the chains is found in the 1986 article by Curély, Georges and Drillon [42].

We note in passing that ferrimagnetic behavior can in principle be observed even in a homometallic chain which contains two distinct sites: the different g-factors corresponding to the two sites will lead to different effective moments and ferrimagnetic behavior, albeit at low relative temperatures.

4.1 Dithiooxaloto Family The first report of ordered bimetallic linear chains was by Gleizes and Verdaguer [43] for $AMn(S_2C_2O_2)_2(H_2O)_3 \cdot 4.5H_2O$, where $S_2C_2O_2$ is dithiooxaloto (DTO) and A has been shown [43, 44] to be Cu, Ni, Pd or Pt. The structure consists of extended ziz-zag chains of the formula

$$A(S_2C_2O_2)Mn(H_2O)_3(O_2C_2S_2)A$$

plus isolated water molecules, Fig. 7. The A atoms are tetracoordinated in a plane to the four sulfur atoms of the two bridging DTO groups while the manganese atoms are heptacoordinated to oxygen atoms located at the vertices of a pentagonal bipyramid. Four of the oxygens come from the DTO groups while the other three are from water molecules. The chains are isolated from one another by the presence of loosely bound lattice waters.

The reported magnetic behavior [44, 45] for this family of compounds is clearly understood in terms of the structure of the chains and the lattice. Firstly, the site of the A ions is square planar which will render diamagnetic a d^8 ion such as Ni^{2+}. The compound then consists of isolated Mn^{2+} ions with no exchange pathway available down the chains. The product χT is consequently constant down to 35 K, below which it rapidly decreases to zero; a Curie-Weiss analysis of this data yields a value for the interchain coupling of $J'/k = -0.1$ K for the [Ni,Mn] compound. The copper 2+ ion will retain its moment in a square planar environment and the [Cu,Mn] chain is therefore a magnetic bimetallic. The product χT is shown in Fig. 8 where it is noted that the product shows a shallow minimum near 130 K before diverging sharply at lower temperatures, reaching a peak value at 7.5 K. It should be noted that the presence of the lattice waters is critical in isolating the chains well enough to observe the ferrimagnetic divergence in χT; the initial report [43] on this compound had studied a partially dehydrated sample and missed the effect. Analysis of the magnetic data in terms of an alternating quantum/classical chain leads to a Cu-Mn exchange strength of -20.4 K. Comparison of this exchange strength to the weak coupling found between the isolated Mn ions in [Pd,Mn] leads to an estimate of the J'/J ratio as a few percent.

Related to the dithiooxalato series described above is the single example known of a bimetallic oxalato chain $CuNi(C_2O_4)_2 \cdot 4H_2O$ [46]. Although not available as single crystals, an EXAFS study indicated that the

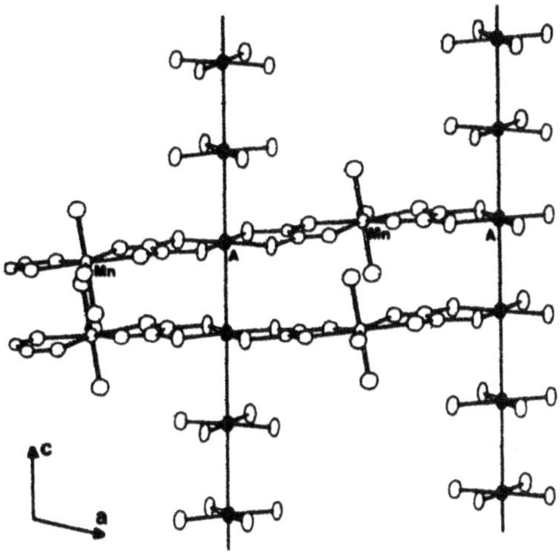

Figure 7. Structure of $AMn(DTO) \cdot 7.5H_2O$. Ref. 44.

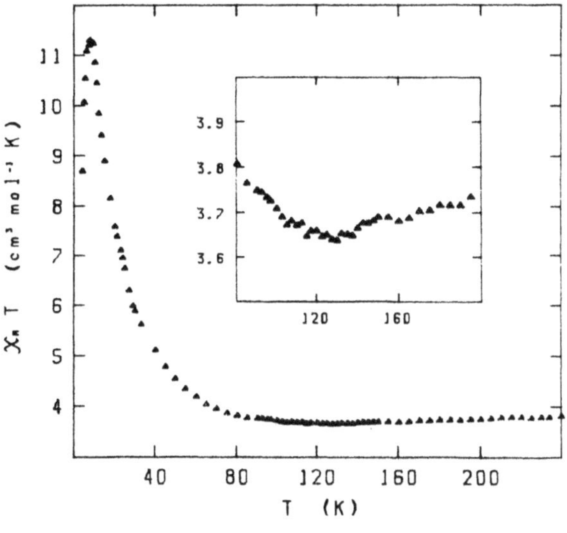

Figure 8. Magnetic susceptibility of $CuMn(DTO) \cdot 7.5\ H_2O$, plotted as the product of susceptibility times temperature versus temperature. Ref. 45.

chains were ordered. The nickel ions were proposed to be coordinated to six oxygens, four from the oxalate groups and two from water molecules in the axial positions; the copper ions were said to be coordinated to only four oxalate oxygens, but also see two other coppers from adjacent chains. The magnetic data demonstrate the existence of antiferromagnetic interactions but the expected minimum in the product of χT was not seen; this failure is probably due to the relatively large coupling between the chains via the Cu-Cu contacts. Comparison of the high-temperature data to the results of finite rings of (S=1, S=1/2) leads to an estimate of the Ni-Cu exchange strength of J/k = -38 K.

4.2 EDTA Family A more versatile bimetallic system was reported [47] in 1982 by Beltran, Escriva and Drillon, $MM'EDTA \cdot 6H_2O$, where EDTA = $[C_2H_4N_2(CH_2COO)_4]^{4-}$, ethylenediamine tetraacedic acid, Fig. 9. The hexadentate EDTA ligand provides a distorted octahedral environment for a metal ion (the "chelated position"), bonding to it with two nitrogen and four oxygen atoms. A second site (the "hydrated position") is provided by the four oxygens from water molecules and two other oxygens from the carboxylate groups from the adjacent anions. The structure then consists of infinite chains of alternating "hydrated" and "chelated" octrahedra bridged by carboxylate groups.

$$-M(H_2O)_4-M'EDTA-M(H_2O)_4-$$

For convenience of notation, members of the $MM'EDTA \cdot 6H_2O$ family will be denoted by the symbol (M,M'), where the metal on the hydrated site will be listed first.

The versatility of this family is due to the regularity of the two metallic sites: unlike the heptacoordinate site in the DTO compounds which are only occupied by Mn ions, both sites in the EDTA family are pseudo-octahedral and can be occupied by many different ions. As an example of the flexibility allowed, consider the case of the cobalt ion. Chains have been prepared in which the cobalt ions occupy only the hydrated sites, (Co,Ni) or (Co,Cu), [47], only the chelated sites, (Mn,Co) [48] or both sites in the same chain, as in the homometallic (Co,Co) [47].

While the EDTA system provides the possibility of investigating a wide variety of ferrimagnetic systems, with Ising or Heisenberg exchange and many choices for the alternating spin values, it is at the same time an alternating chain of the alternating spacing type. For the (Ni,Ni) chain, the distance from the hydrated nickel to the chelated nickel is 5.517 Å while the subsequent distance to the next hydrated nickel is 5.890 Å. The inequivalence of the bond distances is larger when two different metals occupy the sites. So far, most analyses of the magnetic data in the EDTA system have been based upon the assumption of uniform chains, ignoring the alternation in the exchange parameter, although recent studies of the weakly interacting (Co,Co) and (Co,Cu) chains have shown behavior characteristic of dimerized systems [49]. Since the models for ferrimagnetic chains are in early stages of development, it is difficult to estimate the size of the error introduced by this assumption; that it has been made should be borne in mind when comparing exchange strengths.

The members of the EDTA family of ferrimagnetic chains which have been characterized magnetically are listed in Table 2. The cobalt ion appears in four of these six materials but in three of them, (Co,Co), (Co,Cu), and (Co,Ni), the interactions are weak and no characteristic minimum in χT is observed. The (Mn,Ni) member was the first compound in which the minimum was demonstrated [50], appearing at 2.8 K, Fig 10-a. The data have been fit to the results of a finite chain calculation of a Heisenberg chain of alternating S=5/2, S=1 units, with a consequent exchange strength of J/k =

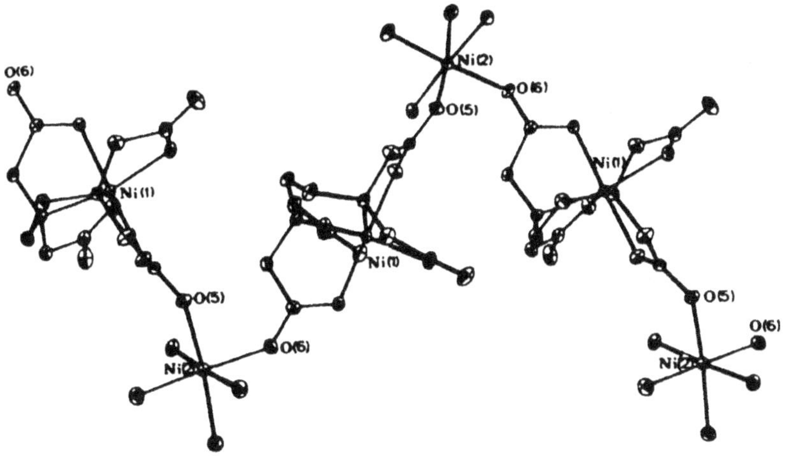

Figure 9. Structure of NiNiEDTA·$6H_2O$. Ref. 47.

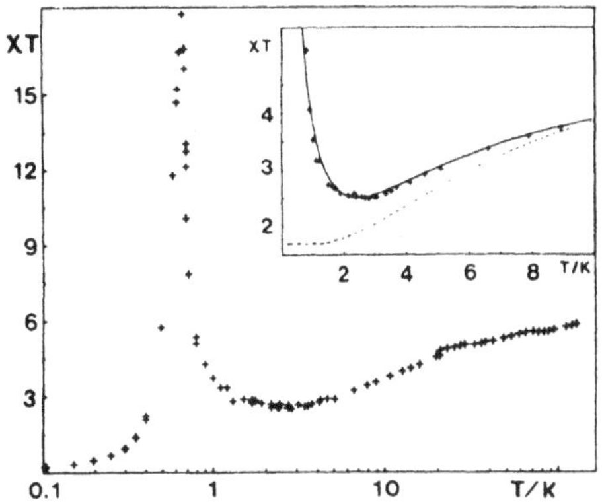

Figure 10a. Product of χT vs. temperature for MnNiEDTA·$6H_2O$. Ref. 48.

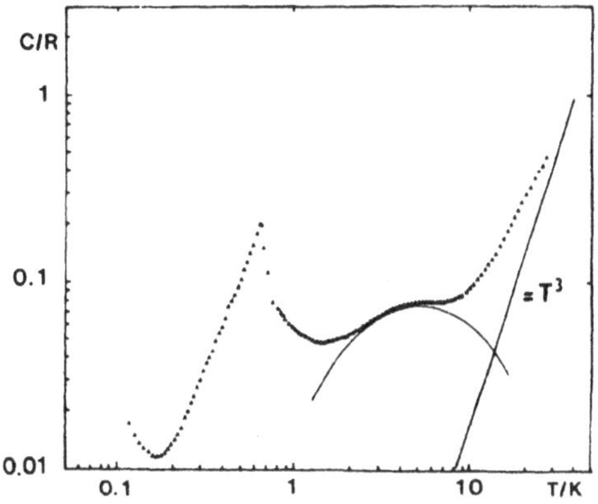

Figure 10b. Specific heat vs. temperature for MnNiEDTA·$6H_2O$. Ref. 48.

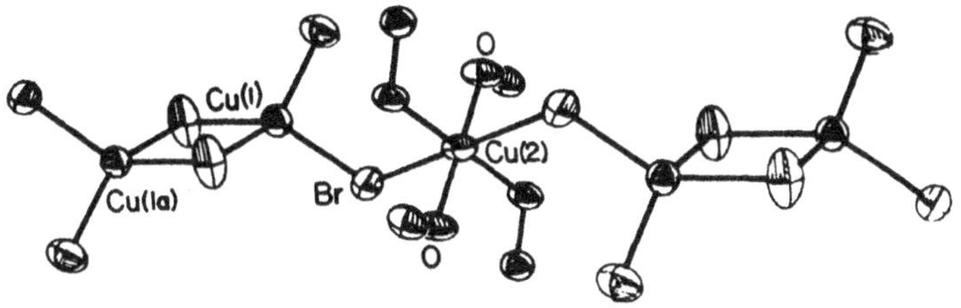

Figure 11. Structure of the chains in $CuBr_2$·4/3TMSO. Ref. 53.

−0.75 K [48]. It is clear that the EDTA groups provide a much weaker exchange between the adjacent spins than do the dithiooxalato group. A measure of the adequate isolation between the chains is the low critical temperature of $T_c = 0.66$ K, as determined by specific heat measurements [48], Fig 10-b. One would expect the isolation of the similar {Mn,Co} chain to be comparable, yet the critical temperature is sixty percent higher, occurring at 1.06 K. The origin of the difference may lie within the nature of the exchange interactions within the respective chains. While the {Mn,Ni} chain was modeled with a Heisenberg interaction, the data for the {Mn.Co} could be fit to the predictions for a S=5/2, S=1/2 Ising chain, as obtained with a transfer matrix technique [42]. While the exchange strength of the cobalt analog was only 26 % larger at $J/k = -0.95$ K, the nature of the Ising interaction is to produce much longer correlation lengths than the Heisenberg interaction at a given reduced temperature. Since the 3-d transition occurs (in the molecular field picture) when correlated blocks of spins are coupled by the interchain interaction J', an Ising system will order at a higher temperature than a corresponding Heisenberg system with an identical J'/J ratio.

Ferrimagnetic behavior is possible even within homometallic materials, provided the local g-factors on the alternating sites are different. Theoretical studies [51,52] show that the susceptibility of an antiferromagnetic chain of identical S=1 ions can diverge at low temperatures due to the non-cancellation of the somewhat different moments. The divergence is pronounced in an Ising model but occurs at very low temperature in the Heisenberg model. The susceptibility of the {Ni,Ni} chain [47] is characterized by a maximum near 10 K; no divergence is seen at lower temperatures. The data have been explained in terms of both Heisenberg and Ising models; the exchange constants obtained are comparable ($J(H)/k = -4.17$ K; $J(I)/k = -4.13$ K) but the Ising model surprisingly gives a better overall fit to the data [52].

4.3 TMSO Family Ferrimagnetic behavior is easier to demonstrate in homogeneous chains when there is an odd number of metal ions per unit cell. This is the origin behind the observed behavior of the isostructural compounds, $CuX_2 \cdot 4/3$TMSO, where X = Cl, Br [53]. The structure of these triclinic compounds, full name tetrakis(tetramethylenesulfoxo)copper(II)hexachlorodicuprate(II), consists of chains of alternating dimers and monomers, Fig. 11. The dimers consist standard $Cu_2X_6^{2-}$ anions built around a center of symmetry; long bonds from terminal halides on two different dimers to a monomeric copper links all the coppers into an alternating chain. The coordination sphere about the monomer consists of four closely bonded oxygens from the TMSO groups, as well as the two long bonds to the terminal halides. The bulk of the sulfoxide groups provides good isolation between the chains.

The magnetic susceptibility of the bromide analog of $CuX_2 \cdot 4/3$TMSO is shown plotted as χT versus T in Fig. 12. For both analogs the data are characterized by a gradual rise in the effective moment which is attributed to the presence of ferromagnetic interactions within the dimers, characterized by the parameter J_d. Below 20 K, the two data sets differ greatly, χT of the bromide undergoing a steep decline as shown while χT of the chloride ascends sharply. This change of behavior reflects the the onset of the weaker coupling J_m between the ferromagnetic dimers and the monomers. The systems are best described as alternating chains in which the alternation pattern involves three exchange strengths, two of which are identical by symmetry:

$$\cdots - J_d - J_m - J_m - \cdots$$

where the temperature dependence of the data indicates that $J_d \gg J_m$. If both

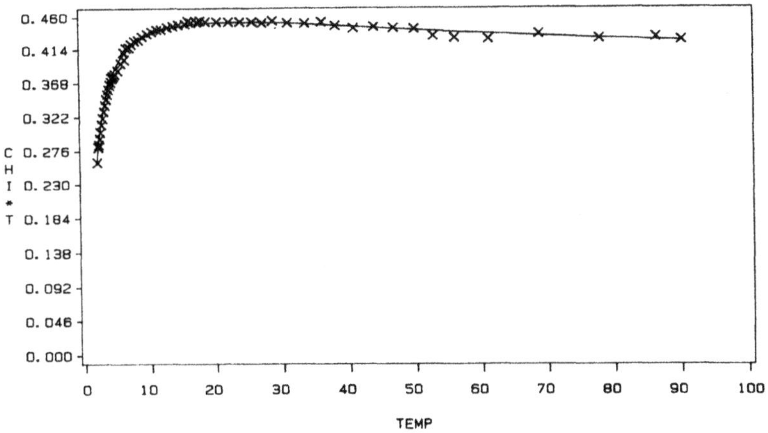

Figure 12. Magnetic susceptibility times temperature vs. temperature for $CuBr_2 \cdot 4/3TMSO$. Ref. 53.

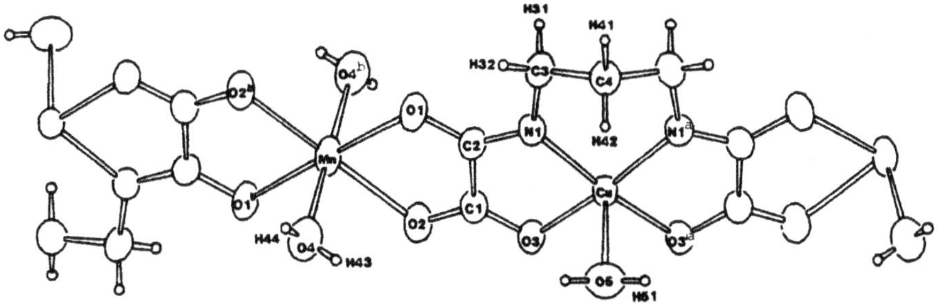

Figure 13. Structure of the chains in $MnCu(pba)(H_2O)_3 \cdot 2H_2O$. Ref. 54.

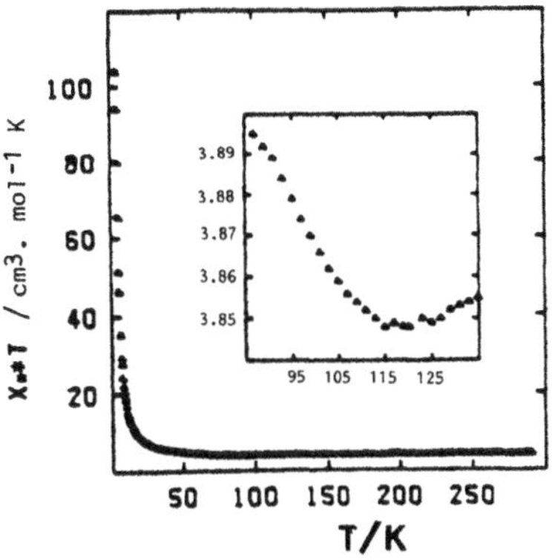

Figure 14a. Product of χT vs. T for $MnCu(pbaOH)(H_2O)_3$. Ref. 56.

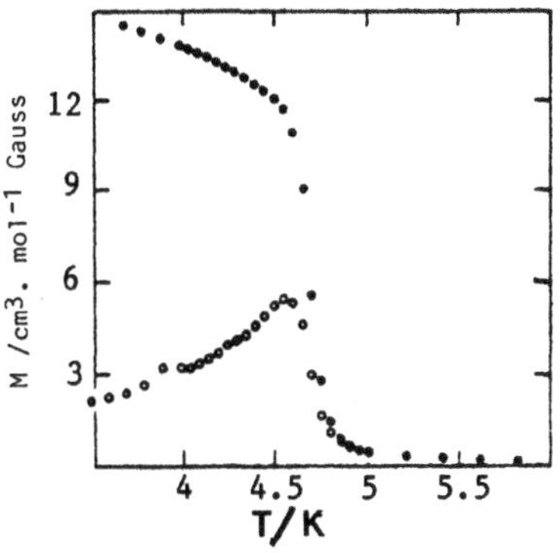

Figure 14b. Magnetic Moment vs. T for $MnCu(pbaOH)(H_2O)_3$. Ref. 56.

the exchange constants are positive, at low temperatures the chain becomes an alternating S=1, S=1/2 ferromagnetic chain, with a consequent divergence in susceptibility. Even if J_m is antiferromagnetic, the material acts as an alternating S=1, S=1/2 antiferromagnet with the subsequent ferrimagnetic behavior at temperatures low compared to J_m.

The data have initially been compared to the predictions for the susceptibility of isolated trimers with two exchange constants, in order to obtain values for the exchange strengths of the ferromagnetic dimers. The results for the bromide are plotted as the solid line in Fig. 12 and show $J_d(Br)/k = 18(4)$ K and $J_m(Br)/k = -1.5$ K. The values obtained from the fit for J_m have been divided by two to adjust for the two exchange paths between the monomers and the adjacent dimers. The corresponding values for the chloride are $J_d(Cl)/k = 16(3)$ K and $J_m(Cl) = 1.3$ K. These values for J_m are only approximate since they are based on the isolated trimer model but they do justify the description of these materials as S=1,S=1/2 ferrimagnetic chains at low temperatures. Note that the characteristic divergence in such an antiferromagnetic chain is not predicted until a temperature of 1.36 $|J|/k$ [46].

4.4 1,3-Propylenebis(oxamato) Family The most recently discovered of the ferrimagnetic systems is based upon polymerization of the mononuclear dianion $[Cu(pba)]^{2-}$, where pba is the title ligand above [54, 55]. The resulting chains are shown below in Fig. 13. The copper ions are located in an approximately square planar site, coordinated to two oxygens and two nitrogens from the pba ligand, plus an oxygen from the water molecule in the axial site. The alternate ion (Mn^{2+} or Ni^{2+}) is situated in a center of symmetry and has an elongated octahedral environment; the equatorial positions are filled by oxygens from the oxamato bridges and the axial positions by water molecules. This structure is uniform, not alternating, and employs the oxamato bridge, an effective superexchange pathway. In addition, the separation between metal ions on adjacent chains is 5.21 Å for the (Mn,Cu) compound, which provides for better one-dimensional isolation than in the corresponding $MnCu(DTO)\cdot 7.5H_2O$ chain.

Both the (Mn,Cu) and (Ni,Cu) pba compounds show magnetic behavior characteristic of the well-isolated ferrimagnetic chains. Minima are seen in χT at 115 K and 83 K, respectively, before the products climb to large values at low temperatures; a maximum in χT appears at 7.3 K for the (Ni,Cu) compound while the corresponding maximum for the (Mn,Cu) is found at 2.3 K. The exchange strength for the Mn compound was found by comparison of the data to the predictions of the model of Verdaguer et al [45] in which one chain is assumed to be classical and the other is quantum. The resulting fit is excellent with a $J/k = -16.9$ K. This value is comparable to that found for the Mn-Cu exchange in the DTO chains [45]. Likewise, the exchange in the (Ni,Cu) pba compound was found to be -60 K, even larger than the value of -38 K determined for the corresponding oxalate chain [46].

It is fitting to close the analysis of the current status of ferrimagnetic chains with a discussion of the first linear chain system which undergoes a transition into a state of 3-d long range order with a spontaneous moment [56]. The compound is $MnCu((pbaOH)(H_2O)_3$, where pbaOH differs from pba only by the replacement of the central methylene group on the propylene chain by CHOH. The character of the antiferromagnetic interactions within the chains is virtually identical to that found in the (Mn,Cu) pba chain, as demonstrated by a minimum in χT seen at the same temperature of 115 K, Fig 14-a. The important difference lies is the relative orientation of adjacent chains. Whereas the pba chains are related by a simple unit cell translation (a), adjacent pbaOH chains are displaced by one-half unit cell along the chain axis (b).

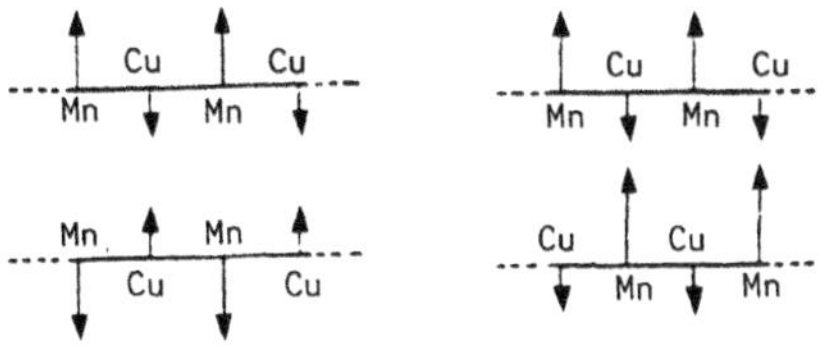

The magnetic effect of this displacement is to insure the antiferromagnetic interchain Mn–Cu interaction aligns the net moments in each chain, giving a spontaneous ordered moment. Indeed, below 4.6 K, spontaneous moments are found, Fig 14-b. This is a lovely demonstration of molecular engineering at work, where a system with only antiferromagnetic interactions has been manipulated to produce a spontaneous ferrimagnetic moment.

ACKNOWLEDGEMENTS

The author wishes to expresses his thanks to his students (particularly Dr. A. Lamas) and colleagues for their assistance in the research and the preparation of this article. The support of the National Science Foundation under grant DMR-8306432 is gratefully acknowledged.

REFERENCES

1. Birgeneau, R. J. and Shirane, G. 1978, Physics Today, December, 32.
2. Dingle, R., Lines, M. E., and Holt. S. L. 1969 Phys. Rev. 187, 643.
3. Smith, T. and Friedberg, S. A. 1968 Phys. Rev. 176, 660.
4. de Jongh, L. J. and Miedema, A. R., 1974, Adv. Phys. 23, 1.
5. "Magneto-Structural Correlations in Exchange Coupled Systems", eds. Willett, R. D., Gatteschi, D. and Kahn, O.,1985, Reidel.
6. Bonner, J. C. 1987 J. Appl. Phys., 61, 3941.
7. See for examples the articles by Kahn, Willett, Hodgson, Hendrickson and Hatfield in reference 5.
8. See, for example, Carlin, R. L, "Magnetochemistry",1987, Springer-Verlag, chapter 7. For ferromagnetic chains, see Willett, R. D., Gaura, R. M., and Landee, C. P. 1983 in "Extended Linear Chain Compounds, Vol. 3", ed. J. S. Miller, Plenum Press, 143.
9. Mennenga, G., de Jongh, L. J., Huiskamp, W. J., and Reedijk, J., 1984, J. Magn. Magn. 44, 89. See also Losee, D. B., Richardson, H. W. and Hatfield, W. E. 1973 J. Chem. Phys. 59, 3600.
10. Losee, D. B., McElearney, J. N., Shankle, G. E., Carlin, R. L, Cresswell, P. J., and Robinson, W. T. 1973 Phys. Rev. 8, 2185.
11. Hoogerbeets, R., Wiegers, S. A. J., van Duyneveldt, A. J., Willett, R. D. and Geiser, U. 1984 Physica B+C 125, 135.
12. Landee, C. P., Greeney, R. E., Reiff, W. M., Zhang, J. H., Chalupa, J., and Novotny, M. A. 1985 J. Appl. Phys. 57, 3343.
13. de Jongh, L. J., 1982 J. Appl. Phys. 53, 8018.
14. Haldane, F. D. M., 1983 Phys. Rev. Lett. 50, 1153.
15. For a review see Bray, J. W., Interrante, L. V., Jacobs, I. S., and Bonner, J. C. in "Extended Linear Chain Compounds, Vol. 2", ed. J. S. Miller, Plenum, 1982.
16. de Jongh, L. J. 1981 "Recent Developments in Condensed Matter Physics, Vol. 1", ed. J. T. Devreese et al, Plenum, p.343.
17. Geiser, U., Willett, R. D., Lindbeck, M. and Emerson, K. 1986, J. Am. Chem. Soc. 108, 1173.
18. Bonner, J. C., Friedberg, S. A., Kobayashi, H., Meier, D. L., and Blöte, H. W. J. 1983 Phys. Rev. B 27, 248.
19. Brown, D. B., Donner, J. A., Hall, J. W., Wilson, S. R., Wilson, R. B., Hodgson, D. J., and Hatfield, W. E. 1979, Inorg. Chem. 18, 2635.
20. Shastry, B. S. and Sutherland, B. 1981 Phys. Rev. Lett. 47, 964.

21. ter Haar, L. W., and Hatfield, W.E. 1985, Inorg. Chem. 24, 1022.
22. Roberts, S. A., Bloomquist, D. R., Willett, R. D. and Dodgen, H. W. 1981, J. Am. Chem. Soc. 103, 2603.
23. Hall, J.W., Marsh, W.E., Weller, R.R., and Hatfield, W. E. 1981, Inorg. Chem. 20, 1033.
24. De Groot, H.J.M., de Jongh, L. J., Willett, R. D., and Reedijk, J. 1982, J. Appl. Phys. 53, 8038.
25. Wolthuis, A. J., Huiskamp, W. J., de Jongh, L. J., and Reedijk, J. 1985, Physica B+C 133, 161.
26. de Jongh, L. J., de Groot, H. J. M., and Reedijk, J. 1982, J. Appl. Phys. 53, 8027.
27. Hatfield, W. E. 1983 Inorg. Chem. 22, 833.
28. Duffy, W. and Barr, K. P.1968 Phys. Rev. 165, 647.
29. O'Brien, S., Gaura, R. M., Landee, C. P., Ramakhrishna, B. L., and Willett, R. D. 1987, submitted for publication.
30. Woenk, J. W. and Spek, A. L., 1976, Cryst. Struct. Commun. 5, 805.
31. Landee, C. P. and Willett, R. D. 1979, Phys. Rev. Lett. 43, 463.
32. Dupas, C. and Renard, J.P. 1981 "Recent Developments in Condensed Matter Physics, Vol. 4", ed. J. T. Devreese et al, Plenum, 213.
33. Olmstead, M. M., Musker, W. K., ter Haar, L. W. and Hatfield, W. E. 1982 J. Am. Chem. Soc. 104, 6627.
34. Hatfield, W. E., ter Haar, L. W., Olmstead, M. M., and Musker, W. K. 1986 Inorg. Chem. 25, 558.
35. Bencini, A., Benelli, C., Fabretti, A. C., Franchini, G., and Gatteschi, D. 1986 Inorg. Chem. 25, 1063.
36. Bencini, A., Benelli, C., Gatteschi, D. and Zanchini, C. 1984 J. Am. Chem. Soc. 106, 5813.
37. Benelli, C., Gatteschi, D., Carnegie, D. W. Jr., and Carlin, R. L. 1985 J. Am. Chem. Soc. 107, 2560.
38. Caneschi, A., Gatteschi D., Laugier, J. and Rey, P. 1987 J. Am. Chem. Soc. 109, 2191.
39. Daoud, A., Ben Salah, A., Chappert, C., Renard, J.P., Cheikhrouhou, A., Duc, T., and Verdaguer, M. 1986, Phys. Rev. B 33, 6253.
40. Smit, J. J., de Jongh. L. J., van Ooijen, J. A. C., Reedijk, J., and Bonner, J. C. 1979 Physica 79B, 229.
41. Chiari, B., Hatfield, W. E., Piovesana, O., Tarantelli, T, ter Haar, L. W., and Zanazzi, P. F. 1983, Inorg. Chem. 22, 1468.
42. Curély, J., Georges, R., and Drillon, M. 1986 Phys. Rev.B 33, 6243.
43. Gleizes, A. and Verdaguer, M. 1981 J. Am. Chem. Soc. 103, 7373.
44. Gleizes, A. and Verdaguer, M. 1984 J. Am. Chem. Soc. 106, 3727.
45. Verdaguer, M., Gleizes, A., Renard, J-P. and Seiden, J. 1984 Phys. Rev. B 29, 5144.
46. Verdaguer, M., Julve, M., Michalowicz, A. and Kahn, O. 1983 Inorg. Chem. 22, 2624.
47. Beltran, D., Escriva, E., and Drillon, M. 1982 J. Chem. Soc., Faraday Trans. 78, 1773.
48. Drillon, M., Coronado, E., Beltran, D., Curély, J., Georges, R., Nugteren, P.R., de Jongh, L.J., and Genicon, J.L. 1986 J. Magn. Magn. 54-57, 1507.
49. Coronado, E., Beltran, D., Drillon, M., 1987, unpublished results.
50. Drillon, M., Coronado, E., Beltran, D., and Georges, R. 1983 Chem. Phys. 79, 449.
51. Georges, R., Curély, J., and Drillon, M. 1985 J. Appl. Phys. 58, 914.
52. Coronado, E., Drillon, M., Fuertes, A., Beltran, D., Mosset, A., and Galy, J. 1986 J. Am. Chem. Soc. 108, 900.
53. Landee, C. P., Djili, A., Willett, R. D., Place, H., this volume.
54. Pei, Y., Kahn, O. and Sletten J. 1986 J. Am. Chem. Soc., 108, 3143.
55. Pei, Y., Verdaguer, M., Kahn, O., Sletten, J. and Renard, J-P. 1987 Inorg. Chem. 26, 138.
56. Pei, Y., Verdaguer, M., Kahn, O., Sletten, J. and Renard, J.-P. 1986 J. Am. Chem. Soc. 108, 7428.

MAGNETIC ORDERING OF $Mn^{II}Cu^{II}$ BIMETALLIC SYSTEMS : DESIGN OF MOLECULAR FERROMAGNETS

Olivier Kahn

Laboratoire de Spectrochimie des Eléments de Transition
U.A. n° 420, Université de Paris Sud, 91405 Orsay, France

INTRODUCTION

For a few years, we have participated to the efforts to design molecular systems ordering ferromagnetically. Molecular in this context means that we use the synthesis methods of the molecular chemistry -we work in solution with mild conditions of temperature and pressure- and that we attempt to build the tridimensional lattice by assembling molecular bricks in a controlable fashion. Our first step along this line has been to find a strategy leading to ferromagnetic interactions between nearest neighbor magnetic centers. This strategy is that of the orthogonality of the magnetic orbitals. When in an AB pair, the magnetic orbitals around A are all orthogonal to the magnetic orbitals around B, the interaction between A and B is purely ferromagnetic and the ground state of the pair has the highest spin multiplicity. We have synthesized several compounds of this kind involving pairs of metal ions[1,2], or a metal ion and an organic radical[3]. This strategy to achieve a ferromagnetic interaction is very efficient in the sense that we have not found any counterexample so far. When the strict orthogonality of all the magnetic orbitals is realized, then the ground state has the highest spin multiplicity. However, the symmetry requirements allowing such an orthogonality are often difficult to obtain. A small distortion with regard to the ideal symmetry generally destroys this orthogonality, so that the interaction becomes antiferromagnetic with a stabilization of the state of lowest spin multiplicity. In a certain sense, this strategy of orthogonality requires going again a natural trend which favors the pairing of the electrons in molecular orbitals of low energy. This is why we have decided to explore an alternative strategy allowing to "throw off the yoke" of the symmetry requirements. The basic idea of this new strategy consists in imposing the parallel alignment of local spins 5/2 (Mn^{II} or Fe^{III}) owing to an antiferromagnetic interaction with local spins 1/2 (Cu^{II}), according to the principle scheme :

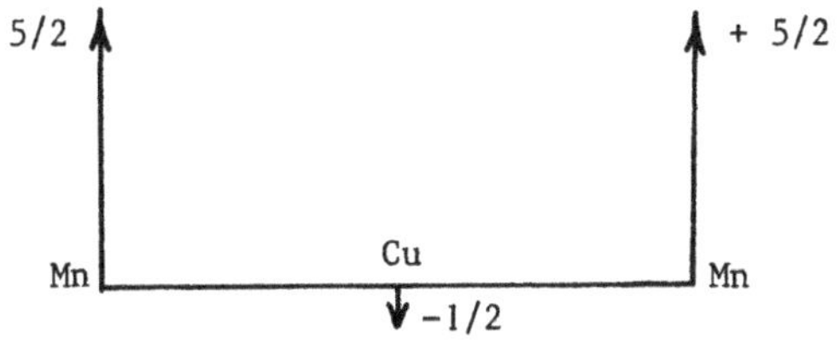

This paper is devoted to the first results obtained along this line and is organized as follows : in the next Section, we describe a $Mn^{II}Cu^{II}Mn^{II}$ trinuclear species and we introduce the concept of irregular spin state structure, then we present a new family of $Mn^{II}Cu^{II}$ ordered bimetallic chains and we discuss both the intrachain and interchain interactions. Finally, we propose some outlooks in the field of the molecular ferromagnets.

IRREGULAR SPIN STATE STRUCTURE IN A $Mn^{II}Cu^{II}Mn^{II}$ TRINUCLEAR SPECIES[4,5].

$\{[Mn(Me_6-[14]ane-N_4)]_2Cu(pba)\}(CF_3SO_3)_2.2H_2O$ (1) with $Me_6-[14]ane-N_4$ = ± 5,7,7,12,14,14-hexamethyl-1,4,8,11-tetra-azacyclotetradecane and pba = propylene-1,3-bisoxamato has been synthesized according the procedure shown in Figure 1 and the molecular skeleton is shown in Figure 2

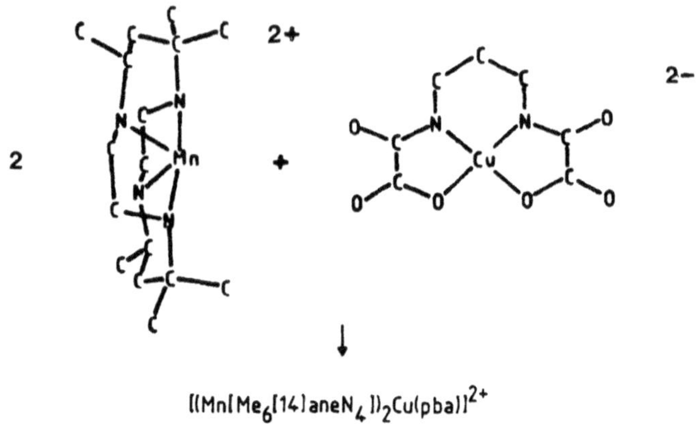

Fig. 1. Schema of the synthesis of 1

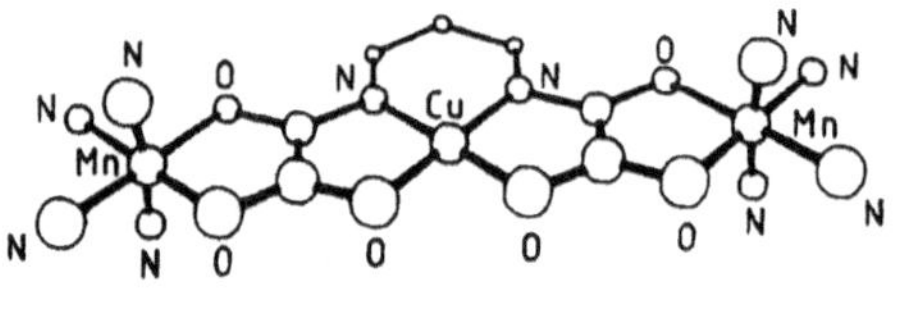

Fig. 2. $Mn^{II}Cu^{II}Mn^{II}$ trinuclear unit in 1

The low lying states arising from the interaction Hamiltonian in zero field :

$$\mathcal{H}^{\circ} = - J(\hat{S}_{Mn1}.\hat{S}_{Cu} + \hat{S}_{Mn2}.\hat{S}_{Cu}) \qquad (1)$$

are given in Figure 3, where each state is indicated by an arrow of which the lenght represents the spin associated with this state. In (1), it is assumed that the interaction between nearest neighbors is purely

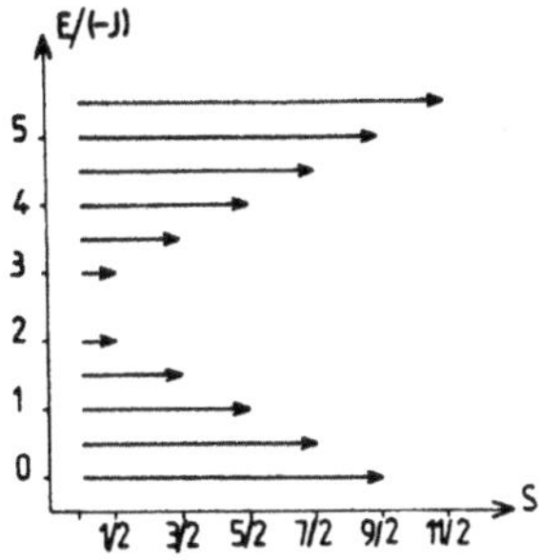

Fig. 3. Spin state structure for a $Mn^{II}Cu^{II}Mn^{II}$ trinuclear species

isotropic and that between terminal centers is negligible. For J < 0, the ground state is S = 9/2. When going up in energy, the spin regularly decreases form 9/2 to 1/2, then increases from 1/2 to 11/2 for the most excited state. Such a spin state structure where the spin multiplicity does not vary monotonically versus the energy is said to be irregular[4]. In the present case, the low energy range of the spin state structure is reminiscent of a ferromagnetically coupled system, in the sense where the lower the energy is, the higher the spin multiplicity. On the other hand, the high energy range (E > - 5J/2) well corresponds to what is implicitely expected for an antiferromagnetically coupled system. The consequence of this irregularity of the spin state structure is the presence of a minimum in the $\chi_M T$ versus T plot, χ_M being the molar magnetic susceptibility and T the temperature. The $\chi_M T$ versus T plot for 1 is shown in Figure 4.

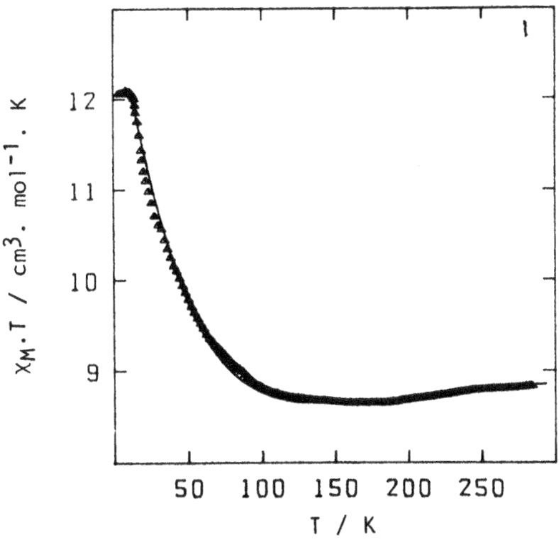

Fig. 4. $\chi_M T$ versus T plot for the $Mn^{II}Cu^{II}Mn^{II}$ trinuclear compound 1.

At high temperature ($kT \gg |J|$), $\chi_M T$ corresponds to what is expected for two Mn^{II} and one Cu^{II}. Upon cooling down, $\chi_M T$ decreases and reaches a minimum around 170 K. Below 170 K, $\chi_M T$ rapidly increases and reaches a plateau in the 2-10 K temperature range with $\chi_M T = 12.1\ cm^3mol^{-1}K$, which is the value expected for a $S = 9/2$ state ($\chi_M T = 33\ N\beta^2 g^2/4k$). To fit the experimental data, we used the theoretical expression deduced from the spin Hamiltonian

$$\mathcal{H} = \mathcal{H}^\circ + \mathcal{H}_{ZE} \qquad (2)$$

with :

$$\mathcal{H}_{ZE} = \beta\left[g_{Mn}(\hat{S}_{Mn1} + \hat{S}_{Mn2}) + g_{Cu}\hat{S}_{Cu}\right] \cdot \vec{H} \qquad (3)$$

and found $J = -36.6\ cm^{-1}$, $g_{Mn} = 2.03$ and $g_{Cu} = 2.10$. The magnitude of the Mn^{II}-Cu^{II} antiferromagnetic interaction confirms that the bis bidentate bridging ligands like oxamato have quite a specific ability to favor such a large interaction, provided that there is a magnetic orbital of xy-symmetry on each ion, as schematized below[6] :

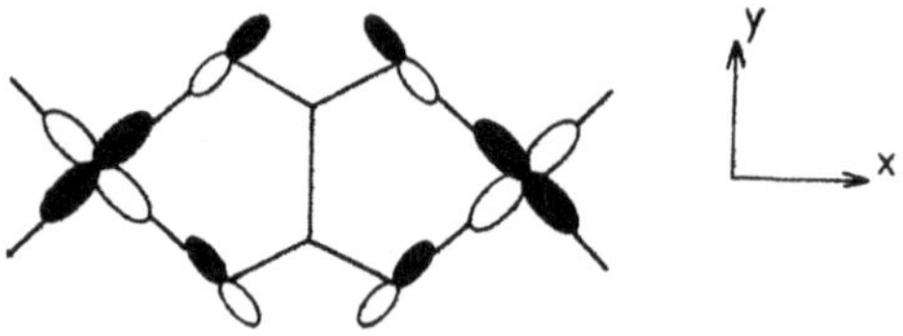

Such a xy-type magnetic orbital is available for both Mn^{II} and Cu^{II}.

$Mn^{II}Cu^{II}$ ORDERED BIMETALLIC CHAINS

The concept of irregular spin state structure is not limited to trinuclear species or other polynuclear molecular entities but can be extended to the ordered bimetallic chains, of which several examples have been recently reported[7-13]. Let us consider a $Mn^{II}Cu^{II}$ chain of this kind with antiferromagnetic interaction between the adjacent $S_{Mn} = 5/2$ and $S_{Cu} = 1/2$ local spins. The ground state in absence of any interchain interaction might be schematized as :

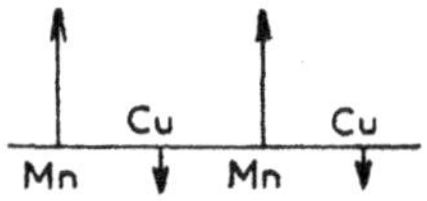

and a 1D magnetic ordering would be expected at 0 K with a divergence of $\chi_M T$, χ_M being here the molar magnetic susceptibility per MnCu unit. This behavior may be defined as the one-dimensional ferrimagnetism. At high temperature, $\chi_M T$ is equal to the sum of what is expected for a Mn^{II} and a Cu^{II} ions and when starting to cool down, the first chain state to be depopulated is that of highest spin with all the 5/2 and 1/2 local spins aligned along a same direction. It follows that upon cooling down, $\chi_M T$ decreases and reaches a minimum before increasing rapidly.

We have synthesized the two compounds, chemically very similar, of formula $MnCu(pba)(H_2O)_3 \cdot 2H_2O$ (2) and $MnCu(pbaOH)(H_2O)_3$ (3). Both are obtained by polymerization of a Cu^{II} mononuclear dianion $[Cu(pba)]^{2-}$ or $[Cu(pbaOH)]^{2-}$, represented here as :

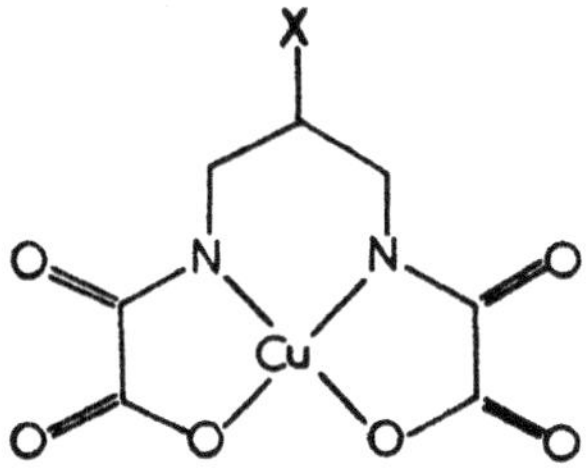

X = H for $[Cu(pba)]^{2-}$

= OH for $[Cu(pbaOH)]^{2-}$

with Mn^{II}. 2 and 3 have very similar chain structures. That of 3 is shown in Figure 5.

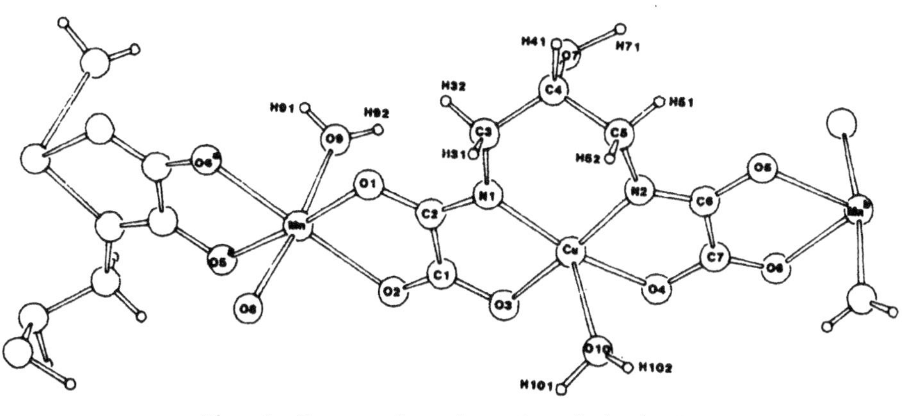

Fig. 5. Perspective view of a chain in 3

In the temperature range 30 < T/K<300, the two compounds have exactly the same magnetic behavior within the experimental uncertainties, with a minimum of the $\chi_M T$ versus T plot at 115 K and a rapid increase of $\chi_M T$ upon cooling down below 115 K. The $\chi_M T$ versus T plot for 2 is shown in Figure 6.

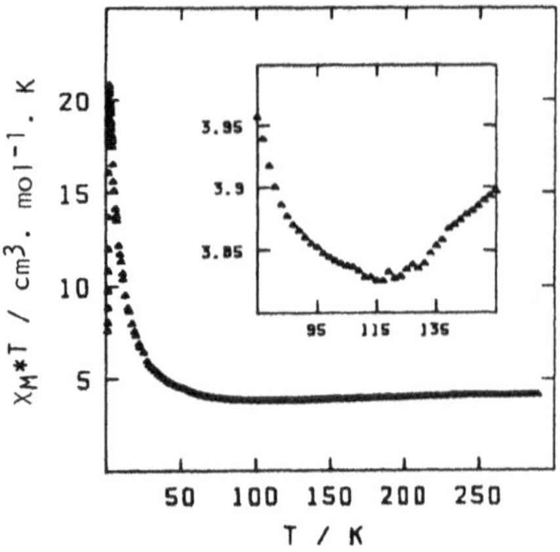

Fig. 6. $\chi_M T$ versus T plot for $MnCu(pba)(H_2O)_3.2H_2O$, (2).

A quantitative model to interpret the magnetic behavior of such a $Mn^{II}Cu^{II}$ chain has been proposed[14]. It consists in taking S_{Mn} as a semi-quantum spin and S_{Cu} as a purely quantum spin. This model leads to numerical results that can be reasonably well fitted by the empirical expression :

$$\chi_M T = (g^2/4)(4.75 - 1.62370\ X + 2.05042\ X^2 - 4.52588\ X^3 - 8.64256\ X^4)/$$
$$(1 + 0.77968\ X - 1.56527\ X^2 - 1.57333\ X^3 - 0.11666\ X^{4.5}) \qquad (4)$$

with :

$$X = |J|/kT$$

(4) is valid for $J < 0$ and $X < 0.752$. In this model, the local g_{Mn} and g_{Cu} factors are assumed to be isotropic and equal ($g_{Mn} = g_{Cu} = g$). The fitting of the experimental data for 2 and 3 leads to $J = -23.4(4)$ cm^{-1} and $g = 1.97(2)$.

INTERCHAIN INTERACTIONS AND MAGNETIC ORDERING IN 2 AND 3

Both 2 and 3 crystallize in the orthorhombic system. The chains run along the *b* axis. The shortest interchain separations along the *a* axis involve metal ions of the same nature in 2 (Cu...Cu = 6.545 Å and Mn...Mn = 6.977 Å) and of different nature in 3 (Cu...Mn = 5.751 and 6.398 Å). In a simplified fashion, one can say that in 3, with regard to 2, every other chain is displaced by slightly less than half of a repeat unit along *b*, as schematized below :

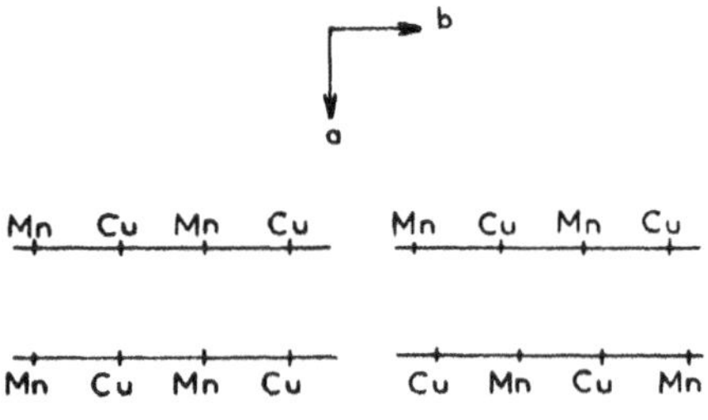

2 3

In both compounds, the chains closest to one another are those related by a unit cell translation along *c*. In 2, we have $Cu...Cu^{c} = Mn...Mn^{c} =$ 5.2105 Å and in 3, $Cu...Cu^{c} = Mn...Mn^{c} =$ 5.073 Å. The crystal packings for 2 and 3 are compared in Figure 7.

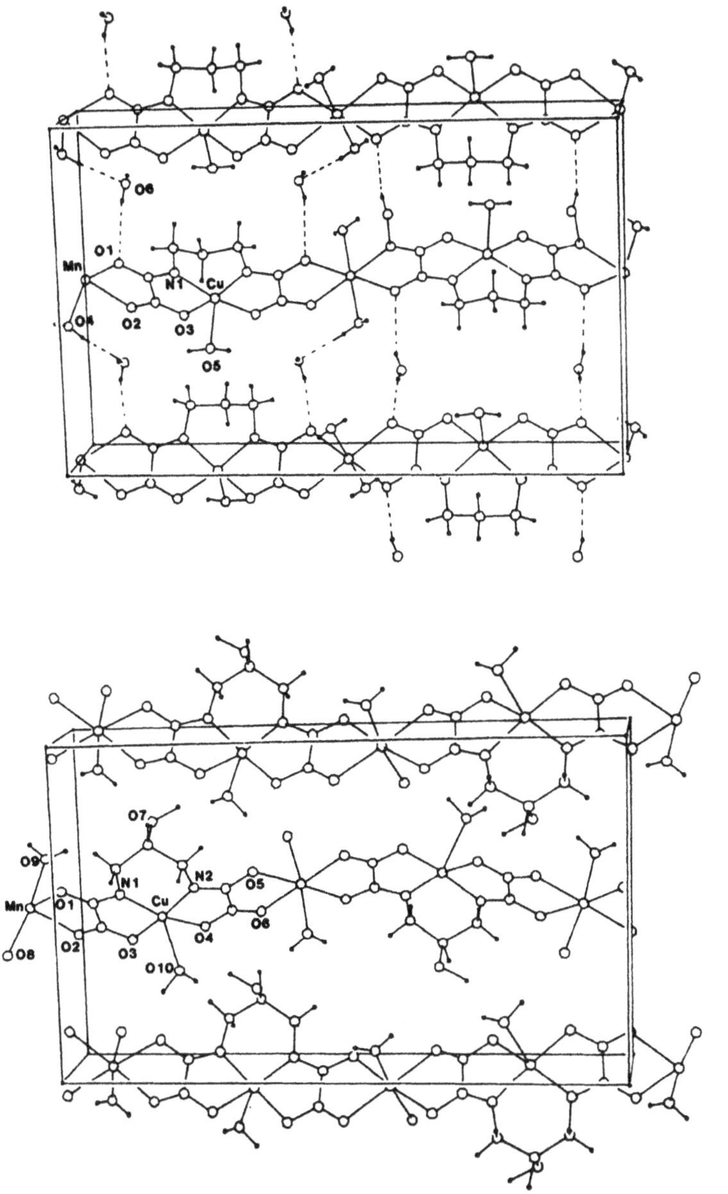

Fig. 7. Perspective view of three neighboring chains in $MnCu(pba)(H_2O)_3 \cdot 2H_2O$ (2) (top) and $MnCu(pbaOH)(H_2O)_3$ (3) (bottom) ; the *a* axis runs top to bottom of page and the *b* axis left to right.

The crystal structure differences involving the relative positions of the chains along the *a* axis have drastic consequences as far as the low temperature magnetic properties are concerned. Upon cooling down below 30 K, $\chi_M T$ increases much faster for 3 than for 2 and below 5 K, the two compounds behave quite differently. 2 exhibits a maximum of $\chi_M T$ at 2.3 K and a maximum of χ_M at 2.2 K, due to the onset of a 3D antiferromagnetic ordering. In contrast, $\chi_M T$ for 3 diverges and becomes strongly field-dependent, which suggests that a ferromagnetic transition takes place. The field cooled magnetization (FCM) and zero field cooled magnetization (ZFCM) curves at 3×10^{-2} G for 3 shown in Figure 8 confirms that such a 3D ferromagnetic ordering occurs. The FCM curve shows the typical feature of a ferromagnetic transition, i.e. a rapid increase of the magnetization M when T decreases and a break in the curve around $T_c = 4.6$ K. If we cut the field at 3.5 K, we observed a remanent magnetization that vanishes at T_c. The ZFCM curve is obtained by cooling down in zero field, then applying the field and heating up. Below T_c, the ZFCM is smaller than the FCM, due to the fact that the applied field is too weak to move the domain walls of the polycrystalline sample. The ZFCM exhibits a maximum at T_c, as expected for a polycrystalline ferromagnet.

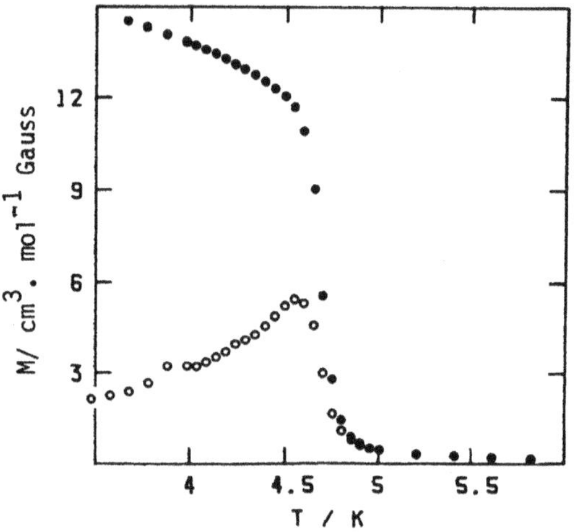

Fig. 8. FCM (•) and ZFCM (o) curves for 3 in the 6-3.5 K temperature range and a field of 3×10^{-2} G.

We also measured the molar magnetization M_a, M_b and M_c along the principal axes of the orthorhombic lattice in the 5-2 K temperature range and found that for any field up to 20 G, we have $M_c \gg M_a \sim M_b$. It follows that the easy magnetization axis is along the *c* axis or very close to the *c* axis. Finally, we investigated the magnetic hysteresis at various temperatures between 4.2 and 1.3 K. At 4.2 K, the hysteresis effect is very weak. It becomes more pronounced when T decreases. The hysteresis loop at 1.3 K is shown in Figure 9. At this temperature, the remanent magnetization is equal to 2.25×10^3 cm^3 mol^{-1} G, i.e. about a tenth of the saturation magnetization and the coercive field is about 50 G.

The main difference between the crystal structures of 2 and 3 concerns the packing of the chains along the *a* axis (see Fig. 7). The Mn^{II} ion has the high-spin d^5 configuration. Consequently, it most likely interacts with any other neighbor ion in a antiferromagnetic fashion[2]. Indeed, whatever the geometry of the system may be, at least one of the five magnetic orbitals centered on Mn^{II} gives a non zero overlap with a magnetic orbital centered on the nearest neighbor ion. It follows that the relative positions of the chains along *a* in 2 lead to a cancellation of the resulting spin in the *ab* plane and to a parallel alignement of the $S_{Mn} = 5/2$ local spins in 3. Since the interactions within the chains are much larger than those between the chains, one can say that the ferrimagnetic chains couple antiferromagnetically along *a* in 2 and ferromagnetically in 3 as schematized in Figure 10. In this Figure, we took into account that in both 2 and 2, the easy magnetization axis has been found along *c*.

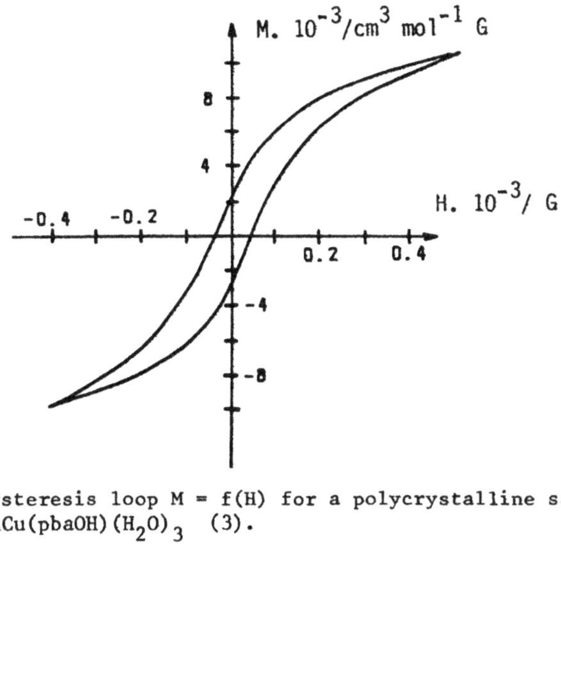

Fig. 9. Hysteresis loop M = f(H) for a polycrystalline sample of $MnCu(pbaOH)(H_2O)_3$ (3).

The last point deserving a discussion is the mechanism of the interaction between the ferromagnetic planes *ab* in 3. In this compound as well as in 2, the shortest separations along *b* are Cu...Cu and Mn-Mn, which would favor an antiferromagnetic coupling of the *ab* planes. This is not consistent with the observed 3D ferromagnetic ordering. We suggest the following interpretation : the ferromagnetic ordering in 3 would be due to the magnetic dipolar interactions between the *ab* planes. These interactions that vary as the square of the local spin momenta would be particularly important owing to the presence of $S_{Mn} = 5/2$ local spins. Moreover, the interchain separation along *c* in 3 is rather short with Mn...Mn = Cu...Cu = 5.073 A.

OUTLOOK AND CONCLUSION

In this last Section, we would like to go beyond our own results and to discuss about the design of molecular ferromagnets in a more general fashion.

At the beginning of this chapter, we emphasized that the strategy of orthogonality of the magnetic orbitals, when applied to a molecular system, leads to the stabilization of the state of highest spin multiplicity. This orthogonality may be strict (symmetry imposed) or accidental. It is strict when the interacting magnetic orbitals transform as different irreducible representations of the symmetry group and it is accidental when some interacting magnetic orbitals have the same symmetry, with however a zero overlap between them owing to some peculiar values of the structural parameters. Examples of accidental and strict orthogonalities are given in Figure 11, where we represented the ϕ_A and ϕ_B magnetic orbitals at the top and the overlap densities $\rho(i) = \phi_A(i)\phi_B(i)$ at the bottom[25]. The first column refers to a $Cu^{II}Cu^{II}$ pair with two bridging angles close to 90°, and the second to a $Cu^{II}VO^{II}$ pair.

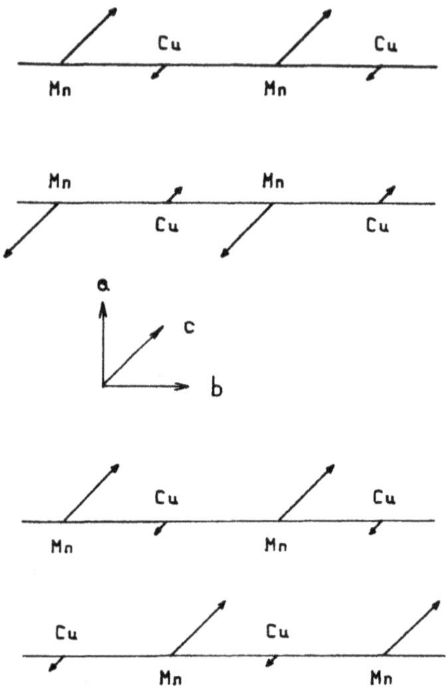

Fig. 10. Schematic representation of the magnetic structure in the *ab* plane for 2 (top) and 3 (bottom).

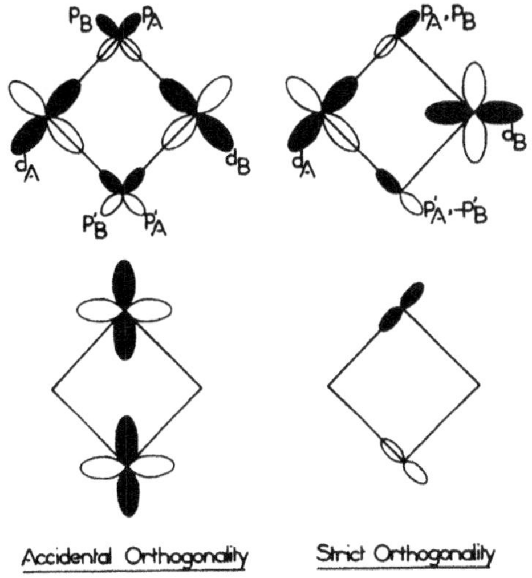

Fig. 11. Schematic representation of accidental orthogonality in a $Cu^{II}Cu^{II}$ pair with bridging angles close to 90° and of strict orthogonality in a $Cu^{II}VO^{II}$ pair. The interacting magnetic orbitals are represented at the top and the overlap densities at the bottom.

It is important to notice that the accidental orthogonality in the $Cu^{II}Cu^{II}$ pair can be distroyed by a very small variation of the bridging angles whereas in the case of strict orthogonality, such a variation has as only consequence to modify slightly the magnitude of the ferromagnetic interaction. In spite of the fragility of the accidental orthogonality, several ferromagnetically coupled Cu^{II} and Ni^{II} dinuclear complexes and linear chains have been obtained in which the nature of the interaction was actually due to this accidental (or quasi accidental) orthogonality[15-23].

The strict orthogonality is more easily realized with pairs of different magnetic centers, either two different metal ions or a metal ion and an organic radical. Examples of such pairs are $Cu^{II}VO^{II}$, $Cu^{II}Cr^{III}$, $Cu^{II}Cr^{II}$, $Ni^{II}Cr^{III}$, Cu^{II} orthosemiquinone etc...[1-3,6]. Therefore, we may imagine to design bimetallic ferromagnetic chains. A difficulty however appears. All the bimetallic chains synthesized so far contain extended bridging ligands with rather long intrachain metal...metal separations (> 5 Å). With such extended bridging networks and in the case of orthogonality, the ferromagnetic interaction is expected to be very weak if not negligible[6,24]. Indeed, each magnetic orbital is then delocalized over a large number of bridge atoms and the overlap densities between pairs of magnetic orbitals are spread over the whole bridge without any pronounced maximum. It follows that the two-electron exchange integrals of the form

$$j = \int_{space} \frac{\rho(i)\ \rho(j)}{r_{ij}}\ d\tau(i)\ d\tau(j) \qquad (5)$$

which govern the magnitude of the ferromagnetic interaction remain very weak. This results has been experimentally confirmed in the case of an oxalato bridged $Cu^{II}VO^{II}$ dinuclear complex ; in spite of the strict orthogonality of the magnetic orbitals, the triplet state is stabilized by less than 1 cm^{-1} with regard to the singlet state[24]. Of course, we can imagine ordered bimetallic chains with monoatomic bridging ligands like halogenato or oxo ions but the synthesis of such systems seems very problematical. To sum up the discussion above, we can say that two kinds of one-dimensional systems are better candidates to exhibit a ferromagnetic-like behavior, either homometallic chains with ferromagnetic interaction between nearest neighbors due to the accidental orthogonality of the magnetic orbitals, or bimetallic chains with antiferromagnetic interaction between nearest neighbors and non compensation of the local spins.

The problem at hand now is to assemble those chains so as to avoid the cancelling of the spins at the scale of the crystal lattice. Concerning the homometallic ferromagnetic chains, we do not see any evident way to impose an interchain ferromagnetic interaction, if not by attempting to realize again an accidental orthogonality between closest magnetic centers belonging to two adjacent chains. In contrast, with bimetallic AB chains, a strategy emerges ; it consists in imposing A...B as the shortest interchain separation instead of A...A and B...B. This situation is obtained in the *ab* planes of $MnCu(pbaOH)(H_2O)_3$ owing to the hydrogen bonds. It is clearly not easy to control the network of hydrogen bonds during the synthesis and we are looking for a more efficient approach.
Several possibilities are investigated. One of them consists in fixing two bulky ligands in the apical positions around Cu^{II} and in imposing a kind of gearing between the chains as schematized in Figure 12. The relative positions of the chains with Mn...Cu as the shortest interchain separation avoid too large holes in the crystal lattice

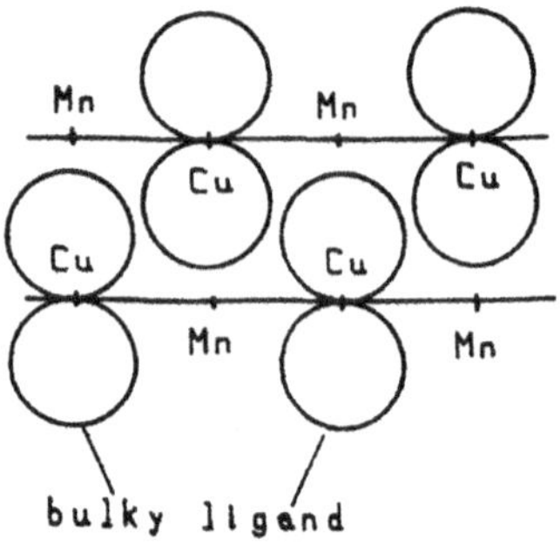

Fig. 12. The toothed gearing strategy to impose Mn...Cu as the shortest interchain separation.

Another way to avoid a cancelling of the spin at the scale of the lattice would be to introduce Cu^{II} mononuclear species either between two Cu^{II} ions or between two Mn^{II} ions belonging to adjacent chains as schematized in Figure 13. The mononuclear species would polarize the chains so as to align the $S_{Mn} = 5/2$ local spins along a same direction.

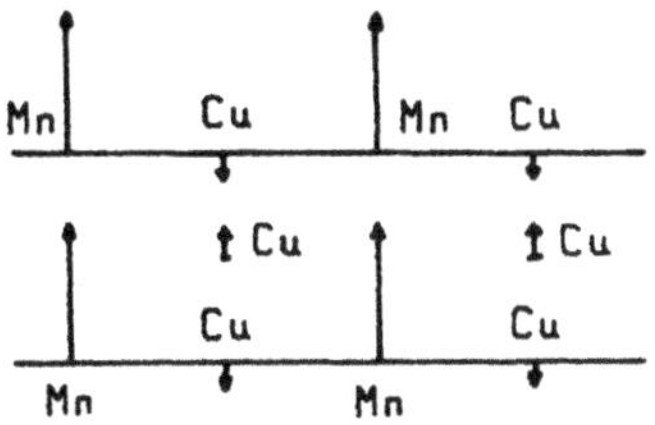

Fig. 13. Ferromagnetic-like polarization of $Mn^{II}Cu^{II}$ bimetallic chains through the presence of Cu^{II} mononuclear species between the chains.

So far, we have only discussed about arrangements of chains without genuine bonds (except hydrogen bonds) between the chains. It is also possible to couple chemically the chains so as to obtain bimetallic planes. These planes will have a non zero resulting moment if all the shortest separations along the x as well as the y directions involve different metal ions. This can eventually be achieved by using dissymmetrical bridging ligands X coupling the chains as schematized in Figure 14.

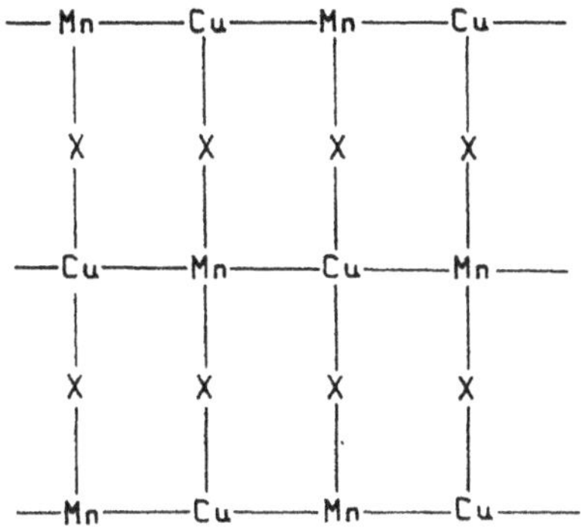

Fig. 14. Ferrimagnetic plane

Another very exciting approach would consist in starting the polymerization process with a Cu^{II} mononuclear brick as that schematized below with a fourfold or a pseudo fourfold symmetry axis.

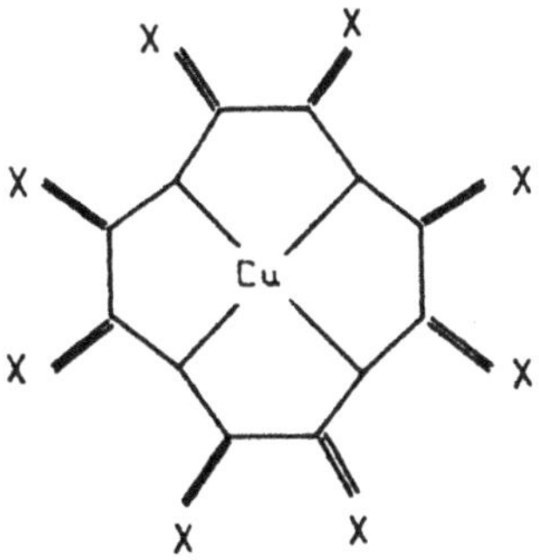

Several groups over the world, belonging to both the organic and inorganic chemistry communities, are working in the two related fields of the molecular species with an high-spin multiplicity within the ground state and of the molecular ferromagnets. To finish this paper, we would like to summarize the main achievements reported so far.

Iwamura et al. synthesized the polycarbene schematized below.

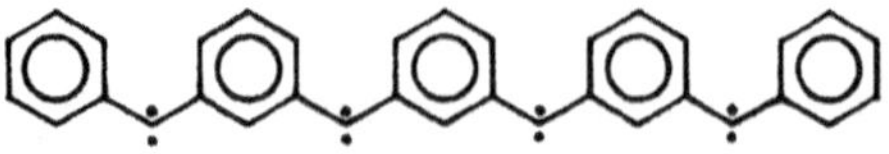

Due to topological symmetry, this molecule has eight quasi degenerate molecular orbitals accommodating eight electrons. According to Hund's rule, the ground state is a nonet, which has been unambiguously proved from magnetic and EPR data[26-28]. An hypothetical one-dimensional system made up of units as those shown above should be a linear ferromagnet. However, actually the polycarbene molecules with a $S = 4$ ground state interact in an antiferromagnetic fashion within the host lattice in which they are dilute.

Torrance suggested a strategy to obtain ferromagnetic interactions in seggregated stacks of molecular units with a three fold symmetry axis[29]. This strategy consists in stabilizing a triplet low lying state by configuration interaction with a triplet excited state arising from the charge transfer configuration in which two electrons occupy two degenerate molecular orbitals as schematized in Figure 14.

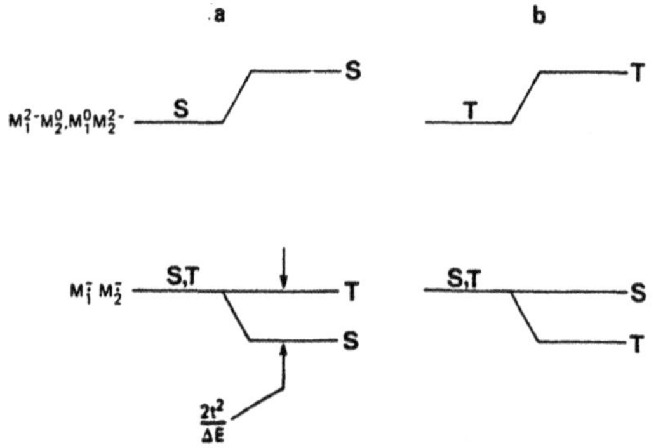

Fig. 14. (a) Anderson's model for antiferromagnetic interaction ;
(b) Torrance's model for ferromagnetic interaction, from Ref. 29.

Previously, a model for ferromagnetic coupling in alternated stacks had been proposed by McConnell[30,31]. In this model also, a ferromagnetic state is stabilized through an interaction with a parallel spin charge transfer excited state. Miller et al. recently reported the charge transfer compound $[Fe(C_5Me_5)_2]^+[TCNE]^-$ in which the ferromagnetic transition observed at 4.8 K can be explained by McConnell's model. Since a Contribution in this book will be devoted to this important result, we will not discuss it further.

It is worthwhile to mention also the report by Ovchinnikov et al. on an organic polymer ferromagnet prepared by thermal polymerization of the biradical shown below[33].

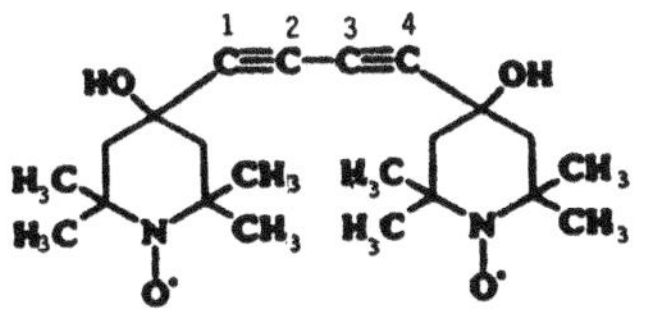

The yield of the reaction in term of ferromagnetic material is only 0.1 % but apparently it is possible to enrich the polymer in particules exhibiting a spontaneous magnetization by magnetic separation. So far, the structure of the magnetically active material is unknown.

Finally, some of the metal ion-nitroxide polymer compounds obtained by Gatteschi, Rey et al. ordering ferromagnetically in the 10-25 K temperature range are particularly interesting[34]. To our knowledge, the nature of the interactions between adjacent metal ion and bridging organic radical is not yet established. The knowledge of the crystal structure of at least one of these compounds would bring important insights on this promising new class of molecular materials.

The area of the molecular ferromagnets is in first infancy. The emulation between the colleagues mentioned in this paper and others should favor a fast development. The main goals are first of all to shift T_c toward higher temperatures, but also to design stable systems, given a well defined hysteresis loop with large remanent magnetization and coercitive field. We think that our approach utilizing both the efficiency of the organic based ligands to transmit the electronic effects on long distances and the specific ability of the d metal orbitals to provide high magnetic moments will allow to report on novel systems of this kind in a near feature.

ACKNOWLEDGEMENT

I want to express my deepest gratitude to my coworkers and colleagues who have participated to the work described in the first four Sections of this paper, particularly Y. Journaux, Y. Pei, J.P. Renard, J. Sletten and M. Verdaguer.

REFERENCES

1. O. Kahn, J. Galy, Y. Journaux, J. Jaud and I. Morgenstern-Badarau, J. Am. Chem. Soc. 104 : 2165 (1982).
2. Y. Journaux, O. Kahn, J. Zarembowitch, J. Galy and J. Jaud, J. Am. Chem. Soc. 105 : 7585 (1983).
3. O. Kahn, R. Prins, J. Reedijk and J.S. Thompson, Inorg. Chem. in press.
4. O. Kahn, Proc. Indian Acad. Sci. (Chem. Sci.) 98 : 33 (1987).
5. Y. Pei, Y. Journaux, O. Kahn, A. Dei and D. Gatteschi, J. Chem. Soc. Chem. Comm. 1300 (1986).
6. O. Kahn, Angew. Chem. Int. Ed. Engl. 24 : 834 (1985).

7. A. Gleizes and M. Verdaguer, J. Am. Chem. Soc. 103 : 7373 (1981) ; 106 : 3727 (1984).
8. D. Beltran, E. Escriva and M. Drillon, J. Chem. Soc. Faraday Trans. 2, 78 : 1773 (1982).
9. M. Drillon, E. Coronado, D. Beltran and R. Georges, Chem. Phys. 79 : 449 (1983).
10. Y. Pei, M. Verdaguer, O. Kahn, J. Sletten and J.P. Renard, J. Am. Chem. Soc. 108 : 7428 (1986) ; Inorg. Chem. 26 : 138 (1987).
11. Y. Pei, O. Kahn and J. Sletten, J. Am. Chem. Soc. 108 : 3143 (1986).
12. M. Verdaguer, M. Julve, A. Michalowicz and O. Kahn, Inorg. Chem. 22 : 2624 (1983).
13. M. Drillon, E. Coronado, D. Beltran, J. Curely, R. Georges, P.R. Nugterer, L.J. de Jongh and J.L. Genicon, J. Magn. Magn. Mater. 54-57 : 1507 (1986).
14. M. Verdaguer, A. Gleizes, J.P. Renard and J. Seiden, J. Phys. Rev. B : Condens. Matter 29 : 5144 (1984).
15. W.H. Crawford, H.W. Richardson, J.R. Wasson, D.J. Hodgson and W.E. Hatfield, Inorg. Chem. 15 : 2107 (1976).
16. C.P. Landee and R.D. Willett, Inorg. Chem. 20 : 2521 (1981).
17. A.P. Ginsberg, R.L. Martin, R.W. Brookes and R.C. Sherwood, Inorg. Chem. 11 : 2884 (1972).
18. D.M. Duggan and D.M. Hendrickson, Inorg. Chem. 13 : 2929 (1974).
19. E.J. Laskowski, T.R. Felthouse, D.N. Hendrickson and G. Long, Inorg. Chem. 15 : 2905 (1976).
20. K.O. Young, C.J. O'Connor, E. Sinn and R.L. Carlin, Inorg. Chem. 18 : 804 (1979).
21. Y. Journaux and O. Kahn, J. Chem. Soc. Dalton 1575 (1979).
22. R.D. Willett, R.M. Gaura and C.P. Landee, "Extended Linear Chain Compounds", ed. J.S. Miller, Plenum, New.York, Vol. 3 : 143 (1983).
23. W.E. Hatfield, W.E. Estes, W.E. Marsh, M.W. Pickens, L.W. Ter Haar, and R.R. Weller "Extended Linear Chain Compounds", ed. J.S. Miller, Plenum, New York, Vol. 3 : 43 (1983).
24. M. Julve, M. Verdaguer, M.F. Charlot, O. Kahn and R. Claude, Inorg. Chim. Acta 82 : 5 (1984).
25. O. Kahn and M.F. Charlot, Nouv. J. Chim. 4 : 567 (1980).
26. H. Iwamura, T. Sugawara, K. Itoh and T. Takui, Mol. Cryst. Liquid. Cryst. 125 : 251 (1985).
27. T. Sagawura, S. Bandow, K. Kimura, H. Iwamura and K. Itoh, J. Am. Chem. Soc. 108 : 368 (1986).
28. Y.Teki, T. Takui, K. Itoh, H. Iwamura and K. Kobayashi, J. Am. Chem. Soc. 108 : 2147 (1986).
29. J.B. Torrance, S. Oostra and A. Nazzal, Synt. Met. in press.
30. H.M. McConnell, J. Phys. Chem. 39 : 1910 (1963).
31. H.M. McConnell, Proc. Robert A. Welch Found. Chem. Res. 11 :144 (1967).
32. J. Miller, J.C. Calabrese, H. Rommelmann, S.R. Chittipeddi, J.H. Zhang, W.M. Reiff and A.J. Epstein, J. Am. Chem. Soc. 109 : 769 (1987).
33. Y.V. Korshak, T.V. Medvedeva, A.A. Ovchinnikov and V.N. Spector, Nature 326 : 370 (1987).
34. D. Gatteschi and P. Rey, private communication.

EXOTIC MAGNETIC SYSTEMS

Cristiano Benelli, Andrea Ganeschi, and Dante Gatteschi

Department of Chemistry
University of Florence
Florence, Italy

Paul Rey

Centre d'Etudes Nucleaires
Grenoble, France

1.INTRODUCTION

Among the various meanings of exotic, the one which applies to the present title, is "strikingly out of the ordinary, rarely met with" according to the Webster's dictionary or perhaps also "excitingly strange, having the appeal of the unknown". A title like this is therefore very dangerous, because, in a field which has been actively investigated for more than three thousand years now, it is certainly fairly difficult to report really exciting new data. However we have accepted with pleasure the invitation of the organizers to make a resume of data with this title, because, in our opinion, at the moment many new developments are reported in the literature which can be collected under this heading.

Indeed molecular magnetism is presently living one of the most exciting periods of its story, and unusual behaviors are currently observed for clusters and infinite arrays of metal centers. A large impetus in the synthesis of new molecular systems exhibiting controlled magnetic properties has started from the studies of heterodinuclear metal pairs, which have shown how behaviors which were extremely rare for homonuclear species can be rather easily obtained.[1] So it has been realized for istance that ferromagnetic coupling is much more common than it was believed up to few years ago, that ferrimagnetism is possible also for molecular systems, and that this can provide a useful strategy to design molecular materials with bulk magnetic properties.

Once it has been realized that by this procedure novel properties can be obtained, the extension to systems containing even more different kinds of spins has been obvious. Indeed paramagnetic systems can be broadly classified into four groups: transition metal ions, lanthanide and actinide ions, main group radicals, organic radicals. Up to a few years ago the spins which were brought into interaction belonged in general to the same group, i.e. transition metal ions with transition metal ions, or organic radicals with organic radicals. These kind of compounds can be considered to give "normal" magnetism, while we like to consider as "exotic" all those materials in which the interacting spins belong to different groups in the above classification. In fact this is the operational definition which we adopt here for exotic. Therefore this Chapter will be organized into two main sections reporting the interactions of transition metal ions with lanthanides, and of transition metal ions with organic radicals, while the interactions of lanthanides with organic radicals are reported in a separate contribution to this book.[2] At the moment the interactions of main group element radicals are still too exotic, if superoxide and nitric oxide complexes are neglected, to justify a mention here. Finally, we want just to advice that in the following the exchange coupling constant which we will use is defined by the spin hamiltonian: $H = \sum_{i,j} J_{ij} S_i . S_j$.

2.TRANSITION METAL IONS AND ORGANIC RADICALS

This is a field of large interest ranging from biological samples to low dimensional magnetic materials. The most intensively studied radicals are semiquinones and nitroxides, but also other anionic radical ligands have been usefully employed.[3] We will refer briefly to the former here, and more at length to the latter, where our contribution has been more relevant.

2.1 Semiquinone complexes

The interest for this class of complexes is mainly due to their ability to undergo redox reactions, not the least important of which being their involvement in bacterial photosynthetic processes.[4] For instance it is known that a broad EPR signal observed at g = 1.8 in purple bacteria is due to a magnetic interaction between a semiquinone and an iron(II) ion.

Several simple complexes containing semiquinones have been synthesized, in which the coupling of the metal ion with the semiquinone ranges from weak to strong antiferromagnetic. For instance tris semiquinone complexes have been reported with iron(III) and chromium(III).[5,6]

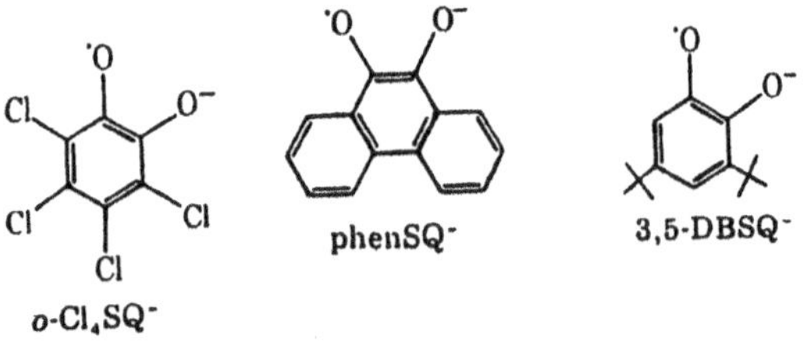

The ligands are ortho-tetrachloro semiquinone, o-Cl_4SQ^-, 9,10-phenantrenesemiquinone, phenSQ^-, and 3,5-di-tert-butyl- semiquinone, 3,5-$DBSQ^-$. The magnetic orbitals are essentially p_z in nature on the oxygen atoms, however EPR data showed that unpaired spin density is spread all over the aromatic ligands.

The $Fe(3,5\text{-}DBSQ)_3$ complex has a temperature independent magnetic moment corresponding to two unpaired electrons, as one can expect for a strong antiferromagnetic coupling between the five unpaired electrons of iron(III) and the three electrons of the semiquinones. The corresponding phenSQ^- and o-Cl_4SQ^- complexes on the other hand have magnetic moments which are temperature dependent: at high temperature the two complexes have moments higher than that expected for two unpaired electrons, showing that the antiferromagnetic coupling is less intense here. On the other hand at low temperature the moment drops below the value for two unpaired electrons, showing that intermolecular interactions are operative. It is interesting to note the difference in behavior of the two least bulky semiquinones. The corresponding chromium(III) complexes also show a moderate antiferromagnetic coupling.

Another interesting class of compounds has been obtained by reacting M(II)SALen with various quinones (SALen is the Schiff base formed by salycilaldehyde and ethylenediamine, M = Fe, Mn, Co).[7] The metal ions are oxidized to the +3 oxidation state, and discrete complexes of formula M(SALen)(SQ) are obtained, where SQ is a semiquinone. No crystal structure is available, but they are believed to be as shown below:

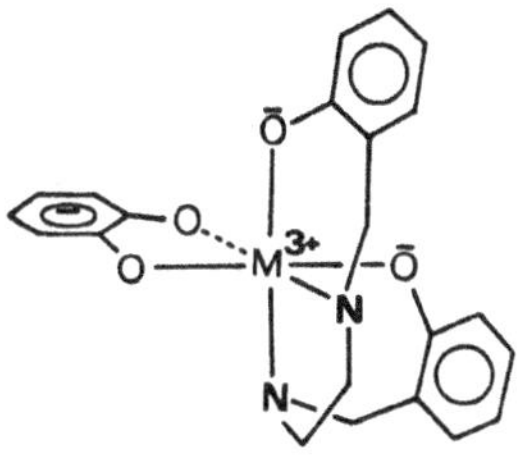

For M = Fe, the ground state is S = 2 showing that a strong antiferromagnetic exchange is operative between $S_{Fe} = 5/2$ and $S_r = 1/2$. J has been estimated to be larger than 1,200 cm^{-1} and strong antiferromagnetic coupling has been observed also with manganese and cobalt. This is readily understood considering that the unpaired electron in the magnetic orbital of the semiquinones can extensively overlap with the orbitals of the metal ions of σ symmetry.

On the basis of simple overlap considerations of this kind a moderate ferromagnetic coupling can be anticipated in octahedral complexes of metal ions which have σ magnetic orbitals, such as copper(II) and nickel(II). Indeed J has been found to be $-220\ cm^{-1}$ in (di-2-pyridylamine)(3,5-di-t-butyl-o-semiquinonato)copper(II)[8] in which the semiquinone ocupies two equatorial positions in the coordination sphere of the metal ion. Relatively strong ferromagnetic coupling has been observed also for both nickel(II) and copper(II) macrocyclic derivatives with dibenzo ortho semiquinone.[9]

Tetranuclear species have been reported [10,11] of formula $M_4(SQ)_8$, where M = manganese(II), cobalt(II), nickel(II). A view of the inner coordination sphere of the $Co_4|O_2C_6H_2(t\text{-}Bu)_2|_8$ molecule is shown in Figure 1.

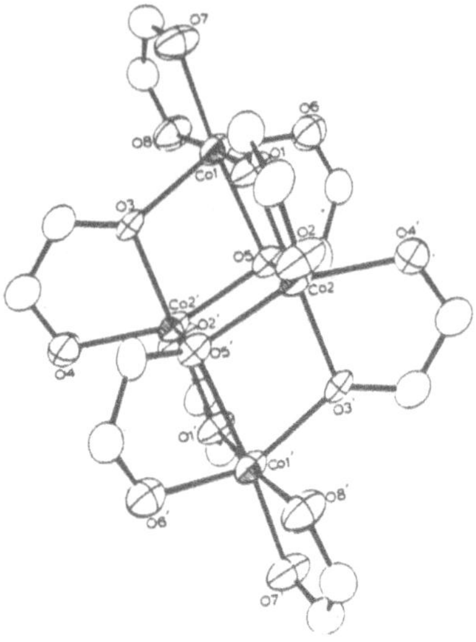

Figure 1. View of the inner coordination sphere $Co_4(SQ)_8$.
After ref. 11

The magnetic moments at high temperature indicate weak spin-spin interactions.

In principle it might be possible to form low dimensional magnetic materials with semiquinones, although, to our knowledge, this has never been obtained. However in several semiquinone complexes intermolecular magnetic interaction have been observed.

2.2 Nitroxide complexes

Several complexes with nitroxides have been studied.[12] The organic radicals are characterized by the functional group N-O.

The unpaired electron which is in a π^* orbital spends roughly 60% of its time on the nitrogen and 40% on the oxygen as shown by EPR spectra and confirmed by MO calculations.[13] Many different types of nitroxides have been synthesized. Some of those which were used to form complexes with paramagnetic ions are depicted in Figure 2.

As far as complexes with transition metal ions are concerned two different classes have been synthesized, one in which the oxygen atom is directly bound to the metal ion, and the other in which the oxygen atom is far from the metal ion. We will not expand further on the latter class of compounds, for which excellent reviews are available[12], because, although some magnetic interactions can be operative, they are generally weak.

When the oxygen atom of the nitroxide is bound to the transition metal ion, a direct exchange interaction is operative. The mechanism is best understood referring to copper(II) complexes, for which several data are available. In fact two structural types can be distinguished, one in which the nitroxide occupies an axial position in a tetragonal pseudosymmetry, and the other in which it enters an equatorial site.[14] For the former a weak to moderate ferromagnetic coupling has been observed, while for the latter the molecule has always been found to be in the ground low spin state, with no evidence of excited higher spin multiplicity states.

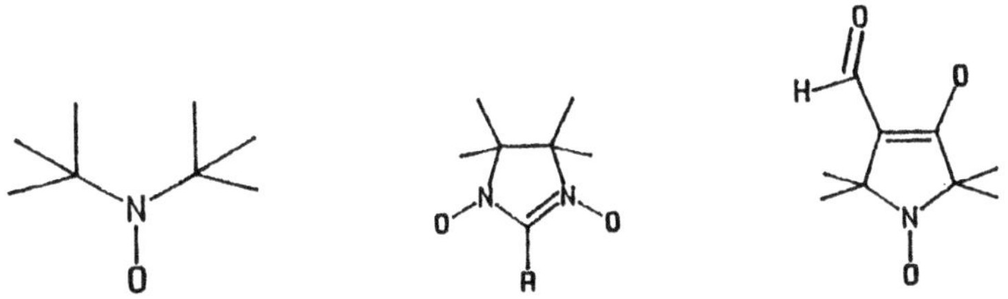

DTBNO NITR proxFORMYL

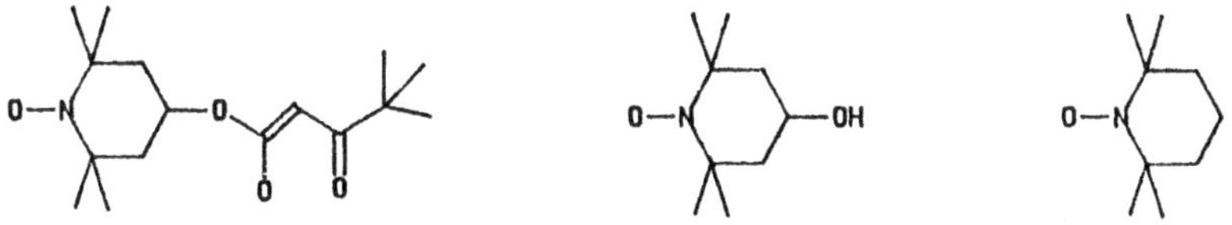

pacTEMPOL TEMPOL TEMPO

Figure 2. Structural formula of some nitroxides used to form complexes with transition metal ions.

These two limit cases are easily justified on the basis of a simple orbital mechanism: in fact for the axial case the two magnetic orbitals, of the radical and of the copper ion respectively, are orthogonal thus yielding parallel alignment of the spins, while for the latter a substantial overlap occurs, with an antiferromagnetic coupling of the spins. Attempts were made to find structural-magnetic correlations, in the case of ferromagnetic coupling, also using Extended Hückel calculations. It was found that the overlap between the metal and radical magnetic orbitals remains substantially zero irrespective of the rotations of the nitroxide relative to the metal. The main factor which is relevant to the interaction appears to be the metal-oxygen distances, as shown in Table 1. Unfortunately it has not been possible up to the moment to obtain complexes with relatively short copper-oxygen distances. As a consequence the largest ferromagnetic coupling observed is 78 cm^{-1}.

Much shorter distances were observed when the nitroxide binds in an equatorial position,[19,21,22,24,25] and in this case EH calculations show that the metal and ligand orbitals overlap quite extensively, and indeed strongly coupled pairs have been found to be formed[14], with no clear evidence of populated triplet states.

Table 1. Metal-oxygen distances(pm) and exchange coupling constants (cm^{-1}) in some axial copper(II) nitroxide complexes[a].

Compound	R	-J	Reference
$Cu(hfac)_2TEMPOL$	243.9	13	15,16
$Cu(pacTEMPOL)_2$	258.3	19*	17
$Cu(pacTEMPOL)_2$	315.7	19*	17
$Cu(proxFORMYL)_2$	260.6	21	18,19
$Cu(hfac)_2(NITMe)_2$	234.1	26*	20
$Cu(hfac)_2(NITMe)_2$	243.1	26*	20
$Cu(hfac)_2(NITPh)_2$	239.3	10	21
$Cu(tfac)_2(NITMe)_2$	231.0	72	14

[a]The names of the radicals are defined in Figure 1. hfac = hexafluoroacetylacetone; tfac = trifluoracetylacetone.
*In the unit cell two non equivalent sites are present. The J value are the averaged ones.

Beyond discrete complexes containing only one copper(II) ion, also a few linear chain compounds were reported. In $Cu(hfac)_2TEMPOL$, whose structure[15] is schematized in Figure 3, the copper(II) ion is

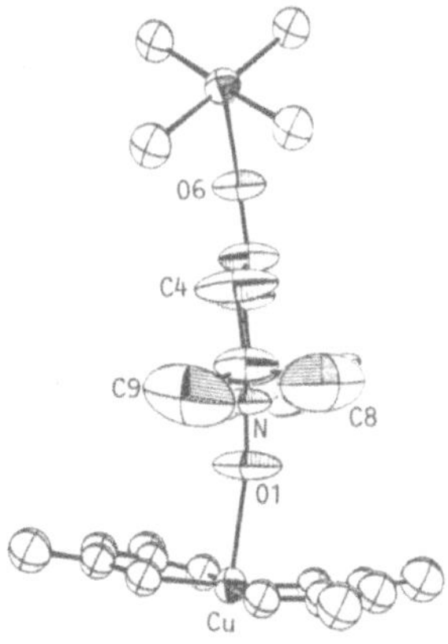

Figure 3. Simplified view of $Cu(hfac)_2TEMPOL$. After ref. 15.

ferromagnetically coupled to the nitroxyl radical[13], $J = -13\ cm^{-1}$. A much weaker interaction is operative between the nitroxide and the copper ion which is bound to the -OH group.[26] This has been estimated to be antiferromagnetic, $J' = 5.4 \times 10^{-2}$ cm^{-1}. Therefore this compound is an example of an alternating chain with ferro- and antiferromagnetic coupling.

$Cu(proxFORMYL)_2$ has been shown to have the structure reported in Figure 4. The spins of the copper(II) ions and of the radicals are roughly in the bc plane of a monoclinic lattice. Each copper ion is bound to four different proxFORMYL ligands, yielding a structure in which two different chains intersect at each copper ion. Two ligands bind to copper with the chelate oxygen atoms and two with the oxygen of the nitroxide group. The magnetic properties were interpreted with a simple model of clusters of three spins, one copper and two nitroxides. However the EPR spectra showed that additional interactions between the clusters are operative. The inter-cluster coupling constant was estimated to be 0.13 cm^{-1} from the ratio of the linewidths at X- and Q-band frequency, in reasonable agreement with the value observed in $Cu(hfac)_2TEMPOL$.

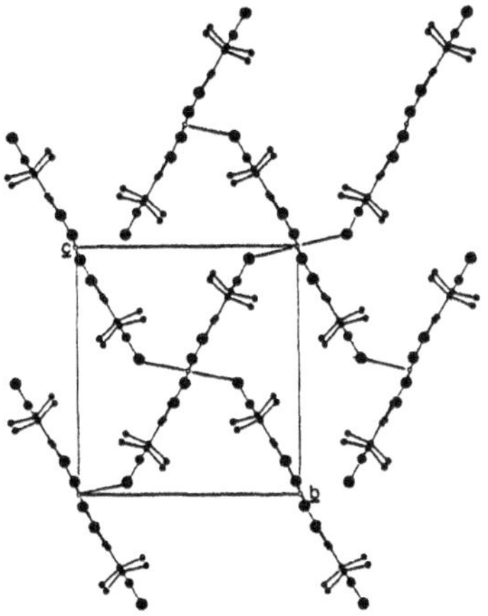

Figure 4. General view of $Cu(proxFORMYL)_2$. After ref. 19.

An unusual linear chain is formed by $Cu(NITPh)_2Cl_2$ whose structure[23] is schematized in Figure 5. Each copper ion is coordinated to two chloride ions and two oxygen atoms of two different NITPh radicals in a square planar coordination environment. Each nitroxide is bound to copper via one of its oxygen atoms, while the oxygen atoms which are not bound to copper have shortest contacts with a nitrogen atom (356 pm) of another NITPh of a neighboring molecule, in such a way that a chain is formed, see Figure 6. From the view point of exchange pathways, this might be described as:

$$\begin{array}{ccccccccccccc} --- & R & — & Cu & — & R & --- & R & — & Cu & — & R & --- \\ & J' & J & & J & & J' & & J & & J & & J' \end{array}$$

including only nearest neighbor interactions. Indeed this would be an interesting case to analyze, but in practice $J > 500$ cm^{-1}, so that magnetically $Cu(NITPh)_2Cl_2$ can be considered as a linear chain of

R--Cu--R units having one unpaired electron. It is interesting to notice however that J' is 15 cm^{-1}, a fairly large value for two paramagnetic centers which are 356 pm apart, with no intervening atom.

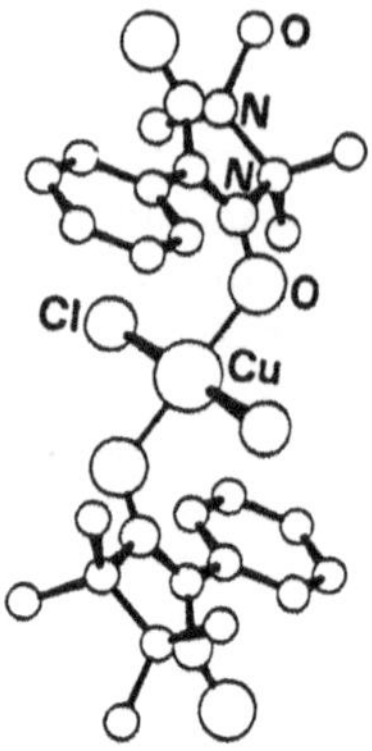

Figure 5. Schematic view of $Cu(NITPh)_2Cl_2$. After ref. 23

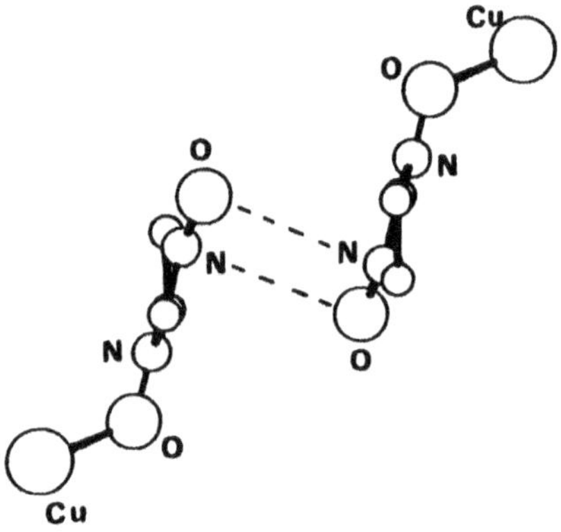

Figure 6. Simplified view of two interacting molecules responsible of the J' interaction. After ref. 23

Ferromagnetic chains could eventually be obtained[20] by reacting $Cu(hfac)_2$ with NITMe. The nitroxide binds to two different copper atoms with its two oxygen atoms, as shown in Figure 7. The χT vs T curve increases smoothly down to 6 K when its reaches 3 emu mol^{-1} K. The data can be fit reasonably well with the Bonner-Fisher equation to yield $J = -25\ cm^{-1}$.

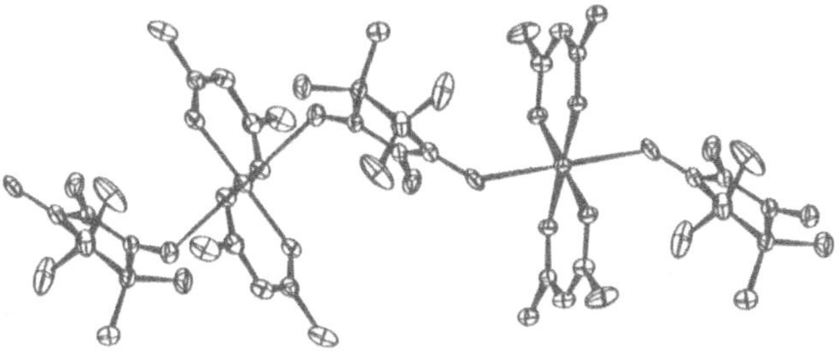

Figure 7. Schematic view of $Cu(hfac)_2NITMe$. After ref. 20.

Also the EPR spectra of this compound appear to be interesting[28]: at room temperature they show a $(3\cos^2\theta-1)^n$ dependence of the linewidth as would be expected for a linear chain with enhanced secular behavior due to spin diffusion affects. θ is the angle of the static magnetic field with the chain axis. On lowering temperature the linewidth initially decreases almost linearly, while approaching 4.2 K it increases. The resonance fields are also found to be temperature dependent below 10 K, a result which has been rationalized as due to the presence of short range order. However it must be stressed that down to 4.2 K the compound does not show any evidence of long range order: apparently the chains are too well shielded to yield three dimensional order.

The nickel and cobalt analogues, which up to now did not yield any crystal suitable for X-ray analysis, are also linear chains, as is shown by the XT vs T curves which at low temperature increase up to 129 emu mol^{-1} K for nickel and to 445 emu mol^{-1} K for cobalt. The magnetic behavior of Ni $(hfac)_2$ NITMe can be accounted for by an Ising model which requires a ferromagnetic coupling constant of ca. 15 cm^{-1}. The analysis of the magnetic properties of $Co(hfac)_2NITMe$ is complicated by the fact that in octahedral coordination cobalt (II) has an orbitally degenerate ground state, which makes the usual spin hamiltonians unsuitable. However the high value of XT observed at low temperature suggests a strong ferromagnetic interaction, and eventually a ferromagnetic phase transition can be suspected. Another octahedral cobalt(II) complex in which the metal ion is bound to two nitroxides was previously characterized.[29] In that case XT decreases on lowering temperature, suggesting an antiferroferromagnetic coupling between the metal ion and the radicals, but no analysis was attempted. The coupling was found to be antiferromagnetic also in a tetrahedral adduct of $CoBr_2$ with di-t-butylnitroxide.[30-32]

Manganese(II)-nitroxide complexes are also rather well characterized now. A few compounds containing one manganese(II) and two radicals have been reported.[29] The geometry of these complexes can be schematized as shown in Figure 8. The magnetic orbitals of the axial nitroxides in these complexes can overlap to the magnetic orbitals of manganese(II), which in general are linear combinations of xz, yz and z^2 orbitals. The remarkable feature is that notwithstanding the overlap,

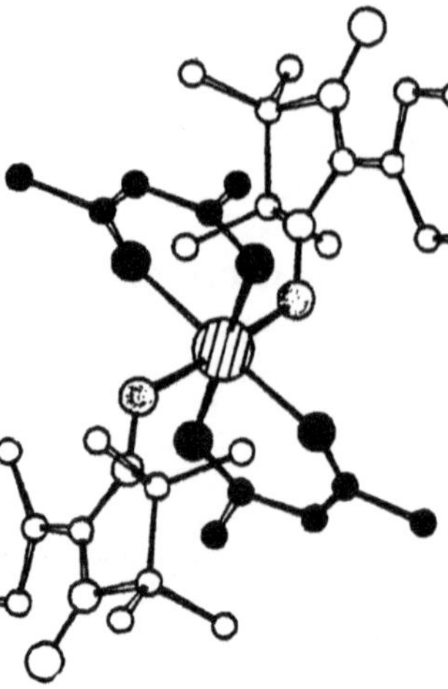

Figure 8. Schematic view of $Mn(hfac)_2(NITPh)_2$

which might suggest a complete pairing of the electrons, this is far from complete. Indeed the configuration with all the spins aligned parallel is only slightly higher in energy than that with two spins reversed.[29] As a consequence the ground state of these bis nitroxide manganese(II) complexes is a quartet, with thermally populated sextets and octets. The magnetic data have been interpreted with J ranging from 130 to 210 cm^{-1}. The EPR spectra of $Mn(hfac)_2(TEMPOL)_2$ and $Mn(hfac)_2(proxyl)_2$ have shown additional evidence of the thermal population of the quartet and sextet states.[33] The EPR spectra have been able to discriminate between two S = 5/2 states formed by the interaction of $S_{Mn} = 5/2$ with two $S_r = 1/2$.

Since the J value in octahedral manganese(II)-nitroxide complexes is so large manganese(II) is an ideal candidate to yield ferrimagnetic chains. Indeed it has been shown that in chains containing alternate S = 5/2 and S = 1/2 spins, the low temperature magnetic behavior corresponds to a chain of ferromagnetically coupled S = 2 spins.[34] The XT vs T curve for a chain like this has a limit value of 4.75 emu mol^{-1} K at high temperature, then decreases to a minimum of 4.10 emu mol^{-1}K at T close to 2.5 J/k, and then increases rapidly, diverging at very low temperature.[34,35] With a large J the minimum occurs at very high temperature, thus giving the opportunity to have a large effective spin along the chains at relatively high temperature. Therefore the interactions between the chains can acquire more importance, and three dimensional order can set in. In fact this is what we have observed in complexes of formula $Mn(hfac)_2NITMe$[36], which are supposed to have a structure similar to that of $Cu(hfac)_2(NITMe)_2$.[20]

When manganese(II)hexafluoroacetylacetonate is reacted with NITMe several different compounds are obtained. More details for this are given in a separate contribution in this book.[37] Of particular interest is the hexagonal $|Mn(hfac)_2NITMe|_6$ complex whose structure[38] shows a ring of six manganese(II) ions bridged by six radicals. From the point of view of the magnetic properties the main reason of interest of this compound is provided by the fact that it has six S = 5/2 and six S = 1/2 spins coupled. Indeed currently the knowledge of the magnetic properties for both small and infinite clusters is rather satisfactory, but much less is known for clusters with a relatively large number of spins. This is a field where active research can be foreseen in the next few years. For instance high nuclearity clusters are being tackled[39] and on another side complexes containing 11 paramagnetic iron(III) atoms have been synthesized and studied[40] as models for understanding the mechanism of metal uptake and release in the iron storage protein ferritin.

The XT vs T curve of $|Mn(hfac)_2NITMe|_6$ increases on decreasing temperature, until it reaches a plateau at 45 K. Presently it is not possible to analyze quantitatively these curves, due to the complexity of the hamiltonian matrix. In fact the basis of spin functions comprises $6^6 x2^6$ states, and even using symmetry arguments, the dimensions of the matrices remain too large to allow a quantitative assessment of the problem. However a qualitative analysis is already able to provide useful information. The room temperature XT value is 5.8 emu mol^{-1} K, higher than that expected for non-interacting S = 5/2 and S = 1/2 spins. This clearly indicates that the high temperature limit is not obtained, suggesting a fairly large value of J. This in principle might be either ferro- or antiferromagnetic, but, given the comparison with the complexes containing one manganese(II), it appears much more probable that the coupling is in fact antiferromagnetic. This result is confirmed by the low temperature limit of XT. For a cluster of N pairs of S = 5/2 and S = 1/2 spins, the low temperature limit corresponds to:

$$(XT)_{lowT} = \frac{g^2 S(S+1)}{8N}$$

where S is the total spin multiplicity and X is calculated for one pair. For antiferromagnetic coupling S = 12, while for ferromagnetic coupling S = 18. Setting $g = g_e$ these two correspond to limit values of 13 and 28.5 emu mol^{-1} K respectively. Since the observed value is 12.6 emu mol^{-1} K it is apparent that the coupling must be antiferromagnetic. Since no minimum in the XT curve is observed up to 300 K, J can be estimated to be larger than 100 cm^{-1}.

EPR spectra are in principle a useful tool for investigating spin dynamics and a cluster like the one present in $Mn|hfac)_2NITPh|_6$ might in principle be expected to show properties intermediate between those of pairs and those of one dimensional materials. In fact the spectra at room temperature show one signal at g = 2, and another one at g = 4, but the latter can be observed only in crystal orientations close to the magic angle. The line width is found to follow a $(3\cos^2\theta - 1)^n$ dependence, θ being the angle with the hexagonal axis, while the line-shape is gaussian, as can be expected for a cluster which has so many thermally populated levels available. In an analogous compound with a chain structure the line shape should be either lorentzian or intermediate between gaussian an lorentzian.[41]

Concluding this section we want to stress how versatile have the radicals proven to be as ligands towards paramagnetic metal ions. One positive characteristics is that they can couple relatively easily in a ferromagnetic way, and, exploiting this, ferromagnetic alternating chains could be obtained. On the other hand they can also couple very effectively in an antiferromagnetic way and, exploiting this, ferrimagnetic chains, similar to those previously reported for copper-manganese systems, but with much larger coupling constants could be obtained. With this potentiality it would be extremely important to design new radicals which can transmit the exchange interaction not only in one dimension, but also in two or three. We feel that much research will be performed along these lines in the next few years.

3. LANTHANIDES AND TRANSITION METAL IONS

Most of the studies on the interaction of lanthanides and transition metal ions have been performed in doped ionic lattices. The main difficulty in the interpretation of the magnetic properties in this field are given by the unquenched angular orbital momentum of the lanthanide ions, which leads to large anisotropy effects. Therefore the effective spin hamiltonian is more complicated than the usual Heisenberg one. A convenient form for pairs is given below[41]

$$H = \sum_{k=0}^{2l} \sum_{q=-k}^{k} \alpha_{kq} T_q^{(k)} . S^{RE} . S^{TM}$$

where α_{kq} are the components of the exchange potential, $T_q^{(k)}$ is the q-th component of the k-th irreducible tensorial operator which acts on the orbital part of the rare-earth wave function, S^{RE} is the spin operator acting on the rare-earth electrons and S^{TM} is the spin operator relative to the transition metal ion. The latter has been assumed to have no first order orbital angular momentum for the sake of simplicity. A notable exception to the use of this hamiltonian is for pairs involving gadolinium(III), which has a ground ^8S state, and consequently no first order orbital momentum contribution. Therefore for that case a simple spin-hamiltonian approach is totally justified.

The systems which have been most studied are the iron garnets, the orthoferrites and the orthochromites doped with rare earth ions. Due to space limitations we will give here only some example, without any attempt to cover completely the area. Europium(III) iron garnet shows anisotropic magnetic properties which were attributed to crystalline field effects around the lanthanide ion and to europium(III)-iron(III) exchange, treated at the molecular field level.[42] The exchange fields of ytterbium(III) in ytterbium iron garnet have been estimated to be 349,000, 611,000 and 678,000 gauss along three orthogonal axis.[43] Detailed studies of holmium(III) in ytterbium iron garnet yielded a large set of exchange parameters,[44] among which we may quote an isotropic part corresponding to 2.5 cm^{-1}, with fourth order terms as large as 84 cm^{-1}.

Gadolinium(III)-chromium(III) interactions have been studied in $GdCrO_3$.[45] The isotropic exchange in this case was found to be negligible, and the observed effects on the magnetization were attributed to high-order anisotropic exchange interactions. f-d-f exchange was shown[46,47] to be relevant to understand the magnetic properties and the nature of the orientational phase transitions in orthoferrites and orthochromites.

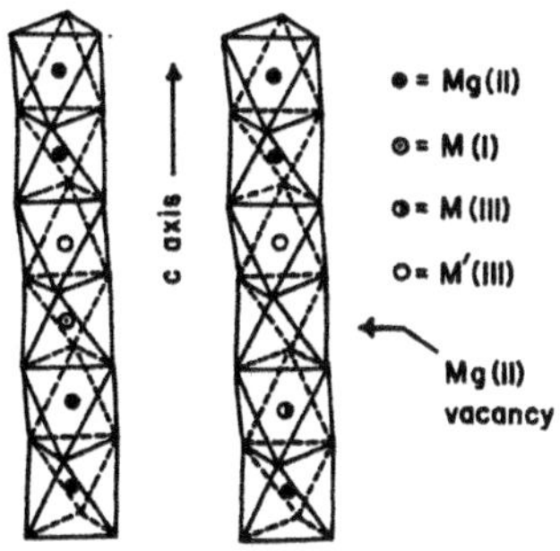

Figure 9. Perspective views of $|MgCl_3^-|_n$ after ref. 48.

EPR spectroscopy was used to characterize[48] gadolinium(III)-ruthenium(III) pairs in $CsMgCl_3$. They were assumed to be due to ions in the lattice as shown in Figure 9.

The exchange parameters defined by

$$H = J_z S_z S_z' + J_{xy}(S_x S_x' + S_y S_y')$$

were determined as $J_z = +0.0055\ cm^{-1}$, $J_{xy} = \pm 0.020\ cm^{-1}$. For two ions at ca. 600 pm the above parameters yield isotropic exchange components of $\pm 0.0144\ cm^{-1}$, according to the sign chosen for J_{xy}.

Discrete molecular species containing lanthanides and transition metal ions are still relatively rare. An interesting pentanuclear species of formula $Ni_3Eu_2(O_2C_2S_2)_6(H_2O)_{10}.xH_2O$ ($10 < x < 12$) has been reported[49] but since the nickel ion is in a square planar environment, which induces a ground singlet state, no magnetic coupling could be observed.

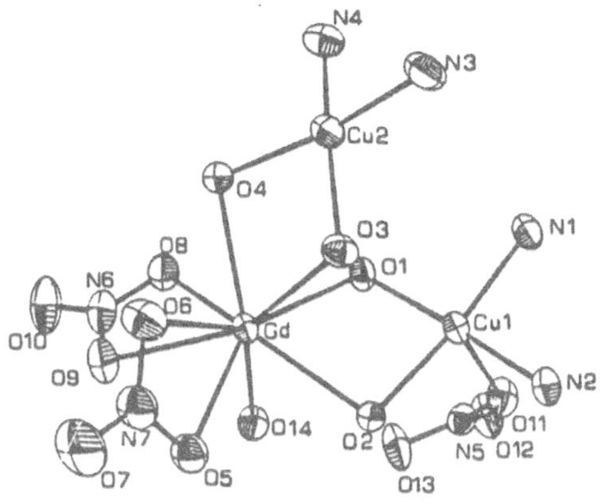

Figure 10. Schematic view of $Gd(NO_3)_3(CuSALTn)_2$ after ref. 55

Compounds containing coupled copper(II) and gadolinium(III) ions were obtained by reacting[50] copper Schiff base complexes of formula shown in Figure 10 with appropriate gadolinium salts[51,52].

The best characterized examples are trinuclear species containing one lanthanide and two transition metal ions, $GdCu_2$[53-55]. The magnetic properties of these complexes show that XT increases at low temperature, then it is stabilized at a plateau corresponding to a ground S = 9/2 state, originated by ferromagnetic coupling between the lanthanide and the copper ion. The values of the isotropic coupling constants range from -1.22 to -7.38 cm^{-1}. Also analogous europium derivatives were reported. The EPR spectra of these, which show a rhombic zero field splitting tensor, indicate a sizeable exchange contribution to the anisotropic spin-spin interaction tensor.

It is apparent that much remains to be done in this area. However the data already existing clearly indicate that interesting behaviors must be expected when transition metal ions and lanthanides are coupled, thus justifying the effort to synthesize new systems.

REFERENCES

1 O. Kahn, Angew. Chemie Int. Ed. Engl. 24: 834 (1985).
2 C. Benelli, A. Caneschi, D. Gatteschi, J. Laugier, P. Rey, this book
3 W. Kaim, Coord. Chem. Rev., 76: 187 (1987).
4 W. W. Parson, R. J. Cogdell, Biochim. Biophys. Acta, 416: 105 (1975).
5 M. Y. Okamura, R. A. Isacson, G. Feher, Proc. Nat. Acad. Sci. USA, 72: 3491 (1975).
6 R. M. Buchanan, S. L. Kessel, H. H. Downs, C. G. Pierpont, D. N. Hendrickson, J. Am. Chem. Soc., 100: 7894 (1978).
7 S. L. Kessel, R. M. Emberson, P. G. Debrunner, D. N. Hendrickson, Inorg. Chem., 19: 1170 (1980).
8 O. Kahn, R. Prins, J. Reedjik, J. S. Thompson, submitted for publication
9 A. Dei, D. Gatteschi, L. Pardi, to be published
10 A. Lynch, D. N. Hendrickson, B. J. Fitzgerald, C. G. Pierpont, J. Am. Chem. Soc., 106: 2041 (1984).
11 R. M. Buchanan, B. J. Fitzgerald, C. G. Pierpont, Inorg. Chem., 18: 3439 (1979).
12 S. S. Eaton, G. R. Eaton, G.R. Coord. Chem. Rev. 26: 207 (1978).
13 B. R. Knauer, J. J. Napir, J. Am. Chem. Soc. 98: 4395 (1976).
14 A. Caneschi, D. Gatteschi, A. Grand, J. Laugier, L. Pardi, P. Rey, submitted for publication
15 D. P. Anderson, T. C. Kuechler, Inorg. Chem. 19: 1417 (1980).
16 A. Bencini, C. Benelli, D. Gatteschi, C. Zanchini, J. Am. Chem. Soc. 106: 5813 (1984).
17 A. Grand, P. Rey, F. Subra, Inorg. Chem. 22: 391 (1983).
18 J. Laugier, R. Ramasseul, P. Rey, J. C. Espie, A. Rassat, Nouv. J. Chim. 7: 11 (1983).
19 C. Benelli, D. Gatteschi, C. Zanchini, J. M. Latour, P. Rey, Inorg. Chem 25: 4242 (1986).

20 A.Caneschi, D. Gatteschi, J. Laugier, P. Rey, J. Am. Chem. Soc. (1987).
21 D. Gatteschi, J. Laugier, P. Rey, C. Zanchini, Inorg. Chem. (1987).
22 M. H. Dickman, R. J. Doedens, Inorg. Chem. 20: 2677 (1981).
23 J. Laugier, P. Rey, C. Benelli, D. Gatteschi, C. Zanchini, J. Am. Chem. Soc. 108: 6931 (1986).
24 L. C. Porter, M. H. Dickmann, R. J. Doedens, Inorg. Chem. 22: 1962 (1983).
25 L. C. Porter R. J. Doedens, Inorg. Chem. 211: 1006 (1985).
26 J. J. Linn, R. S. Drago, J. Am. Chem. Soc. 93: 891 (1971).
27 C. Benelli, D. Gatteschi, D. W. Carnegie, R. L. Carlin, J. Am. Chem. Soc. 107: 2560 (1985).
28 A. Caneschi, D. Gatteschi, P. Rey, C. Zanchini, J. Chem. Soc. Faraday Trans. I in press.
29 M. H. Dickmann, L. C. Porter, R.J. Doedens, Inorg. Chem. 106: 5813 (1986).
30 W. Beck, K. Schmidter, H. S. Keller, Chem. Ber. 100: 503 (1967).
31 D. G. Brown, T. Maier, R. S. Drago, Inorg. Chem 10: 2804 (1971).
32 C. Benelli, D. Gatteschi, C. Zanchini, Inorg. Chem. 23: 798 (1984).
33 C Benelli, D. Gatteschi, C. Zanchini, R. J. Doedens, M. H. Dickman, L. C. Porter, Inorg. Chem. 25: 3453 (1986).
34 Y. Pei, M. Verdaguer, O. Kahn, J. Sletten, J.P. Renard, Inorg. Chem. 26: 138 (1987).
35 M. Drillon, E. Coronado, D. Beltran, R. Georges, J. Appl. Phys. 57: 3353 (1985).
36 to be published
37 A. Caneschi, D. Gatteschi, R. Sessoli, C. Zanchini, J. Laugier, P. Rey, this book
38 A. Caneschi, D. Gatteschi, R. Sessoli, C. Zanchini, J. Laugier, P. Rey, submitted for publication
39 L. J. de Jongh, this book
40 S. Gorun, S. J. Lippard, Nature(London)319: 666 (1986).
41 J. E. Drumheller, Magn. Reson. Rev. 7: 123 (1982).
42 P. M. Levy, Phys. Rev. 135: A155 (1964).
43 P. M. Levy, Phis. Rev. 147: 320 (1966).
44 K. A. Wickersheim, Phys. Rev. 122: 1376 (1961).
45 H. Kamimura, T. Yamaguchi, Phys. Rev. 1: 2902 (1970).
46 A. H. Cooke, D. M. Martin, M. R. Wells, J. Phys. C 7: 3133 (1974).
47 E. I. Golovenchtis, V. A. Sanna, Sov. Phys. Solid State 23: 977 (1981).
48 G. L. Mc Pherson, L. A. Martin, J. Am. Chem. Soc.106: 6884 (1984).
49 J. C. Trombe, A. Gleizes, J. Galy, Compt. Rend. Acad. Sc. Paris 294: 1369 (1982).
50 S. J. Gruber, C. M. Harris, E. Sinn, J. Inorg. Nucl. Chem. 30: 1805 (1968).

51 A. Chisani, A. Musumeci, M. Vidali, A. Seminara, Inorg. Chim. Acta 81: L19 (1984).

52 G. Condorelli, I. Fragalà, S. Giuffrida, A. Cassol, Z. Anorg. Allg. Chem. 251: 412 (1975).

53 A. Bencini, C. Benelli, A. Caneschi, R. L. Carlin, A. Dei, D. Gatteschi, J. Am. Chem. Soc. 107: 8128 (1985).

54 A. Bencini, C. Benelli, A. Caneschi, A. Dei, D. Gatteschi, J. Magn. Magn. Mater. 54-57: 1485 (1986).

55 A. Bencini, C. Benelli, A. Caneschi, A. Dei, D. Gatteschi, Inorg. Chem. 25: 572 (1986).

MAGNETIC PHASE TRANSITIONS IN LOW-DIMENSIONAL SYSTEMS

Jean-Pierre Renard

Institut d'Electronique Fondamentale, CNRS UA 022
Bâtiment 220, Université Paris-Sud, 91405 Orsay
Cédex, France

1. INTRODUCTION

Phase transitions are of current interest in many areas of condensed matter physics. They occur in many systems, among them we can mention liquid-gas transition, superfluidity, binary alloys (such as β-brass), structural phase transitions and magnetic phase transitions (MPT). The MPT are especially interesting because of the wide variety of available magnetic compounds with different lattices, spins and interactions. The simplest case of an MPT is the ferromagnetic transition which is usually of second order and only occurs in zero magnetic field. The case of two sublattices antiferromagnets is richer since they undergo phase transitions in a certain range of applied magnetic fields, which can be of first order (metamagnetism; spin-flop phases). Besides these simple ferromagnets and antiferromagnets (F and AF), other magnetic compounds which exhibit a more complex class of phase transitions are found such as: systems with a non magnetic ground state induced either by zero field splitting, or in the case of magnetic chains, by alternating AF interactions; magnetic chains with a large magnetoelastic coupling inducing a spin-Peierls transition; frustrated antiferromagnets with more than two sublattices or with incommensurate magnetic ordering.

An enormous amount of theoretical and experimental work has been devoted to MPT. An exhaustive review is out of the scope of this paper. The reader is refered to the review papers by Kadanoff et al [1] and de Jongh and Miedema [2], and for the transparent magnetic materials to the more recent paper by Ferré and Gehring [3]. Our aim is to introduce a few simple physical ideas, which will be useful for the understanding of MPT, in low dimensional compounds.

The outline of the paper is the following : In Section 2, we discuss the characteristic features of second order MPT and the role of the relevant parameters : lattice dimensionality d and order parameter dimensionality n; Section 3 is devoted to the correlation length in simple low dimensional models; In

Section 4 we discuss crossovers induced by small deviations to ideal models exhibited by real magnetic compounds, and critical temperature. Sections 5 and 6 are respectively devoted to the effects of impurities on MPT and to the field-temperature phase diagram of low dimensional antiferromagnets. Finally in section 7, we discuss a few complex cases with non magnetic ground state, and especially the S = 1, Heisenberg AF chain which exhibits the quantum gap conjectured by Haldane [4].

2 - GENERAL FEATURES OF SECOND ORDER MAGNETIC PHASE TRANSITIONS-CRITICAL EXPONENTS

Figure 1 illustrates the simple second order phase transition from paramagnetic to ferromagnetic phase achieved by lowering the temperature in zero magnetic field.

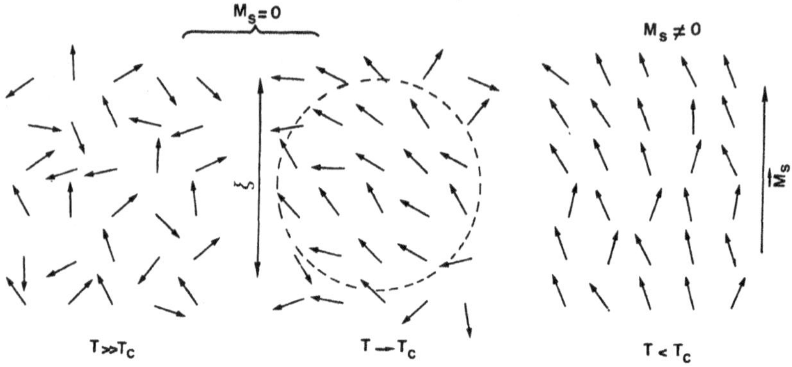

Figure 1 . Microscopic picture of a ferromagnetic transition.

At temperatures much higher than the critical temperature T_c, the orientation of a spin is independent from the orientation of neighbouring spins. Consequently, pair correlations $\langle \vec{S}_i \ \vec{S}_j \rangle$ for $i \neq j$ are close to zero. At temperatures slightly above T_c, the spin orientation is almost the same over a certain distance which is the correlation length ξ, but the magnetization integrated over a macroscopic volume $\gg \xi^3$ and the time average spin value are equal to zero. The correlation length diverges at $T = T_c$, and below T_c the average magnetization has a finite value i.e. the spontaneous magnetization, which is the order parameter of the transition. For isotropic interactions, the magnetic system has a spherical symmetry above T_c, while it has a cylindrical symmetry around the direction of the spontaneous magnetization below T_c. This symmetry breaking at $T = T_c$ is a very general feature of phase transitions. The divergence of the correlation length at $T = T_c$ is the keystone of 2nd order MPT. It insures that the magnetic system is invariant by expansion at $T = T_c$. An important consequence is the homogeneity of the Gibbs function with respect to the reduced temperature t and to the magnetic field H .

$t = (T-T_c)/T$ for $T>T_c$; $t = (T_c-T)/T_c$ for $T < T_c$,

This homogeneity first assumed by Widom [5], was given on a rigourous basis by Wilson [6] within the frame of the renormalization group (RG) theory. It leads to the following singular behaviour of thermodynamical quantities in the critical region, described by the power laws as following :

correlation length	$\xi = \xi_0 . t^{-\nu}$
magnetic susceptibility	$\chi = \Gamma t^{-\gamma}$
specific heat	$C = (A/\alpha)\ (t^{-\alpha} - 1) + B$
spontaneous magnetization	$M_s = B\ t^{\beta}$
magnetization at $T = T_c$	$M = D\ H^{1/\delta}$

Each critical exponent can be expressed as a linear function of two others through the scaling relations :

$$\alpha + 2\beta + \gamma = 2 \quad ; \quad \gamma = \beta\ (\delta - 1)$$
$$\gamma = (2 - \eta)\nu \quad ; \quad d\nu = 2 - \alpha$$

The existence of university classes for the critical behaviour is an important feature, pointed out by Kadanoff et al [1] and firmly established by RG theory. Indeed, the critical exponents depend only on a few relevant parameters which are the lattice dimensionality d; the dimensionality of the order parameter n and the range of the interactions. They do not depend on the spin value S and on the detailed lattice structure. For magnetic compounds the possible values of d are 1 (chains), 2 layers or surfaces) and 3. For n, the common values are 1 (Ising model), 2 (XY model) and 3 (Heisenberg model), respectively described by the following hamiltonians.

$n = 1$, Ising $= - \Sigma\ J_{ij}\ S_i^z\ S_j^z$
$n = 2$, XY $= - \Sigma\ J_{ij}\ (S_i^x\ S_j^x + S_i^y\ S_j^y)$
$n = 3$, Heisenberg $= - \Sigma\ J_{ij}\ \vec{S_i}\ \vec{S_j}$

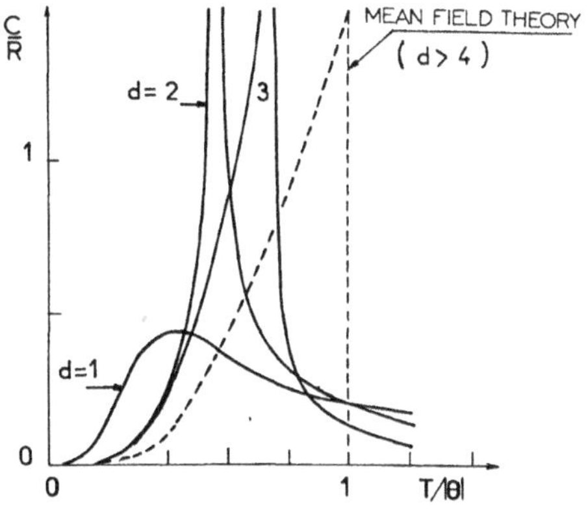

Figure 2 . Theoretical magnetic specific heat for the S = 1/2 Ising model for d = 1,2 and 3; θ = 2zJS(S+1)/3k. (from ref.2).

It should be noticed that the addition of a large single ion anisotropy term $\Sigma\ D(S_i^z)^2$ to the Heisenberg interaction induces an Ising type behaviour for D < 0 or an XY type behaviour for d > 0, at low temperatures, and especially near

T_c. values of n larger than 3 can be found in complex phase transitions which involve an increase of the unit cell in one or more directions as shown by Mukamel and Krinsky [7]. Some physical realizations such as Mn O (n=8) [7,8] are known, but generally the phase transitions of these systems are of first order.

In Table 1, for short range interactions, we classify the values of the critical exponents β and γ for various values of d and n.

TABLE I : Values of the critical exponents β and γ for various models with short range interactions . For d = 3, the critical exponents were obtained from field theoretical methods [9].

Increasing stability →

	d = 1	d = 2	d = 3	d > 4
n = 3	NO LONG RANGE ORDER	NO LONG RANGE ORDER	β = 0.3647 γ = 1.3866	MEAN FIELD THEORY β = 1/2 γ = 1
n = 2	AT T≠0	Kosterlitz Thouless γ = ∞ No LRO	0.346 1.3160	
n = 1		β = 1/8 γ = 7/4	0.325 1.2402	

Table I shows a few important features :

i - For a given value of n, the stability is increased, and consequently, the critical fluctuations decreased, as the lattice dimensionality d is increased. This is illustrated by the Fig. 2, which shows the specific heat of the Ising model (n=1) for different lattice dimensionalities.

ii - For the dimensionalities larger than 4, which can be considered in theoretical models, the mean, or molecular, field theory, which neglects the critical fluctuations, is valid for all values of n.

iii - For a given value of d, the stability is increased by reducing n. This is very clear for the two-dimensional lattice for which there is no PT at finite temperature for n = 3, whereas for n=2 there is a special transition described by Kosterlitz and Thouless (KT) [10] which has an exponential divergence of the correlation length at a finite temperature T_{KT} without LRO and specific heat anomaly, and finally for n = 1 (standard d = 2 Ising model) the critical behaviour was obtained by Onsager [11].

Finally, we consider the effect of dipolar interactions which are present in all magnetic compounds. These interactions which slowly decrease with the distance r as $1/r^3$, are relevant to the critical behaviour. The most interesting case is the d = 3 Ising dipolar ferromagnet which has been shown to have the same critical behaviour as the d = 4 short range system [12]. In this case, the critical behaviour is fully described by RG theory [13,14] as a mean field behaviour with logarithmic corrections. For example, the susceptibility is expressed as $\chi = t^{-1}|Lnt|^{1/3}$. Such critical behaviour has been observed in the rare earth dipolar ferromagnets $LiTbF_4$ and $LiHoF_4$ [15-18].

3. CORRELATION LENGTH OF SIMPLE LOW-DIMENSIONAL MODELS

Most of the magnetic properties of a system of interacting spins can be obtained from a knowledge of the pair correlation functions (PCF) $\langle S_i^\alpha(0)\, S_j^\alpha(t)\rangle$, where α = x, y, z. The static behaviour is determined from the static PCF at t = 0 and the dynamic behaviour for $t \neq 0$. In the following, we describe only the static properties. The spin dynamics of 1d magnetic systems was previously discussed in the review papers by Steiner et al. [19] and Renard [20]. Let us first consider the case of magnetic chains (d=1) with neareast-neighbour exchange interaction J. The corresponding Hamiltonian is :

$$\mathcal{H} = -2J\sum_i [a\, S_i^z S_{i+1}^z + b(S_i^x S_{i+1}^x + S_i^y S_{i+1}^y)]$$

where a = 1, b = 0 corresponds to Ising model, n = 1; a = 0, b = 1 to the XY model, n = 2 and a = 1, b = 1 to the Heisenberg model, n = 3.

The correlation length ξ^α is defined from the long range exponential decrease of the static PCF [19].

$$\langle S_i^\alpha S_{i+1}^\alpha\rangle = (\mathcal{E})^{|l|} \langle (S_i^\alpha)^2\rangle \exp(-K^\alpha|l|)$$

where $\mathcal{E} = +1$ for J > 0 and $\mathcal{E} = -1$ for J < 0, and $K^\alpha = a/\xi^\alpha$ where a is the distance between neighbouring spins. The susceptibility (staggered susceptibility for AF) is related to the PCF by :

$$\chi^\alpha = [N(g\mu_B)^2/kT]\sum_l (\varepsilon)^l \langle S_i^\alpha S_{i+l}^\alpha\rangle$$

For $T\to\infty$, this general expression yields the Curie law since $\langle S_i^\alpha S_{i+l}^\alpha\rangle$ = 1/3 for l = 0 and = 0 for $l \neq 0$. In the low temperature range, as $T\to 0$, χ^α is proportional to ξ^α/kT.

In 1d systems, the development of short range order (SRO) ($\xi/a \gg 1$), occurs when lowering the temperature below the Curie-Weiss value $\Theta = 4S(S+1)|J|/3k$.

The correlation length has been calculated using different models. Some results for $T\to 0$ are given in the table 2.

The S = 1/2, XY and Heisenberg 1d systems are still controversial. From calculations on finite chains, a T^{-x}

TABLE 2. Low temperature correlation length for some low dimensional magnetic systems.

	Model	n	ξ/a	ref.
	Classical Heisenberg (S=∞)	3	$\simeq 2|J|S(S+1)/kT$	[19]
d=1	" XY (S=∞)	2	$\simeq 4|J|S(S+1)/kT$	"
	Ising S = 1/2	1	$\simeq 2\exp(|J|/kT)$	"
	Heisenberg S = 1/2	3	$\simeq \exp(2|J|S^2/kT)$	[21]
d=2	(square lattice)			
	XY	2	$\sim \exp[b(T/T_{KT}-1)^{-1/2}]$	[10]

divergence of as $T\to 0$ been suggested with x = 0.6 - 0.8 [22-23]

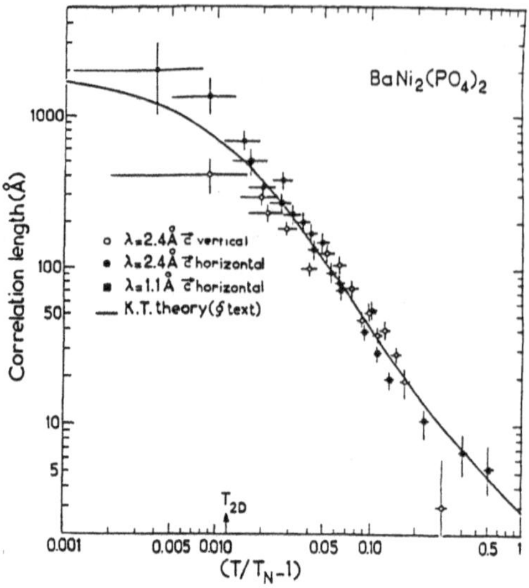

Figure 3 . In plane correlation length of Ba $Ni_2(PO_4)_2$ as a function of temperature. The solid line is a fit to the KT theory. (from ref. 28).

For 2d magnetic systems, recent calculations [21] on the S = 1/2 Heisenberg ferromagnet with a square lattice suggest an exponential divergence of the correlation length at T = 0 while for the XY model, a special kind of exponential divergence at the Kosterlitz-Touless temperature T_{KT} is predicted [10] (table 2). The associated divergence in χ is given by $\xi^{2-\eta}$ with η = 1/4. For both these 2d systems, Mermin and Wagner [24] have demonstrated the absence of LRO at T = 0. The Kosterlitz-Thouless transition has been depicted as the transition of a gas of free vertices to bound pairs of vortex-antivortex. This picture has been confirmed by Monte-Carlo studies [25]. There is little experimental data on correlation lengths in low dimensional magnetic systems. The first data concerned the quasi-1d AF $(CH_3)_4$ N Mn Cl_3 (TMMC), in which the correlation length was measured by quasi elastic magnetic neutron scattering in the temperature range 1-40 K [26]. The results were found to be consistent with the prediction for the classical Heisenberg model with J/k = - 7.7 K. Recent neutron studies [27] show a more complex ξ(T) behaviour, which is mainly of XY type at the lowest temperatures (see further sections). For d=2, a good fit to the theory was obtained for the quasi-2d XY AF Ba $Ni_2(PO_4)_2$ and Ba Ni_2 $(As\ O_4)_2$ [28], especially for the ξ(T) behaviour determined by quasi elastic neutron scattering (Fig.3).

4. CROSSOVER EFFECTS AND T_c IN LOW DIMENSIONAL MAGNETIC COMPOUNDS

4.1. Crossover: The real magnetic compounds are rarely fully described by the ideal models presented before. They show deviations which become important as T decreases. In a general way, their Hamiltonian, $\mathcal{H}$, is the sum of a main term $\mathcal{H}_0$ of dimensionalities d_0, n_0 and of perturbation terms $\mathcal{H}'$, much smaller than $\mathcal{H}_o$, of dimensionalities d,n. If d > d_o, or n < n_o, the perturbation terms leads to a more stable critical behaviour (CB) than the CB for (d_o, n_o). As T decreases, a crossover from the (d_o, n_o)CB the (d,n) CB occurs at a certain crossover temperature T*. Different possible crossovers are illustrated in table 3.

Let us first consider the simplest cases. One of them is the lattice dimensionality crossover of Ising systems which corresponds to the lowest line in table 3. This situation has been studied in details by Navarro and de Jongh [29] in the Ising cobalt compounds $CoCl_2.2H_2O$ and $Co\,Cl_2\;2NC_5H_5$ (crossover from d = 1 to d = 3), and $(CH_3)_3$ NH Co $Cl_3.2H_2O$

TABLE 3 : Different possible crossovers for magnetic systems with with short range interactions : O NO LRO ●LRO.

		chains d = 1	layers 2	3
Heisenberg,	n = 3	O	O	●
XY ,	2	O	KT	●
ISING ,	1	O	●	●

(crossover from d = 1 to d = 2). This Ising system is very attractive since the 2d array of Ising chains was exactly solved by Onsager. Indeed the measured temperature dependence of the specific heat fits very well the theoretical dependence as shown in Fig.4. In this Ising system, the LRO is clearly induced by the 2d interchain interaction.

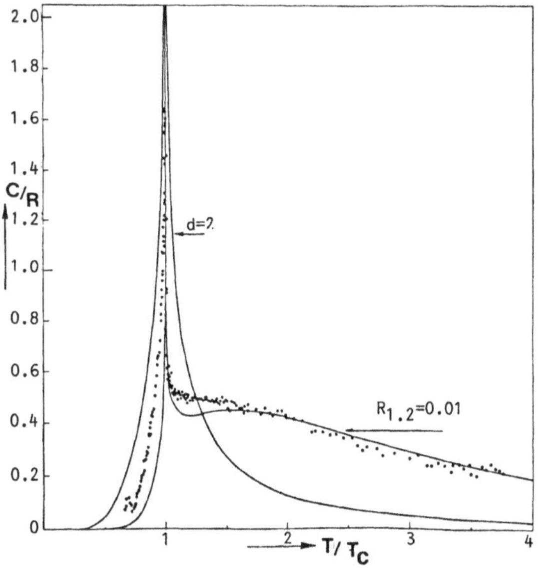

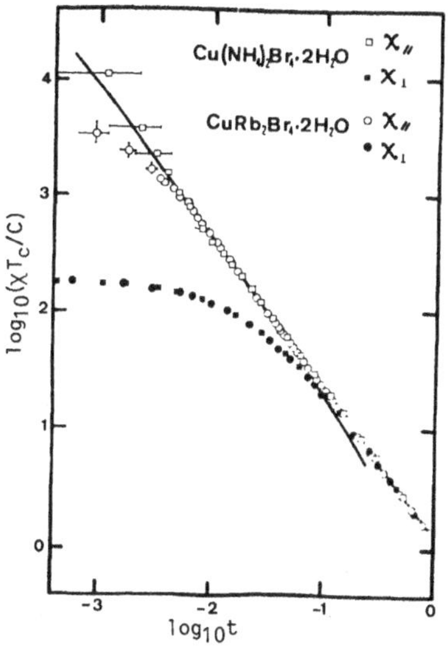

Figure 4 . Experimental specific heat for $[(CH_3)_3\,NH]\,CoCl_3.2H_2O$ compared with the Onsager result for d=1→2 crossover.

Figure 5 . Parallel and perpendicular susceptibility of two 3d ferromagnets in the critical region. The solid curve is the theoretical curve for n=3→1 anisotropy crossover (from ref. 31).

An other simple case is the anisotropy crossover for d = 3 which corresponds to the third column of the table 3. The crossover from n = 3 to n = 1 in the critical region, has been studied theoretically studied by Pfeuty et al [30] and experimentally in the $Cu\ M_2\ Br_4 . 2H_2O$ ($M=NH_4$, Rb), quasi Heisenberg bcc ferromagnets, with a small relative Ising exchange interaction of about 10^{-2} [31]. The anisotropy crossover can be easily observed by comparing the T dependence of the susceptibilities parallel, $\chi_{//}$, and perpendicular, $\chi_{\perp}$, to the easy axis. Below the crossover temperature, these susceptibilities separate from each other, $\chi_{//}$ diverging at T_c while $\chi_{\perp}$ reaches a plateau as shown in Fig.5. An excellent agreement between theory and experiment is observed for the $\chi_{//}$ variation [31]. In principle the critical exponent would shift from the value for n = 3 far from T_c to the value for n = 1 close to T_c. In the limited experimental range $10^{-3} < t < 10^{-1}$, an intermediate γ value is observed.

In many magnetic compounds, more complex situations are met. For example, the critical behaviour of the quasi-2d Heisenberg ferromagnets shows the series of crossovers from d = 2, n = 3 $\rightarrow$ d = 2, n = 2 $\rightarrow$ d = 3, n = 2 for $K_2\ Cu\ F_4$ [32], and from d = 2; n = 3 $\rightarrow$ d = 2, n = 2 $\rightarrow$ d = 2, n = 1 $\rightarrow$ d = 3, n = 1 for $(C_n H_{2n+1}\ NH_3)_2\ Cu\ Cl_4$ [2,33].

4.2. T_c in quasi-1d magnetic compounds : In many systems such as AMX_3 (A = Cs, Rb, NH_4..., M = 3d(II) metal ion), the magnetic chains parallel to z axis are isotropically coupled in the x, y directions. The LRO is achieved by the interchain interaction J' which insures the crossover from d = 1 to d = 3 and LRO below $T_c \neq 0$. A useful intuitive relation for T_c has been given by Villain and Loveluck [34]. They assumed that 3d LRO is achieved when the thermal energy is comparable to the interaction energy between blocks of correlated spins of the z' neighbouring chains :

$$kT_c \simeq z'|J'|S(S+1)\,\xi(T_c)/a \qquad (1)$$

This relation is consistent with a more rigorous relation based on the susceptibility per spin $\chi(T_c) = 1/2z'|J'|$ [35]. A graphical solution of eq. 1 is shown in Fig.6. One can easily see that for given values of J and J', T_c is a decreasing function of n, which is consistent with the general trends mentionned before (increasing stability with decreasing n). For XY or Heisenberg chains, the T^{-1} behaviour of ξ leads to a proportionality relation given $T_c \propto (JJ')^{1/2}$. It should be noticed that in real quasi 1d compounds, successive crossovers may complicate the preceeding scheme. For example, in TMMC which is the archetype of 1d AF, the small XY anisotropy due to dipolar interactions between Mn II spins leads to a crossover from n = 3 to n = 2 at about 10 K. Below this temperature, TMMC behaves like a XY chain and finally has a transition to 3d LRO at T_c = 0.85 K [36]. TMMC, and more generally the anisotropic chain of classical spins, has been the subject of theoretical calculations. The temperature dependence of the correlation length has been determined for XY anisotropy [37] and Ising anisotropy [38]. Boersma et al [39] have performed detailed numerical calculations of T_c using the matrix transfer technique.

Except for a few simple cases, the relation between T_c and the interchain interaction J' is not straightforward. Thus direct determinations of J' are very interesting. For ratios of $|J'/J|$ which are not too small, J' can be extracted from the spin wave dispersion curve for $\vec{q}$ perpendicular to the chain axis [19]. For ferromagnetic chains with AF interchain coupling, J' can be obtained from the metamagnetic transition field $H_c(0) = 2z'|J'|S/g\mu_B$. Non magnetic substitution in bimetallic chains can also provide J' [40].

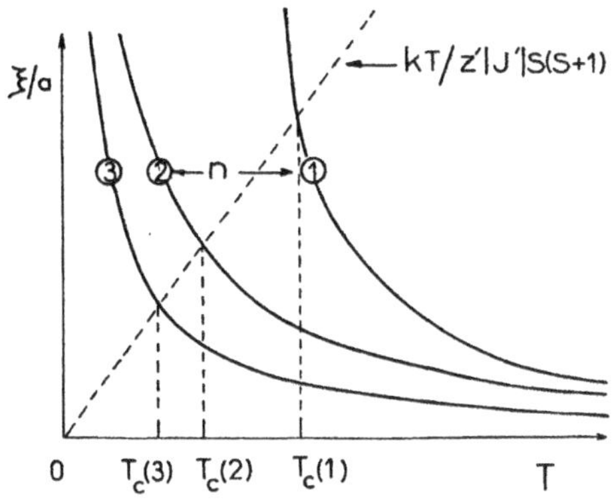

Figure 6 . Comparison between the critical temperatures of Heisenberg, XY and Ising quasi-1d systems determined from eq. (1).

5. CRITICAL TEMPERATURE OF IMPURE QUASI-1d SYSTEMS

1d magnetic compounds with nearest neighbour exchange interactions are rather sensitive to impurities. Indeed, non magnetic ions substituted for magnetic ions break the chains into independent segments. Instead of diverging as $T\rightarrow 0$, the correlation length is limited to the average distance between impurities. At low atomic concentration of impurities, x, $\xi(x)$ is approximately given [41] by:

$$a/\xi(x) = x + a/\xi(0) \qquad (2)$$

where $\xi(0)$ is the correlation length of the pure chain. From eq.1 and 2 and knowing the T dependence of $\xi(0)$, $T_c(x)$ can be easily determined. For $x<<1$, the T_c decrease induced by the impurities, can be written as $(T_c(0)-T_c(x)/T_c(0) = Kx$. The constant K is equal to 1 in the mean field approximation.

For 2d and 3d magnetic compounds, K is slightly larger than 1 [42], whereas in 1d systems, K may have fairly large values as shown in table 4. For smaller J', the decrease of T_c is larger. The extreme case is TMMC diluted by cadmium where K is equal to 200 [43] in very good agreement with the theory for XY chains with weak interchain interaction $|J'/J|\simeq 10^{-4}$.

The effect of magnetic impurities in quasi-1d systems is less drastic than for non magnetic impurities, since there exists an exchange interaction J_{ih} between impurity and host ion. The absolute value $|J_{ih}|$ can be deduced from the $T_c(x)$ data in some cases such as quantum impurities (S=1/2) in chains of classical spins (TMMC doped by Cu [43]) or classical impurities in quantum chains ($(CH_3)_4NCuCl_3$ doped by Mn [44]). In some cases such as $CsMn_{1-x}Co_xCl_3.2H_2O$, the orthogonal competing anisotropies of Co and Mn which favour different spin easy axes gives rise to a new magnetic phase (oblique AF phase) for a certain range of x. [52-54]. The high degree of disorder induced by impurities in quasi-1d systems is also favourable to the occurence of spin glass phases at low

Table 4. T_c decrease in some quasi-1d impure compounds.

compound	J/k	|J'/J|	Imp.	K	ref.
			Cd	200	
$(CH_3)_4NMnCl_3$	-6.7	$<10^{-4}$	Cu	20	[43]
			Ni	10	
$(CH_3)_4NCuCl_3$	+45	10^{-3}	Mn	3	[44]
$Cs\ Ni\ F_3$	+10	10^{-3}	Mg	5.6	[45]
$Cs\ Co\ Cl_3$	-80	10^{-2}	Mg	10	[46]
$(CH_3)_2\ NH_2MnCl_3$	-6,5	10^{-3}	Cd	20	[47]
$CsMnCl_3 . 2H_2O$	-3.2	10^{-2}	Cu	3.4	[48]
			Co	1	
$(CH_3)_3NHCoCl_3 . 2H_2O$	+15	10^{-2}	Mn	4	[49]
Cs_2CoCl_4	-35		Zn	4.15	[50]
CuMn(dto)	-45	3.510^{-3}	Pd	8.5	[51]

temperatures, as shown for example in $C_6H_{11}NH_3Cu_{1-x}Mn_xCl_3$ [55].

Finally, we mention measurements, by quasi-elastic neutron scattering, of the correlation length in TMMC : Cu, which are in good agreement with the theoretical predictions [56,57].

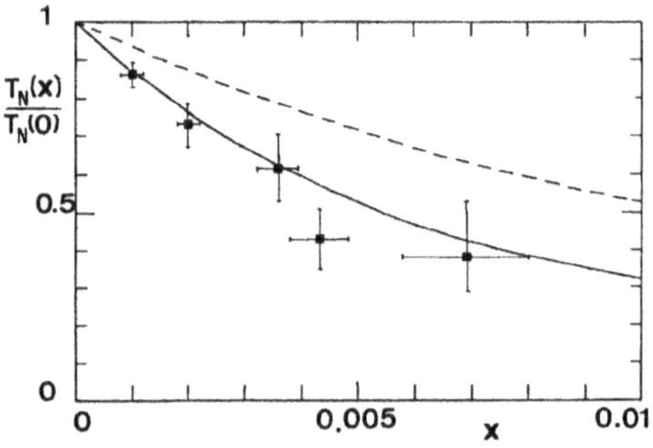

Figure 7 . Critical temperature of $(CH_3)_4$ N $Mn_{1-x}Cd_xCl_3$ (TMMC : Cd). The solid curve and the dashed curve are calculated respectively for XY and Heisenberg model (from ref.43).

6. PHASE DIAGRAM OF LOW-DIMENSIONAL ANTIFERROMAGNETS

As mentionned in section 1, antiferromagnetic-paramagnetic MPT is not destroyed by applying magnetic field H. Mean field calculations based on the development of the Brillouin function in powers of H/T show that $T_c(H)$ decreases in proportion to H^2. However the first experiments showed that in the quasi-1d AF, α-Cu NSal (S=1/2) [58], Cs Ni Cl_3 (S=1) [59] and TMMC (S=5/2) [60], T_c is an increasing function of H. An elegant explanation of this phenomenon was developped by Villain and Loveluck [34]. Since the parallel susceptibility is smaller than the perpendicular susceptibility in AF compounds, the Zeeman energy is lowered for a configuration with spins perpendicular to the field. Thus the field suppresses the parallel spin fluctuations and decreases the spin dimensionality from n to n-1. For a purely Heisenberg chain, the reduction of n from 3 to 2 leads to an increase of T_N, as shown in Fig.6, which can reach the maximum relative value $\sqrt{2}$. The case of XY chain in a field parallel to the easy plane is more spectacular because it changes the T^{-1} divergence of ξ for n = 2, to an exponential divergence for n=1. The relative T_c increase may have large values in this case. For example, in the quasi 1d-XY AF TMMC, T_c increases by a factor 4 for H = 120 kOe [61] as shown in Fig.8.

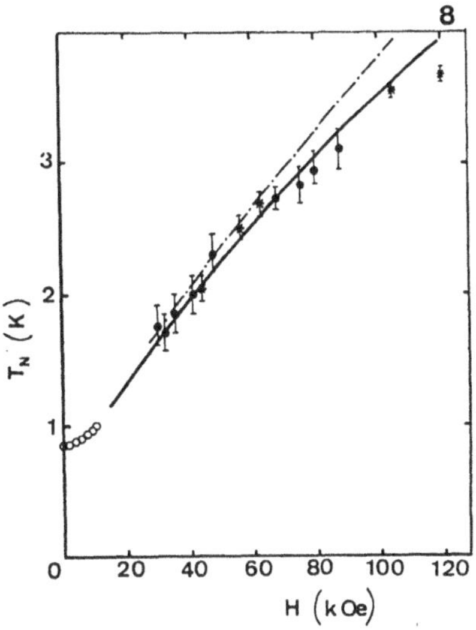

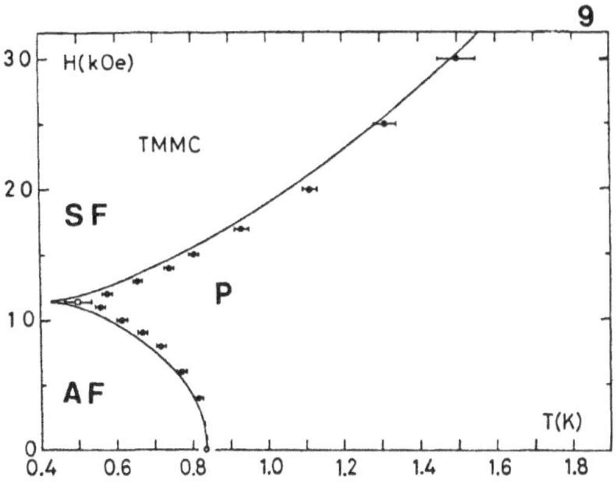

Figure 8 . Field dependence of the 3d ordering temperature of TMMC. The dot-dashed curve corresponds to the soliton model for a pure crystal. The full curve is for impurity concentration $x = 210^{-4}$. (from ref. 61).

Figure 9 . Magnetic phase diagram of TMMC for field parallel to the easy axis. (from ref. 63).

For a more detailed analysis of the (H,T) phase diagram for H parallel to the easy axis x, the small Ising anisotropy which imposes the spin easy direction must be considered. This is done in the following Hamiltonian :

$$\mathcal{H} = \sum_i \left[-2J\, \vec{S}_i \cdot \vec{S}_{i+1} + D(S_i^z)^2 - E(S_i^x)^2 - g\mu_B H\, S_i^x \right]$$

D = 0, E > 0 correspond to axial anisotropy; D >> E > 0 correspond to biaxial anisotropy.

An important feature is the competition between the Ising anisotropy $-E(S_i^x)^2$ which tends to align the spins along x and the Zeeman term $-g\mu_B H S_i^x$ which favours the spin orientation perpendicular to x. For small values of H, the spins lie along x and for $H > H_{SF} = 2S(EJ)^{1/2}/g\mu_B$ at T = OK, the spin direction changes from x to y. The corresponding variation of the spin dimensionality n with H is shown below.

	$H < H_{SF}$	$H = H_{SF}$	$H > H_{SF}$
unixial AF	n = 1	3	2
biaxial AF	n = 1	2	1

A consequence of this increase of n at H_{SF} is a marked minimum in T_c. The resulting singular point in the (H,T) phase diagram is called the bicritical point. At the bicritical point the three phases : paramagnetic (P), antiferromagnetic (AF) and spin-flop (SF) coexist as shown in Fig.9. Detailed calculations of $T_c(H)$ based on the transfer matrix [39] and on the model of magnetic solitons [61,62] have been performed. They are in good agreement with the experimental data in various quasi-1d AF, TMMC [61,63], see Fig. 8 and 9, $CsMnCl_3.2H_2O$ [64,65], $CsMnBr_3.2H_2O$, $(CH_3)_2NH_2MnCl_3$ and $CuCl_2.2NC_5H_5$ [64]. The same physical ideas were successfully applied by de Groot and de Jongh [62] to interpret the phase diagram of quasi-2d AF.

To be complete, we will briefly discuss the phase diagram of impure antiferromagnets. Fishman and Aharony [66] have shown that a field applied to an AF system with random exchange (RE) generated random fields (RF). The effects of RF in a magnetic system can be simply described by assuming that they reduce the lattice dimensionality d to an effective dimensionality $\bar{d}=d-1$ [67-69]. They are sketched for a 3d impure AF in Fig. 10. Most experimental studies were performed on the AF↔P transition far from the bicritical point in Mn(II) and Fe(II) fluorides strongly diluted by Zn or Mg [70-74] in order to check the theoretical predictions for the Ising model in random fields. There are still few experimental studies of the effect of impurities on the transition lines near the bicritical point [65,75-77]. These experiments are consistent with the schematic picture of Fig.10.

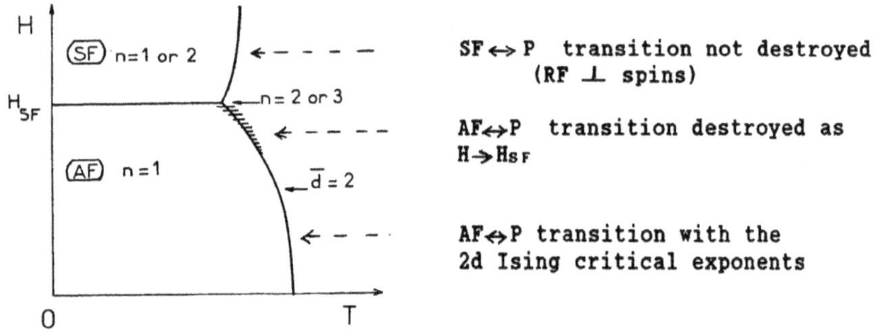

Figure 10 . Effect of random fields on the AF phase diagram

7 . SINGLET GROUND STATE MAGNETIC SYSTEMS.

Among the low dimensional systems which exhibit a singlet ground state, the alternating AF chain with S=1/2 has been extensively studied [78,79]. This system is characterized by two different exchange interactions J_1 and J_2, the degree of alternation being defined by $\alpha = J_2/J_1 < 1$. The energy gap between the singlet ground state and the continuum of excited states has been calculated as a function of J_1 and α [80]. The gap has important consequences on the low temperature thermodynamic properties. It inhibits the increase of the correlation length at low temperature and consequently suppresses the MPT to LRO, if the interchain interaction is sufficiently weak with respect to the intrachain interaction. In this case, the magnetic susceptibility drops down to zero for all field orientations.

In an applied magnetic field H, the alternating AF chain exhibits the interesting behaviour [81] sketched in Fig. 11. For a certain field range $H_{c1}-H_{c2}$ which depends on J_1 and α, the energy gap vanishes. This allows the development of spin correlations and the establishment of LRO induced by interchain interactions. The phase diagram illustrated in Fig.11, has been observed in the alternating quasi-1d AF with $\alpha = 0.27$, $Cu(NO_3)_2.2.5H_2O$ [82]. However it is surprising that no systematic neutron measurements of gap and correlation length were performed in the alternating AF-chains.

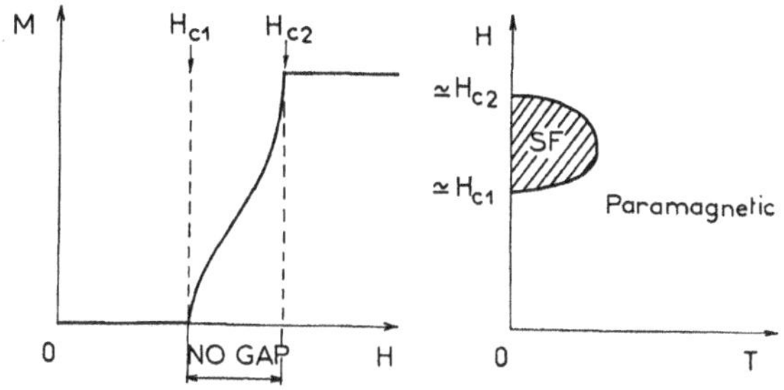

Figure 11 . Field dependence of magnetization and magnetic phase diagram of the quasi-1d alternating antiferromagnet.

The spin-Peierls chains were also largely investigated. These systems behave as a uniform AF chain at high temperature and undergo a progressive dimerization below the spin-Peierls transition, accompanied by a weak specific heat anomaly. The (H,T) phase diagram is not fully understood to date. The reader is refered to the paper by de Jongh [78] for a recent review.

The most recent development in 1d magnetism concerns the Heisenberg AF chains with integer spin value, for which Haldane [4] has predicted a singlet ground state separated by an energy gap E_G from the excited states. This prediction was supported by numerical calculations [83,84] on S=1, AF chains which yield $E_G \simeq 0.4|2J|$. Since the Haldane conjecture and its experimental check on the S=1, quasi-1d AF, Cs Ni Cl_3 and NENP, are discussed in the present book by Steiner [85] and by Renard et al [86], we will focus here on the possible phase transition to LRO in these systems. Following the simple idea developped in section 3, we can define a critical number of correlated spins n_{cr} by comparing the effective energy of interchain interaction, to E_G : $n_{cr} \simeq 0.4|J/z'J'|$. n_{cr} is typically a few x10 spins for Cs Ni Cl_3 and a few $x10^2$ spins for NENP. For NENP the correlation length is limited by the gap before attaining n_{cr} and LRO cannot establish in contrast with Cs Ni Cl_3 where LRO is achieved at $T_c = 0.29|J|$.

Kosevich and Chubukov [87] have recently shown, by calculating the quantum corrections to the Green's function of 1d-Heisenberg AF, that there exists a limiting value $|J'/J|_{cr}$ below which LRO cannot establish. $|J'/J|_{cr}$ is approximately given by $\exp(-\pi S)$, i.e. 4.10^{-2} for S = 1. This critical value which is larger than the $|J'/J|$value of Cs Ni Cl_3, is not consistent with the occurence of LRO. In fact, the expression of $|J'/J|_{cr}$ is probably incorrect for S=1, since it was established for values of S >> 1.

Further experiments on the phase diagram of S = 1, AF chains versus $|J'/J|$ and H would be of great interest.

ACKNOWLEDGMENTS

I would like to thank my coworkers P. Beauvillain, S. Clément, C. Dupas, R. Mégy and E. Vélu; Drs. J.P. Boucher and L.J. de Jongh for figures and reprints; Dr. A.V. Chubukov for a reprint; Dr. Bennett for helpful suggestions in writing the manuscript; B. Lesquer for drawing some figures and Mrs F. Genet for typewriting the manuscript.

REFERENCES

1. Kadanoff, L.P., Götze, W., Hamblen, D., Hecht, R., Lewis, E.A.S., Palciauskas, V.V., Rayl M. and Swift, J. 1967, Rev. Mod. Phys. 39, 395.
2. De Jongh, L.J. and Miedema, A.R. 1974, Adv. Phys. 23, 1.
3. Ferré, J. and Gehring, G.A. 1984, Rep. Prog. Phys. 47. 513.
4. Haldane, F.D.M. 1983, Phys. Lett. A 93, 464 and Phys. Rev. Lett. 50, 1153.
5. Widom, B. 1965, J. Chem. Phys. 43, 3892 and 3898.
6. Wilson, K.G. 1971, Phys. Rev. B4, 3174.
7. Mukamel, D. and Krinsky, S. 1976, Phys Rev. B 13, 5065 and 5078.
8. Bloch, D., Hermann-Ronzad, D., Vettier, C. Yelon, W.B. and Alben, R. 1975, Phys. Rev. Lett. 35, 963.
9. Le Guillou, J.C. and Zinn-Justin, J. 1977, Phys. Rev. Lett. 39, 95.
10. Kosterlitz, J.M. and Thouless, D.J. 1973, J. Phys. C6,1181. Kosterlitz, J.M. 1974, J. Phys. C7, 1047.
11. Onsager, L. 1944, Phys. Rev. 65, 117.
12. Larkin, I.A. and Khmel'Nitskii, D.E. 1969, Sov. Phys. JETP. 29, 1123.
13. Brézin, E. 1975, J. Physique Lett. 36, L51.
14. Brézin, E. and Zinn-Justin, J. 1976, Phys. Rev. B13, 251.
15. Ahlers, G. Kornblit, A. and Guggenheim, H.J. 1975, Phys. Rev. Lett. 34, 1227.
16. Als Nielsen, J. 1976, Phys. Rev. Lett. 37, 1161.
17. Beauvillain, P., Chappert, C. and Laursen, I. 1980, J. Phys. C 13, 1481.
18. Beauvillain, P., Renard, J.P., Laursen, I. and Walker, P.J. 1978, Phys. Rev. B 18, 3360.
19. Steiner, M., Villain, J. and Windsor, C.G. 1976, Adv. Phys. 25, 87.
20. Renard J.P. 1985, Magnetostructural correlations in

exchange coupled systems, edit. Willett, R.D, Gatteschi, D. and Kahn O., NATO ASI Series, D. Reidel Publ. Co., 140, 207.
21. Takahashi, M. 1987, Phys. Rev. Lett. 58, 168.
22. Bonner, J.C. and Fisher, M.E. 1964, Phys. Rev. A 135, 640.
23. Baker, G.A.J., Rushrooke, G.S. and Gilbert, H.E. 1964, Phys. Rev. A 135, 1273.
24. Mermin N.D. and Wagner H. 1966, Phys. Rev. Lett. 17, 1133.
25. Tobochnik, J. and Chester, G.V. 1979, Phys. Rev. B 20, 3761.
26. Birgeneau, R.J., Dingle, R., Hutchings, M.T., Shirane, G. and Holt, S.L. 1971, Phys. Rev. Lett. 26, 718; Hutchings, M.T., Shirane, G., Birgeneau, R.J. and Holt, S.L. 1972, Phys. Rev. B 5, 1999.
27. Boucher, J.P., Regnault, L.P., Rossat-Mignod, J., Villain, J. and Renard J.P. 1979, Solid State Commun. 31, 311.
28. Regnault, L.P. and Rossat-Mignod, J. "Magnetic properties of layered transition metal compounds", Chapter 6 (to be published); Regnault, L.P., Rossat-Mignod, J. and Henry J.Y. 1983, J. Phys. Soc. Japan 52, 1.
29. Navarro, R. and de Jongh, L.J. 1978, Physica 94 B, 67.
30. Pfeuty, P., Jasnow D. and Fisher M.E. 1974, Phys. Rev. B 10, 2088.
31. Velu, E., Renard, J.P. and Lécuyer, B. 1976, Phys. Rev. B 14, 5088.
32. Dupas, A. and Renard, J.P. 1976, J. de Physique 37, C1-213.
33. de Jongh, L.J. 1976, Physica 82B, 247.
34. Villain. J. and Loveluck, J.M. 1977, J. de Physique Lett. 38, L77.
35. Imry, Y. 1976, Phys. Rev. B13, 3018.
36. Dupas, C. and Renard, J.P. 1973, Phys. Lett. 43A, 119.
37. Hone, D. and Pires, A. 1977, Phys. Rev. B15, 323.
38. Faria, A.C. and Pires, A.S.T. 1979, J.Phys. C 12, 2637.
39. Boersma, F. de Jonge, W.J.M. and Kopinga, K. 1981, Phys. Rev. B 23, 186.
40. Verdaguer, M., Gleizes, A., Renard, J.P. and Seiden, J. 1984, Phys. Rev. B 29, 5144.
41. Thorpe, M.F. 1975, J. Physique 36, 1177.
42. de Jongh, L.J. 1983, Magnetic Phase Transitions, Springer Series in Solid-State Sciences, Volume 48, 172.
43. Dupas, C. and Renard, J.P. 1978, Phys. Rev. B 18, 401.
44. Cheikh-Rouhou, A., Dupas, C., Renard, J.P. and Seiden, J. 1982, Phys. Rev. B 25, 3261.
45. Steiner, M. and Axmann, A. 1976, Solid State Commun. 19, 115.
46. Ferré, J., Jamet, J.P., Renard, J.P. and Briat, B. 1983, J. Phys. C 16, 108.
47. Schouten, J.C., Boersma, F., Kopinga, K. and de Jonge, W.J.M. 1980, Phys. Rev. B 21, 4084.
48. Velu, E., Renard, J.P., and Corbel, G. 1978, J. Physique. Colloque, 39, C6-717.
49. Takeda, K. 1976, J. Phys. Soc. Japan 40, 1781.
50. Bartholome, J., Algra, H.A., de Jongh, L.J. and Huiskamp, W.J. 1977, Physica 86-88 B, 707.
51. Frasse, C. 1985 thesis, Toulouse (unpublished).
52. Yamamoto, I. 1980, J. Phys. Soc. Japan 49, 74.
53. Nagata, K., Iio, K., Igarashi, M. and Tazuke, Y. 1986, J. Mag. Mag. Mat. 54-57, 33.
54. Kubo, H., Uryû, N., Hamasaki, T., Hidaka, M., Nishihara, H. and Yasuoka, H. 1986, J. Mag. Mag. Mater. 54-57, 35.

55. Cheikh-Rouhou, A., Dupas, C. and Renard, J.P. 1983, J. Physique-Lettres 44, L.777.
56. Boucher, J.P., Dupas, C., Fitzgerald, W.J., Knorr, K. and Renard, J.P. 1978, J. Physique-Lettres 39, L-86.
57. Endoh, Y., Heilmann, I.U., Birgeneau, R.J., Shirane, G., Mc Gurn, A.R. and Thorpe, M.F. 1981, Phys. Rev. B 23, 4582.
58. Clark, W.G., Azevedo, L.J. and Mc. Lean, E.O., 1975, Proceedings of LT 14 Conference, Otaniemi, 4, 369.
59. Almond, D.P. and Rayne, J.A. 1975, Phys. Lett. 55A, 137.
60. Dupas, C. and Renard, J.P. 1976, Solid State Commun. 20, 581.
61. Boucher J.P., Regnault, L.P., Rossat-Mignod, J., Renard, J.P. Bouillot, J. and Stirling, W.G., 1981, J. Appl. Phys. 52, 1956.
62. de Groot, H.J.M. and de Jongh, L.J. 1986, Physica 141B, 1.
63. Takeda, K., Koike, T., Tonegawa, T. and Harada, I. 1980, J. Phys. Soc. Japan 48, 1115.
64. de Jonge, W.J.M., Hijsmans, J.P.A.M., Boersma, F., Schouten, J.C. and Kopinga, K. 1978, Phys. Rev. B 17, 2922.
65. Vélu, E., Mégy, R. and Seiden, J. 1983, Phys. Rev. B 27, 4429.
66. Fishman, S. and Aharony, A. 1979, J. Phys. C12, L 729.
67. Grinstein, G. and Ma, S.K. 1982, Phys. Rev. Lett. 49, 685.
68. Imbrie, Z. 1984, Phys. Rev. Lett. 53, 1747.
69. Villain, J. 1985, Scaling Phenomena in disordered systems, ed. R. Pynn and A. Skjeltorp, Plenum Publ. Co., 423.
70. Belanger, D.P., King, A.R., Jaccarino, V. and Cardy, J.L. 1983, Phys. Rev. B 28, 2522.
71. Shapira, Y., Oliveira, Jr., N.F. and Foner, S. 1984, Phys. Rev. B30, 6639.
72. Cowley, R.A., Yoshizawa, H., Shirane, G. Hagen, M. and Birgeneau, R.J. 1984, Phys. Rev. B 30, 6650.
73. King, A.R., Mydosh, J.A. and Jaccarino, V. 1986, Phys. Rev. Lett. 2525.
74. de Groot, H.J.M. and de Jongh, L.J. 1986, Physica Scripta T13, 219.
75. Rohrer, H. 1981, J. Appl. Phys. 52, 1708.
76. Shapira, Y. 1982, J. Appl. Phys. 53, 1931.
77. Mégy, R. 1987, thesis, Orsay (unpublished).
78. de Jongh, L. J. 1985, Magnetostructural correlations in exchange coupled systems, NATO ASI series, D. Reidel Publ. Co, 140, 1.
79. Landee, C.P., in this book.
80. Fields, J.N., Blöte, H.W.J. and Bonner, J.C. 1979, J. Appl. Phys. 50, 1807.
81. Mohan, M. and Bonner, J.C. 1982, J. Appl. Phys. 53, 8035.
82. Bonner, J.C., Friedberg, S.A., Kobayashi, H., Meier, D.L., and Blöte, H.W. 1983, Phys. Rev. B 27, 248 and references therein.
83. Botet, R. Jullien, R. and Kolb, M. 1983, Phys. Rev. B 28, 3914.
84. Nightingale, M.P. and Blöte, H.W.J. 1986, Phys. Rev. B 33, 659.
85. Steiner, M., paper in this book.
86. Renard, J.P., Verdaguer, M. Regnault, L.P. Erkelens, W.A.G., Rossat-Mignod, J. and Stirling, W.G., Paper in this book.
87. Kosevich, Yu. A. and Chubukov, A.V. 1986, J.E.T.P. 91, 1105 (in (Russian); English translation to appear in Sov. Phys. JETP.

CHAIN-LIKE MAGNETIC SYSTEMS:

NON-LINEAR EXCITATIONS AND QUANTUM EFFECTS

K. Kopinga and W.J.M. de Jonge

Department of Physics
University of Technology
P.O. box 513
56000 MB Eindhoven
The Netherlands

1. INTRODUCTION

The presence of non-linear excitations in one-dimensional magnetic systems was already indicated about 50 years ago[1]. It was found that, apart from the usual spinwaves, a solution of the Hamiltonian describing a ferromagnetic $S = \frac{1}{2}$ Heisenberg system is formed by "bound" complexes of two or more spinwaves. During the last decades, there has been a renewed interest in the static and dynamic properties of this type of systems. The reason for this is twofold. First, $S = \frac{1}{2}$ spin systems with a ferromagnetic nearest neigbor interaction form one of the most simple non-trivial infinite ensembles of interacting particles. Secondly, a number of experimental systems have been synthesized[2], which are very good approximations of the "idealized" theoretical model systems.

Estimates of the behavior of these model systems can be obtained from brute force numerical computations[3] or, for instance, Quantum Monte Carlo simulations[4]. During the last years, the availability of increasingly large computer systems has led to a significant improvement of the accuracy of these estimates, but the results in general hardly yield any information about the specific elementary excitations. In the limit of high uniaxial anisotropy (Ising), it is well established[5] that the bound complexes are localized clusters of adjacent spins that are reversed with respect to the ground state. However, for systems with a small uniaxial (Ising-Heisenberg) or easy-plane (XY) anisotropy, or systems with an isotropic nearest neighbor interaction (Heisenberg), only the low-energy part of the excitation spectrum has been solved up till now[6]. In the presence of a symmetry-breaking external magnetic field, exact results have, to our knowlegde, not been reported.

A large theoretical and experimental effort was prompted by the observation that the equation of motion of a chain of classical spins with easy-plane anisotropy in a symmetry-breaking field can, under certain approximations, be mapped to a sine-Gordon (sG) equation[7]. Apart from the usual linear excitations (magnons), this equation has non-linear solutions which are known as kink solitons. In ferromagnetic chains, such a soliton can be viewed as a more or less local 2π-twist of the spins within a chain with respect to the ground state.

A variety of experiments on quasi one-dimensional compounds with easy-plane anisotropy have been interpreted in terms of the sG model. Among these we mention neutron scattering studies on the ferromagnetic $S=1$ chain system $CsNiF_3$[8] and the antiferromagnetic $S=\frac{5}{2}$ chain system $(CD_3)_4NMnCl_3$ (TMMC)[9], as well as nuclear spin-lattice relaxation time[10] and heat-capacity measurements[11] on $CsNiF_3$, TMMC[12,13] and the ferromagnetic $S=\frac{1}{2}$ chain system $(C_6H_{11}NH_3)CuBr_3$ (CHAB)[14].

Although a fair qualitative description of most of the data could be obtained, the validity of the sG approximation to real magnetic systems has been questioned. In some cases, the observed effects can also be explained in terms of other elementary excitations whereas, on the other hand, it is not clear to what extent the various approximations involved with the mapping of the original magnetic system to the sG system affect a meaningful comparison of the latter model with the experimental data.

In this paper the effect of linear and non-linear excitations on various magnetic properties of CHAB will be reviewed. We focus our attention to the magnetic heat capacity, the nuclear spin-lattice relaxation time, and the magnetization. The effects of the approximations involved with the description of CHAB in terms of the sG model will be analyzed. We will show that this model, if suitably modified, yields a consistent description of all data on CHAB that has been collected in the presence of an external field within the XY plane.

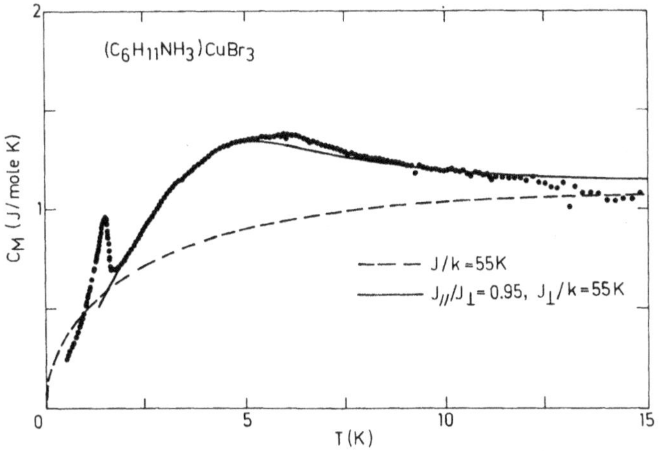

Fig.1. Zero-field magnetic heat capacity of CHAB, together with numerical results for some ferromagnetic $S=\frac{1}{2}$ chains.

2. CRYSTALLOGRAPHIC AND MAGNETIC PROPERTIES

The crystallographic structure of $(C_6H_{11}NH_3)CuBr_3$ is othorhombic, spacegroup $P2_12_12_1$, with a = 19.84 Å, b = 8.78 Å, and c = 6.44 Å, and four formula units in the crystallographic unit cell[15]. It consists of bibridged linear chains of $CuBr_3^-$ ions running parallel to the c axis, which are effectively isolated in the a and b directions by the cyclohexylammonium complexes. The zero-field magnetic heat capacity of CHAB[16] is shown in Fig. 1. The λ-anomaly shows the onset of three-dimensional long-range order below T_c = 1.50 K. The data in the paramagnetic region can be described rather well by the numerical estimate[17] for a Heisenberg-XY chain system with J/k = 55 ± 5 K and 5 % easy-plane anisotropy. The observed "bump" in the heat capacity can be associated with a cross-over from Heisenberg behavior above ~ 12 K to XY behavior at low temperatures. Some typical results of ferromagnetic resonance (FMR) experiments[18] on CHAB are shown in Fig. 2. The field-dependence of the observed signals at T = 1.2 K can be fully explained by associating their frequencies with the zero wavenumber magnon energy. The splitting of the resonance signals for external fields along one of the crystallographic axes is due to the interactions between the chains, which are smaller than J/k by three orders of magnitude. From these experiments and the heat-capacity measurements it was deduced that the individual chains in CHAB can be described by the Hamiltonian

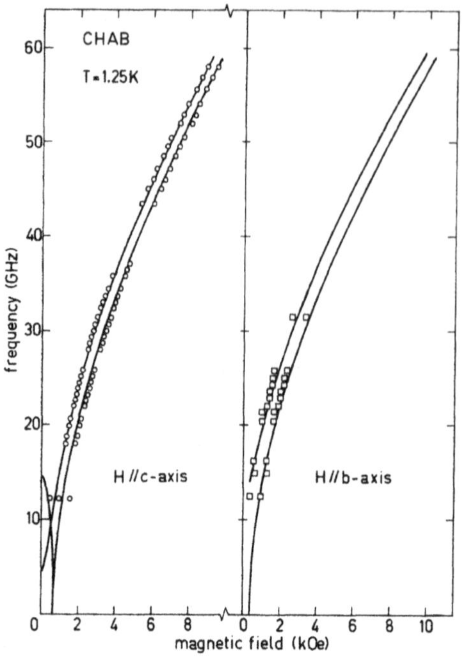

Fig.2. FMR frequencies in CHAB for fields along the c and b axis, together with the corresponding theoretical predictions.

$$\mathcal{H} = -2 \sum_i \left[J^{xx}S_i^x S_{i+1}^x + J^{yy}S_i^y S_{i+1}^y + J^{zz}S_i^z S_{i+1}^z \right], \qquad (1)$$

with $J^{xx}/k = 55 \pm 5$ K, $J^{zz}/J^{xx} = 0.95$, and $(1 - J^{yy}/J^{xx}) \simeq 5 \times 10^{-4}$. The y axis coincides with the crystallographic c axis, whereas the x axis is located within the ab plane at an angle φ from the b axis. Two symmetry-related types of chains are present, with $\varphi = +25^0$ and -25^0, respectively. It is obvious that CHAB is a very good realization of a one-dimensional (1d) $S = \frac{1}{2}$ ferromagnetic model system. Mapping to the sG system is formally possible for external fields applied along the crystallographic c axis, which is located in the easy (XY) plane for both types of chains.

3. MAGNETIC EXCESS HEAT CAPACITY

The mapping of the spin Hamiltonian (1) to the sG model involves the assumptions that the spins can be considered as classical vectors, their motion is confined to the XY plane, and the limit of zero lattice spacing (continuum limit) can be taken. Apart from this, the experimental results are generally compared with predictions based upon the ideal-gas phenomenology, in which the solitons are assumed to behave as a dilute gas of noninteracting quasiparticles[19]. Within this approximation, the free energy per spin can be written as $F = F_m + F_s$, where F_m reflects the contribution of linear excitations (magnons) and F_s is proportional to the soliton density n_s:

$$F_s = -kTn_s \qquad \text{with } n_s = \left[\frac{8}{\pi}\right]^{\frac{1}{2}} m\,a\,t^{-\frac{1}{2}} e^{-1/t} \ . \tag{2}$$

In this equation m denotes the soliton mass, a is the lattice spacing between adjacent spins, and $t = kT/E_s^o$, where E_s^o is the soliton rest-energy. For a system described by Eq. (1) with **B** in the XY plane, m and E_s^o are given by

$$m = \frac{1}{a}(g\mu_B B/2JS)^{\frac{1}{2}} \ , \ E_s^o = 8S(2g\mu_B BJS)^{\frac{1}{2}} \tag{3}$$

In comparing the sG model with experimental data usually only the excess heat capacity, i.e., the heat capacity in the presence of a field **B** minus that in zero field, is considered. The reason for this is that, because of the lack of discrete energy levels, a classical model fails to describe the total heat capacity of a real (quantum) system at low temperatures. On the other hand, the kink-solitons occuring in the sG system behave as quasiparticles with a finite rest-energy. Since these solitons are not present in zero field, it is not *a priori* impossible that the classical sG model may give a fair description of the excess heat capacity ΔC, at least in that field and temperature region where the solitons have sufficient statistical weight.

The soliton contribution ΔC_s to the excess heat capacity per spin can be obtained by straightforward differentiation of Eq. (2) as

$$\Delta C_s = \left[\frac{8}{\pi}\right]^{\frac{1}{2}} kma\, t^{-\frac{5}{2}}\, e^{-1/t}(1 - t - \frac{1}{4}t^2) \ . \tag{4}$$

The magnon contribution to ΔC, which in the sG model is represented by a power series in t multiplied by kma, that should be added to Eq. (4), is often replaced by its quantummechanical counterpart based on linear spin-wave theory.

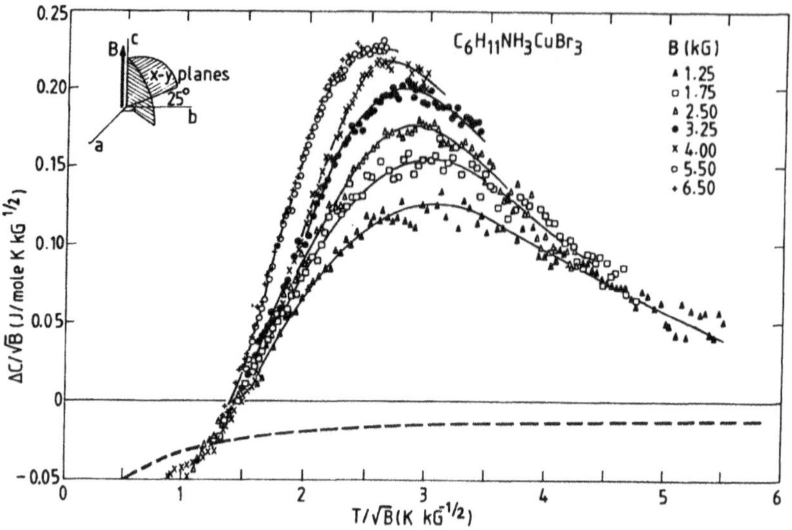

Fig.3. Excess heat capacity of CHAB for **B**∥**c** *plotted in reduced form. Drawn curves are guides to the eye. The dashed curve is the prediction for B = 4 kG based on linear spin-wave theory.*

Some static properties of the sG model have been obtained more directly[20] from a numerical solution of the transfer integral equation of a discretized chain. These results indicate that the original ideal gas phenomenology according to Eq. (4) leads to incorrect results for $t > 0.15$, in which region the majority of the experimental data has been collected. More recently, the thermodynamic properties of the sG model were calculated[21] using a modified WKB method. The resulting expansion for the free energy can be associated with a magnon part, given by a power series in t, and a soliton part $-kTn_s$ with

$$n_s = \left[\frac{8}{\pi}\right]^{\frac{1}{2}} m\, a t^{-\frac{1}{2}} e^{-1/t}\left(1 - \frac{7}{8}t - \frac{59}{128}t^2 + \,.....\right) . \qquad (5)$$

Comparison of this equation with Eq. (2) reveals the presence of finite-temperature corrections to the density of kink-solitons. The heat capacity obtained from this approach is in much better agreement with the numerical results than Eq. (4).

In order to compare the various theories outlined above with experimental data on an $S=\frac{1}{2}$ ferromagnetic chain system we measured the specific heat of a single-crystal of CHAB of 2.5 g by means of a heat-pulse calorimeter[14] with an accuracy of 2%. Temperature readings were obtained from a germanium thermometer and were corrected for magnetic field effects

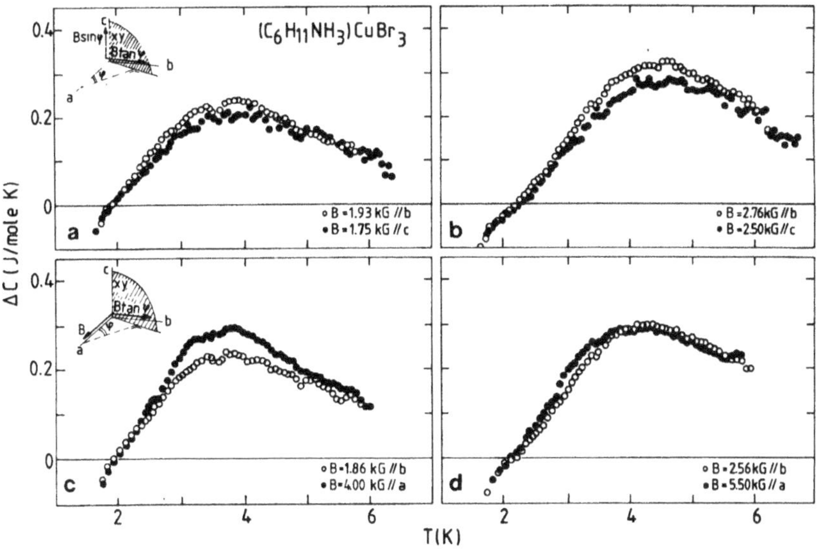

Fig.4. Excess heat capacity of CHAB measured in inclined fields.

(< 30 mK), if appropriate. The excess heat capacity ΔC in the region $1.5 < T < 6.5$ K, $B \lesssim 6.5$ kG was obtained by subtracting the zero-field heat capacity from the data obtained for **B** // **c**. No demagnetization corrections were applied since these amount to at most 50 G. In Fig. 3 we plotted the corresponding excess heat capacity in reduced form, i.e., $\Delta C/\sqrt{B}$ against $T/\sqrt{B}$.

Although ΔC certainly includes the effect of **B** on the linear spin-wave excitations, this contribution on itself cannot explain the observed behavior of the excess heat capacity, since it is negative and monotonically increasing, as indicated by the dashed curve in Fig. 3. This strongly indicates that the bump-like behavior of ΔC originates from non-linear excitations. The solid curves in the figure are obtained by smooth interpolation of the data. For sake of clarity only these curves will be depicted in some of the following figures.

Measurements of ΔC were also performed with **B** along the a and b axis. For **B** // **a**, the field can be decomposed into a component $B\sin\varphi$ within the XY plane, whereas for **B** // **b** the in-plane component is equal to $B\cos\varphi$. In Fig. 4 four representative sets of data are plotted, arranged in such a way that for each set the in-plane component of the field is the same. It appears that the location of the maximum of ΔC can almost completely be explained by the effect of the in-plane component of the field. This observation supports an interpretation of ΔC in terms of solitons, since, as long as the out-of-plane component of the field $g\mu_B B$ is smaller than

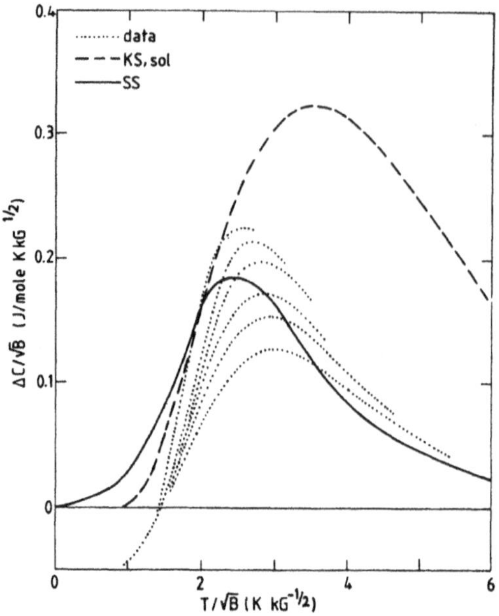

Fig.5. Excess heat capacity of CHAB for **B**∥**c** *plotted in reduced form together with the soliton contribution (Eq. 4) based upon the ideal-gas phenomenology (dashed curve) and the numerical results for the sG model (solid curve).*

$(J^{xx}-J^{zz})/k$, only the in-plane component induces the symmetry-breaking required for soliton excitations. We will therefore continue with an interpretation of our data collected for **B**∥**c** in terms of the sG model.

In Fig. 5 the data is represented by dotted curves. The dashed curve denotes the soliton contribution to ΔC according to Eq. (4) for $S = \frac{1}{2}$, $J/k = 55$ K, and $g = 2$. The corresponding soliton rest energy amounts to $10.95\sqrt{B}$ K (B in kG). Although this curve gives a fair qualitative description of the data, the corresponding prediction is formally only valid up to $T/\sqrt{B} \simeq 1.7$. A much better quantitative description is given by the numerical results for the sG model[20], which are represented by the solid curve. Although these calculations, in which all elementary excitations are implicitly included, give a nice description of the average data on ΔC, some systematic deviations remain, which will now successively be discussed.

First, the experimental excess heat capacity at low values of $T/\sqrt{B}$ becomes negative, in contrast to the various classical model predictions depicted in Fig. 5. This is a direct consequence of the shift of entropy to higher temperatures induced by an external field, and is characteristic for a quantum system. In this respect we note that both recent Quantum

Monte Carlo simulations[4] and extrapolations of numerical results on finite $S=\frac{1}{2}$ chains[3] yield a negative excess heat capacity at low T. Secondly, if ΔC could be perfectly described by the sG model, the data plotted in Fig. 3 should all collapse onto a single curve, since in the sG model $\Delta C/m$ is a universal function of kT/E_s^o, and for a ferromagnetic chain system both E_s^o and m are proportional to $\sqrt{B}$ [cf. Eq. (3)]. Inspection of Figs. 3 and 5, however, shows that for higher fields the magnitude of $\Delta C/\sqrt{B}$ increases significantly, whereas the location of the maximum shifts to somewhat lower values of $T/\sqrt{B}$. Since the numerical results for the sG model are in principle exact, it is evident that the discrepancies between data and theory must originate from the various approximations involved with the mapping of the original equation of motion of the spins to the sG equation. The effect of each individual approximation on ΔC, however, is far from obvious.

To obtain additional information about these effects we compared the observed excess heat capacity with exact numerical results for ΔC obtained from transfer-matrix calculations on a discrete chain of classical spins in a transverse field[22]. In these calculations the set of exchange and anisotropy-parameters appropriate for CHAB are used. The results are represented by the solid curves in Fig. 6, in which we also plotted the smoothed experimental data (dashed curves). Inspection of this figure shows that the magnitude of ΔC obtained from the transfer-matrix calculations is too high by about a factor of 2. Nevertheless, the systematic deviations of the data from universal behavior are predicted correctly, and hence we conclude that these features very likely originate from the effect of out-of-plane spin-components. The shaded area in Fig. 6 reflects the results of similar transfer-matrix calculations on a discrete chain of classical spins which are fully confined to the XY plane ($n=2$). The numerical results for the sG model are located within this area, which indicates that the influence of the continuum limit on ΔC is rather small in the present field region.

The bare quantum effects in CHAB can be inferred from Fig. 6 by comparing the $n=3$ transfer-matrix results with the data. It is obvious that these effects tend to *decrease* the magnitude of the excess heat capacity in the experimental field and temperature region. On the other hand, a comparison of the results of the $n=2$ transfer-matrix calculations with those for $n=3$ reveals that the presence of spin-components out of the XY plane gives rise to an *increase* of ΔC. This suggests that the fair description of the average data on CHAB by the classical sG model most likely arises from a compensation of both effects. This conclusion is

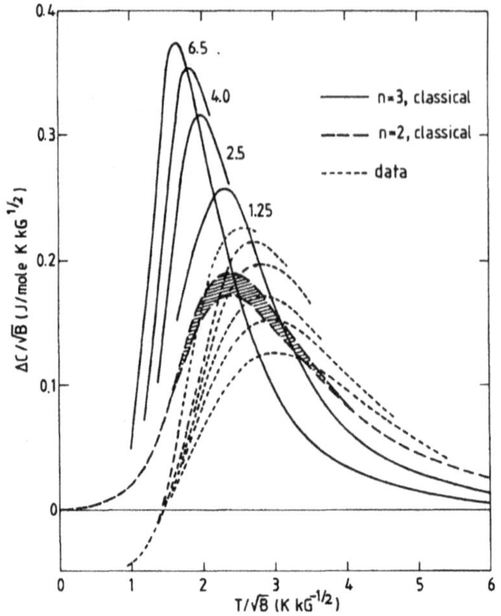

Fig.6. Excess heat capacity of CHAB for **B** // **c** *plotted in reduced form together with the results of transfer-matrix calculations on several discrete classical spin-systems. The curves are labelled by the value of the field.*

corroborated by a number of studies which consider corrections on the sG model resulting from out-of-plane spin-components[23] or quantum effects[24]. An analysis of the excess heat capacity of two other potentially soliton-bearing systems, $CsNiF_3$ and TMMC, similar to the approach outlined in this section, can be found in Ref. 25.

4. NUCLEAR SPIN-LATTICE RELAXATION

In Sec. 3 we have shown that for a proper description of the excess heat capacity of CHAB the inclusion of non-linear soliton-like excitations is necessary. To investigate to which extent these excitations manifest themselves in the dynamic properties of this compound, we studied the spin-lattice relaxation of the hydrogen nuclei for $1.2 \text{ K} < T < 6.5 \text{ K}$ and external fields $0 < B < 70$ kG. The nuclear spin-lattice relaxation time T_1 was determined from the recovery of the magnetization of the nuclear spin system after demagnetization, which was measured by means of a spin-echo technique, using a sequence of incoherent RF pulses matching the Larmor frequency of the nuclear spins of interest. Since the relaxation of the nuclear spin system towards equilibrium is induced by fluctuations in the electron spin system, it can be expressed in terms of the various elementary excitations. We have interpreted the experimental data by assuming that the dominant excitations for **B** // **c** are magnons and solitons. To enable a quantitative comparison of the effect of these excitations on T_1, we have calculated their respective contributions by exploiting the relation between T_1^{-1} and the dynamic structure factors $\mathcal{S}^{\alpha\beta}(q,\omega)$[26]. This approach has the advantage that the contribution of the solitons can be evaluated without approximating their actual shape by, for instance, a rectangular profile[10]. In our case this relation can be written as

$$T_1^{-1}(j) = \sum_{\alpha}\sum_{\beta}\sum_{m=0}^{\infty} G^{\alpha\beta}(j,m)\ \frac{2\pi}{N}\sum_{q}\mathcal{S}^{\alpha\beta}[q,\omega_N(j)]\cos(mqa)\ . \tag{6}$$

In this equation j refers to a set of nuclear spins, α and β run over the coordinates x,y,z, and $G^{\alpha\beta}(j,m)$ are the so-called geometrical factors describing the interaction between the set of nuclear spins and electron spin system. $\omega_N(j)$ represents the Larmor frequency of the nuclear spins j, which depends on both the external and the internal (dipolar) field.

The experimental results are presented in Fig. 7. It can be seen from this figure that the data exhibit a remarkably large field and temperature dependence; T_1 varies over almost five decades in the experimental region. The solid curves in the figure represent calculations based on the two-spinwave or Raman process, i.e., processes in which one magnon is annihilated and another is created or *vice versa* as the nuclear spin flips. These calculations are based upon the Hamiltonian (1) and the set of parameters appropriate to CHAB. In the present system the direct process is not possible, since ω_N is much smaller than the magnon energy gap. It is obvious that the overall behavior of the data collected in fields along the a and b axis can be described satisfactorily by this process. Only for **B** // **c** significant deviations occur at $T \gtrsim 4.2$ K. These deviations suggest that, especially at low fields and high temperatures, other relaxation mechanisms contribute significantly. To elucidate this point we calculated the contribution of three-spinwave relaxation processes and that of soliton-like excitations, which may be important for this particular field direction.

It appeared that for **B** // **a**,**b** the three-spinwave processes can be neglected with respect to the Raman process. For **B** // **c** the Raman process is about a factor 10 less effective, and three-spinwave processes yield a significant contribution at low B or high T. In Fig. 7 the relaxation rate for **B** // **c** calculated on basis of both Raman and three-spinwave processes is

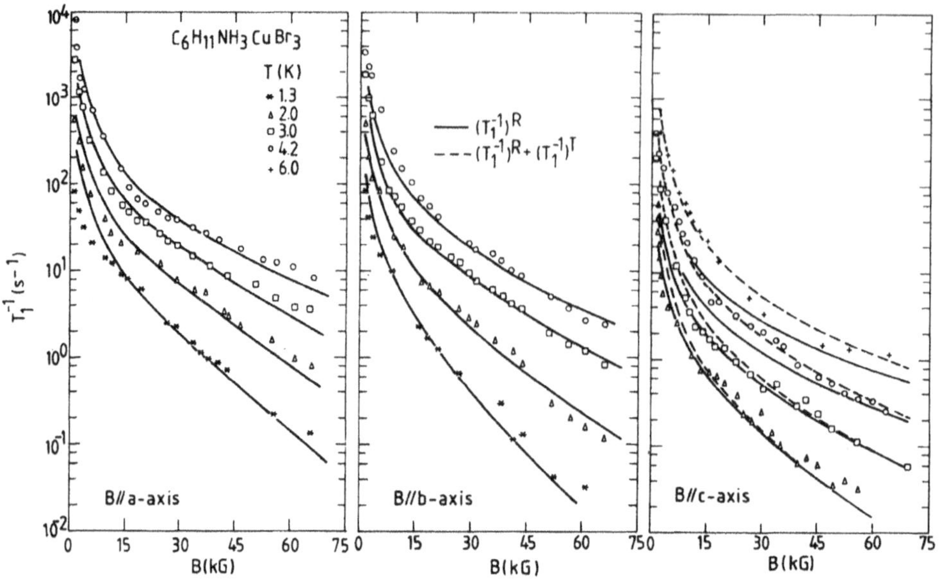

Fig.7. Field dependence of the spin-lattice relaxation rate of the hydrogen nuclei in CHAB together with the theoretical predictions based on the Raman process (solid curves) and three-spinwave processes (dashed curve).

reflected by the dashed curve. It is obvious that this curve describes the observed relaxation rate rather well, which might indicate that the soliton-like excitations have only a minor contribution. This contribution was calculated by substituting the known expressions[27] for the dynamic structure factors in Eq. (6). The resulting expression shows that if the relaxation rate would be solely due to the presence of solitons, it would be given by

$$T_1^{-1} \sim \frac{1}{T}\, e^{-E_s^o/kT} \tag{7}$$

where $E_S^o/k = 10.95\sqrt{B}$. This equation results in a universal linear relation between $\ln(T\,T_1^{-1})$ and $\sqrt{B}/T$ for all fields and temperatures. In Fig. 8 the experimental data collected for **B** // **c** are represented in such a way. This figure reveals that the data almost perfectly collapse onto one universal curve. However, the quantitative contribution of solitons, which is reflected by the dashed line in the figure, is very small. This is more clearly illustrated in the inset of the Fig. 8, where the various contributions to T_1^{-1} are plotted separately for low values of $\sqrt{B}/T$. The labels

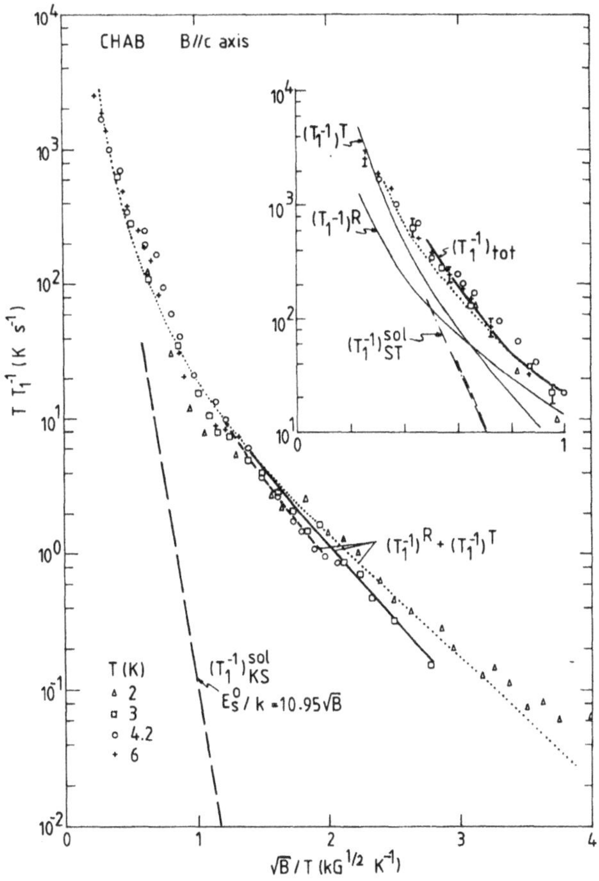

*Fig.8. Spin-lattice relaxation rate of the hydrogen nuclei in CHAB for **B // c** plotted in reduced form. The meaning of the various theoretical predictions is explained in the text.*

KS and ST refer to the soliton density used in the calculations (Eqs. (2) and (5), respectively). It appears that the inclusion of a soliton contribution yields a slightly better description in the appropriate $\sqrt{B}/T$ region [solid curve labelled "$(T_1^{-1})_{tot}$"], but the resulting improvement is not very significant and does therefore not yield solid support for the presence of soliton-like excitations in CHAB.

5. MAGNETIZATION

The magnetization (M) was measured with a commercial (PAR) magnetometer in applied fields B up to 50 kG along the crystallographic c direction. Sample temperatures between 1.4 and 10 K could be achieved with a typical stability of 50 mK. The accuracy of the measured value of M did

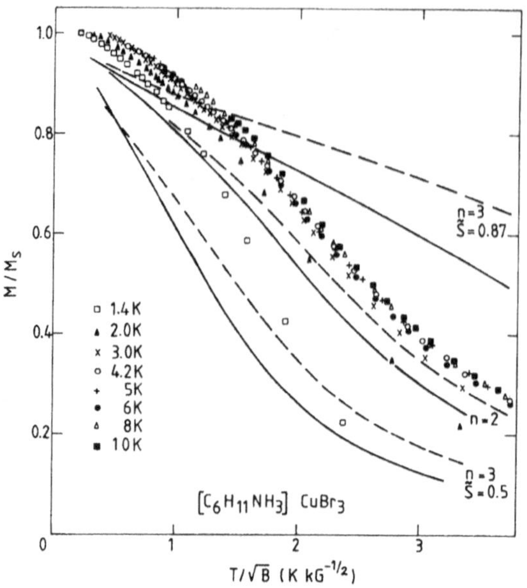

Fig.9. Magnetization of CHAB for **B**∥**c** *plotted in reduced form together with results of transfer-matrix calculations on various discrete classical spin systems. Solid and dashed curves correspond to T = 10 K and 4.2 K, respectively.*

amount to 2%, whereas saturation of M could be reached for T < 3 K. From the saturation value M_S a value $g^{cc} = 2.01 \pm 0.02$ was deduced, which is in nice agreement with the value $g^{cc} = 2$ inferred from the FMR experiments presented in Sec. 2.

Since for several magnetic model systems, including the sG model, M/M_S has been reported to be a universal function of $T/\sqrt{B}$ [20,28], we have plotted our results accordingly in Fig. 9. From this figure it appears that the data collected above T = 3 K almost perfectly collapse onto a single curve, suggesting universal behavior. At lower T systematic deviations occur, which are due to the small interchain interactions. In the figure we have included several theoretical predictions, based on transfer-matrix calculations on discrete chains of classical spins. The curves labelled "n = 3, $\tilde{S}$ = 0.5" denote the results from calculations based on the Hamiltonian (1). The solid and dashed curve correspond to T = 10 K and T = 4.2 K, respectively. The curves labelled "n = 3, $\tilde{S}$ = 0.87" reflect similar calculations based on a semiclassical spin length $\tilde{S}$ equal to $\sqrt{S(S+1)}$ instead of S.

It is obvious that these results systematically deviate from the experimental data and, apart from this, do not show universal behavior. For comparison, we also included the results of transfer-matrix calculations on a chain of classical spins having only two non-zero (XY) spin-components ($n=2$, cf. Sec. 3). Although this prediction agrees much better with the data than the $n=3$ transfer-matrix calculations, significant deviations occur at low values of $T/\sqrt{B}$.

We will now turn to an interpretation of M in terms of the sG model. First, we compare our data with numerical results on the magnetization of this model[20], since such calculations for ΔC yielded a fair description of the corresponding data. The resulting prediction for M/M_S, which is a universal function of $T/\sqrt{B}$, is plotted in Fig. 10, together with the data. Inspection of this figure shows that the overall agreement between this prediction and the data is fairly good, but the qualitative behavior at low $T/\sqrt{B}$ is essentially different from the experimental results. Possibly, this is caused by the presence of spin-components out of the easy plane, which are not taken into account in this model. However, the prediction lies outside the region covered by the results of $n=2$ transfer-matrix calculations, such in contrast to the excess heat capacity (cf. Fig. 6). To obtain more information about the origin of these deviations, we have described the magnetization with the same model that we used in the interpretation of the T_1 data, i.e., a model in which the basic excitations are magnons and solitons.

Both magnons and solitons cause the magnetization at finite T to decrease with respect to its saturation value M_s. The decrease ΔM_{sw} caused by magnons was calculated using linear spinwave theory. The results obtained from the Hamiltonian (1) and the appropriate parameters for CHAB are reflected by the dashed curve. Inspection of the figure shows that this prediction agrees nicely with the data up to $T/\sqrt{B} \simeq 1.2$. At higher values of $T/\sqrt{B}$ the predicted decrease of M is too small, suggesting that other excitations may contribute significantly in this region.

The decrease of the magnetization ΔM_{sol} caused by solitons was calculated by straightforward differentiation of the soliton free energy corresponding to Eq. (5) with respect to B, yielding

$$\Delta M_{sol} = -4Ng\mu_B S\left[\frac{8}{\pi}\right]^{\frac{1}{2}} t^{-\frac{1}{2}} e^{-1/t}\left[1 - \frac{19}{8}t - \frac{3}{128}t^2 - \frac{59}{256}t^3\right]. \qquad (8)$$

The resulting prediction is denoted by the dashed-dotted curve in Fig. 10. It appears that ΔM_{sol} is neglegible for $T/\sqrt{B} < 1$, whereas it has a maximum

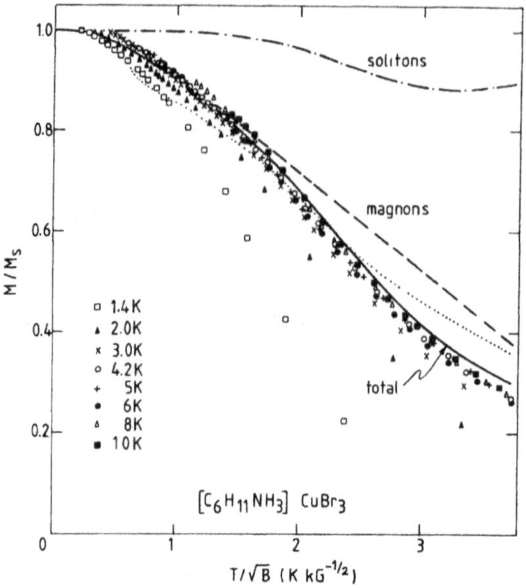

*Fig.10. Magnetization **M** of CHAB for **B**∥**c** plotted in reduced form together with the decrease of **M** resulting from magnons and that resulting from kink-solitons.*

at $T/\sqrt{B} \simeq 3.2$. Within the framework of the sG model, the free energy of a system can be written as $F = F_m + F_{sol}$ (cf. Sec. 3). Following a similar approach as that underlying the interpretation of the T_1 data we replace, in first order, the classical magnon contribution F_m by its quantummechanical counterpart and, correspondingly, add ΔM_{sol} to ΔM_{sw}. The result is reflected by the solid curve in Fig. 10. It is obvious that this approach yields a very good description of the data, especially since one has to recall that no adjustable parameters have been used. A possible explanation for this nice agreement will be presented in the next section.

6. DISCUSSION

In concluding we would like to remark that a consistent description of various magnetic properties of CHAB for **B**∥**c** seems possible if the linear excitations are described by conventional spinwave theory and the non-linear excitations by the classical sG model.

With respect to the total magnetic heat capacity, one should note that classical model predictions for $n = 2$ and $n = 3$ are dominated by a magnon contribution, having a non-physical zero-temperature limiting value of Nk/2 or Nk, respectively. In contrast to the heat capacity itself, the excess heat capacity of CHAB is dominated by non-linear excitations. Within the framework of the sG model, this can be inferred from a comparison of the soliton contribution ΔC_s, following from Eq. (5), with the numerical results for this model, but it is demonstrated directly by a straightforward calculation of ΔC_m using linear spinwave theory (cf. Fig. 3). Partly, this may explain the fact that the observed excess heat capacity agrees rather well with the numerical results for the sG model.

The variation of the nuclear spin-lattice relaxation rate in CHAB with field and temperature can be described satisfactorily by linear spinwave theory alone. This can be understood from the observation that in the experimental range of B and T the soliton contribution to T_1^{-1} is smaller than the magnon contribution by at least one order of magnitude. Hence the measurements of T_1^{-1} neither confirm nor rule out the presence of solitons in this compound.

Attempts to describe the magnetization of CHAB by a classical model, i.e., numerical results for the sG model or transfer-matrix calculations on discrete classical spin systems, did not yield a satisfactory agreement with the data. Deviations between these calculations and the data appeared to persist for $T/\sqrt{B} < 1$, where soliton-like excitations have a neglegible statistical weight. This suggests that the inadequacy of these classical models mainly originates from a poor description of the linear excitations.

REFERENCES

1. F. Bloch, Z. Physik 61: 206 (1931).
   H. Bethe, Z. Physik 71: 205 (1932).
2. C.P. Landee and R.D. Willett, Phys. Rev. Lett. 43: 463 (1979).
3. G. Kamieniarz and C. Vanderzande, Phys. Rev. B35: 3341 (1987).
4. G.M. Wysin and A.R. Bishop, Phys. Rev. B34: 3377 (1986).
5. J.B. Torrance and M. Tinkham, Phys. Rev. 187: 587, 595 (1969), L.A. Bosch, G.J.P.M. Lauwers, K. Kopinga, C. van der Steen, and W.J.M. de Jonge, J. Phys. C20: 609 (1987).
6. See, for instance, J.D. Johnson and J.C. Bonner, Phys. Rev. B22: 251 (1980), P. Schottman, Phys. Rev. B33: 4880 (1986), T. Schneider and E. Stoll, Phys. Rev. B25: 4721 (1982), and T. Schneider, E. Stoll, and U. Glaus, Phys. Rev. B26: 1321 (1982).
7. H.J. Mikeska, J. Phys. C11: L29 (1978); C13: 2913 (1980).

8. M. Steiner, K. Kakurai, and J.K. Kjems, Z. Physik B53: 117 (1983).
9. L.P. Regnault, J.P. Boucher, J. Rossat-Mignod, J.P. Renard, J. Bouillot, and W.G. Stirling, J. Phys. C15: 1261 (1982).
10. T. Goto, Phys. Rev. B28: 6347 (1983).
11. A.P. Ramirez and W.P. Wolf, Phys. Rev. Lett. 49: 227 (1982); Phys. Rev. B32: 1639 (1985).
12. J. Boucher and J.P. Renard, Phys. Rev. Lett. 45: 486 (1980).
13. F. Borsa, Phys. Rev. B25: 3430 (1982).
14. K. Kopinga, A.M.C. Tinus, and W.J.M. de Jonge, Phys. Rev. B29: 2868 (1984).
15. G.C. de Vries, E. Frikkee, R.B. Helmholdt, K. Kopinga, W.J.M. de Jonge, and E.F. Godefroi, J. Phys. Chem. Solids (1987).
16. K. Kopinga, A.M.C. Tinus, and W.J.M. de Jonge, Phys. Rev. B25: 4685 (1982).
17. H.W.J. Blöte, Physica 79B: 427 (1975).
18. A.C. Phaff, C.H.W. Swüste, W.J.M. de Jonge, R. Hoogerbeets, and A.J. van Duyneveldt, J. Phys. C17: 2583 (1984).
19. J.A. Krumhansl and J.R. Schrieffer, Phys. Rev. B11: 3535 (1975).
20. T. Schneider and E. Stoll, Phys. Rev. B22: 5317 (1980).
21. K. Sasaki and T. Tsuzuki, Solid State Commun. 41: 521 (1982).
22. F. Boersma, K. Kopinga, and W.J.M. de Jonge, Phys. Rev. B23: 186, (1981).
23. See, for instance, K.M. Leung and A.R. Bishop, J. Phys. C16: 5893 (1983), H.C. Fogedby, P. Hedegård, and A. Svane, Phys. Rev. B28: 2893, (1983), and G. Wysin, A.R. Bishop, and P. Kumar, J. Phys. C17: 5975 (1984).
24. See, for instance, K. Maki, Phys. Rev B24: 3991 (1981), H.J. Mikeska, Phys. Rev. B26: 5213 (1982), P. Riseborough, Solid State Commun. 48: 901 (1983), and H.C. Fogedby, K. Osano, and H.J. Jensen, Phys. Rev. B34: 3462 (1986).
25. A.M.C. Tinus, W.J.M. de Jonge, and K. Kopinga, Phys. Rev. B32: 3154 (1985).
26. S.W. Lovesey, "*Theory of neutron scattering from condensed matter*", Clarendon Press, Oxford (1984).
27. K. Maki, "*Progr. Low Temp. Phys. VIII*", North Holland, Amsterdam (1982), E. Allroth and H.J. Mikeska, Z. Physik B43: 209 (1981).
28. M.G. Pini and A. Rettori, Phys. Rev. B29: 5246 (1984).

FERROMAGNETIC EXCHANGE IN MOLECULAR SOLIDS

Joel S. Miller and Arthur J. Epstein

Central Research and Development Department, Experimental Station E-328, E. I. du Pont de Nemours & Co., Inc., Wilmington, DE 19898 and Department of Physics and Department of Chemistry, The Ohio State University, Columbus, OH 43210-1106.

INTRODUCTION

Ferromagnetism requires spin alignment throughout the bulk and albeit rare for a molecular solid[1] the quest for molecular based ferromagnets is the subject of increasing contemporary interest.[3-7] Several mechanisms have been proposed for achieving ferromagnetic coupling in a molecular solid and in the past few years research directed toward this end has started. In the 1960's Harden M. McConnell proposed two models, namely, Heitler-London spin exchange between positive spin density on a radical and negative spin density on another[8] and admixing of a virtual triplet excited state with the ground state for a chain of alternating radical cation donors and radical anion acceptors.[9] Noboru Mataga[10] (1968) and later Alexandr A. Ovchinnikov[11] (1978) essentially suggested that very high spin multiplicity alternate hydrocarbon molecules may have ferromagnetic domains.

Experimental evidence for ferromagnetic behavior in a molecular compound, however, has been limited to the charge transfer salt of decamethylferrocene, (η^5-$C_5Me_5)_2Fe^{II}$, **1**, with tetracyanoethylene, TCNE, **2**.[2,3,4,12-13] The mechanisms that govern the stabilization of ferromagnetism in this class of linear chain alternating radical cation donors, D, and radical anion acceptors, A, *i. e.*, $\cdots D^{\cdot+}A^{\cdot-}D^{\cdot+}A^{\cdot-}D^{\cdot+}A^{\cdot-}\cdots$, molecular charge transfer complexes are not firmly established. However, it is attractive to consider[2-4,13] the admixing of a virtual triplet excited state with the ground state model originally proposed by McConnell[9,14] as this complex possesses both the crystal and electronic structures described required by this model.

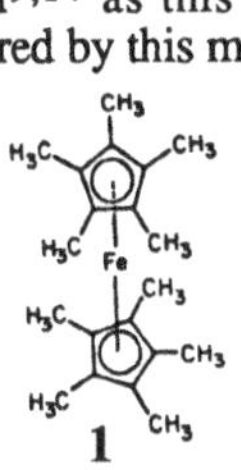

1

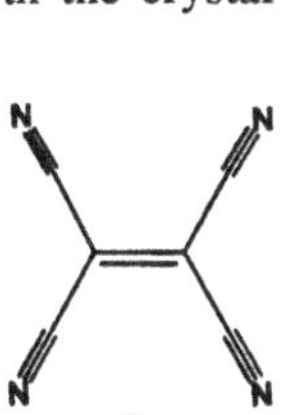

2

HEITLER-LONDON SPIN EXCHANGE

In 1963 McConnell stated[8] that radicals with "...large positive *and* negative atomic π-spin densities ... [that] pancake ... so that atoms of positive spin density are exchanged coupled ... to atoms of negative spin density in neighboring molecules ... give a ferromagnetic exchange interaction." Here ferromagnetic exchange results from the incomplete cancellation of opposite spin components. This is a mechanism for pairwise ferromagnetic exchange,[15] not bulk ferromagnetism. Ferromagnetic exchange in three dimensions is necessary for bulk ferromagnetic behavior. Since free radicals, *e. g.*, R·, readily dimerize, bond formation must be avoided to achieve ferromagnetic exchange via this mechanism. Stable free radicals, *e. g.*, $[TCNE]^{\cdot -}$ [16] or $[TCNQ]^{\cdot -}$ [17], possess atoms with unequal positive and negative spin densities such that it is conceivable that they might form a structure that complies with McConnell's requirement. However, of the numerous structures reported none have the overlap described by McConnell.[18] Even with these stable radicals dimerization to form diamagnetic π-dimers occurs.[19] Examples of C-C σ-bond formation between $-C(CN)_2$ moieties[20] on adjacent TCNQ's as well as of more complex chemical reactions[21] have also been reported. Hindered stable radicals, for example allyl or nitroxide radicals, *e. g.*, **3**[22] and **4**,[23] respectively, might be suitable. However, where known they exhibit Curie or antiferromagnetic behavior. High spin multiplicity radicals, *e. g.*, S = 1 O_2[24] and $C_6(NCH_2)_6$[7,14] or S = 4 polycarbenes,[5,25] studied to date likewise exhibit either antiferromagnetic coupling or independent spin Curie behavior.

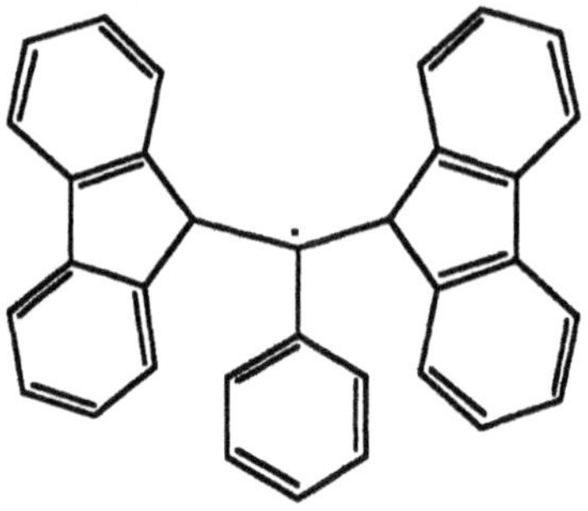

3

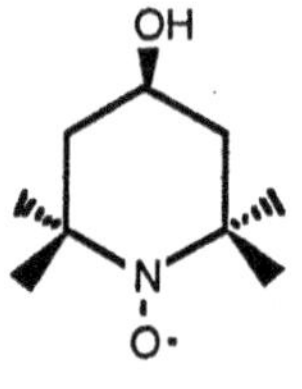

4

Cleverly designed stable radicals with the proper solid state packing are sought to test this model. Recently the spin multiplicity of dicarbenes incorporated into rigid [2.2] cyclophanes have been shown to be controlled by the overlapping modes of the spins in the aromatic C_6 rings. McConnell's model[8] suggests that the pseudo*meta* isomer should possess a singlet ground state.[5,26] This is in contrast to the behavior of the quintet pseudo*ortho* and pseudo*para* isomers of the bis(phenylmethylenyl)[2.2]cyclophanes. ESR[5,26] and computational data[27] on this system are consistent with these results suggesting that ferromagnetic exchange can be achieved by this McConnell model. Preliminary results on intermolecular systems are also consistent with McConnell's model;[28] however, a bulk ferromagnet has yet to reported.

HIGH SPIN MULTIPLICITY MOLECULES AND POLYMERS

Based upon the ground state triplet behavior of diphenylcarbene in 1968 Mataga suggested[10] that according to Hund's rule large planar alternate hydrocarbons comprised of *meta* substituted triplet diphenylcarbene moieties will have a high spin (ferromagnetically coupled) ground state. Bulk ferromagnetic behavior requires spin alignment throughout the solid (*i. e.*, inter- as well as intramolecular ferromagnetic coupling). A mechanism for achieving intermolecular ferromagnetic coupling was not discussed. McConnell's Heitler-London spin exchange mechanism might suffice; however, intermolecular ferromagnetic coupling may not be necessary if the high spin multiplicity radical was its own ferromagnetic domain. Examples of high spin multiplicity polymers that Mataga suggested are **5**, **6** and **7**.

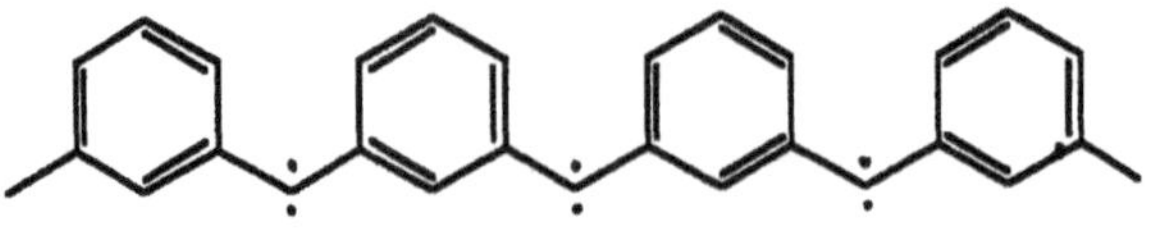

5

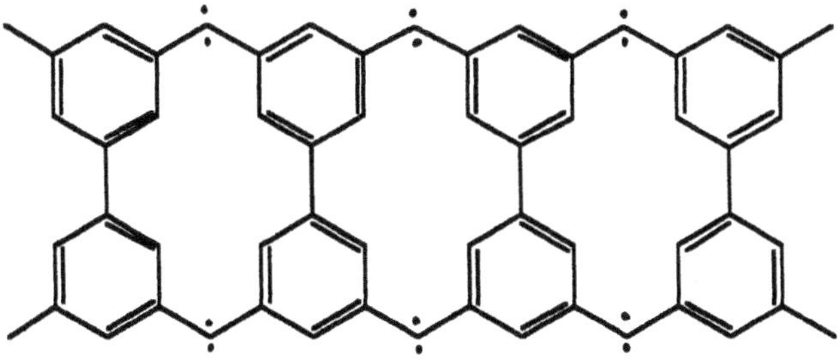

6

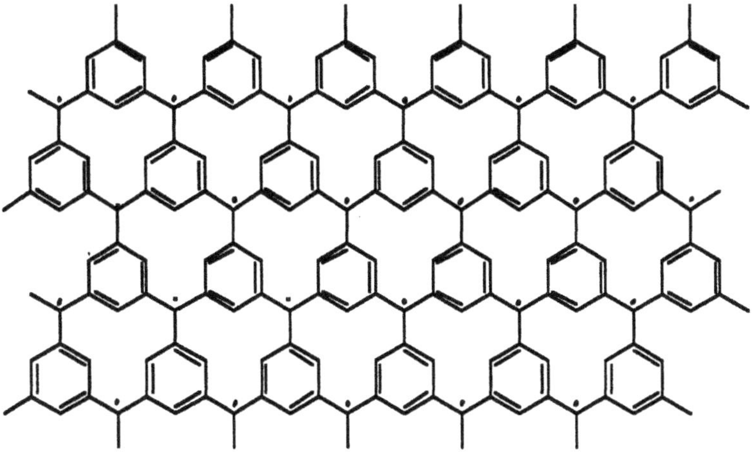

7

With guidance from McConnell's Heitler-London spin exchange model Hiizu Iwamura[5,25,29] and coworkers recently have prepared several high spin polycarbenes via a route that Mataga suggested, namely, the *in situ* photolysis of polydiazoderivatives. The best characterized compound is a nonet (S = 4) tetracarbene, **8**, eq. [1]. The paramagnetic susceptibility of **8** diluted in either a 2-methyltetrahydrofuran glass or a benzophenone crystal confirms the nonet ground state ($\mu_{eff} = 9.08\ \mu_B$ vs. a calculated spin only value of $8.49\ \mu_B$), however, the temperature dependence of the susceptibility exhibits only independent spin or antiferromagnetic behavior.[25] Iwamura has also proposed additional structures that comply with the Mataga prescription.[5]

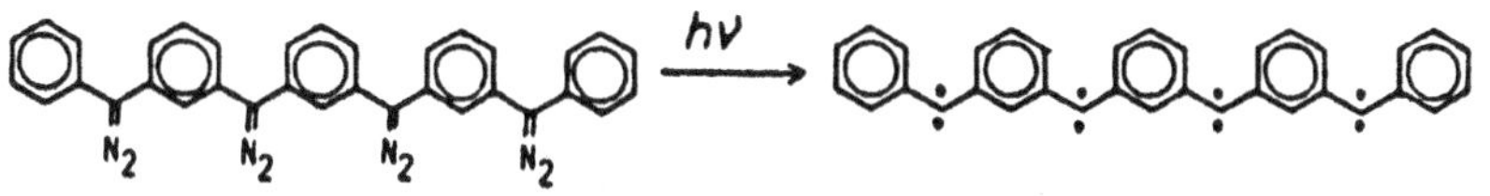

[1]

8

Preparation of an organic ferromagnet via Mataga's prescription is problematic. Mataga points out that as the molecule becomes larger the energy levels forms bands and the bonding/high spin nonbonding and nonbonding/antibonding band gaps becomes smaller. If these gaps become comparable to k_BT, then ferromagnetic spin alignment may no longer be possible due to thermal population of higher excited states. However, bond alternation increasing the bandgap might occur. Additionally he states that correlation lowers the energy of the nonmagnetic states making it more difficult to realize ferromagnetism. Stable triplet carbenes are required. Carbenes can be further stabilized via greater sp hybridization of the carbene carbon.[30] However, the low temperature photolytic loss of dinitrogen leaves sp^2 hybridized carbons. Routes to more sp -$C(N_2)$- carbons, *e. g.*, via addition of steric bulk, may lead to greater stability of the triplet. Other chemical problems that must be overcome prior to the realization of an organic ferromagnet via Mataga's concept include minimization of bond formation or spin pairing via distortions or rehybridization, *e. g.*:

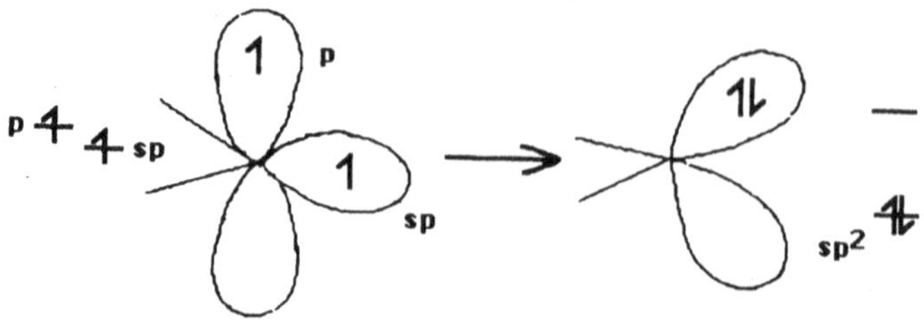

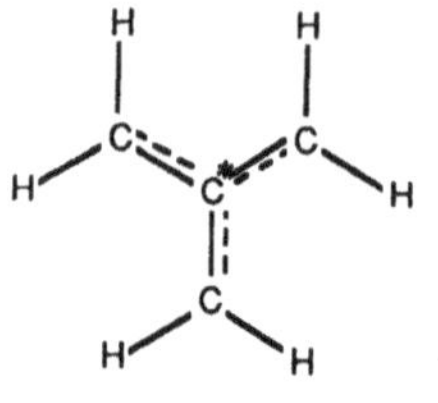

9 S = 1

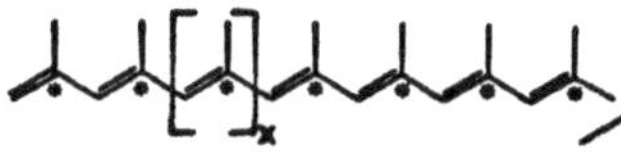

11 S = x/2

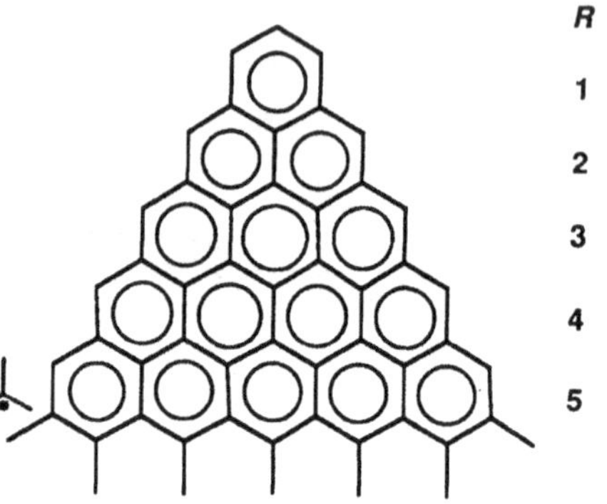

10 $S = (R - 1)/2$

Ovchinnikov[11] in 1978 stated that the spin, S, for planar alternate hydrocarbons can be calculated by $S = |n^*-n|/2$ [where adjacent atoms are respectively starred (*) and nonstared and identically denoted atoms are not adjacent to each other and n is the number of nonstarred atoms and n* is the number of starred atoms]. Thus, trimethylenemethane[31], **9**, is predicted to have an S = 1. Likewise, large molecules should have large S's. The ground state of fused C_6 rings, **10**, are also predicted to have high spin ground states via the formula $S = (\boldsymbol{R}-1)/2$ where $\boldsymbol{R}$ is the number of rows.[11] Three and six fused rings possess S = 1/2 and S = 1 ground states, respectively.[32] Polymers such as **11** are predicted to have a large value for S, *i. e.*, x/2.[11,33] Boron substituted graphite (BC_7), **12**, is also predicted to have a high spin state.[11] Recently, BC_3 has been prepared,[34] however, other compositions as well as their magnetic properties have yet to be reported.

Chemical instability arising from the unpaired elections on **9** and **10** ($\boldsymbol{R}$ = 2,3) suggest that stable high spin materials are going to be difficult to prepare. Distortions as well as intermolecular coupling which pair up electrons must also be avoided.

Recently attempts to prepare a polymer similar to **7** based on the iodine oxidation of 1,3,5-triaminobenzene has been reported, although its reproducible chemical or physical characterization has not been achieved.[35] Recent attempts to prepare an organic polymer ferromagnet via the formation of a polydiacetylene from a stable nitroxy biradical monomer have been disclosed.[36] The susceptibility obeys the Curie-Weiss expression with $\theta = -2$ K and thus indicates that the three-dimensional polymer has dominant antiferromagnetic coupling. These authors report a magnetization corresponding to only ~0.1 to 2% of the maximum theoretical magnetization. The low value and marked variation between samples suggest that the ferromagnetic behavior in these polymers may not be intrinsic.

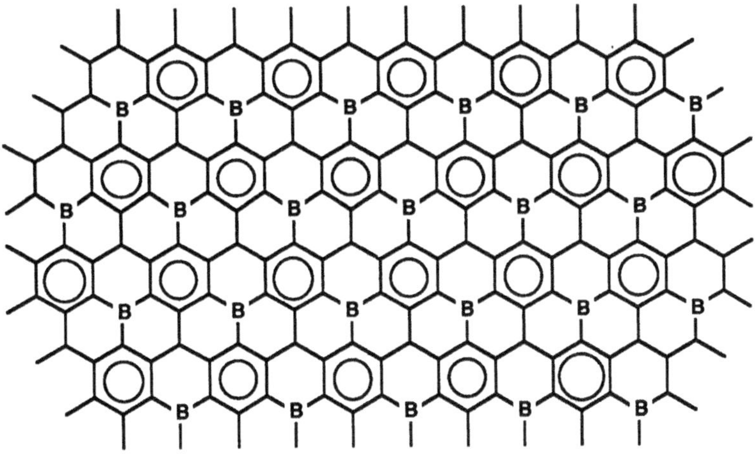

12 S = ∞

VIRTUAL TRIPLET EXCITED STATE MIXING WITH THE GROUND STATE

In 1967 McConnell proposed a configuration interaction model for stabilizing ferromagnetic coupling in a $\cdots D^{\cdot+}A^{\cdot-}D^{\cdot+}A^{\cdot-}\cdots$ structure where either ion possesses a triplet charge transfer excited state. The essence of this McConnell model is that if an excited state, ES, with $m_s = 1$ resulting from either a $m_s = 1$ donor, D[37], or $m_s = 1$ acceptor, A[37], *but not both*[14], formed by either virtual retro ($D^o + A^o \leftarrow D^{+\cdot} + A^{-\cdot}$) or virtual forward ($D^{2+} + A^{2-} \leftarrow D^{+\cdot} + A^{-\cdot}$) charge transfer or virtual disproportionation (*e. g.*, $D^o + D^{2+} \leftarrow 2\ D^{\cdot+}$)[2,3,13] admixes with the $M_s = 1$ ground state, GS, then ferromagnetic coupling will be stabilized.[9,14,15] Using Hund's rule[38,39] the lowest energy excited state most likely to virtually admix with the ground state can be identified enabling the prediction of the type of magnetic coupling. This can be illustrated for a $D^{+\cdot}$ with a doubly degenerate, *d*,[40] partially occupied molecular orbital, POMO, containing three electrons, *i. e.*, d^3, and an $A^{-\cdot}$ with a nondegenerate, *s*,[40] POMO, containing one electron, *i. e.*, s^1.[41] For this $D^{+\cdot}/A^{-\cdot}$ pair two ground states, GS_{FO} and GS_{AF}, Fig. 1, exist. In the absence of spin interactions between these moieties, GS_{FO} and GS_{AF} are of equal energy leading to simple paramagnetic behavior. GS_{FO} with $M_s \equiv m_sA + m_sD = 1$ is ferromagnetically coupled; whereas GS_{AF} with $M_s = 0$ is antiferromagnetically coupled, Fig. 1.

If $A^{-\cdot}$'s valence electron is in a degenerate orbital (*e. g.*, *d* not *s*), then two ground states akin to those discussed above for the d^3/s^1 electron configuration case exist. Admixing the lowest excited state formed from virtual retro charge transfer by the above argument leads to the stabilization of antiferromagnetic coupling. Likewise, forward charge transfer also leads to stabilization of antiferromagnetic coupling as the easiest excited $D^{+\cdot}$ electron has $m_s = -1/2$. The electron excited from the $D^{\cdot+}$ is promoted to the vacant orbital on $A^{-\cdot}$. Although the unpaired electron on $A^{-\cdot}$ can have $m_s = \pm 1/2$, the lowest excited state to maximize spin multiplicity requires a $m_s = -1/2$ electron. Thus, unlike the d^3/s^1 case antiferromagnetic coupling is stabilized regardless of the direction of virtual charge transfer. This is the basis for the aforementioned '*but not both*' statement.[14]

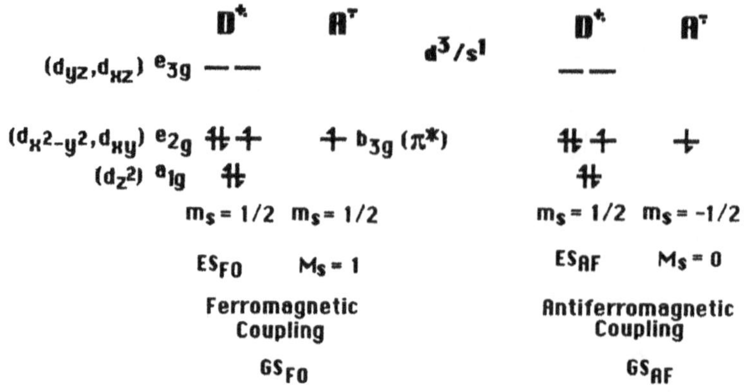

Figure. 1. Paramagnetic ground states assuming no spin interactions between radical ions.

Spin interactions can lead to lowering of either GS_{FO} or GS_{AF} and ferro- or antiferromagnetic coupling. The GS lowest in energy is the one with the greatest probability of admixing with the lowest energy virtual charge transfer excited state. Hund's rule can be used to predict the lowest charge transfer excited state for each direction of charge transfer. Thus, magnetic coupling (ferro-, ferri-, or antiferromagnetic) can be predicted for a specific direction of charge transfer. Retro charge transfer for the above case, Fig. 2, is an informative example. Since the $D^{+\cdot}$ can only accept a $m_s = -1/2$ electron via virtual charge transfer from $A^{-\cdot}$ and GS_{AF}, not GS_{FO}, has a $m_s = -1/2$ electron, the admixture of the $ES_{AF}{}^{D\leftarrow A}$ excited state will lower the energy of GS_{AF} to GS'_{AF} and stabilize pairwise antiferromagnetic coupling.

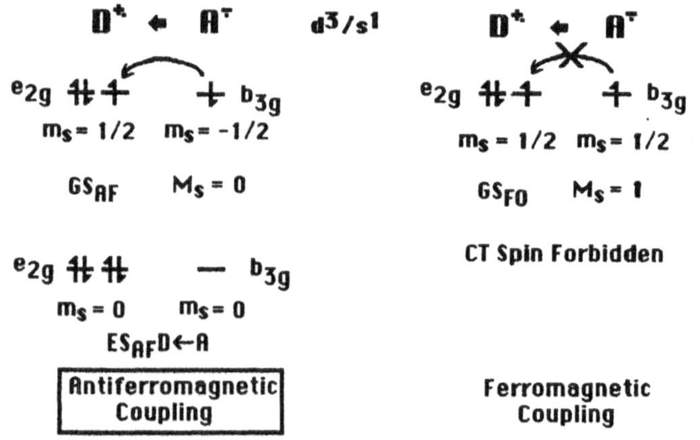

Figure. 2. Schematic illustration of stabilization of antiferromagnetic coupling via retro charge transfer from a s^1 $A^{-\cdot}$ to d^3 $D^{+\cdot}$.

For virtual forward charge transfer excited states $ES_{FO}{}^{A\leftarrow D}$, $ES_{AF}{}^{A\leftarrow D}$ and $ES'_{AF}{}^{A\leftarrow D}$ are possible, Fig. 3. From Hund's rule the $D^{+\cdot}$ electron with the lowest ionization energy has $m_sD = -1/2$ and can only be virtually transferred to the ferromagnetically coupled $A^{-\cdot}$, *i. e.*, $ES_{FO}^{A\leftarrow D}$. Transfer of a $m_s^D = +1/2$ electron to an antiferromagnetically coupled $A^{-\cdot}$ requires admixture of higher excited states, *i. e.*, $ES_{AF}^{A\leftarrow D}$ or $ES'_{AF}^{A\leftarrow D}$. Thus, for forward charge transfer the ferromagnetically coupled ground state, GS_{FO}, leads to pairwise ferromagnetic coupling and may lead to bulk ferromagnetic behavior as observed for d^3/s^1 $[Fe^{III}(C_5Me_5)_2]^{+\cdot}[TCNE]^{-\cdot}$.[3,4,12-13]

This pair of examples for the d^3/s^1 electron configuration leads to differing magnetic coupling depending on the direction of virtual charge transfer; however, an *a priori* prediction of the direction of charge transfer cannot be made. Nonetheless, if McConnell's model is appropo and the direction of charge transfer is known, then the magnetic coupling can be predicted.

Table 1. Magnetic Coupling for Homospin Systems [2,3,a,b]

Spin 1/2 Systems

D(A)	A(D)	D->A	A->D	Example
s^1	s^1	AF	AF	[TMPD][TCNQ],[c] [TTF][$Pt(S_2C_4F_6)_2$],[d] $V(C_6H_6)_2^{\cdot}$,[e] [$Cr(C_6H_6)_2$]I [f]
d^1	s^1	AF	FO	$[Ni^{III}(C_5Me_5)_2]^{+\cdot}[TCNE]^{-\cdot}$ [43]
d^3	s^1	FO	AF	$[Fe^{III}(C_5Me_5)_2]^{+\cdot}[TCNE]^{-\cdot}$ [12,13]
t^1	s^1	AF	FO	
t^5	s^1	FO	AF	
d^1	d^1	FO	FO	$Co^{II}(C_5H_5)_2^{\cdot}$ [42], NO· [g]
d^3	d^1	AF	AF	
t^1	d^1	FO	FO	
t^5	d^1	AF	AF	
d^3	d^3	FO	FO	$[Fe^{III}(C_5Me_5)_2]^{+\cdot}[BF_4]^-$
t^1	d^3	AF	AF	
t^5	d^3	FO	FO	
t^1	t^1	FO	FO	
t^5	t^1	AF	AF	
t^5	t^5	FO	FO	

Spin 1 Systems

D(A)	A(D)	D->A	A->D	Example
d^2	d^2	AF	AF	O_2,[21] $[Ru(OEP)]_2$ [h]
d^2	t^2	FO	AF	
d^2	t^4	AF	FO	
t^2	t^2	FO	FO	
t^2	t^4	AF	AF	
t^4	t^4	FO	FO	

[a] AF refers to antiferromagnetic coupling and FO to ferromagnetic coupling. [b] POMO orbital degeneracy (intrinsic or accidental): *s* = singly (a or b), *d* = doubly (e), *t* = triply (t), *q* = quadruply, or *p* = quintuply. [c] Ohmasa, M.; Kinoshita, M.; Sano, M.; Akamatu, H. *Bull. Chem. Soc. Jap.* **1968**, *41*, 1998. [d] Bray, J. W.; Interrante, L. V.; Jacobs, I. S.; Bonner, J. C. "Extended Linear Chain Compounds," J. S. Miller, ed. Plenum Pub. Corp., *3*, **1983**, 353-415. [e] Fischer, E. O.; Joos, G.; Meer, W. Z. *Naturforsch.*, **1958**, *13b*, 456-457. [f] Karimov, Yu. S.; Chibrikin, V. M.; Shchegolev, I. F. *J. Chem. Phys. Sol.*, **1963**, *24*, 1683-1685. [g] Collman, J. P.; Barnes, C. E.; Sweptson, P. N.; Ibers, J. A. *J. Am. Chem. Soc.* **1984**, *106*, 3500-3510. [h] Skaaup, S.; Skante, P. N.; Boggs, J. E. *J. Am. Chem. Soc.* **1976**, *98*, 6106.

Table 2. Magnetic Coupling for Heterospin Systems [2,3,a,b]

Spin 1/2 - 1 Systems

D(or A)	A(or D)	D -> A	A -> D	Example
s^1	d^2	FI	FI	
s^1	t^2	FO	FI	
s^1	t^4	FI	FO	
d^1	d^2	FI	FO	
d^1	t^2	FO	FO	
d^1	t^4	FI	FI	
d^3	d^2	FO	FI	
d^3	t^2	FI	FI	
d^3	t^4	FO	FO	
t^1	d^2	FI	FO	
t^1	t^2	FO	FO	
t^1	t^4	FI	FI	
t^5	d^2	FO	FI	
t^5	t^2	FI	FI	
t^5	t^4	FO	FO	

Spin 1/2 - 3/2 Systems

D(or A)	A(or D)	D -> A	A -> D	Example
s^1	t^3	FI	FI	$[Cr^{III}(C_5Me_5)_2][TCNE]$[43]
d^1	t^3	FI	FO	
d^3	t^3	FO	FI	
t^1	t^3	FI	FO	
t^5	t^3	FO	FI	

[a] Since $S_D \neq S_A$, FI refers to ferrimagnetic coupling and FO to ferromagnetic coupling. [b] POMO orbital degeneracy (intrinsic or accidental): *s* = singly (a or b) and *d* = doubly (e).

Generalized Hubbard Model

The competition among the various excited states in the lowering of the total ground state energy in principle can be evaluated by a generalized Hubbard model.[42] This model is useful in examining the competition between delocalization, as indicated by the nearest neighbor charge transfer integral, β, and the energy of transferring an electron between the neighboring sites, ΔE. For ΔE >> β, the energy difference between singlet (antiferromagnetic) and triplet (ferromagnetic) alignments is $\Sigma J_{ij}\mathbf{S}_i \cdot \mathbf{S}_j$ where $\mathbf{S}_i$ and $\mathbf{S}_j$ are the spins on adjacent sites *i* and *j*, and J_{ij} is an effective exchange interaction. From perturbation analysis it can be shown that the effective exchange interaction resulting from each virtual excitation is $J_{ij} \sim \beta^2/\Delta E$. The sum of all of the possible exchange interactions between pairs of neighboring sites is $J_{ij} \sim \Sigma\beta_n{}^2/\Delta E_n$ where n refers to each allowed virtual excitation. Thus, the resulting J_{ij} is the sum of the contributions from each possible excitation weighted by the square of the overlap integral (which varies with excitation) and is inversely proportional to the excitation energy.

The magnetic coupling of the solid is dependent upon the total sum of pairwise interactions among the spins, $\Sigma J_{ij}\mathbf{S}_i \cdot \mathbf{S}_j$. Hence, the generalized Hubbard model can be used as a guide to identify the chemical and physical modifications of the solid to enhance the ferromagnetic exchange. For example, chemical modifications of the molecular units to decrease ΔE_n and/or increase β_n should increase the ferromagnetic exchange interaction and lead to the stabilization of ferromagnetic coupling. Application of pressure should increase the overlap and consequently the relative and absolute magnitudes of the β_n, again improving the magnetic properties. Application of this generalized Hubbard model to the permethylmetallocenium radical anion salts is in progress.

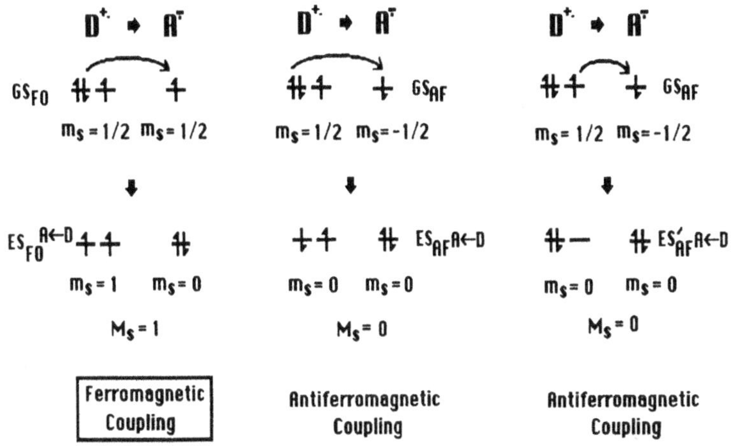

Figure. 3. Schematic illustration of stabilization of ferromagnetic coupling via forward charge transfer from a d^3 $D^{+\cdot}$ to a s^1 $A^{-\cdot}$.

Generalization to Systems with Other Electron Configurations

More generally D's and A's with different electron configurations can be anticipated to support ferromagnetic coupling via the McConnell mechanism and treated with a generalized Hubbard model. The results of the evaluation of the magnetic stabilization (antiferromagnetic, AF, or ferromagnetic, FO) are summarized in Table 1 for homospin (m_s = 1/2, 1) systems ($m_sD = m_sA$). For heterospin (m_s = 1/2-1, 1/2-3/2) systems ($m_sD \neq m_sA$) due to incomplete spin cancellation antiferromagnetic coupling should lead to ferrimagnetic behavior, FI. The predicted FO and FI interactions are summarized in Table 2. It is assumed that the ground state D and A each possess one or more unpaired electrons (*i. e.*, m_sD, $m_sA \geq 1/2$) and they form a chain composed of alternating D and A, *i. e.*, ···DADA···. Radicals representative of specific electron configurations (intrinsic or accidental) which may serve as components for the preparation of a molecular/organic ferromagnet are listed in Table 3.

In contrast to the McConnell model[2,9] ions may not be necessary, but stable radicals with strong admixing of virtual charge transfer excited states are required. Homomolecular species (D ≡ A) in principle are sufficient as long as virtual disproportionation (*e. g.*, D^{2+} + $D^o \leftarrow 2\ D^{+\cdot}$)[13] dominates and one of species formed via disproportionation has $m_s > 1/2$, *vide infra.* Since the key point is mixing of an excited state ($m_s^{ES} \geq 1/2$) with a ground state with spin conservation, a chain structure as proposed by McConnell[9] may not be requisite, but designing materials with strong state mixing is important. Organic or inorganic based polymer chain and network structures[6] (albeit not molecular solids[1]) as well as nonchain structures with the proper admixture of excited and ground states should suffice.

Homospin systems For an A with the ground state spin residing in an *s* orbital, ferromagnetic coupling can be stabilized by forward charge transfer from a D with an d^3 or t^5 electron configuration. For retro charge transfer although antiferromagnetic coupling is predicted for the d^3/s^1 electron configuration, ferromagnetic coupling can result if the D possesses a d^1 or t^1 electron configuration. The complex $[Ni^{III}(C_5Me_5)_2]^{+\cdot}[TCNE]^{-\cdot}$ which is isomorphous to the orthorhombic phase of $[Fe^{III}(C_5Me_5)_2]^{+\cdot}[TCNE]^{-\cdot}$ possesses the d^1/s^1 electronic configuration and analysis of the susceptibility data is consistent with antiferromagnetic behavior.[43]

Several homomolecular $m_s = 1/2$ electron configurations (*i. e.*, d^1/d^1, d^3/d^3, t^1/t^1, and t^5/t^5) stabilize ferromagnetic coupling via virtual disproportionation. This is illustrated for the d^3/d^3 case in Fig. 4. For ferromagnetically coupled GS_{FO} the electron with $m_s = -1/2$ can be virtually transferred to the adjacent site which can only accept a $m_s = -1/2$ electron. For the antiferromagnetically coupled GS_{AF} the site can only accept a $m_s = +1/2$ electron; however, excitation of such an electron from the donor would result in the formation of a higher energy excited state. Thus, ES_{FO} is lower energy than ES_{AF} and ferromagnetic coupling is stabilized. Virtual disproportionation between $[Fe(C_5Me_5)_2]^{+\cdot}$'s has been suggested as a mechanism for stabilizing ferromagnetic coupling between in-registry chains and provides an additional mechanism for establishing bulk ferromagnetic behavior as observed for $[Fe^{III}(C_5Me_5)_2]^{+\cdot}[TCNE]^{-\cdot}$.[3,4,12,13] The t^1/d^1 and t^5/d^3 configurations should stabilize ferromagnetic coupling regardless of the direction of virtual charge transfer. Examples of such systems, however, have yet to be identified.

Heterospin systems Heterospin systems, *i. e.*, systems with different spin magnitudes on the donors and acceptors, provide an opportunity to obtain ferrimagnetic as well as ferromagnetic solids. Within this simple nearest neighbor model antiferromagnetic coupling of adjacent spins of the D and A sublattices can only produce incomplete cancellation of the total spin and ferrimagnetic behavior.

$m_sD = 1/2$; $m_sA = 1$. For the lower symmetry *s* and *d* electron configurations two combinations support ferromagnetic coupling, *i. e.*, d^3/d^2 with forward charge transfer and d^1/d^2 with retro charge transfer. Several combinations (d^1/t^2, d^3/t^4, t^1/t^2, and t^5/t^4) are ferromagnetic invariant of direction of electron transfer. Illustrative systems have yet to be identified for these electron configurations.

$m_sD = 1/2$; $m_sA = 3/2$. Several electron configurations, depending on the direction of virtual charge transfer, can stabilize ferromagnetic coupling. The t^3/s^1 configuration expected for $[Cr(C_5Me_5)_2]^+[TCNE]^{\cdot-}$ due to an accidental degeneracy of the cations e_{2g} and a_{1g} orbitals[44] is predicted to exhibit ferrimagnetic coupling for either retro or forward charge transfer. Preliminary magnetic susceptibility data show high susceptibility that is characteristic of either ferri- or ferro-, but not paramagnetic, behavior.[43] Detailed analysis is required to distinguish between these two magnetic states.

Table 3. Electron Configuration of Representative Radicals [a]

EC[b]	S	Radical	Acronym	Symmetry
s^1	1/2	p-$(NC)_2CC_6H_4C(CN)_2]^{-\cdot}$	$[TCNQ]^{-\cdot}$	D_{2h}
		$[C_4(CN)_6]^{-\cdot}$		C_{2v}
		p-$[Cl_2C_6(CN)_2O_2]^{-\cdot}$	$[DDQ]^{-\cdot}$	C_{2h}
		$[CC(CN)_2]_3^{-\cdot}$		D_{3h}
		$[H_2C_2S_2C]_2^{+\cdot}$	$[TTF]^{+\cdot}$	D_{2h}
		$[F_2CCF_2]^{-\cdot}$		C_{2h}
		$[V(C_6H_6)_2]^{\cdot}$		D_6
		p-$[Me_2NC_6H_4NMe_2]^{+\cdot}$	$[TMPD]^{+\cdot}$	D_{2h}
		$[M(S_2C_2R_2)_2]^{z-}$		D_{2h}
		$R_2NO\cdot$		C_{2v}
		$[M(phthalocyanine)]^{+\cdot}$	$[MPc]^{+\cdot}$	D_{4h}

		$[C(NMe_2)]_3{}^{\cdot}$		D_{3h}
		$[Tc_2Cl_8]^{3-}$		D_{4h}
		$[Re_2Cl_4(PR_3)_4]^{+\cdot}$		D_{4h}
		$[(NC)_2CC(CN)_2]^{-\cdot}$	$[TCNE]^{-\cdot}$	D_{2h}
		$[Cr(C_6H_6)_2]^{+}$		D_6
d^1	1/2	$NO^{\cdot}$		$C_{\infty v}$
		$Co(C_5R_5)_2{}^{\cdot}$		D_5
		$[Ni(C_5R_5)_2]^{+\cdot}$		D_5
		[Ru(octaethlyporphyrin)]$^{+\cdot}$	$[Ru(OEP)]_2{}^{+\cdot}$	D_{4d}
		$[O_2]^{+\cdot}$		$D_{\infty h}$
		[M(octacyanophthalocyanine)]$^{-\cdot}$		D_{4h}
		$[Fe(C_6R_6)_2]^{+\cdot}$		D_6
		$[Co(S_2C_4R_4)_2]^{2-}$		D_{3h}
d^2	1	[Ru(octaethlyporphyrin)]	$[Ru(OEP)]_2$	D_{4h}
		O_2		$D_{\infty h}$
		$[C_6(NEt(CH_2))_6]^{2+}$		C_{3h}
		$[C_6(NC_2H_4)_6]^{2+}$		C_{3h}
		$[C_5R_5]^{+}$		D_{5h}
		$[Co(C_6R_6)_2]^{+}$		D_6
		$Ni(C_5H_5)_2$	$NiCp_2$	D_5
		$[Co(S_2C_4R_4)_2]^{-}$		D_{2h}
		$[Fe(S_2C_2(CN)_2)_2]^{2-}$		D_{2h}
d^3	1/2	$[Fe(C_5R_5)_2]^{+\cdot}$		D_5
		$[C_6(NEt(CH_2))_6]^{+\cdot}$		C_{3h}
		$[C_6(NC_2H_4)_6]^{+\cdot}$		C_{3h}
		[Ru(octaethlyporphyrin)]$^{-\cdot}$	$[Ru(OEP)]_2{}^{-\cdot}$	D_{4h}
		$[C(NMe_2)]_3{}^{2+}$		C_{3h}
		$[C_6H_2(OMe)_2]_3{}^{+\cdot}$		C_{3h}
		$[(Me_2N)_2CC(NMe_2)_2]^{+\cdot}$		
		$[(RS)_2CC(SR)_2]^{+\cdot}$		

[a] From reference 2. [b] Electronic configuration.

STABILIZATION OF BULK FERROMAGNETISM

The admixing of a virtual triplet excited state with the ground state for stabilization of ferro- or antiferromagnetic coupling model is limited to the pairwise interactions among the repeat unit of a linear chain, *e. g.*, $[Fe^{III}(C_5Me_5)_2]^{\cdot +}[TCNE]^{\cdot -}$. Bulk ferromagnetism requires inter- *and* intrachain spin alignment.[2,13] Reiterating, mixing of the ES_{FO} with GS_{FO} leads to lowering of the energy to GS'_{FO}, Figs. 3 and 5a. Since the cation is essentially equidistant to a $[TCNE]^{\cdot -}$ above and below it within a chain, virtual transfer of an e_{2g} electron forming the admixable $M_s = 1$ excited state with either $[TCNE]^{\cdot -}$ may occur. Thus, two excited states can mix with the GS_{FO} to further lower the energy to GS''_{FO}, Fig. 5b, leading to intrachain spin alignment. Intrachain spin alignment leads to energy lowering; however, even with complete intrachain spin alignment (*i. e.*, ferromagnetically coupled), if spins on adjacent chains are aligned in the opposite sense, then bulk antiferromagnetic behavior should dominate. If adjacent chains are out-of-registry by one-half the chain axis repeat length, then $[TCNE]^{\cdot -}$'s residing in adjacent chains may be comparably separated from a Fe^{III} site as are the intrachain $[TCNE]^{\cdot -}$'s. Thus, the ES_{FO} on adjacent chains can additionally mix with GS''_{FO}, Fig. 5c, to further lower the energy of the system to ES'''_{FO} leading to the spin alignment throughout the bulk necessary for bulk ferromagnetism.[2,3,12] For in-registry chains the s^1/s^1 interaction (*e. g.*, $[TCNE]^{\cdot -}/[TCNE]^{\cdot -}$) will lead to antiferromagnetic coupling between the chains. The relative magnitude of the resulting interchain ferro- and antiferromagnetic exhange terms will determine if a particular salt will form a 3-D ordered ground state.

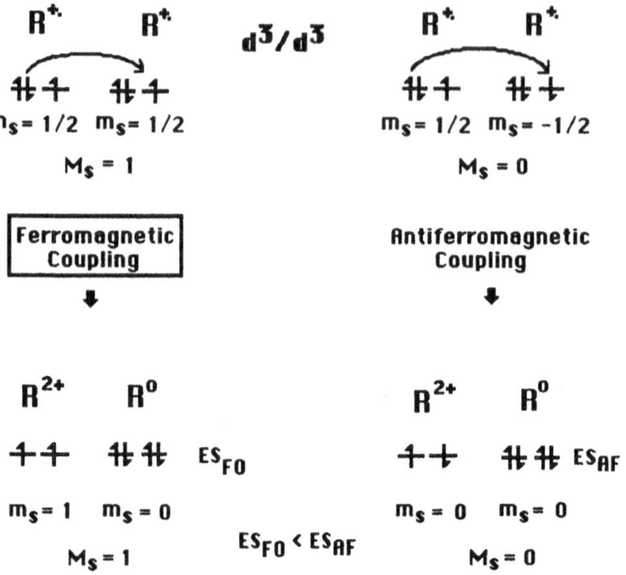

Figure. 4. Schematic illustration of stabilization of ferromagnetic coupling via disproportionation between d^3 radicals, $R^{+\cdot}$. ES_{FO} is stabilized $ES_{FO} < ES_{AF}$.

For in-registry chains virtual disproportionation between interchain Fe^{III}'s sites (*vide supra*) may provide an additional mechanism to align the spins throughout the bulk and contribute to the stabilization of bulk ferromagnetism.[12] These in-registry and out-of-registry arrangements are present in the $[Fe^{III}(C_5Me_5)_2]^{\cdot +}[TCNE]^{\cdot -}$.[13]

GENERAL CONSIDERATIONS

Assuming the virtual charge transfer involves only the highest energy POMO,[45] stabilization of ferromagnetic coupling via the McConnell mechanism requires that the *stable radicals possess a nonhalf-filled degenerate POMO*. Thus, for homomolecular systems ferromagnetic stabilization requires radicals with a d^1, d^3, t^1, t^2, t^4, or t^5 POMO. These radicals must avoid structural/electronic distortions that lower the symmetry and break the degeneracies, *e. g.*, the Jahn-Teller effect; however, accidentally degenerate systems (*e. g.*, high spin transition, lanthanide, and actinide metal coordination complexes) are sufficient. The chains need not be comprised of charged radical donors and acceptors as originally suggested by McConnell,[3] but may be homomolecular radicals or radical ions. For example, chains of radicals [or radical ions with small diamagnetic counterions (*e. g.*, Na^+ or Cl^-)]

possessing a nonhalf-filled degenerate POMO could via virtual disproportionation exhibit ferromagnetic behavior via the McConnell mechanism. These specific combinations of electron configurations are necessary but not sufficient for bulk ferromagnetic behavior[2] via the McConnell mechanism. Opposing effects (*e. g.*, retro vs. forward virtual charge transfer) or magnitude of the stabilization (*e. g.*, inversely proportional to distance and energy difference between the mixing states) may obscure the effect and lead to other phenomena, *e. g.*, para-, meta-, or ferrimagnetism. Additionally, other mechanisms[8,10,11] for molecular based ferromagnetic behavior may be operative.

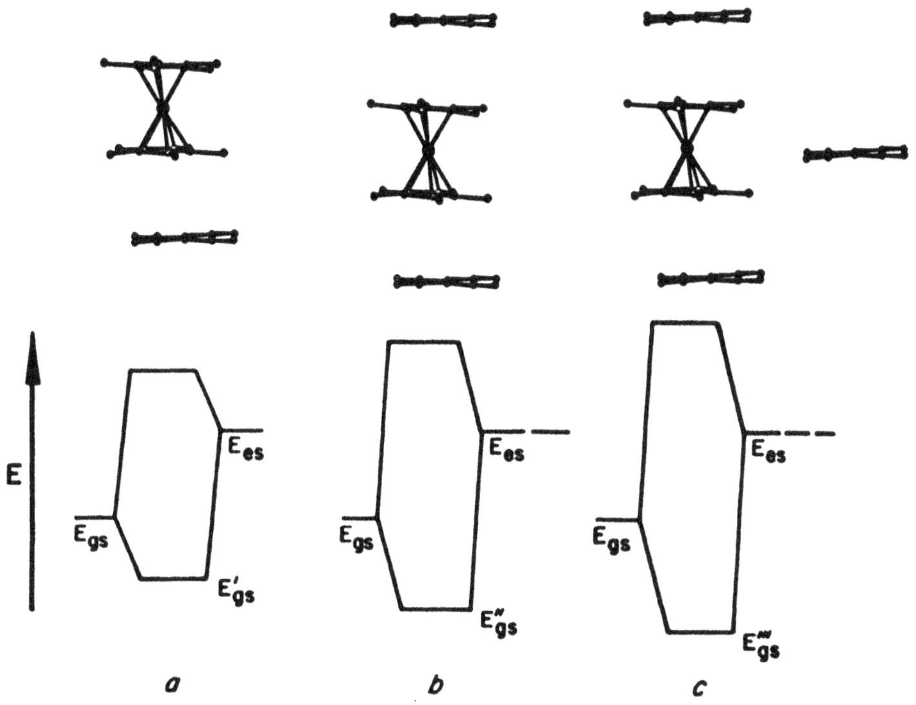

Figure. 5. Schematic illustration of (a) mixing of the GS_{FO} with $m_s = 1$ ES_{FO} to lower the energy to GS'_{FO}; (b) mixing of the GS_{FO} with a pair $m_s = 1$ ES_{FO}'s arising from intrachain interactions to lower the energy to GS''_{FO}; (c) mixing of the GS_{FO} with additional $m_s = 1$ ES_{FO}'s arising from interchain interactions to further lower the energy to GS'''_{FO} which may lead to the macroscopic spin alignment necessary for bulk ferromagnetic behavior.

For radicals with intrinsic POMO degeneracies the relatively high symmetry required by either the D or A, if not both, restricts the choice of compounds to those which belong to specific point groups. For doubly degenerate orbitals the molecule or ion must belong to the D_{2d}, C_3 or higher point groups. For triply degenerate systems the point groups are further restricted to the cubic (*T* and *O*) and I_h groups. As noted above the effective orbital symmetry need not be rigorous and can be accidental. However, unpaired electrons residing in orbitals of different local symmetry may lead to partial or full cancellation of the net magnetic moment and lead to ferri- or antiferromagnetic if not more complex magnetic behavior.

ORGANIC FERROMAGNETS

The desire to prepare an organic ferromagnet remains of academic interest. Many electron configurations that support ferromagnetic coupling via the McConnell mechanism require partial occupancy of triply degenerate orbitals which necessitates a cubic space group or an accidental orbital degeneracy. Both situations are well characterized for transition metal based coordination complexes, but are rare for organic compounds. An *a priori* prediction of accidentally degenerate orbitals is difficult to make except for large low symmetry radicals with several pseudo equivalent radical sites. Except, for example, complex poly-$R_2NO\cdot$ radicals these radicals are neither thermally nor chemically stable. Consequently, stable D_{2d}, C_3 or higher symmetry $m_s = 1/2$ organic radicals with a degenerate POMO[46] are necessary to prepare a ferromagnet based on the McConnell mechanism. Breslow[1b,8] previously pointed out the necessity of a triplet state and has focused research toward the synthesis of stable C_3 or higher symmetry organic triplets.[7,14] Radicals possessing D_{2d} symmetry, albeit rare, in principle may possess a POMO of *d* (e) symmetry and be a suitable candidate as a component of an organic ferromagnet. Hexacyanotrimethylenecyclopropane, **13**, is predicted to have an e^4 HOMO and oxidation should lead to an e^3, d^3, radical cation.[47] Since neutral **13** has never been isolated, derivatives with electron donating groups, *e. g.*, NR_2, to stabilize the radical cation, *e. g.*, **14**, would have to prepared and studied. Alternatively, with the goal of building a solid where adjacent chains are out-of-registry by one-half of a unit cell length as noted for the $[Fe(C_5Me_5)_2]^{\cdot+}[A]^{\cdot-}$ system per*meta*-substituted multilayer cyclophanes,[48] *e. g.*, [2.2.2], **15**, or [3.3.3], are challenging molecules to synthesize. These bulky materials may provide the proper solid state structure enabling bulk ferromagnetic behavior.

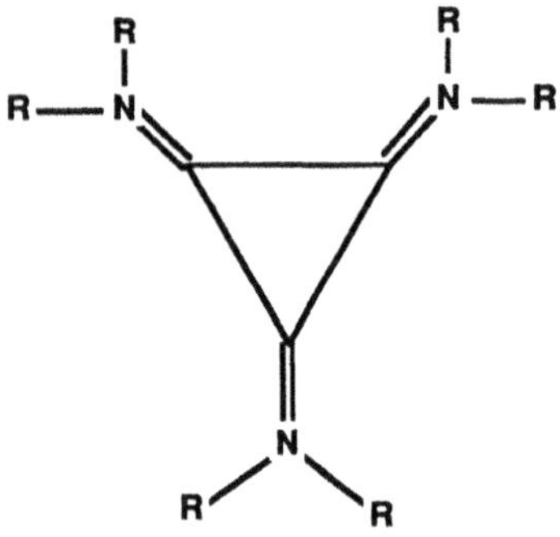

13

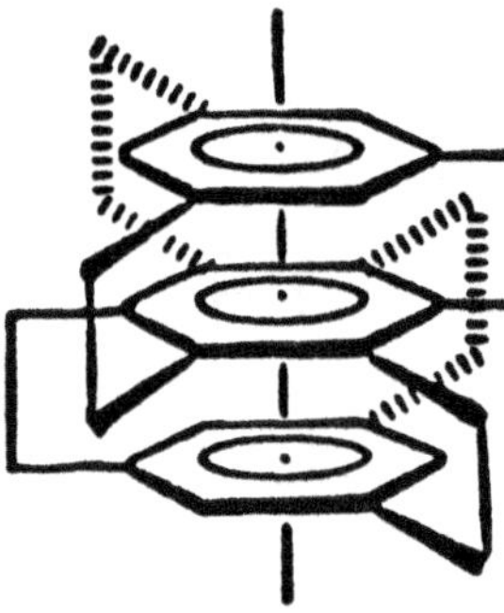

14

SUMMARY

The extended McConnell model and its mathematical embodiment as the generalized Hubbard model offer a convenient guide for the synthetic chemist to prepare new materials for exploration of ferro-, antiferro- and ferrimagnetic phenomena in molecular systems. For a compound to exhibit bulk ferromagnetism via the McConnell mechanism ferromagnetic coupling is necessary. To achieve ferromagnetic coupling in a molecular material (organic, organometallic, main group, polymer, and/or inorganic coordination complex) via the McConnell mechanism at least one type of radical (neutral, radical cations/radical anions, or radical ions with small diamagnetic counterions) must possess a degenerate POMO that is not half-filled and the lowest excited state formed via virtual charge transfer (retro or forward) possesses the same spin multiplicity and mixes with the ground state to stabilize the ferromagnetically coupled ground state. The requirement for a radical to possess an intrinsically degenerate POMO limits the structure of a radical to D_{2d}, C_3 or higher symmetry where symmetry lowering distortions do not occur. Intrinsic doubly or triply degenerate orbitals are not necessary as accidental degeneracies suffice. Since ferromagnetism is a bulk phenomenon, to be achieved the aforementioned ferromagnetic coupling must be established throughout the solid. Extensive chemical syntheses of cleverly designed radicals, as well as physical, experimental and theoretical insight are necessary to test these concepts and establish a deeper understanding of cooperative phenomena in molecular solids.

ACKNOWLEDGMENT

We gratefully acknowledge partial support by the Department of Energy Division of Materials Science (Grant No. DE-FG02-86ER45271.A000).

REFERENCES

1. A molecular solid is comprised of low molecular weight molecules or ions that do not possess extended covalent bonding in the solid state. However, shorter than van der Waal separations may be present in the solid. Dissolution into conventional aqueous or organic solvents leads to solvation of the individual ions or molecules that were used to prepare the molecular solid.[2]
2. Miller, J. S.; Epstein, A. J. *J. Am. Chem. Soc.*, **1987**, *109*, 0000.
3. Miller, J. S., Epstein, A. J., Reiff, W. M. submitted.
4. Miller, J. S.; Epstein, A. J.; Reiff, W. M. *Isr. J. Chem.*, **1987**, in press.
5. Itoh, K.; Takai, K. *Mol. Cryst. Liq. Cryst.*, **1985**, *125*, 251. Iwamura, H. *Pure & Appl. Chem.*, **1986**, *58*, 187.
6. Kahn, O. *Angew. Chem. Int. edit.*, **1985**, *24*, 834-850.
7. Breslow,, R. *Mol. Cryst. Liq. Cryst.* **1985**, *125*, 261-267; Breslow, R.; Maslak,P.;Thomaides, J. S. *J. Am. Chem. Soc.* **1984**, *106*, 6453.
8. McConnell, H. M. *J. Phys. Chem.*, **1963**, *39*, 1910.
9. McConnell, H. M. Proc. *R. A. Welch Found. Chem. Res.*, **1967**, *11*, 144.
10. Mataga, N. *Theor. Chim. Acta*, **1968**, *10*, 372.
11. Ovchinnikov, A. A. *Dok. Nauk Akad. SSSR*, **1977**, *236*, 957-961; *Theor. Chim. Acta*, **1978**, *47*, 297-304.
12. Miller, J. S.; Calabrese, J. C.; Bigelow. R. W.; Epstein, A. J.; Zhang, R. W.; Reiff, W. M. *J. Chem. Soc.,Chem. Commun.*, **1986**, 1026-1028.
13. Miller, J. S.; Calabrese, J. C.; Rommelmann, H.; Chittapeddi, S. R. ; Zhang, R. W.; Reiff, W. M.; Epstein, A. J.,*J. Am. Chem. Soc.* **1987**, *109*, 769-781.
14. Breslow, R. *Pure Appl. Chem.*, **1982**, *54*, 927-938; Breslow, R.; Juan, B.; Klutz, R. Q., Xia, C-Z. *Tetrahedron*, **1982**, *38*, 863-867.
15. Ferromagnetic exchange coupling in a molecular solid is essentially equivalent to stabilization of a high spin, *e. g.*, triplet, state. This is a necessary but insufficient condition for bulk ferromagnetism.
16. Dixon, D. A.; Miller, J. S. *J. Am. Chem. Soc.*, **1987**, *109*, 0000.
17. Miller, J. S.; Zhang, J. H.; Reiff, W. M.; Dixon, D. A.; Preston, L. D.; Reis, A. H. Jr.; Gebert, E.; Extine, M.; Troup, J., Epstein, A. J.; Ward, M. *J. Phys. Chem.* **1987**, *91*, 000.
18. *e. g.*, Herbstein, F. H. *Prespect. Struct. Chem.*, **1977**, *B33*, 2744-2750. Shibaeva, R. P.; Atovmyan, L. O. *J. Struct. Chem.* **1972**, *13*, 514-531.
19. *e. g.*, Reis, A. H., Jr.; Preston, L. D.; Williams, J. M.; Peterson, S. W.; Candela, G. A.; Swartzendruber, L. J.; Miller, J. S. *J. Am. Chem. Soc.*, **1979**, *101*, 2756.
20. Hoffmann, S. K.; Corvan, P. J.; Singh, P.; Sethulekshine, C. N.; Metzger, R. M.; Hatfield, W. E. *J. Am. Chem. Soc.*, **1983**, *105*, 4608; Harms, R. H.; Keller, H. J.; Nothe, D.; Werner, M. *Mol. Cryst. Liq. Cryst.*, **1981**, *65*, 1979.
21. Miller, J. S., Calabrese, J. C. in preparation.
22. Duffy, W., Jr.; Dubach, J. F.; Pianetta, P. A.; Deek, J. F.; Strandburg, D. L.; Miedema, A. R. *J. Chem. Phys.*, **1972**, *56*, 2555-2561. Yamauchi, J. *J. Chem. Phys.* **1977**, *67*, 2850-2855.
23. Kumano, M.; Ikegama, Y.; Sato, T.; Saito, S. *Chem. Phys. Lett.* **1976**, *41*, 354-356. Boucher, P. D.; Nechtschein, M.; Saint-Paul, M. *Phys. Lett.*, **1973**, *42A*, 397-398.
24. Triplet oxygen, O_2, and triplet carbenes have this d^2/d^2 electronic structure (intrinsic and accidental, respectively) and are antiferromagnetic at lower temperature, but this is due to dimerization to $(O_2)_2$. "Comprehensive Inorganic Chemistry," **1973**, *2*, 705.
25. Sugawara, T.; Murata, S.; Kimura, K.; Iwamura, H. *J. Am. Chem. Soc.*, **1985**, *107*, 5293-5294. Sugawara, T.; Bandow, S.; Kimura, K.; Iwamura, H. *J. Am. Chem. Soc.*, **1984**, *106*, 6449-6450. Sugawara, T.; Bandow, S.; Kimura, K.; Iwamura, H.; Itoh, K. *J. Am. Chem. Soc.*, **1986**, *108*, 368-371.

26. Izoka, A.; Murata, S.; Sugawara, J.; Iwamura, H. *J. Am. Chem. Soc.*, **1985**, *107*, 1786-1787.
27. Yamaguchi, K.; Fukui, H.; Fueno, T. *Chem. Lett.*, **1986**, 625-628; Yamaguchi, K.; Fueno, T.; Nakasuji, K.; Murata, S. *Chem. Lett.*, **1986**, 629-637.
28. Sugawara, T.; Tukada, H.; Izuoka, A.; Iwamura, H. *J. Am. Chem. Soc.*, **1986**, *108*, 4272-4278.
29. Marata, S.; Sugawara, T., Iwamura, H. *J. Am. Chem. Soc.* **1987**, *109*, 66-126.
30. Hoffmann, R.; Zeiss, G. D.; Van Dine, G. W. *J. Am. Chem. Soc.* **1968**, 901, 1485-1499. Humphreys, R. W. R.; Arnold, D. R. *Can. J. Chem.*, **1976**, *55*, 2286-2291.
31. Dowd, P. *Acc. Chem. Res.*, **1972**, *5*, 242.
32. Clar, E.; Kemp, W.; Stewart, D. G. *Tetrahedron*, **1958**, *3*, 325-333.
33. Klein, D. J.; Nelin, C. J.; Alexander, S.; Matsen, F. A. *J. Chem. Phys.* **1982**, *77*, 3101-3108.
34. Kouvetakis, J., Kaner, R. B., Sattler, M. L.; Bartlett, N. *J. Chem. Soc., Chem. Commun.*, **1986**, 1758-1759.
35. Torrance, J. B.; Oostra, S.; Nazzal, A. *Syn. Met.* **1987**, *19*, 708.
36. Korshak, Yu. V.; Ovchinnikov, A. A.; Shapiro, A. M.; Medvedeva, T. V.; Specktor, V. N. *Pisma Zh. Eksp. Teor. Fiz*, **1986**, *43*, 309-311. Korshak, Yu. V.; Medvedeva, T. V.; Ovchinnikov, A. A.; Specktor, V. N. *Nature*, **1987**, *326*, 370-372.
37. Throughout the paper the arguments hold for either D/A or the reverse A/D. For convenience D and A are assumed to be neutral; however, they may be ionic.
38. Purcell, K. F.; Kotz, J. C. "Inorganic Chemistry," W. B. Saunders & Co., **1977**, (a) p51-52; (b) p147-148.
39. Hund's rule was originally proposed for atoms (*e. g.*, McWeeny, R."Coulson's Valence," Oxford Univ. Press, **1979**, p102-103); however, it was later successfully applied to molecules and biradicals (*e. g.*, Lahti, P. M.;Rossi, A.; Berson, J. A. *J. Am. Chem. Soc.* **1985**, *107*, 4362-4363; Seeger, D.E.; Lahti, P. M.; Rossi, A.; Berson, J. A. *J. Am. Chem. Soc.* **1986** *108*, 1251-1265.
40. Throughout the paper the singly and doubly orbital degeneracies are denoted as *s* (a or b symmetry), or *d* (e symmetry).
41. The ferromagnetic charge transfer complex $[Fe^{III}(C_5Me_5)_2]^{+\cdot}[TCNE]^{-\cdot}$ [3,4,12,13] possesses the d^3/s^1 electron configuration.
42. Hubbard, J. *Proc. Roy. Soc. (London)* **1963**, *A276*, 238; **1963**, *A277*, 237; **1964**, *A281*, 401; Kanamori, J. *Prog. Theor. Phys. (Kyoto)* **1963**, *30*, 275; Beni, G.; Pincus, P.; Hone, D. *Phys. Rev. B* **1973**, *B8*, 3389; Lyon-Caen, C.; Cyrot, M. *J. Phys. C*, **1975**, *8*, 2091.
43. Miller, J. S.; Epstein, A. J. unpublished.
44. Robbins, J. L.; Edelstein, N.; Spencer, B.; Smart, J. C. *J. Am. Chem Soc.*, **1982**, *104*,1882-1893. Cauletta, C.; Green, J. R.; Kelly, M. R.; Powell, P.; van Tilborg, J.; Robbins, J.; Smart, J. C. *J. Elect. Spectro. Rel. Phen.* **1980**, *19*, 327-353. Evans, S.; Green, M. L. H.; Jewitt, B.; King, G. H.; Orchard, A. F. *J. Chem. Soc., Faraday Trans. II*, **1974**, *70*, 356-376. Gordon. K. R.; Warren, K. D. *Inorg. Chem.* **1978**, *17*, 987-994.
45. Circumstances where virtual excitation from a lower lying filled orbital dominate the admixing exciting state are conceivable. For example, $[Cr^{I}(C_6Me_6)_2]^{+\cdot}[TCNE]^{\cdot-}$ has a s^1/s^1 ground state electronic configuration and preliminary susceptibility data can be fit by the Curie-Weiss expression with $\theta = +12$ K suggesting dominant ferromagnetic interactions, O'Hare, D. M.; Miller, J. S. unpublished results.
46. Previous work[2,3,13] suggests that a stable triplet may not be a necessary component for an organic ferromagnet; a stable doublet with a virtually accessible triplet capable of admixing with ground state as observed for $[Fe^{III}(C_5Me_5)_2]^{+\cdot}[TCNE]^{-\cdot}$ should suffice.
47. Fukunaga, T. *J. Am. Chem. Soc.* **1976**, *98*, 610-611; Fukunaga, T.; Gordon, M. D.; Krusic, P. J. *J. Am. Chem. Soc.* **1976**, *98*, 611-613.
48. Huber, A. J. *J. Chem. Soc.* (C), **1967**, 13-14. Toyoda, T.; Tatemitsu, H.; Sakata, Y.; Kasai, N.; Mitsumi, S. *Chem. Soc. Jap.*, **1986**, *59*, 3994-3996.

THEORY OF THE ONE- AND TWO-DIMENSIONAL ELECTRON GAS

V. J. Emery*

Physics Department
Brookhaven National Laboratory
Upton, New York 11973

INTRODUCTION

The study of organic conductors and superconductors has always been intimately associated with low-dimensional physics: polymers or compounds involving planar molecules have anisotropic structure and associated anisotropic physical properties. Frequently, there is a temperature range over which the behavior is essentially one-dimensional, since thermal broadening is sufficient to mask any departure from the planar Fermi surface which is the hallmark of one-dimensionality. Below a crossover temperature T_x there may be a region of two-dimensionality before all aspects of the crystal structure finally make themselves felt. As a consequence, the past ten to fifteen years have seen an extensive development of the theory of the one-dimensional electron gas[1-4]: by now we have a detailed understanding of a wide variety of mathematical models, and the analytical or numerical techniques are available for the study of new models, as they become relevant. It has steadily become clear that, although quasi one-dimensional systems are well-suited to the study of structural phase transitions, it is much easier to find superconductivity, and particularly high-temperature superconductivity, in higher dimensions. The arguments leading to this conclusion involve structure in time, momentum space and real space: in other words retardation, nesting and angular momentum.

In most models of superconductivity, pairing is produced by an attractive interaction which comes about because the charge carriers polarize their surroundings--the (relatively large) organic molecules, the lattice or the spin or charge density of the other carriers. Typically, the induced interaction is weaker than the Coulomb repulsion but it can be effective because the polarization takes time to relax. In three dimensions, this "retardation" leads to superconductivity but in one dimension it serves to enhance the formation of charge-density waves[3,4]. In particular phonon exchange[3] and Little's exitonic mechanism[4,5] both suffer from this effect. The upshot is that, for retardation to lead to superconductivity, the crossover temperature T_x should be greater than the Debye energy of the phonons or the analogous characteristic energy of other modes responsible for the effective interaction. In this sense, superconductivity always is a two- or three-dimensional phenomenon. (This is quite apart from the fact that there is no long-range order at finite temperatures in one dimension.)

The charge- or spin-density wave states which compete with superconductivity have a characteristic wave vector q which should span a significant portion of the Fermi surface in order to provoke an electronic instability: the Fermi surface should be "nested." This condition is relatively easy to satisfy for the nearly flat Fermi surface of a quasi one-dimensional system but otherwise is quite exceptional. Thus higher dimensionality gives superconductivity a considerable edge in the competition with structural phase transitions.

Finally there is the possibility of new mechanisms of superconductivity or new pairing states when the constrictions of one-dimensional motion are removed. One example is the exchange of spin-density wave fluctuations, which may be relevant for organic superconductors[6]. The effective interaction oscillates in real space, and the motion of two electrons must have a specific impact parameter in order to take advantage of the regions of attraction. This may be accomplished by pairing electrons or holes on different organic stacks or by constructing a pair state with finite angular momentum[6]. Neither possibility is available in one-dimension. All of these considerations may come together when the exchange of soft modes related to a nearby structural phase transition produce a retarded interaction which leads to pairing with a finite impact parameter. This is the case for spin fluctuations in organic superconductors[6].

Given all of this, it is hardly surprising that the organic superconductors[7] are much less anisotropic than the previously studied organic conductors, and that the ET compounds with a T_c of 8K are more two-dimensional than the TMTSF compounds for which T_c is about 1K. This situation alone makes it clear that, in the future, the two-dimensional electron gas will receive much more attention from theorists. However, the recent discovery of high T_c superconductivity in oxides, with their quasi two-dimensional CuO_2 planes, guarantees that this will be the case.

In line with these remarks, the remainder of these lectures will be concerned with two problems (a) the competition between $2k_F$ and $4k_F$ charge-density waves in a one-dimensional electron gas and (b) a two-dimensional model of high T_c superconductivity in the oxides.

$2k_F$ and $4k_F$ CHARGE-DENSITY WAVES

Over the past ten years, there has been an extensive investigation of materials which show $4k_F$ charge-density waves, sometimes in competition with the more conventional variety which have wave vector $2k_F$. This work is surveyed in the lectures of Dr. Pouget at this workshop[8]. There is general agreement that $2k_F$ charge-density waves are a consequence of a Fermi surface instability, a weak coupling effect, whereas $4k_F$ charge-density waves are essentially the first Bragg peak of a Wigner crystal and require strong coupling[1,2,3]. However, there is relatively little understanding of the crossover from one to the other as the coupling constants or other parameters of the system are changed. This is not a purely academic question: some systems, such as TTF-TCNQ display both effects—$4k_F$ appears to dominate at high temperature, but $2k_F$ is the first to condense into an ordered state[8]. Clearly, an understanding of the competition between the two phenomena is essential for a full description of the phase diagram of such materials.

In order to study this problem[9], first imagine an extended Hubbard model with transfer integral t, on-site interaction U, and interactions V_1 and V_2 between holes on first and second neighboring sites. As a first step, we shall consider the $U \to \infty$ limit, which essentially removes the spin degrees of freedom from active consideration, since it is no longer possible to have two electrons on the same site[1]. This is an advantage, since it allows us to focus on the charge-density which is our immediate concern. As is well known[1], the resulting Hamiltonian can be rewritten as a spin chain

$$H = \sum_n \{-t(S^x_n S^x_{n+1} + S^y_n S^y_{n+1}) + V_1 S^z_n S^z_{n+1} + V_2 S^z_n S^z_{n+2}\} \quad (1)$$

where the $\underset{\sim}{S}_n$ are Pauli matrix vectors and, physically, a spin is up ($S^z_n = +1$) if site n is occupied or down ($S^z_n = -1$) if site n is unoccupied. Equation (1) describes a Heisenberg-Ising model with second neighbor interactions. In the realm of spin chains, the notation (J_{xy}, J_1, J_2) would be more appropriate for (t, V_1, V_2), but the latter is a better reflection of the Hubbard model convention. The repulsive interactions V_1 and V_2 correspond to antiferromagnetic coupling, since they discourage occupancy of the two sites that they couple--i.e. they discourage parallel spins. They compete since V_1 would like all even sites or all odd sites to have the same spin, whereas V_2 favors opposite spins. In the absence of a magnetic field, the mean magnetization is zero, which corresponds to an average of half an electron per site. The the spin analog of a $4k_F$ charge-density wave is antiferromagnetic order (with periodicity two lattice sites) and a $2k_F$ charge-density wave is equivalent to (2,2) order whose extreme form is (+,+,-,-,+,+,-,-,....) (with periodicity four lattice sites). Indeed, when $t = 0$, it is easy to show that the ground state of H is antiferromagnetic for $J_1 > 2J_2$ and a (2,2) phase for $J_1 < 2J_2$. The model can be solved exactly along another boundary[10], $V_2 = 0$. There it is the usual Heisenberg-Ising model which has long-range antiferromagnetic order for $V_1 > t$ but power law correlation functions (as at a critical point) otherwise. The problem then is to work out the behavior when both t and V_2 are nonzero. We expect that for larger t, there is a region of values of V_1 and V_2 with critical point behavior. This is indeed the case. Although, so far as we know, there is no exact analytical solution, it is possible to work out the characteristic (fixed point) Hamiltonian which describes the critical region. But to match this to original Hamiltonian and hence to find critical exponents and phase boundaries, it is necessary to do a numerical calculation. The most appropriate method is to diagonalize finite chains of N sites and to use renormalization group arguments to take the $N \to \infty$ limit. A detailed description of this technique and of the method of exploring the ordered phases is given in reference 9. The important conclusion of the $4k_F$ - $2k_F$ crossover is that there are two kinds of $4k_F$ order. At small t, $2k_F$ and $4k_F$ density waves have their maximum amplitude on the lattice sites. But there may also be a shift of phase by π so that the peak lies on a bond, midway between two sites. What happens is that a region of $4k_F$ bond order separates the $2k_F$ site-ordered phase from the $4k_F$ site-ordered phase and from the critical region. The $4k_F$ bond order extends into the $2k_F$ phase where they coexist until the latter finally dominates at $t = 0$.

There is a natural physical interpretation of this picture, which is easy to visualize by drawing the electron lattices as shown in reference 3. A direct transition from $4k_F$ to $2k_F$ site order would require a uniform shift of half of the electrons, all even ones or all odd ones. What actually happens as t decreases is that a bond $4k_F$ charge-density wave is unstable to a (period doubling) dimerization which produces the $2k_F$ site order.

HIGH T_c SUPERCONDUCTIVITY IN OXIDES

The discovery of high T_c superconductivity in doped La_2CuO_4, $YBa_2Cu_3O_{7-\delta}$ and related compounds[11-13] has provoked an unprecedented burst of research into the properties of these materials and a search for new materials of even higher transition temperature. They are relatively easy to make (although good samples are not so easy to come by) and laboratories all over the world are equipped to make some kind of measurement of relevance for superconductivity.

For theorists, the challenge is to understand the mechanisms--the source of the attractive interaction and the details of pairing with such a large binding energy. BCS theory[14] is well-established as a good description of superconductivity in virtually all metals. Over the years, this has come to mean two things: (a) pairing is produced by electron-phonon coupling (b) the BCS approximation to many-body theory, in which the electrons condense at T_c into pairs with zero total momentum, is a valid description. Although (a) has been questioned on a number of occasions, it has always proven to be relevant for metals. At the present time, there is reason to believe that another mechanism may be appropriate for organic superconductors[6] and heavy fermion superconductors[15], but this has yet to be established to the satisfaction of all concerned. Another interesting system is liquid He^3, which is a fermion superfluid very much like a superconductor except that the He^3 atoms are uncharged[16]: phonon exchange is irrelevant because there is no lattice, and pairing is a consequence of the combined effect of the Van der Waals interaction between He^3 atoms and the exchange of ferromagnetic spin fluctuations. The latter, which are important because He^3 is an almost ferromagnetic Fermi liquid, have the distinctive consequence that the pairs are spin triplets rather than singlets as in metals.

The second element of BCS theory, Cooper pairing, is valid for liquid He^3 but has been questioned for the high T_c oxides because Δ/E_F (the gap divided by the Fermi energy) is not exceptionally small. This ratio can be as low as 10^{-4} in metals and 10^{-2} in He^3. However much larger gaps have been reported for the oxides[17], although Δ/E_F should still be a good expansion parameter in most cases. If BCS theory proved to be inapplicable, the most likely alternative is that the attractive interaction is strong enough to produce real pairs (with negative energy) at a relatively high temperature. The pairs are bosons which undergo a separate superfluid transition at T_c. As we shall see, this could happen at sufficiently low carrier density in the oxides.

The common feature of doped La_2CuO_4 and $YBa_2Cu_3O_{7-y}$ are quasi one-dimensional CuO_2 planes, shown in Fig. 1. Experiments suggest that the superconductivity is mainly associated with these planes and that the surroundings mainly serve to control the carrier density. The simplest representation is to start with a basic state (or vacuum) consisting of Cu^+, with all 3d states occupied and O^{2-} which has filled 2p states. The number of holes per unit cell will be denoted by $1 + \delta$, where δ is determined by doping, oxygen defects and the charge states or band structure of the environment. We shall mainly be concerned with doped La_2CuO_4. Simple valence arguments suggest that the undoped material contains La^{3+}, Cu^{2+} and O^{2-} so that $\delta = 0$. This picture is much too simple to describe what actually happens, but it is a good starting point for estimating the carrier density. The Hamiltonian is assumed to be an extended Hubbard model with the following parameters[18]:

$\varepsilon_p, \varepsilon_d$ are single site energies for O(2p) and Cu(3d) holes

U_p, U_d are interaction energies of two holes on an O(2p) or Cu(3d) site

V is the interaction between holes on neighboring Cu,O sites

t is the hopping parameter characterizing hybridization of the $Cu(d_{x^2-y^2})$ and $O(2p_x)$ or $O(2p_y)$ orbitals.

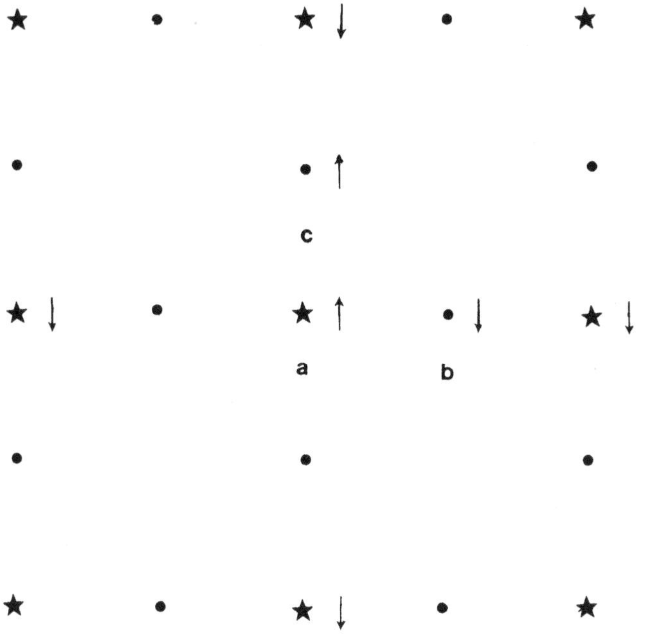

Fig. 1 The structure of the CuO_2 planes. The stars indicate Cu sites and the dots oxygen sites. The particular configuration of spins is discussed in the text.

The values of the parameters are not well-known but to be specific, it will be assumed[18] that $t = 1.3 - 1.5$ eV, $\varepsilon \equiv \varepsilon_p - \varepsilon_d = 1$ eV, $U_p = 1\text{-}3$ eV, $U_d = 5\text{-}6$ eV and $V = 1\text{-}2$ eV.

Consider first the case $\delta = 0$. When $t = 0$, there is exactly one hole per Cu site (Cu^{2+}) since $\varepsilon > 0$. The holes move by hopping via an oxygen atom with an effective hopping parameter $t_d \equiv t^2/(\varepsilon+V)$ to give an effective Cu Hubbard model:

$$H_d = t_d \sum_{\substack{(i,j)\\ \sigma}} d^+_{i\sigma} d_{j\sigma} + U_d \sum_i d^+_{i\sigma} d_{i+} d^+_{i-} d_{i-} \tag{2}$$

where $d^+_{i\sigma}$ creates a hole with spins $\sigma = \pm$ in the $d_{x^2-y^2}$ level of cell i, and the sum over (i,j) is restricted to neighboring cells. There have been several studies of this model, the most recent a Monte Carlo calculation by Hirsch[19]. The ground state is an antiferromagnetic insulator but at finite temperature the system has long range order only if there is weak hopping between the planes. At $|t_d/U_d|$ increases, the holes delocalize and their effective moment is diminished. Recently it has been established[20,21] that there is indeed antiferromagnetic order below 250K in La_2CuO_{4-y}, which lends support to the essential correctness of the approach[22].

The band picture regards hopping as the most important process: three sites per cell lead to three bands of which the lowest is half-filled when $\delta = 0$. The system is then susceptible to electronically-driven structural phase transitions, particularly an oxygen breathing mode, which is the most effective for varying the value of t. This is an essential ingredient in obtaining high T_c by phonon exchange[23]. Experimentally, a number of structural phase transitions have been documented but none so far has been clearly attributable to interaction with the conduction electrons, and there is little support so far for this aspect of the phonon exchange mechanism. Indeed, the existence of antiferromagnetic order may be regarded as evidence to the contrary. With repulsive interactions, the charge-density wave related to the structural distortion has its maximum amplitude on the bonds rather than the sites[4], as described in Sec. 2, and the situation is qualitatively like a spin-Peierls transition[4], as described in the Coulon's lectures[24]. But this evidently has been preempted by antiferromagnetism, which might be expected because the exchange integral, which should be a few hundred degrees, is smaller than the Debye energy of the oxygen modes[24]. An oxygen breathing mode does not modulate t_d so, in the strong-coupling picture which leads to Eq. (2), it is more natural to rephrase the whole argument in terms of Cu phonons[25]--but the conclusion is the same.

Now suppose that holes are added by doping with Sr or Ba. They will go into $2p_x$ or $2p_y$ states of the oxygen, provided the site energy ε_p lies within the gap in the spectrum of H_d. For t = 0 and a finite range of values of t, this requires that $U_d > \varepsilon + 2V$, since the gap is proportional to U_d for intermediate and strong coupling. Experiments on doped La_2CuO_4 strongly support this picture. In particular, an X-ray absorption edge study[26] of $La_{2-x}(Ba,Sr)_xCuO_4$ found that the copper remained in the Cu^{2+} state for all x in the range $0 \leq x \leq 0.3$ and that changes in the La L_3 edge with doping were consistent with variation of the oxygen charge. Thus, the charge carriers which are responsible for superconductivity are holes in the O(2p) states produced by doping. Their density and Fermi energy are quite low.

The O(2p) holes move by hopping via the Cu sites, and it is straightforward to obtain the effective Hamiltonian to second order in t, treating the Cu(3d) holes exactly. Assuming translational invariance on the Cu, it is found that[18]:

$$H_p = \sum_{k\sigma} t^2 e_{\tilde{k}} G_{\tilde{k}}(\varepsilon + 2V - \mu) b^{+}_{\tilde{k}\sigma} b_{\tilde{k}\sigma} \tag{3}$$

where $G_{\tilde{k}}(\omega)$ is the one-particle Green's function and μ the chemical potential of the Cu(3d) holes,

$$e_{\tilde{k}} = 2(2 - \cos k_x a - \cos k_y a) \tag{4}$$

$\underline{a}$ is the lattice spacing and $b_{\tilde{k}\sigma}$ annihilates a hole in an oxygen state of momentum $\tilde{k}$ and spin σ. The energy spectrum is partly determined by the shape of $G_{\tilde{k}}$. For weak to intermediate coupling, $G_{\tilde{k}}$ varies rapidly near the ideal Fermi surface S_F on a scale determined by the mean free path or localization length. In that case, $e_k G_k$ has a minimum just inside S_F. For large U_d, G_k varies slowly with $\tilde{k}$ and the minimum in the spectrum occurs at the zone corners, where e_k is a maximum. Experiment supports the intermediate coupling picture[18], although this may change as more evidence comes in.

Within this picture, there is a natural way of obtaining a strong attraction between O(2p) holes. In perturbation theory in t, there is an attractive exchange interaction between opposite spins, coming from a process in which one spin hops onto the other's site and one or other hops back[1]. The interaction is $O(t^{2n})$ where n is the number of bonds between the two sites. The essential point is that the O(2p) - Cu(3d) exchange

J_{CO} (which is $O(t^2)$) is much stronger than the Cu(3d) - Cu(3d) exchange J_{CC} (which is $O(t^4)$). The first consequence is that antiferromagnetism is rapidly destroyed by doping. The O(2p) holes act as a set of annealed, random, strongly coupled magnetic impurities, and remove the long-range order when $J_{CO}\delta \gtrsim J_{CC}$. Consider now the situation shown in Fig. 1. The O(2p) holes at b and c may be exchanged by first interchanging the holes at a and b and then interchanging those at a and c (which now have opposite spin). The net result is that the spins at b and c have been reversed by exchange of a spin-1 excitation on the Cu, which is a local, strong-coupling version of spin-density wave exchange[6,15]. It is important to note that this process did not require two holes to be on the same O(2p) site, so there is no restriction in the motion of O(2p) holes which may increase their energy.

It is important to notice that each oxygen hole is coupled to six neighbors by this exchange process which helps to obtain[18] attractive interaction of the order of 1 eV. Then the form of pair wave function depends upon the value of U_p and the location of the Fermi surface[18]. Large values of U_p favor d-state pairing[15] for which the gap $\Delta_{\underset{\sim}{k}}$ is given by

$$\Delta_{\underset{\sim}{k}} = \Delta_0(\cos k_x a - \cos k_y a) \qquad (5)$$

But if U_p is smaller than about 1 eV, a uniform gap, independent of $\underset{\sim}{k}$, would give a lower energy. This is particularly true when U_d is large and the Fermi surface is at the zone corners. (Note that the extended s-state, for which $\Delta'_{\underset{\sim}{k}} \sim (\cos k_x a + \cos k_y a)$ is not very different from a uniform gap, since $\Delta'_{\underset{\sim}{k}}$ is constant on the Fermi surface). Another possibility is to construct a wave function pairing holes on different oxygen sites within a cell, to give a uniform gap. In any event, it is not difficult to choose the parameters of the model to obtain transition temperatures of the required order of magnitude for the high T_c oxides.

The nature of the many-body state depends on the density of carriers. In two dimensions even for a spectrum $\varepsilon_k \sim k^2$, the density of states is constant for small k, and an arbitrarily weak interaction will produce a bound state of two holes[27]. As the density increases, the ground state should first be a Bose condensate of these "real" pairs[27] and ultimately will cross over to the BCS state[14] which is a condensate of Cooper pairs. The presence of real pairs should be clearly visible in the normal state properties--for example, the susceptibility should be activated. Up to the present, the most extensive investigations have been concerned with samples displaying the highest T_c: $La_{2-x}Sr_xCuO_4$ with x ~ .15 and $YBa_2Cu_3O_{7-y}$ with y close to zero. In these cases, the evidence favors the BCS picture. However, it would be interesting to study materials with lower carrier density (smaller x or larger y) to seek evidence of crossover to "real" pairing.

In summary, the model of high T_c superconductors, presented here, has a low density low Fermi energy system of change carriers, which are holes with most of their weight on the oxygen sites. The high transition temperature is a consequence of a strong coupling produced by exchange of spin fluctuations, largely on Cu atoms. The strength of this coupling is usually small[6,15]. But the copper oxide superconductors are exceptional because of their electronic structure ($\varepsilon_p \approx \varepsilon_d$) and lattice structure (oxygen on bridge sites between the Cu atoms).

---

*Supported the by Division of Materials Sciences, U.S. Department of Energy, under contract DE-AC02-76CH00016.

REFERENCES

1. V. J. Emery in Highly Conducting One-dimensional Solids, edited by J. T. DeVreese, R. P. Evrard and V. E. Van Doren (Plenum, New York 1979).
2. J. Solyom, Adv. Phys. 28, 201 (1979).
3. V. J. Emery, NATO Advanced Study Institute, Magog, Canada, edited by L. Caron and D. Jérome (Plenum, New York, 1987).
4. J. E. Hirsch, ibid.
5. W. A. Little, Phys. Rev. 134, A1416 (1964); J. E. Hirsch and D. J. Scalapino, Phys. Rev. B32, 117 (1985).
6. V. J. Emery, J. Phys. (Paris) Colloq. 44, C3-977 (1983); Synth. Met. 13, 21 (1986).
7. D. Schweitzer, this volume.
8. J-P. Pouget, this volume.
9. V. J. Emery and C. Noguera, to be published.
10. J. D. Johnson, S. Krinsky and B. M. McCoy, Phys. Rev. A8, 2526 (1973) and references therein.
11. J. G. Bednorz and K. A. Müller, Z. Phys. B64, 189 (1986).
12. S. Uchida, H. Takagi, K. Kitazawa and S. Tanaka, Jpn. J. Appl. Phys. Lett. 26, L1 (1987); C. W. Chu, P. H. Hor, R. L. Meng, L. Gao, Z. J. Huang and Y. Q. Wang, Phys. Rev. Lett. 58, 405 (1987); R. J. Cava, R. B. van Dover, B. Batlogg and E. A. Rietman, Phys. Rev. Lett. 58, 408 (1987).
13. M. K. Wu, J. R. Ashburn, C. J. Tomg, P. H. Hor, R. L. Meng, L. Gao, Z. H. Huang, Y. Q. Wang, and C. W. Chu, Phys. Rev. Lett. 58, 908-10 (1987); R. J. Cava, B. Batlogg, R. B. Van Dover, D. W. Murphy, S. Sunshine, T. Siegrist, J. P. Remeika, E. A. Rietman, S. Zahurak and G. P. Espinosa, Phys. Rev. Lett. 58, 1676 (1987).
14. J. Bardeen, L. N. Cooper and J. R. Schrieffer, Phys. Rev. 108, 1175 (1957).
15. D. J. Scalapino, E. Loh, Jr. and J. E. Hirsch, Phys. Rev. B34, 8190 (1986); K. Miyake, S. Schmitt-Rink and C. M. Varma, Phys. Rev. B34, 6554 (1986); M. T. Béal-Monod, C. Bourbonnais and V. J. Emery, Phys. Rev. B34, 7716 (1986).
16. A. J. Leggett, J. C. Wheatley, Rev. Mod. Phys. 47, 331 (1975).
17. M. Naito, D. P. E. Smith, M. D. Kirk, B. Oh, M. R. Hahn, K. Char, D. B. Mitzi, J. Z. Sun, D. J. Webb, M. R. Beasley, O. Fischer, T. H. Geballe, R. H. Hammond, A. Kapitulnik and C. P. Quate, to be published.
18. V. J. Emery, Phys. Rev. Lett. (to be published).
19. J. Hirsch, Phys. Rev. B31, 4403 (1985).
20. R. L. Greene, H. Maletta, T. S. Plaskett, J. G. Bednorz and K. A. Mueller (to be published).
21. D. Vaknin, S. K. Sinha, D. E. Moncton, D. C. Johnston, J. Newsam and H. King (to be published); T. Freltoft, G. Shirane, J. Remeika, A. S. Cooper, G. Aeppli, D. Harshman, D. E. Moncton, S. K. Sinha, J. E. Fischer and D. Vaknin (to be published).
22. La_2CuO_{4-y} is superconducting for small y and becomes antiferromagnetic as y increases. Possible explanations are that there are lanthanum defects or that the La(5d) level dips below the Fermi level[18], either of which would give rise to holes in the half-filled band expected for y = 0 on the basis of simple valence counting arguments. In that case removal of oxygen would add electrons to the CuO_2 planes, removing the excess of holes and restoring the half-filled band necessary for antiferromagnetism. This situation does not spoil the picture that the density of carriers is $n=1+\delta$ with δ small but it does complicate the attempt to make a correspondence between theory and experiment. (I thank P. Day and J. Torrance for a discussion of the role of lanthanum defects.)
23. W. Weber, Phys. Rev. Lett. 58, 1371 (1987).
24. C. Coulon, this volume.

25. S. Barišić, I. Batistić and J. Friedel, to be published.
26. J. Tranquada, S. M. Heald, A. Moodenbaugh and M. Suenaga, Phys. Rev. B (May 1987).
27. The role of real pairs has been discussed in the context of nuclear physics (V. J. Emery, Nuclear Physics 12, 69 (1959)) and one-dimensional conductors (reference 1, above).

ANION PERIODICITY AND ORDERING EFFECTS IN ONE DIMENSIONAL ORGANIC CONDUCTORS

J.P. Pouget

Laboratoire de Physique des Solides, associé au C.N.R.S.
Université Paris-Sud, F 91405 Orsay

INTRODUCTION

These last years, with the discovery of the organic superconductivity, a considerable number of studies have been devoted to the series $(TMTSF)_2X$ and $(BEDTTTF)_2X$, where X is a monovalent anion. During them, it was found that a sizeable coupling occurs between the anion sublattice and the organic stack. This coupling can be very effective when the charge periodicity of the anion induces a $2k_F$ or $4k_F$ charge density wave (CDW) response of the organic stack. The nature and strength of this response condition the electronic ground state of the 1D metallic chain. Especially, interesting situations occur when the anion sublattice possesses additional degrees of freedom whose ordering in temperature or pressure either change the strength of a preexisting coupling or add new critical periodicities in the system.

The purpose of this lecture is to give a brief summary of such effects with a special attention to their structural counterpart. We shall first review in part I the basic features of the CDW instability of an organic stack. Then, in part II, the effect of the charge periodicity of the anion sublattice will be considered. Finally, the effect of additional periodicities due to the ordering of extra degrees of freedom, like the orientation of non centrosymmetrical anions, will be the object of part III.

I. CDW INSTABILITIES OF THE ORGANIC STACKS

Among all the instabilities shown by 1D conductors, the charge density wave (CDW) instability is the most frequently observed. It occurs particularily in charge transfer salts composed of two electronic (donor and acceptor) subsystems like TTF-TCNQ. These salts always exhibit a low temperature periodic lattice distortion (at $T_P \sim 50K$) due to the electron-phonon coupling of the lattice with the CDW instability of the 1D electron gas. This transition is announced by a sizeable regime of 1D structural fluctuations (extending over a temperature range of more than 100K) which allows a detailed study of the intrastack CDW instability[1]. The relevant physical parameters at the basis of the CDW instability are summarized by fig. 1 : the displacement of the molecules from their steady position (described by U_q, where q is the in chain wave vector of the deformation

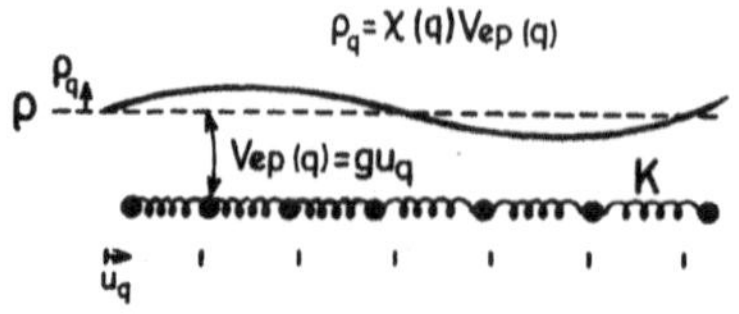

Figure 1. Basic interactions involved in the CDW instability.

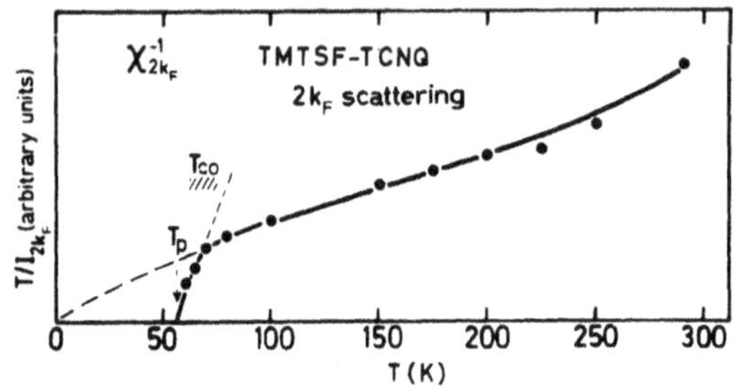

Figure 2. Temperature dependence of $\chi_\rho^{-1}(2k_F)$ for TMTSF-TCNQ (T_p and T_{co} are indicated).

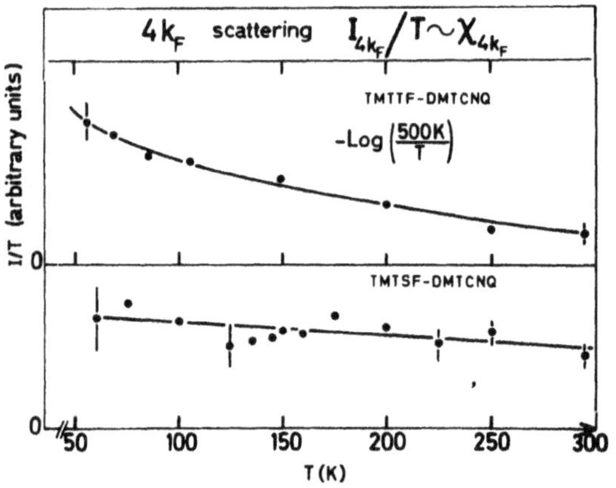

Figure 3. Temperature dependence of $\chi_\rho(4k_F)$ for TMTCF-DMTCNQ (C = S or Se)

wave) provides a q dependent potential $V_{ep}(q)$ which couples to the 1D electron gas. This coupling induces a modulation of the charge density, according to :

$$\rho_q = \chi_e(q)V_{ep}(q), \tag{1}$$

where $\chi_e(q)$ is the q dependent electron-hole polarizability of the electron gas, including the electron-electron interactions. In 2nd order in U_q and ρ_q, the free energy of the organic stack can be written as :

$$F_{org}(\rho_q, U_q) = \frac{1}{2\chi_e(q)} |\rho_q|^2 + \frac{K}{2} |U_q|^2 - g\, U_q\, \rho_q, \tag{2}$$

where the 1st term represents the electronic contribution, the 2nd term the cost of energy due to elastic deformation and the 3rd term the electron phonon coupling (a 1st order coupling where $V_{ep}(q) = g\, U_q$ is assumed). The minimization of (2) with respect to U_q gives: $U_q=(g/K)\rho_q$. This quantity reported in (2) gives :

$$\tilde{F}_{org}(\rho_q) = \frac{1}{2}\, \frac{1}{\chi_\rho(q)} |\rho_q|^2, \tag{3}$$

with

$$\frac{1}{\chi_\rho(q)} = \frac{1}{\chi_e(q)} - \frac{g^2}{K} \quad . \tag{4}$$

In this derivation, based on the RPA approximation of the electron phonon coupling, the CDW lattice instability occurs at the wave vector q for which $\chi_\rho(q)$ diverges (i.e. for which $\chi_e(q)$ shows a maximum). Depending on the strength and on the range of the electron electron interactions $\chi_e(q)$ is maximum either at $q = 2k_F$ (weak interactions) or at $q = 4k_F$ (strong interactions) or even shows two maxima at $q = 2k_F$ and $4k_F$ for intermediate coupling[2,3] (k_F is defined in such a way that $4k_F$, expressed in reciprocal space unit, is equal to the number of conduction electron, ρ, per repeat unit). In the case of strong electron-electron interactions, the $4k_F$ CDW can be viewed as the 1st Fourier component of the lattice of localized charges. Its formation requires, in addition to strong intrasite Coulomb interactions, U, substantial nth intersite interactions, V_n, to establish a periodicity of equidistant charges[4a]. The $4k_F$ CDW is enhanced near the band filling $\rho = 1$ and $\rho = 1/2$, for which periodic ordering of charges necessitates only U and (U, V_1), respectively.

The CDW response function $\chi_\rho(q)$ can be extracted from X-ray diffuse scattering experiments. In the classical approximation, the diffuse intensity peaking at the wave vector $2k_F$ or $4k_F$ amounts to[1] :

$$\delta Id(q) \alpha <|U_q|^2> = \frac{g^2}{K^2} <|\rho_q|^2> = \frac{g^2}{K^2} k_B T \chi_\rho(q) \quad . \tag{5}$$

However, the temperature dependence of χ_ρ certainly differs from that given by (4)[42] especially in the vicinity of T_p^{MF} (defined by the temperature at which χ_ρ diverges in (4)) because the RPA approximation does not treat correctly the 1D lattice fluctuations which are particularily important around T_p^{MF}. Their calculations require the presence of higher order terms[5] in the free energy (2). Also the presence of interchain coupling terms have to be included in (2) to describe the cross over (at T_{co}) towards a regime of 3D fluctuations and the Peierls transition, which occurs at finite temperature, T_p, in real systems.

These concepts are illustrated by figure 2 which shows the temperature dependence of χ_ρ^{-1} for the $2k_F$ instability of TMTSF-TCNQ ($T_{co} \sim 75$K and $T_p = 57$K). In this salt, the $2k_F$ scattering originates mainly from the donor

(TMTSF) stack. There is no 4 k_F instability. The charge transfer of $\rho=0.57$ in TMTSF-TCNQ is reduced to 0.50 by the substitution TCNQ → DMTCNQ. With this value of ρ, the donor stack of TMTSF-DMTCNQ now exhibits the $4k_F$ instability. It shows also more divergent 1D fluctuations at $2k_F$ than TMTSF-TCNQ(1). However, its $4k_F$ fluctuations does not diverge at low temperature (fig.3). $\chi_\rho(4k_F)$ diverges in the sulfur analogue, TMTTF-DMTCNQ, which has about the same charge transfer. This shows that the TMTTF stack has a stronger tendency at the dimerization than the TMTSF stack. The same observation shall be done in $(TMTTF)_2X$ and $(TMTSF)_2X$ where the monovalent anion leaves the same number of hole ($\rho=0.5$) per donor (see part II). This effect is believed to be the consequence of a stronger electronic polarizability at $4k_F$, $\chi_e(4k_F)$, due to larger Coulomb interactions on the TMTTF stack. From (4) it can be enhanced by a stronger effective electron-phonon coupling constant, $\lambda ep=g^2/K$, of the TMTTF sublattice(K lower?).

Finally, let us mention that the analysis of the structure factor of the X-ray diffuse scattering, observed in the regime of 1D fluctuations, shows that the phonon mode coupled to the $2k_F$ or $4k_F$ electronic instability is mainly due to translations of rigid molecules(1). The main effect of such translations is to modulate the intrastack transfer integrals (bond modulation).

II. ANION PERIODICITY

In organic conductors composed of only one kind of metallic stack, the charge transfer is carried out by counter ions forming ionic chains parallel to the metallic stack. These ionic chains provide an external potential on the organic stack, which first Fourier component is $q^A = 2\pi/a$, where a is the ionic periodicity. If the salt has the stoichiometry $D\,A_x$ and is composed of monovalent anions A, the $4k_F^D$ wave vector of the donor (D) stack is of $x\ 2\pi/d$, where d is the intrastack periodicity. With r more A chains than D stacks, one has $rd = x\,a$ and

$$q^A = 4k_F^D/r \tag{6}$$

Table I expresses the relationship (6) for some salts which will be considered in the text.

TABLE I

salts	$x = \rho$	r	relationship (6)
$(TMTSF)_2X$ } $(TMTTF)_2X$	$\frac{1}{2}$	1	$q^A = 4k_F^D$
DIPS $\phi_4(I_3)_{0.76}$	0.76	2	$q^A = 2k_F^D$
DIPS $\phi_4(I_3)_{1.1}$	1.1	3	$q^A = \frac{4}{3}k_F^D$
$TTT_2\ I_{3+\delta}$	$\frac{3+\delta}{6}$	1	$q^A = 4k_F^D$

The electrostatic coupling with the anion sublattice must thus be added at the free energy of the CDW considered in part I ;

$$F = \frac{1}{2\chi_\rho(q)} |\rho_q|^2 - V^A(q)\, \rho_{-q} \quad . \tag{7}$$

$V^A(q)$ is the potential produced by the anion sublattice. It induces a sizeable response of the organic stack if it provides a wave vector $n\, q^A$ (n integer) for which $\chi_\rho(q)$ shows a maximum (i.e. if $n\, q^A = 2k_F$ and/or $4k_F$). The induced CDW is thus given by minimisation of (7) :

$$\rho_q = \chi_\rho(q)\, V^A(q) \ . \tag{8}$$

A gap Δ_q defined by :

$$\rho_q = \chi_e(q)\Delta_q \tag{9}$$

appears in the band structure. (8) and (9), together with (4) show that, as expected, the external potential and the induced CDW contribute at Δ_q according to :

$$\Delta_q = V^A(q) + \lambda_{ep}\, \rho_q \quad . \tag{10}$$

Because of $V^A(q)$, the gap Δ_q is created whatever the temperature. Thus, the CDW response function χ_e^q will depend explicitely on Δ_q on all the temperature range. Also Δ_q is given by a self consistant equation[6]. From (9) and (10), it is:

$$\frac{\Delta_q - V^A(q)}{\Delta_q} = \lambda_{ep}\, \chi_e(\Delta_q) \quad . \tag{11}$$

a) Commensurate case : $(TMTSF)_2PF_6$ and $(TMTTF)_2PF_6$

It was noticed very soon that the $4k_F$ potential created by the anion sublattice (see table I) plays a key role in the physics of (TMTCF)$_2$X [7] (C = S, Se) through the formation of a $4k_F$ gap Δ_{4k_F} (fig. 4) which gives rise to an additional electron-electron Umklapp scattering process (g_3)[8]. In these salts the tendancy at a $4k_F$ localization of a charge (a hole) in the bond near the anions (see fig. 5) is enhanced by the dimerization of the stack resulting from its response at the anion potential. Thus, the gap Δ_{4k_F} in agreement with (10), includes two components : the $4k_F$ bond potential($V^A(4k_F)$; all the molecules being related by inversion symmetry,the electrons experience only a $4k_F$ bond potential from the anion sublattice) and the difference of intrastack transfer integrals ($t_1 - t_2 \propto \frac{\partial t}{\partial U_{4k_F}} \cdot U_{4k_F} = \lambda_{ep}\, \rho_{4k_F}$). Crystallographic data[9] show that the amplitude of dimerization of the organic stack is smaller in $(TMTSF)_2X$ than in $(TMTTF)_2X$. With respect to the selenium compounds, the strongest tendancy of the sulfur analogues to undergo a high temperature dimeric distortion is corroborated by vibrionic intensity measurements[10]. This behaviour is expected from the study of the $4k_F$ CDW response function, χ_ρ, (fig. 3) which shows a more strongly divergent instability in TMTTF-DMTCNQ than in TMSF-DMTCNQ[11]. The activated behaviour of the electrical conductivity observed below about 220K (T_ρ)[12] is a consequence of the strong $4k_F$ charge localization effect which occurs in the $(TMTTF)_2X$ series.

The $4k_F$ charge localization on bonds has several consequences for the electronic ground state. In the limit of strong localization of one hole per dimer, it is easy to see that the instabilities are those of a half filled band (1 site $\simeq$ 1 dimer). In the strong coupling limit, where

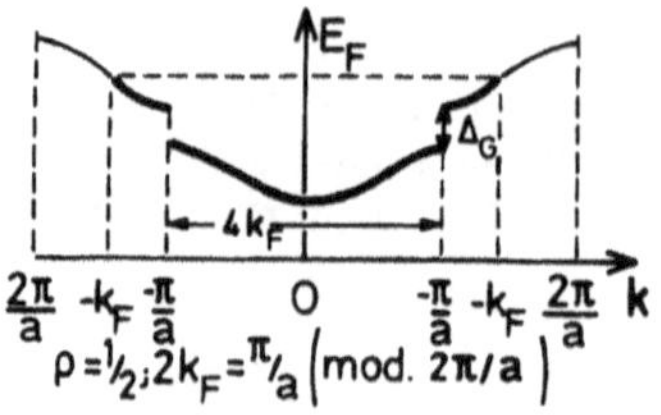

Figure 4. Schematic one electron band structure of $(TMTSF)_2X$. Δ_G corresponds at Δ_{4k_F} defined in the text.

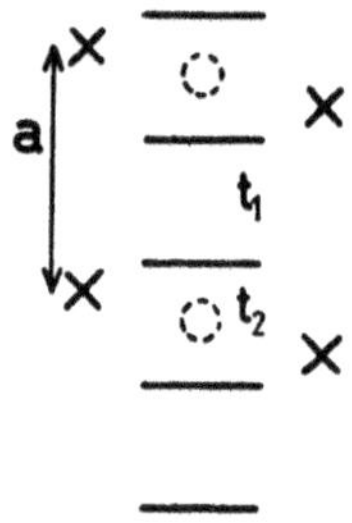

Figure 5. Schematic position of the anions, X, with respect to the organic TMTCF stack. The bonds where the holes are localized by the $4k_F$ anion potential are indicated.

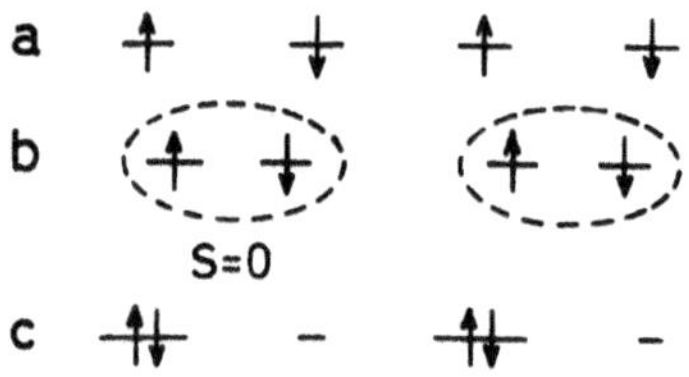

Figure 6. Schematic representation, in the strong coupling limit, of the $2k_F$ antiferromagnetic (a), the $2k_F$ non-magnetic spin Peierls (b) and of an extreme form of the $2k_F$ site CDW (c) ground states. In (c) the $2k_F$ intradimer distortion is represented by segments of different length.

the energy required to put a second hole on the dimer is larger than the interdimer transfer integral, it is also easy to see that the instabilities are either towards antiferromagnetism ($2k_F$ spin density wave (SDW) - fig. 6a) or towards spin Peierls pairing ($2k_F$ bond CDW - fig. 6b). The $2k_F$ site CDW which modifies the charge distribution per site is energetically defavored (an extreme form of it is shown by fig. 6c) [13]. The choice between antiferromagnetism and spin Peierls pairing depends on the electronic parameters and on the lattice rigidity. These two ground states are observed in $(TMTTF)_2X$ and related materials with localized electrons [14]. As illustration, the temperature dependence of the $2k_F$ CDW response function, χ_ρ, driving the spin Peierls instability of $(TMTTF)_2PF_6$ is shown in fig. 7 This figure reports also, for comparison, the same quantity measured in $(TMTSF)_2PF_6$ which shows a low temperature SDW instability (which cannot be described in the strongly localized limit). $(TMTSF)_2PF_6$ shows a weak regime of quasi 1D CDW fluctuations at $2k_F$, which remains always of very smallintensity and spatial extent ($\xi_{//} \sim 10Å$, $\xi_\perp < b$). It is interesting to remark that the vanishing of $\chi_\rho(2k_F)$ below 30 K corresponds at the onset of critical SDW fluctuations at $2k_F$ [15].

b) Incommensurate case : triodide salts

In triodide organic salts like those indicated in Table I, the organic matrix delimits a lattice of channels which are filled with I_3^- anions (fig. 8). The average I_3^- - I_3^- distance, a, is incommensurate with the organic periodicity, d, in chain direction. Thus, for the organic sublattice the anion potential furnishes an infinite set of Fourier components mq^A ($|m|$ integer). In fact, because of the incommensurability the iodine chains are weakly coupled at the cage, and they behave, at high temperature, as nearly isolated 1D systems. Thus they show no long range order in channel direction and no order between channels. Experimentally, it is found that the I_3^- positions are relatively well correlated in channel directions [16], which means that the q dependence of the position-position correlation function :

$$S^\circ_A(q) = <|\eta_q|^2> \tag{12}$$

where η_q is the Fourier transform of the I_3^- density in channel direction, is clearly peaked at mq^A, for low values of m (ℓI_3 on the left side of fig. 9). Because of the disorder between channels, the organic sublattice experiences no net potential from the anion sublattice. A potential with the in chain wave vector mq^A, will arise with the establishment of a transverse coupling between the in chain Fourier density components $\eta_{mq}A$. At the lowest order in $\eta_{\vec{Q}}(=\sum_j \eta^j_q\, e^{i\vec{q}_\perp \vec{R}_j}$, with j running on the 2D lattice of channels; $\vec{Q} = \{\vec{q}_\perp, q \simeq mq^A\}$) the potential $V^A(Q)$ which enters in (7) is given by :

$$V^A(Q) = h(Q)\eta_Q . \tag{13}$$

The anion sublattice contributes also at the free energy under the form [17]:

$$F^A = \frac{1}{2} \quad \frac{1}{\chi_A(Q)} |\eta_Q|^2, \tag{14}$$

with :

$$\chi_A(Q) = \frac{\chi^\circ_A(q)}{1-v(Q)\chi^\circ_A(q)} . \tag{15}$$

$\chi^\circ_A(q) = S^\circ_A(q)/kT$ is the intrachain susceptibility and $v(Q)$ is the 2D Fourier transform of the interchannel coupling term $v_{ij}(q)$. The most important direct coupling mechanism between the I_3^- density waves is Coulombic in nature and illustrated by fig. 10.

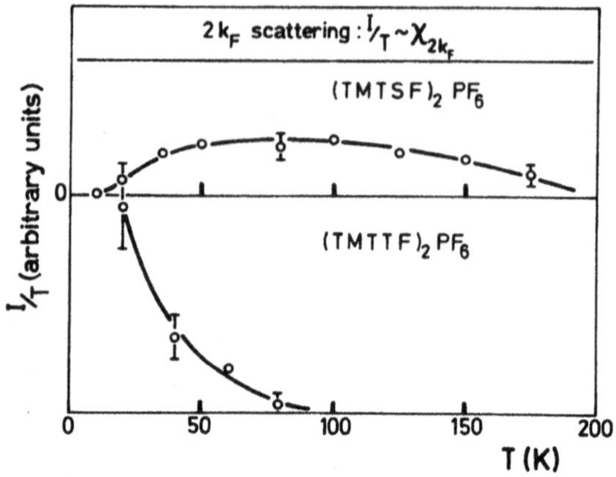

Figure 7. Temperature dependence of $\chi_\rho(2k_F)$ for $(TMTCF)_2PF_6$ (after data of ref. 36)

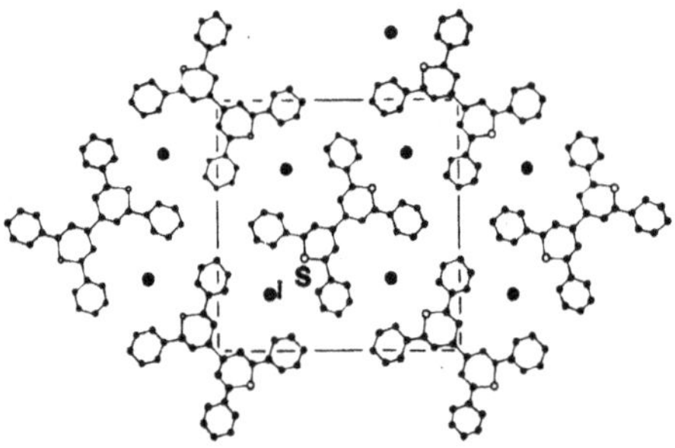

Figure 8. (001) projection of the structure of DIPS $\emptyset_4(I_3)_{0.76}$ showing the 2D lattice of iodine channels.

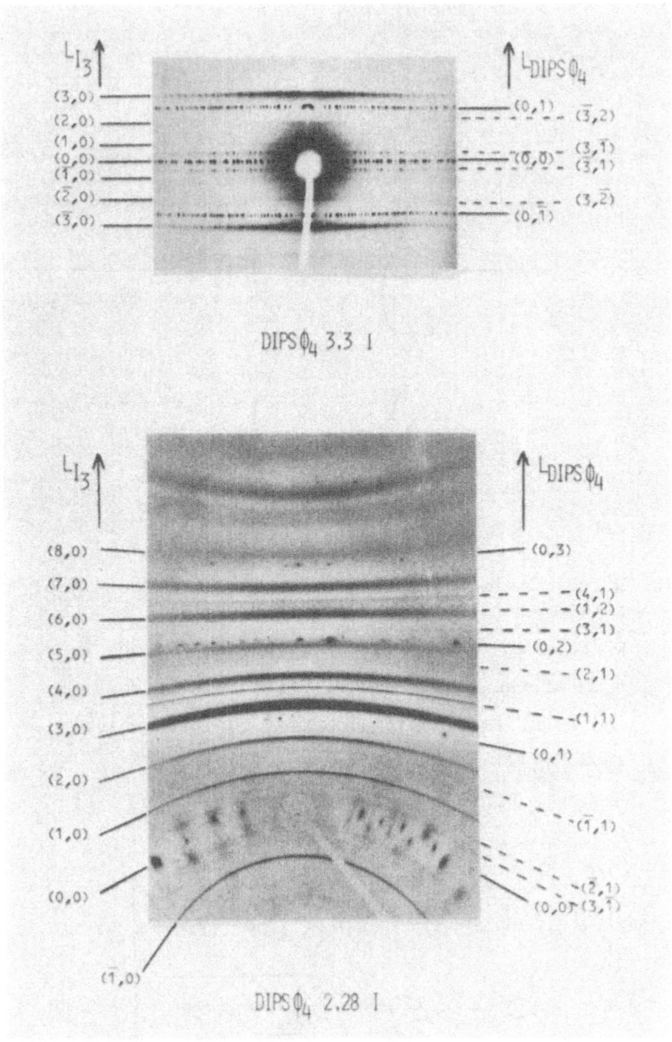

Figure 9. Monochromatic X-ray patterns of $\text{DIPS}\emptyset_4(\text{I}_3)_{0.76}$ and $\text{DIPS}\emptyset_4(\text{I}_3)_{1.1}$ at room temperature showing the 1D nature of the scattering from the iodine sublattice (the channel direction is vertical). The iodine (ℓ_{I_3}) and organic ($\ell_{DIPS\emptyset_4}$) Miller indices in channel direction are indicated. Additional satellite lines, of couple of indices (ℓ_{I_3} , $\ell_{DIPS\emptyset_4}$), and related at the modulation of the iodine density by the organic periodicity, are also shown. In the lower pattern (fixed crystal-fixed film) the ℓ_{I_3} = 1,2, and 3 main lines and associated satellite lines show a short range order. In the upper pattern (oscillating crystal) the ℓ_{I_3} = 3 main line and associated satellite lines show a long range order.

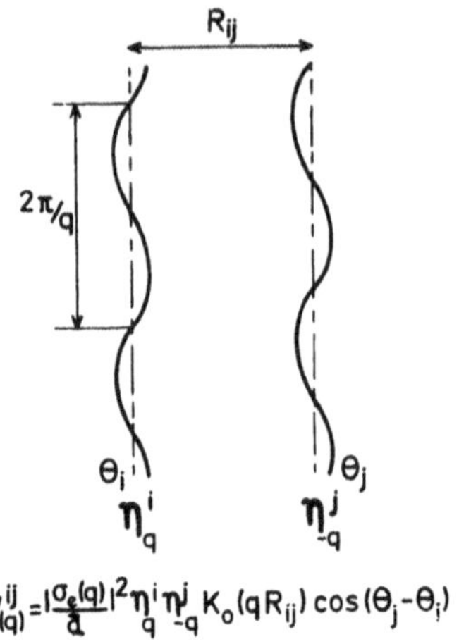

Figure 10. Direct Coulomb coupling between chains i and j of I_3^- density η_q :
$W^{ij}(q) = v_{ij}(q)\eta^i_q\eta^j_{-q}$. In $v_{ij}(q)$, $\sigma_e(q)$ is the Fourier transform of the charge distribution on a I_3^- unit, θ_{ij} is $\vec{q}_\perp\vec{R}_{ij}$, and $K_o(x)$ is a modified Bessel function which decays very rapidly for large $x(K_o(x) \sim \frac{1}{\sqrt{x}} e^{-x})$.

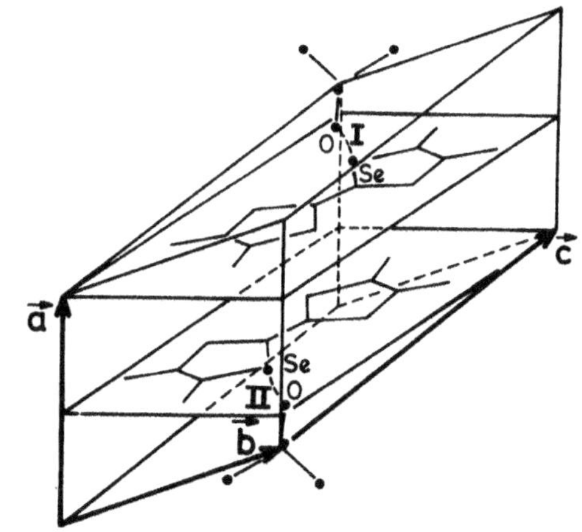

Figure 11. Perspective view of the P $\bar{1}$ structure of $(TMTSF)_2X$ showing for tetrahedral anion the two short contact distances Se–O (I and II).

The minimisation of (7) + (14) with respect to ρ_q gives an effective coupling energy [21] :

$$\tilde{F}_A(\eta_Q) = \frac{1}{2} [\frac{1}{\chi_A(Q)} - |h(Q)|^2 \chi_\rho(q)] |\eta_Q|^2 . \qquad (16)$$

It includes, as 2nd term, a mediated (attractive) interaction, through a CDW distortion of the organic sublattice, which favors the in chain component q for which the CDW response $\chi_\rho(q)$ is maximum. The direct coupling term favors the q component giving a maximum of $v(Q)\chi_A^\circ(q)$. For direct Coulomb coupling, it occurs for $|m| = 1$, because of the exponential decay of $K_o(qRij)$. Both terms favor the critical ordering of the 1st Fourier components of the I_3^- density in DIPS $\emptyset_4(I_3)_{0.76}$, because $q^A = 2k_F$ (see table I). In agreement with this observation, the critical lateral ordering of this component of the iodine density (at 182 K) is accompanied by the opening of a Peierls gap on the organic sublattice [18,19]. Table I shows that the $4k_F$ response of the organic stack could explain the ordering of the m = 3 Fourier component of DIPS $\emptyset_4(I_3)_{1.1}$. Such an ordering already occurs at room temperature (fig. 9). In this compound, the $4k_F$ CDW response could be particularily strong because the charge ρ per molecule (assuming only the presence of I_3^- units) is close to 1 (see part I). In contrast, $\chi_\rho(4k_F)$ is apparently not very effective in $TTT_2I_{3+\delta}$ for the ordering of the m = 1 component which coincides with $4k_F$ of the TTT stack (experimental evidences are that the ordering of iodine chains at $T_o \sim 185K$ is not accompanied by a $4k_F$ electron localization [37] and that short range lateral couplings stronger for the m = 2 and 3 components than for the m = 1 component of the I_3^- density are observed above T_c [20]).

Until now, we have only considered the electrostatic coupling between the organic and iodine sublattices. Mediated interactions between channels can also occur via an elastic deformation of the organic stacks (this will be considered more explicitly in the next section in the case of the anion ordering transitions). Also, via Coulomb interactions and steric effects, the organic stack can modulate the neighbouring columns of I_3^- units. This gives additional components at the iodine density, at the wave vectors $q = mq^A + n\frac{2\pi}{d}$. Some of them are shown on the right side of fig. 9 by the couple of indices (m,n). It is easy to see that small q wave vectors are for (2,1) and (3,1). For them, the direct Coulomb coupling, shown by fig. 10, could be strong, if the associated amplitude of the iodine density, η_q, is large enough [21]. By this Umklapp process involving the organic sublattice, the substantial lateral order observed above T_c in several salts, for the m = 2 and 3 components of the iodine density, could be explained.

III. ANION ORDERING TRANSITIONS

The anion ordering (A.O.) transitions are observed in the $(TMTCF)_2X$ series when X is a non centrosymmetrical anion like ReO_4, BF_4, ClO_4. These anions are placed in cavities delimited by the zig zag stacking of the TMTCF molecule (fig. 11). At room temperature the anions are statistically disordered with a least two inverse orientations equally occupied, leading respectively to short contacts I or II with the Se (S) atom of the TMTCF molecule, as shown by fig. 11. Their low temperature orientational ordering, described by the wave vector Q, introduces new Fourier components of the anion potential, which must be added at the $4k_F$ bond potential (due to the average charge periodicity) already considered in part IIa. We shall not review here the various periodicities $2\pi/Q$ stabilized [22] nor the specific features of the A.O. transitions [23]. Fig. 12 gives only the change in the site and bond potentials associated with the 3 main ordering wave vectors observed for tetrahedral anions. These potentials clearly

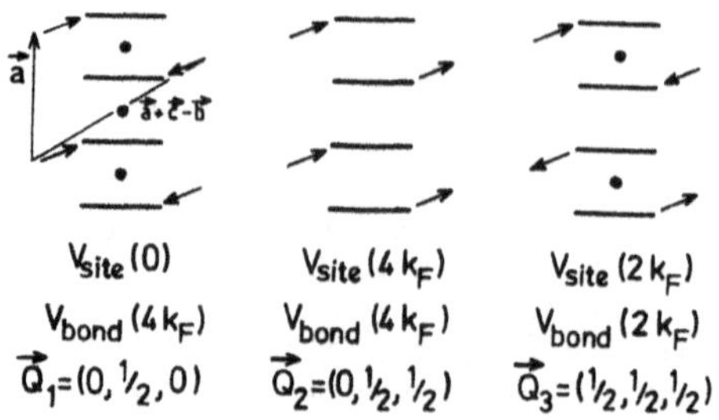

Figure 12. $\vec{Q}_1$, $\vec{Q}_2$, $\vec{Q}_3$ anion ordering patterns in the $(\vec{a}, \vec{c}-\vec{b})$ plane. The arrows indicates the direction of short contact distance of the anion with the TMTCF stack. The change of anion potential for each ordering is indicated.

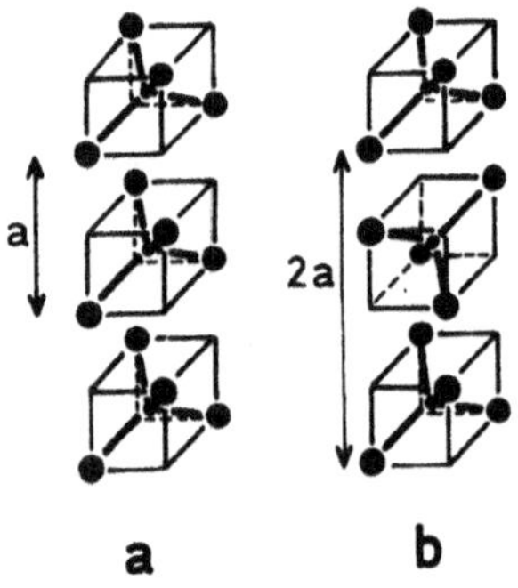

Figure 13. Stacking of regular tetrahedra in uniform (a) and alternate (b) orientations, giving respectively a minimum and a maximum of Coulomb repulsion between charges.

influence the electronic ground state of the $(TMTCF)_2X$ salts [12,24], and contribute to Δq defined by (10).

We shall discuss below, on phenomenological grounds, the nature of the interactions involved in the A.O. process. For that, we start with the harmonic part of the free energy:

$$F(\eta_q, U_q, \rho_q) = \frac{1}{2\chi_A(q)} |\eta_q|^2 + \frac{1}{2\chi_e(q)} |\rho_q|^2 + \frac{1}{2} K|U_q|^2 + gU_q\rho_{-q}$$

$$- h\eta_q\rho_{-q} + k\eta_q U_{-q} \quad . \tag{17}$$

All, the terms have been considered before except the last one which couples the deformation of the cavity with the orientation of the anion. Such a coupling can be sizeable because the anions are in close contact with the methyl groups of the TMTCF molecules[25] (anions of large volume modify even the structure of the organic stack![26]). In (17) the order parameter η_Q is the Fourier transform of η_i, defined by the probability for the anion i to take the orientation I :

$$P_i^I = \frac{1+\eta_i}{2} \; ; (P_i^{II} = 1 - P_i^I) \quad .$$

After minimisation of (17) with respect to U_q and ρ_q, the following effective free energy is thus obtained :

$$\tilde{F}_A(\eta_Q) = \frac{1}{2} \left[\frac{1}{\chi_A(Q)} - \lambda_{ap}(Q) - (h_Q+\mu_Q)^2\chi_\rho(q)\right]|\eta_Q|^2, \tag{18}$$

where $\lambda_{ap} = k^2/K$ is the dimensionless orientation-deformation coupling and $\mu = \sqrt{\lambda_{ep}\lambda_{ap}}$.

$\chi_A(Q)$, related to the direct (electrostatic) interaction energy within the anion sublattice,[43] depends strongly on Q. Fig. 13 shows clearly that in the chain direction the direct Coulomb Coupling between tetrahedra is minimum for $q_a = 0$ and maximum for $q_a = 1/2$. This is probably the Q dependence of this term which imposes the value and the quasi isotropic character of the correlation lengths of the pretransitional fluctuations above T_{AO}[22]. The $Q_3 = (1/2,1/2,1/2)$ order is thus stabilized by the gain of free energy brought by the coupling with the organic stack (3rd term of (18)). $\chi_\rho(2k_F)$ can be estimated from the measurement of the $2k_F$ scattering in salts with octahedral anions such as PF_6(fig. 7).

In the $(TMTSF)_2X$ series $\chi_\rho(2k_F)$ is very weak in intensity and weakly temperature dependent (fig. 7). This suggests that the Q_3. A.O. is due to strong coupling between the anion and organic sublattices [27]. In (18) this coupling is electrostatic (direct electron-anion coupling term : h_Q) and mechanical (electron-anion coupling term via the deformation of the cage: μ_Q). The charge distribution in the anion (related to its electronegativity), as well as the contact distance Se - anion, determine h_Q. λap and thus μ_Q are determined by the volume of the anion. It was previously noticed that the A.O temperature, $T_{A.O}$, for the Q_3 ordering in the $(TMTSF)_2X$ series is correlated with the anion electronegativity[28] or with the anion volume[29]. ($T_{A.O.}$ slightly differs from the temperature at which, in (18), the coefficient of $|\eta_Q|^2$ vanishes, because the Q3 A.O. transition is weakly of 1st order[38]. Table II shows clearly that the two mechanisms are important. The decrease of $T_{A.O.}$ by a factor 4.5 between the ReO_4 and BF_6 salts can be understood by a volume effect (ReO_4 and BF_4 have a comparable electronegativity[41]. The absence of the Q_3

ordering in the ClO_4 salts can be understood by an electronegativity(41) smaller in ClO_4 than in BF_4 (ClO_4 and BF_4 have a comparable volume). Such comparisons are meaningful in the $(TMTSF)_2X$ series because $\chi_\rho(2k_F)$ does not vary appreciably with the anion ($2k_F$ diffuse lines of similar intensity are observed for X = PF_6, AsF_6, ClO_4). This is not the case for the $(TMTTF)_2X$ family where apparently the spin Peierls CDW response function $\chi_\rho(2k_F)$ increases with the degree of $4k_F$ charge localization (in ref. (30) it increases as the temperature T_ρ at which the electrical conductivity shows shows a maximum value). Figure 7 shows for $(TMTTF)_2PF_6$ that $\chi_\rho(2k_F)$ has a low temperature divergence which probably causes the Q_3 A.O of $(TMTTF)_2ClO_4$, even for a weak electron anion coupling(40). Table II shows that $(TMTTF)_2ClO_4$ has now a higher $T_{A.O}$ than $(TMTTF)_2BF_4$. This paradoxal result can be understood if the BF_4 salt has a weaker $\chi_\rho(2k_F)$, due to a lower degree of charge localization, than the ClO_4 salt. This assertion is corroborated by the observation of a lower amount of dimerization of the TMTTF stack (0.02/0.09 Å)(31), a lower value of T_ρ (200K/230K)(12) and a lower intensity of the vibronic absorption, which quantity is, according to ref.(32), related to the degree of localization.

Table II

$X=AB_4$	B-A Pauling electronegativity	Anion Volume(26)	$T_{A.O.}$ $(TMTSF)_2X$	$T_{A.O.}$ $(TMTTF)_2X$
ReO_4	2.04	92 Å^3	177K(Q_3)	153K(Q_3)
BF_4	2.09	67 Å^3	39K(Q_3)	40K(Q_3)
ClO_4	0.67	74 Å^3	24K(Q_1) no Q_3 A.O.(39)	70K(Q_3)

The Q_3 anion ordering which corresponds to a sizeable electron-anion coupling is thus associated with a sizeable distortion U_{Q_3} of the organic stack(33). In $(TMTSF)_2X$, $V^A(Q_3)$ as well as U_{Q_3} contribute at the electrical and magnetic gaps observed below $T_{A.O}$(24a). In the $(TMTTF)_2X$ series, the electrical gap is due to the $4k_F$ charge localization process considered in IIa. The A.O. gap at $2k_F$ affects mainly the spin degrees of freedom (as does the $2k_F$ spin Peierls transition). Thus it appears below $T_{A.O}$ on magnetic measurements(12). The Q_1 A.O of $(TMTSF)_2ClO_4$ occurs below 24K without a distortion of the organic stack(34), which agrees with the expected very weak coupling between the ClO_4 and TMTSF sublattices.

The above considerations allow a qualitative understanding of the basic features of the A.O. A more quantitative interpretation requires the very difficult calculation of the couplings which enter in (18). Also the knowledge of their pressure dependence is necessary to explain the P-T phase diagram of these salts, such as that of $(TMTSF)_2ReO_4$ which shows above about 10Kbar a transition from the Q_3 to the Q_2 ordering wave vectors(34).

The experimental work presented in this paper has been performed in collaboration with P.A. ALBOUY, R. COMES, R. MORET and S. RAVY. Very useful discussions with K. BECHGAARD, C. BOURBONNAIS, C. COULON, V.J. EMERY and C. NOGUERA are recognized.

REFERENCES

1. For a recent experimental review see J.P. Pouget, in "Quasi One Dimensional Conductors" edited by E.M. Conwell (Academic Press) to be published
2. V.J. Emery in "Highly Conducting One Dimensional Solids" edited by J.T. Devreese, R.P. Evrard and V.E. Van Doren (Plenum New York 1979) p. 247
3. J. Solyom, Adv. Phys. 28, 101 (1979)
4. J. Hubbard, Phys. Rev. B17, 494 (1978)
5. W. Dieterich, Adv. Phys. 25, 615 (1976)
6. M.J. Rice, Phys. Rev. Lett. 37, 36 (1976)
7. V.J. Emery, R. Bruinsma, and S. Barisic, Phys. Rev. Lett. 48, 1035 (1982)
8. S. Barisic and S. Brazovskii in "Recent Developments in Condensed Matter Physics" vol. 1, edited by J.T. Devreese (Plenum Press 1981) p. 327
9. B. Liautard, S. Peytavin, G. Brun and M. Maurin, J. Physique Colloque 44, 951 (1983)
10. R. Bozio, C. Pecile, J.C. Scott and E.M. Engler, Mol. Cryst. Liq. Cryst. 119, 211 (1986)
11. Structural and optical studies show that donor stacks are very similar in the TMTCF-DMTCNQ and $(TMTCF)_2X$ series based on the same "C". Thus, close values of the $4k_F$ electron-hole polarizability $\chi_e(4k_F)$ are expected on the TMTCF stack for both categories of salts. However the lattice rigidity K, which also enters in χ_ρ (see (4)), could slightly increase from the two chains to the one chain salts because the charge is more localized on the anions than on the acceptors, and thus closer to the donor stacks (F. Wudl JACS 103, 7064 (1981)).
12. C. Coulon, P. Delhaes, S. Flandrois, R. Lagnier, E. Bonjour and J.M. Fabre, J. Physique 43, 1059 (1982
13. L.G. Caron and C. Bourbonnais, Phys. Rev. B 29, 4230 (1984)
14. C. Coulon, this volume
15. F. Creuzet, C. Bourbonnais, L.G. Caron, D. Jérome and A. Moradpour, Synthetic Metals 19, 277 (1987)
16. P.A. Albouy, J.P. Pouget and H. Strzelecka, Phys. Rev. B35, 173 (1987)
17. V.J. Emery and J.D. Axe, Phys. Rev. Lett. 40, 1507 (1978)
18. P.A. Albouy, P. Le Guennec, J.P. Pouget and C. Noguera, Synth. Metals 19, 687 (1987)
19. Because of the warped nature of the DIPS $\emptyset_4$ molecule, the response of the organic sublattice at the $2k_F$ potential of the iodine sublattice, taken into account of the phase shifts θ_i determined in ref. (16), could result in a twist distortion of the stack around its 2 fold symmetry axis.
20. P.A. Albouy, J.P. Pouget and H. Strzelecka, Synth. Metals (to be published)
21. C. Noguera and P. Le Guennec, Phys. Rev. B (submitted)
22. R. Moret and J.P. Pouget in "Crystal Chemistry and Properties of Materials with Quasi One Dimensional Structures" edited by J. Rouxel (Reidel 1986) p 87
23. J.P. Pouget in "Low Dimensional Conductors and Superconductors" edited by D. Jérome and L.G. Caron (Plenum 1987)
24. (a) C.S. Jacobsen, H.J. Pedersen, K. Mortensen, G. Rindorf, N. Thorup, J.B. Torrance and K. Bechgaard, J. Phys. C : Solid State Physics 15, 2651 (1982)
    (b) S. Tomic, D. Jérome and K. Bechgaard in ref. (23)
25. J.M. Williams, Progress in Inorg. Chem. 33, 183 (1985)
26. P. Batail and L. Ouahab, Mol. Cryst. Liq. Cryst. 125, 205 (1985)

27. R. Bruinsma and V.J. Emery , J. Physique Colloque 44, C3-1115 (1983)
28. F. Wudl , Acc. Chem. Res. 17, 227 (1984)
29. T.J. Kistenmacher, Solid State Comm. 50, 729 (1984)
30. C. Bourbonnais in ref. (23)
31. (a) J.L. Galigne, B. Liautard, S. Peytavin, G. Brun, M. Maurin, J.M. Fabre, E. Torreilles and L. Giral, Acta Cryst. B 35, 1129 (1979)
(b) B. Liautard, S. Peytavin, G. Brun, D. Chasseau, J.M. Fabre and L. Giral, Acta Cryst. C 40, 1023 (1984)
32. R. Bozio, M. Meneghetti and C. Pecile, J. Chem. Phys. 76, 3785 (1982)
33. (a) R. Moret, J.P. Pouget, R. Comès and K. Bechgaard, Phys. Rev. Lett 49, 1008 (1982)
(b) G. Rindorf, H. Soling and N. Thorup, Acta Cryst. C40, 1137 (1984)
(c) T.J. Emge, H.H. Wang, M.R. Beno, J.M. Williams, M.M. Whangbo and M. Evain, J.A.C.S. 108, 8215 (1986)
34. (a) R. Moret, J.P. Pouget, R. Comès and K. Bechgaard, J. Physique 16, 1521 (1985)
(b) B. Gallois, A. Meresse, J. Gaultier and R. Moret, Mol. Cryst. Liq. Cryst. 131, 147 (1985)
35. R. Moret, S. Ravy, J.P. Pouget, R. Comès and K. Bechgaard, Phys. Rev. Lett. 57, 1915, (1986)
36. J.P. Pouget, R. Moret, R. Comès, K. Bechgaard, J.M. Fabre and L. Giral Mol. Cryst. Liq. Cryst. 79, 129 (1982)
37. V.F. Kaminskii, M.L. Khidekel, R.B. Lubovskii, I.F. Shegolev, R.P. Shibaeva, E.B. Yagubskii, A.V. Zvarykina and I.T. Zverena, Phys. Status Solidi, 44a 77 (1977)
38. There is probably a link between the occurence of a strong electron anion coupling and the 1st order character of the Q_3 transition, which requires further investigations.
39. A regime of very local ($\xi \sim 10$Å) Q_3 fluctuations is observed below about 150K. It vanishes at low temperature when the Q_1 A.O. is established (J.P. Pouget, R. Moret, R. Comès, G. Shirane, K. Bechgaard and J.M. Fabre, Journal de Physique Colloque 44, C3-969 (1983))
40 $(TMTTF)_2ClO_4$ and $(TMTTF)_2PF_6$ have the same value of T_ρ, nearly the same amplitude of dimerization of the organic stack at room temperature and the same temperature dependence of the infrared vibronic intensity. Thus, χ_ρ $(2k_F)$ of the ClO_4 salt is expected to be very close to $\chi_\rho(2k_F)$ of the PF_6 salt, which is shown in fig. 7.
41. This is in fact the electronegativity of the outer atoms of the anion with respect to the central one (as in the Table II) which must be considered.
42. In (4), the temperature dependence of $\chi_\rho(q)$ is due to $\chi_e(q)$, which varies as $T^{-\gamma}$, with γ depending of the electron-electron interactions[(2,3)].
43. χ_A^{-1} (Q) is the direct electrostatic coupling energy between tetrahedra and $S_A^\circ = kT\chi_A^\circ$ is an intrasite quantity independent of the temperature ($S_A^\circ = < m_i^2 > = 1$).

MAGNETIC PROPERTIES OF BECHGAARD SALTS AND RELATED COMPOUNDS

ROLE OF THE ELECTRONIC LOCALIZATION

Claude Coulon

Centre de Recherche P. Pascal, CNRS
Domaine Universitaire de Bordeaux I
33405 Talence cedex, France

INTRODUCTION

The $(TMTSeF)_2X$ salts, often known as the Bechgaard salts, have been extensively studied in the past few years because of their superconducting properties [1]. First discovered under pressure for the PF_6 salt[2], superconductivity (Sc) was then reported at ambient pressure for the ClO_4 salt[3] and later observed for several members of the TMTSeF series around a critical pressure which depends on the anion X,[4]. Moreover, superconductivity is not the only remarkable property of these compounds: an antiferromagnetic (AF) ground state can also be observed at low temperature [5] and structural phase transitions due to anion ordering may be present[6].

After these important discoveries, it became quickly obvious that many other compounds present similar properties. Although superconductivity is only observed in a "narrow window", i.e. when several parameters reach a critical value[4], this ground state has also been found in the BEDT series, first under pressure for the ReO_4 salt[7] and more recently at higher temperature with several linear anions [8,9]. Even more convincing is the observation of an antiferromagnetic ground state in many salts prepared with various molecules[10,11]. Among them is TMTTF, the sulfur analog of TMTSeF. In the TMTTF series, the AF magnetic ordering competes with a Spin-Peierls (SP) ground state and a unified description of all these compounds should also include, besides superconductivity, these two electronic instabilities.

Since organic superconductivity is already well documented[8,12], we will be essentially interested in this paper by the AF and SP ground states. The first part will be devoted to a presentation of the different salts. The corresponding molecules are given in table I. In a second part, we compare the transport properties of these salts to introduce and discuss the electronic localization which characterizes several of these compounds. The rest of the paper is devoted to magnetic properties which are used to describe the Spin-Peierls and antiferromagnetic ground states. The competition between these two instabilities is finally discussed in connexion with the strength of the electronic localization and the nature of the interchain couplings.

Table 1. List of the organic molecules mentioned in the text

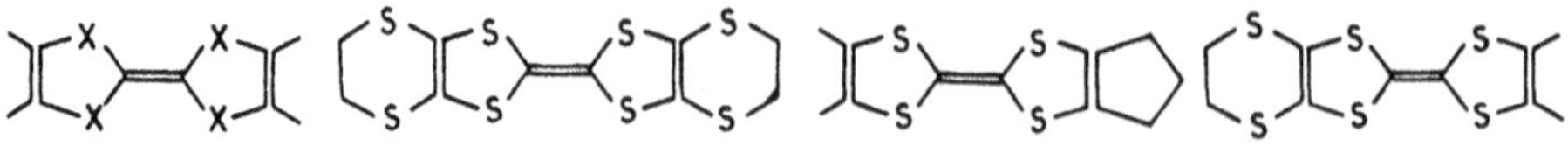

TMTSeF (X = Se) BEDT-TTF DMtTTF DIMET
TMTTF (X = S)

PRESENTATION OF THE DIFFERENT SAMPLES

As mentioned in the introduction, different salts can be compared with the Bechgaard salts to be classified in the same series of compounds. The most obvious analogy between them is their structural organization. The room temperature crystal structure of the $(TMTSeF)_2X$ salts is triclinic (space group $P\bar{1}$) and consists of planes of TMTSeF molecules separated by the anions[13]. Within the planes the molecules are organized in chains and packed in "zig-zag" (see figure 1). Consequently there are two molecules in the unit cell and the stacks are composed of diads even if the dimerization is sometimes very weak. Note that this organization implies the 2:1 stoichiometry which is a common feature for all the materials described in this paper. The anions are located at the center of symmetry of the structure. They can be either centrosymmetrical (like Br, PF_6, AsF_6, SbF_6) or not (like BF_4, ClO_4, ReO_4). When non-centrosymmetrical they should be statistically disordered to preserve the crystal symmetry. In this case, order-disorder phase transitions are generally observed when decreasing temperature and a superstructure corresponding to a larger unit cell is stabilized[6].

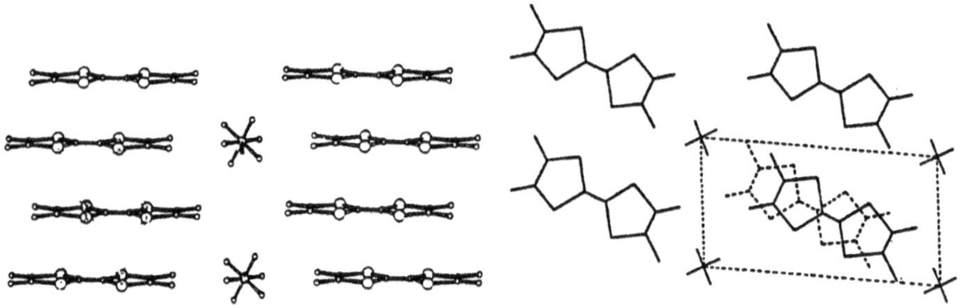

Fig. 1. Crystal structure of $(TMTSeF)_2PF_6$ from ref. 13

Other compounds, including the TMTTF salts[14], $(DIMET)_2AsF_6$,[15] and several DMtTTF salts[16] present the same crystal structure. The local organization of the organic molecules is given in figure 2.

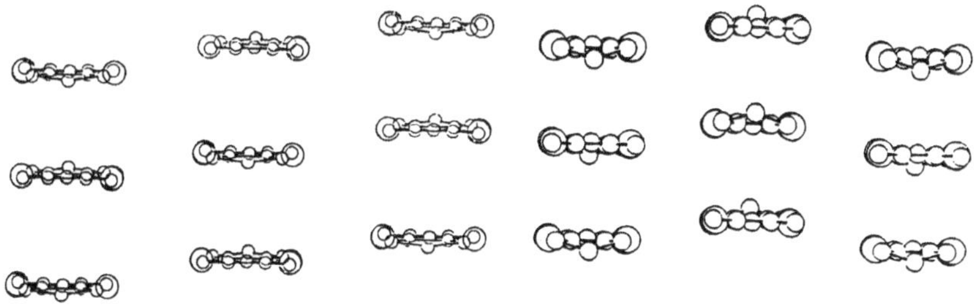

Fig. 2. Local organization of different compounds:
(left) TMTTF like structure,
(right) monoclinic structure in the DMtTTF series.

Other salts present a closely related crystal structure. $(DIMET)_2SbF_6$,[15] and $(BEDT)_2ICl_2$,[17] also have a triclinic structure derived from that of the Bechgaard salts, one essential difference being a larger dimerization. Finally, monoclinic structures are also found in the DMtTTF series with a few anions (ClO_4, ReO_4 and FSO_3),[18,19]. They were described at room temperature in the $C_{2,c}$ space group which implies regular stacks. However, an alternative description is not excluded in the C_c group for which this condition does not exist, indicating that a weak dimerization of the chains may be present. As shown in figure 2, the local environment of the organic molecule presents strong analogies in the triclinic and monoclinic structures, suggesting that one can make a unified description of both compounds. For this reason, in the following, we consider all these materials as belonging to the same series.

Last but not least, it should be noted that the anions can be ordered in the monoclinic structure whatever their symmetry since they are not any more on a center of symmetry. For this reason, monoclinic salts with non-centrosymmetrical anions will be also considered in the following.

TRANSPORT PROPERTIES - EVIDENCE FOR ELECTRONIC LOCALIZATION

Presentation of experimental results

Although the TMTSeF and TMTTF salts have similar structural organizations, their transport properties (at least at ambient pressure) are very different. The electrical conductivity σ of the TMTSeF salts is typically 500 $\Omega^{-1}cm^{-1}$ at 300 K and increases when decreasing temperature to reach a value of about 10^5 $\Omega^{-1}cm^{-1}$ at 20 K[20]. On the other hand, a broad maximum of $\sigma(T)$ is observed at a given temperature T_ρ for the TMTTF salts for which the room temperature conductivity is smaller[21] ($\sigma_{rt} \approx 50\ \Omega^{-1}cm^{-1}$ for most of the salts with the exception of the Br salt for which $\sigma_{rt} \approx 250\ \Omega^{-1}cm^{-1}$). Values of T_ρ range between 250 K to about 80 K (for the Br salt). Below T_ρ the electrical resistivity is activated. Fig. 3a summarizes typical data obtained in the TMTTF series at ambient pressure.

The increase of resistivity below T_ρ should usually be seen as a continuous electronic localization and is generally not the signature of a phase transition. This point is clearly demonstrated by several experimental results:

- Around T_ρ there is no anomaly in the magnetic properties (described in the following section).
- Increasing pressure, T_ρ decreases continuously and the behavior of the TMTTF salts under high pressure becomes similar to that of the TMTSeF salts at ambient pressure[22].
- The same result is obtained by alloying TMTTF salts with TMTSeF since the conductivity of these alloys becomes continuously more metallic when the amount of TMTSeF is increased[23,24].

Other salts described in this paper present similar characteristics. T_ρ goes from 100 K to 270 K in the DMtTTF series in which the magnitude of the localization depends strongly on the anion[25]. Finally, some compounds present a larger localization and the maximum of conductivity is not observed below 300 K for $(DIMET)_2AsF_6$ and $(DIMET)_2SbF_6$,[26]. In this latter case a very strong localization is present as shown by the low value of the conductivity ($\sigma_{rt} \approx 10^{-3}\ \Omega^{-1}cm^{-1}$) and the large value of the gap ($\Delta_\rho \approx 1100$ K) already present at 300 K. Experimental results on DMtTTF and DIMET salts are given in Figure 3b.

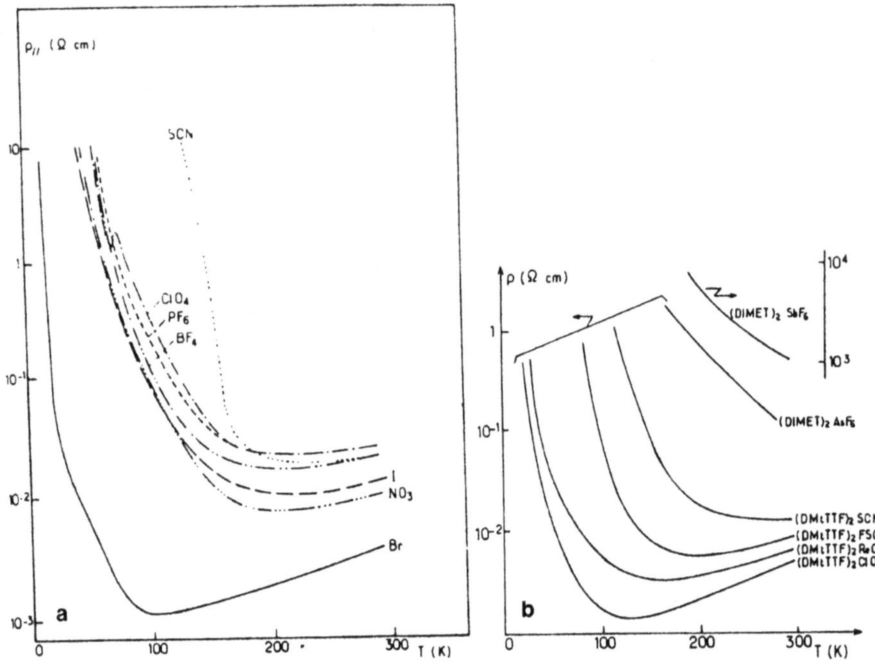

Fig. 3. (a) electrical resistivity of various TMTTF salts
(b) electrical resistivity of some DMtTTF and DIMET salts.

We give in Table 2 a summary of the experimental parameters obtained from conductivity measurements.

Table 2. Typical parameters obtained in the series

	TMTTF			DMtTTF				DIMET		BEDT	
	Br	AsF_6	SbF_6	ClO_4	ReO_4	FSO_3	SCN	AsF_6	SbF_6	ICl_2	AuCl2
σ $(\Omega^{-1}cm^{-1})$	250	20	10	250	170	130	60	10	10^{-3}	3.10^{-2}	5.10^{-2}
T_ρ	~80	220	170	130	150	190	280	-	-	-	-
Δ_ρ	70	600	600	200	200	1000	900	700	1000	1000	1000
Low T gr. st.	AF	SP	AF	AF	AF	AF	-	SP	AF	AF	AF
Ref	21,44 45	21 38	28 38	18 25	18 25	25	25	26	26 53	17 58	58

From Fig. 3a, it is clear that the behavior of the $(TMTTF)_2SCN$ salt is singular. A critical increase in the resistivity is seen at Tc ≈ 160 K. Although there is no anomaly of the magnetic properties at Tc, a structural phase transition has been detected at this temperature probably corresponding to the ordering of the anions with a superstructure (a, 2b, 2c)[27]. The main consequence of this phase transition is a localization of the charges. A quite similar behavior is found for $(TMTTF)_2SbF_6$ where $\rho(T)$ increases critically around 155 K[28]. However, the ordering of the centrosymmetrical anion is neither expected nor observed experimentally. This "structureless" anomaly is also observed for $(TMTTF)_2(SbF_6)_x(AsF_6)_{1-x}$ and $(TMTTF)_2(SbF_6)_x(PF_6)_{1-x}$ alloys and for $(TMTTF)_2ReO_4$ around 230 K, i.e. for various anions[29] . Therefore, the anomaly may result from quick variations of the structure around a given temperature rather than from a real phase transition. Low temperature structural data are still missing to bring more information on that problem.

Thermopower (S) also brings information on charge degrees of freedom. Data are essentially available for TMTSeF and TMTTF salts[30,24,29]. Thermopower of TMTSeF salts is a decreasing function of temperature in the metallic regime. However, a simple proportionality of S with temperature as predicted by a band model is not found: above T ≈ 100 K, one finds $S = S_0 + \alpha T$ where the slope α seems to be correlated with the conductivity of the material[31] (the larger the conductivity, the smaller α). Electronic localization is seen as a plateau or even a sligthly negative slope of S versus T in the TMTTF series. The result obtained for $(TMTTF)_2Br$ at high temperature is intermediate between this behavior and that of the TMTSeF salts. In this case a minimum of S is observed at a temperature close to T_ρ. A change if the slope of S is clearly detected for the SbF_6 and ReO_4 salts at the "structureless" anomaly described previously.

Comments on the Electronic Localization

As already mentioned, the electronic localization only affects transport properties, with no important perturbation on the magnetic susceptibility. This result implies correlation effects since a band model cannot predict a gap for the charge degrees of freedom with no perturbation on the spin excitation spectrum.

The most consistent explanation for this behavior is based on the structure of these materials. One general property of the triclinic samples is the occurence of a dimerization of the organic chains. Moreover, the dimerization may also be present in the monoclinic samples, in particular at low temperature. This dimerization can be seen as a staightforward consequence of the structure since the anions sublattice acts as an external potential on organic chains and implies non regular stacks. The dimerization in turns leads to a "$4k_f$" modulation of the electron density along the chain axis (due to the charge transfer, $4k_f$ is the wave vector corresponding to the periodicity of the diads). This $4k_f$ CDW implies the localization of the charge degrees of freedom [32]. This description was first applied to the TMTTF salts by Emery et al[33] then soon used for many salts including TMTTF/TMTSeF alloys[23,34].

The relation between the actual lattice dimerization and the amplitude of the $4k_f$ CDW depends on parameters such as the electron phonon coupling[35]. However, it is clear that caracterizing the lattice distortion for example by the dimerization gap $\Delta_d = |t_1 - t_2|$ (where t_1 and t_2 are the transfer integrals within a diad along the stacks) gives an approximate estimation of the localization. Alternatively, this information can also be obtained from the value of T_ρ (temperature of the resistivity minimum) and from the amplitude of the gap Δ_ρ measured from conductivity data below T_ρ. It is clear from figure 3 that the samples described in this paper can be classified in several groups:

-(a) For the TMTSeF salts, no localization is observed and the $4k_f$ potential due to the anions is not relevant. Other compounds present a weak localization with a minimum of resistivity at low temperature ($T_\rho \lesssim 100$ K). For these salts, like $(TMTTF)_2Br$ or $(DMtTTF)_2ClO_4$, the room temperature conductivity is larger than 100 $\Omega^{-1}cm^{-1}$ and the dimerization gap is weak.

-(b) The localization is larger for most of the TMTTF salts or for $(DMtTTF)_2SCN$. In these cases T_ρ is slightly below room temperature. The room temperature conductivity of these salts typically ranges between 10 and 100 $\Omega^{-1}cm^{-1}$. The dimerization of their stacks remains weak.

-(c) Finally a few compounds present a larger localization[17,26]. In these cases T_ρ is not observed (is well above 300 K) and the room temperature conductivity is about 10^{-1} - 10^{-3} $\Omega^{-1}cm^{-1}$. Typical samples of this class are $(DIMET)_2SbF_6$, $(BEDT)_2ICl_2$ and $(BEDT)_2AuCl_2$. The dimerization of their lattice is large. These salts will be described in the following using the strongly localized limit.

In conclusion, a wide variety of transport properties is seen among the series. We describe in the next part how they are correlated to the magnetic properties.

MAGNETIC PROPERTIES IN THE SERIES

Magnetic properties of Bechgaard salts and related compounds have been extensively studied either by static or ESR measurements. In both cases, the $q = 0$ susceptibility is probed but complementary data on the staggered susceptibility $\chi(q = 2k_f)$ can be obtained by other techniques like NMR[36]. For clarity, we will first review and comment typical experimental results. More details on low temperature phase transitions are given separately.

Weakly Dimerized Salts

All the samples presented in this paper are paramagnetic at high temperature. Figure 4 gives typical results obtained in the TMTSeF and TMTTF series[37,38].

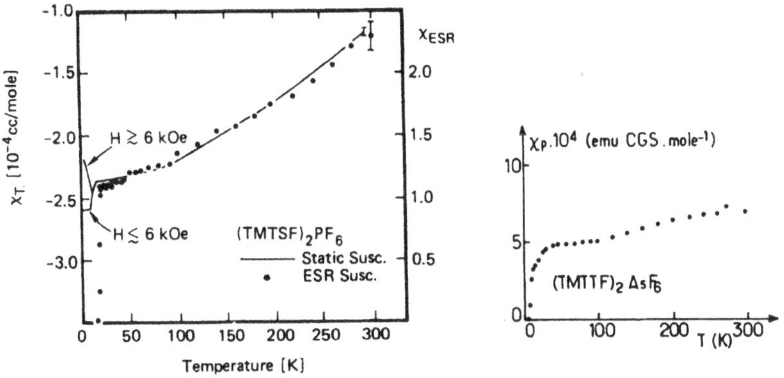

Fig. 4. Spin susceptibility of $(TMTSeF)_2PF_6$ (ref 37) and $TMTTF_2AsF_6$ (ref 38)

Similar temperature dependences of the spin susceptibility χ_p are observed, namely a slight decrease of χ_p with T. This behavior is typical of compounds presenting a weak dimerization (group a and b described above). In all cases χ_p is larger than the estimation obtained in the frame of a band theory (Pauli susceptibility)[21]. The enhancement is typically $\chi_p/\chi_{po} = 4$ (where χ_{po} is the Pauli susceptibility). This characteristic is usually taken as an evidence for important electron-electron interactions[39]. However, $\chi_p(T)$ is far from indicating a strong coupling limit in which the charge degrees of freedom are completely frozen and the experimental data rather suggest the presence of correlated itinerant electrons. As already mentioned, no anomaly occurs at T_ρ for $(TMTTF)_2AsF_6$ indicating a decoupling between charge and spin degrees of freedom.

At low temperature, phase transitions are detected from anomalies in the magnetic behavior. In the case of $(TMTSeF)_2PF_6$ the difference between static susceptibility (which remains finite) and the ESR intensity (which vanishes at 12 K) characterizes the condensation of an antiferromagnetic ground state. Below the magnetic ordering, antiferromagnetic resonance (AFMR) has been observed in closely related salts $(TMTSeF)_2AsF_6$,[40] and $(TMTSeF)_2ClO_4$,[41]. Independently, anisotropy of the static susceptibility indicates an AF long range order[42,43].

The occurence of a low temperature AF ground state is also characteristic of weakly localized salts like $(TMTTF)_2Br$,[44,45] or $(DMtTTF)_2ClO_4$,[18]. In these cases, AFMR is observed using a conventional X band spectrometer. This technique has been systematically applied to many sulfur organic compounds and gives a convincing evidence of the long range magnetic order[11].

In contrast, both ESR and static susceptibility measurements indicate a gap below 11 K in $(TMTTF)_2AsF_6$. This behavior is characteristic of the condensation of a Spin-Peierls (SP) ground state. Similar phase transitions are found in several other samples of the group b (corresponding to $T_\rho \approx$ 200-300 K, i.e. to intermediate localization). Among them are $(TMTTF)_2PF_6$,[34,46] and $(DIMET)_2AsF_6$, [26].

They are however a few cases of samples of the group b for which the SP ground state is not observed. Among them is $(DMtTTF)_2SCN$,[25] which remains paramagnetic down to 2.5 K and $(TMTTF)_2SCN$ or $(TMTTF)_2SbF_6$ which present a low temperature AF ground state[27,47]. For these two TMTTF salts a large anomaly of the electrical conductivity is detected around 160 K and this distinctive behavior might be correlated to the nature of the low T magnetic ground state.

Strongly Dimerized Compounds

A few compounds like $(DIMET)_2SbF_6$, $(BEDT)_2ICl_2$ and $(BEDT)_2AuCl_2$ are characterized by a strong dimerization and localization. In these cases, conductivity is weak and already activated at 300 K (with a gap larger than 1000 K). ESR data for two of these samples are given in Figure 5.

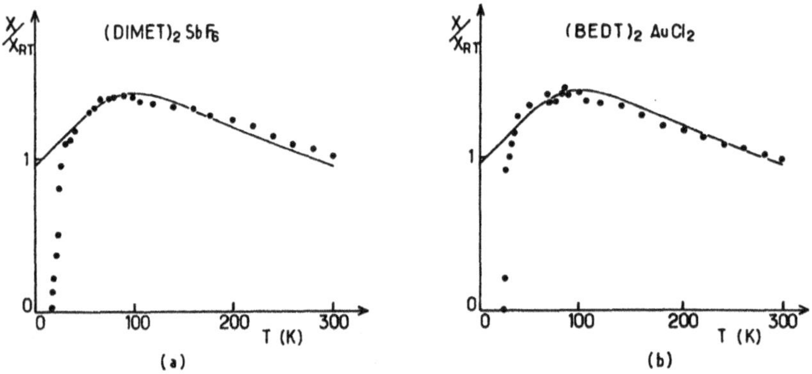

Fig. 5. ESR susceptibility of $(DIMET)_2SbF_6$ (a) and $(BEDT)_2AuCl_2$ (b). The fit for a spin 1/2 Heisenberg AF chain is also given (continuous line) (from ref 49).

In both cases the temperature dependence of χ_p is typical of a 1D quantum spin chain above 30 K (continuous line gives the fit for J = 78 K and χ_p(300 K) = 10^{-3} emu cgs mole^{-1})[17,48,49]. A similar behavior is observed for $(BEDT)_2ICl_2$ with a slightly higher J. In this case larger crystals allow a determination of the absolute value of χ_p in agreement with the fit given above[49]. Below 30 K departure from the 1D behavior is observed. Finally, the condensation of an AF ground state is found at $T_N = 12$ K for $(DIMET)_2SbF_6$ and 27 K for $(BEDT)_2AuCl_2$. Besides static susceptibility, direct proof of the magnetic order has been obtained by AFMR. Particularly striking is the fact that the Neel temperature differs by a factor of two for these two salts despites similar intrachain interactions. This result characterizes quasi-1D materials and should be attributed to different interchain couplings.

To summary, the nature of the magnetic ground state of the different compounds is also given in Table 2.

LOW TEMPERATURE PHASE TRANSITIONS

In this section we give more details on the low temperature magnetic phases (SP and AF) and discuss how they compete in the different samples.

The Antiferromagnetic Instability

The AF ground state is observed for many compounds, either with weak or very large localization.

Strongly localized salts The simple fit of the paramagnetic susceptibility in the case of strongly localized salts suggests that these samples are good examples to introduce the mechanism of the magnetic ordering. The fit in the 1D regime gives an estimate of the intrachain AF exchange: $J_{\parallel} \approx 80$ K. The fact that the 3D magnetic order is only obtained at T_N much lower than $J_{\parallel}$ is easily explained in the frame of a quasi 1D theory: interchain couplings (weaker than $J_{\parallel}$) are necessary to establish the long range order. In terms of 1D staggered susceptibility $\chi_{1D}(q = 2k_f, T)$ and interchain interaction $J_{\perp}$, T_N is given by[50]:

$$1 - J_{\perp}\, \chi_{1D}(q = 2k_f, T_N) = 0$$

As a consequence, fluctuations remain 1D down to a crossover temperature T_x^2 close to T_N[51]. Experimentally, T_x^2 is reached when departure from a single chain behavior is observed: for example, T_x^2 is about 30 K for the samples shown in Figure 5.

The interchain coupling can be estimated from a detailed analysis of the transverse transfer integrals. As shown in Figure 6a, at least three different energies, t_+, t_-, t_δ should be introduced besides t and t_a which describe the intrachain overlaps[52]. In the strong dimerized limit (t_a, t_δ, t_+, $t_- \ll t$) each dimer is equivalent to a supersite with one spin 1/2 in the ground state and exchange between neighboring supersites can be obtained from a perturbation theory. Apart from $J_{\parallel}$ which describes the intrachain interaction, two transfer exchange energies J_δ and J_b describe the interchain couplings (cf Fig. 6b). As usual, one obtains: $J_\delta \sim t_\delta^2/U$ since only one overlap links two neighboring dimers in the δ direction (U is the on-site repulsion energy between two electrons).

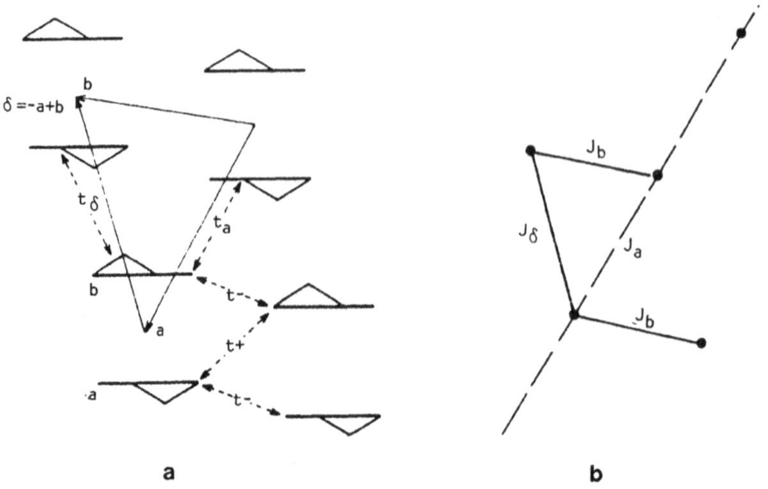

Fig. 6. (a) nature of the transfer integrals between organic molecules, (b) schematization in terms of exchange interactions. The dashed line gives the direction of the chain axis.

The situation is more complex in the b direction because of the competition between positive t_+ and negative t_- transfer integrals and one shows that J_b strongly depends on the ratio t_-/t_+[53]. When this ratio approaches 1/2, J_b becomes ferromagnetic and the resulting magnetic superstructure is expected to be (2a, b). Then, the total transverse interaction is $J_\perp = J_\delta - J_b$ (exchange is taken positive for AF interactions). Although very low temperature structures are missing to estimate the ratio t_-/t_+, extrapolations from room temperature data strongly suggest that we are in this case for the three samples described in this section. Moreover, the increase of T_N going from $(DIMET)_2SbF_6$ to $(BEDT)_2ICl_2$ and $(BEDT)_2AuCl_2$ can be understood in terms of increasing transverse interactions.

Other salts The situation is more complex for weakly dimerized salts (groups a and b). In this case, the intrachain interaction cannot be described by an Heisenberg Hamiltonian and a more elaborated theory is required to determine $\chi_{1D}(q,T)$. However, interchain exchange can still be introduced since the different transverse transfer integrals are still small energies and the same physical description applies to the AF phase transition[51]. In particular, T_N is still given by the same relation $(J_\perp \chi_{1D}(q = 2k_f, T_N) = 1)$ and 1D AF fluctuations should be present at high temperature. Experimental evidence for these fluctuations have been obtained by NMR[36,54].

The wave vector of the magnetic superstructure is also determined by interchain exchange rather than by details of the Fermi surface. Again, a realistic description of the transverse interactions is necessary. Going away from the strong coupling limit, incommensurate superstructures can probably be favored. Experimentally, incommensurate magnetic ordering has been found in $(TMTSeF)_2ClO_4$ and $(TMTSeF)_2PF_6$ from NMR experiments on single crystals[55,56].

Description of the AF Phase

Below T_N, the anisotropy of the static susceptibility and AFMR bring more information on the characteristics of the AF ground state. Both experiments indicate a weak anisotropy with a Spin-Flop field H_{SF} of the order of a few KGauss[11]. In contrast with quasi 1D inorganic systems[57], T_N is almost independant of field for $0 < H < 50$ KGauss[43,58]. This is probably due to the larger interchain interactions of the organic salts ($J_\perp/J_{/\!/}$ is about 5-10% for organics instead of 10^{-4} - 10^{-2} for inorganic compounds[57]). Moreover, H_{SF} is much smaller than the exchange field and no saturation is observed for $H \sim H_{SF}$.

AFMR shows an orthorhombic anisotropy and gives the position of the magnetic axes. This technique is easily performed on one single crystal of reasonable size when no selenium atom is present in the molecule. Working at a fixed frequency, rotation patterns giving the resonance field as a function of the orientation are obtained. We give in Figure 7 the results for $(DIMET)_2SbF_6$,[53] and $(BEDT)_2AuCl_2$,[59]. In both cases the crystal is rotated around simple crystallographic directions.

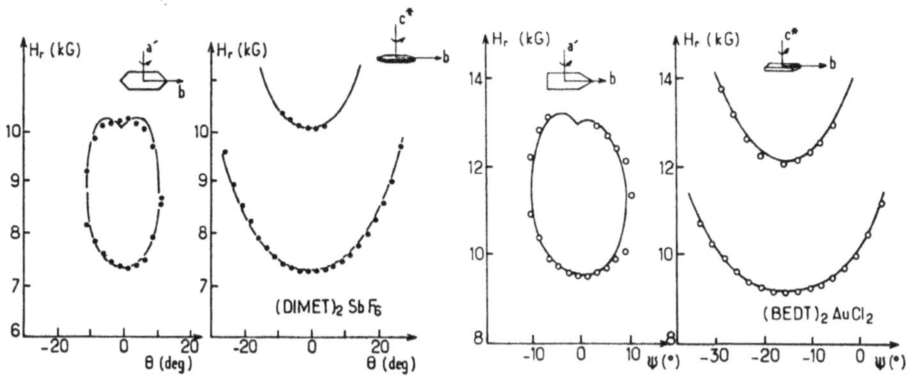

Fig. 7. AFMR rotation patterns for $(DIMET)_2SbF_6$ and $(BEDT)_2AuCl_2$. Rotation axes are shown in insert (a' is defined by a'= b x c*) Continuous line gives the fit using the AFMR theory.

The fit gives the values of Ω_- and Ω_+, zero field frequencies and the position of the rotation axes in the magnetic frame. Ω_- and Ω_+ are related to the anisotropy energy through the relation:

$$\hbar\,\Omega_- = \mu/\mu_B \sqrt{J_{/\!/}\,W_{ei}} \qquad \hbar\,\Omega_+ = \mu/\mu_B \sqrt{J_{/\!/}\,W_{eh}}$$

where W_{ei} and W_{eh} are respectively the easy-intermediate and easy-hard energy difference. μ is the magnetic moment, μ_B the Bohr magneton.

From that information, the position of the magnetic axes relative to crystallographic directions is deduced. As expected for triclinic salts, the magnetic axes are generally not along simple directions. Table 3 summarizes the results obtained for the strongly localized salts.

Table 3. Antiferromagnetic parameters among localized salts. θ et φ are the polar angles of the axes in the cristallographic frame (b, c*, a').

	Easy axis θ	φ	Int. axis θ	φ	Hard axis θ	φ
$(DIMET)_2SbF_6$						
from AFMR	90	0	90	90	0	-
calculated	89	3	92	93	2	-
$(BEDT)_2ICl_2$						
from AFMR	87	7	79	98	13	-113
calculated	86	5	74	95	17	-100
$(BEDT)_2AuCl_2$						
from AFMR	80	0	79	92	15	-130
calculated	86	4	72	96	18	-100

Similar results are found for the three salts with however small differences in the position of the magnetic axes.

For both of them, we have proposed a simple (2a, b) magnetic superstructure and therefore, the characteristics of the anisotropy can be theoretically determined analyzing the spin interactions in more details. Spin-orbit coupling being small for sulfur compounds, it is reasonable to consider only dipolar interactions. Figure 8 shows that, with the given superstructure, the easy axis should be close to the short axis of the molecule since dipolar interactions are in this case attractive in all directions. From similar considerations, the intermediate and hard axes are respectively close to the long axis and to the normal of the molecule.

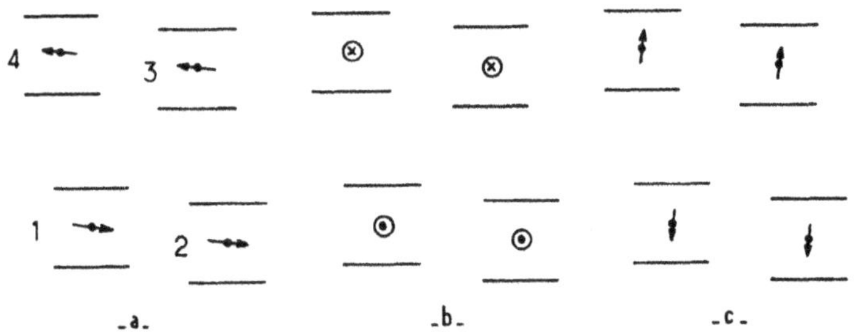

Fig. 8. Hand waving argument to give the position of the magnetic axes:
(a) easy axis: the dipolar interactions are attractive along a, b and δ directions,
(b) intermediate: repulsion along b, attractive along a and δ,
(c) hard: repulsion in all directions.

Accurate determination can be obtained considering the actual distribution of charge on the molecules. The deduced positions of the magnetic axes are also given in Table 3 and are in excellent agreement with experimental results. In particular, the calculation accounts for the small differences observed between the samples. The calculation also gives anisotropy energies in agreement with the obtained zero field frequencies considering the exchange energy obtained from ESR[49].

More generally, similar zero field frequencies are found for all the materials of the series, indicating rather similar magnetic energies. There are however differences in the position of the magnetic axes between sulfur and selenium compounds. This difference has been recently explained including spin-orbit anisotropy for selenium salts[60].

The Spin-Peierls Ground State

A SP ground state is essentially observed in compounds with intermediate localization, i.e. for example $(TMTTF)_2PF_6$, $(TMTTF)_2AsF_6$ or $(DIMET)_2AsF_6$.

The Spin-Peierls instability is known to be specific of quasi 1D materials. In the strong coupling limit it can be described as a dimerization of the spin chain leading to a chain of singlets (i.e. to a gap in the spin excitations spectrum) below the phase transition. This mechanism requires a rather soft lattice and important spin-lattice coupling. Due to the anisotropy of the spin degrees of freedom, 1D fluctuations are expected to be observed above the condensation of the 3D order. A close examination of experimental results supports this prediction. For example, it is clear from Fig. 4b that the susceptibility of $(TMTTF)_2AsF_6$ presents three different regimes:

- Above To = 40 K, χ_p is weakly T dependent and even saturates below 100 K.
- Between To and Tc = 11 K, χ_p decreases with T, indicating the presence of a pseudo gap in the spin excitations spectrum.
- Finally a gap is developed below Tc where the long range SP order appears. Similar behavior is observed for $(TMTTF)_2PF_6$ with Tc = 15 K and To $\approx$ 40 K,[54]. In this case, NMR experiments have been done to probe directly the $2k_f$ fluctuations. 1D AF fluctuations are seen above To. Between To and Tc, the temperature dependence of the NMR relaxation time supports the existence of a pseudo gap due to the coupling between spins and phonons. In the same temperature range, 1D $2k_f$ fluctuations of the lattice are seen from diffuse X-ray scattering which are the direct evidence of these 1D SP fluctuations[61]. At the condensation of the long range order, a commensurate superstructure appears with the wave vector ($a^*/2$, $b^*/2$, $c^*/2$). This scenario is indeed predicted in the frame of a quasi 1D theory but is not so clearly followed in other systems presenting a SP ground state. However, it seems quite general in samples issued from the TTF series and for example, $(DIMET)_2AsF_6$ has a similar behavior with Tc = 17 K and To = 30 K. A pseudo gap is also observed for TMTTF alloys made with AsF_6, PF_6 or SbF_6 even when no long range SP order occurs[62] and the competition between the AF and SP instabilities should be discussed in more details.

Competition Between the AF and SP Ground States

Both AF and SP ground states originate from 1D AF fluctuations and it is interesting to study what conditions favor one or the other instability.

This problem has been extensively discussed for the TMTTF and TMTSeF salts[36,51,54]. The conclusion is that the electronic localization is a relevant parameter to discuss this competition. From this theory, weakly localized salts are AF whereas a stronger localization favors a SP ground state. The resulting phase diagram is given in Figure 9.

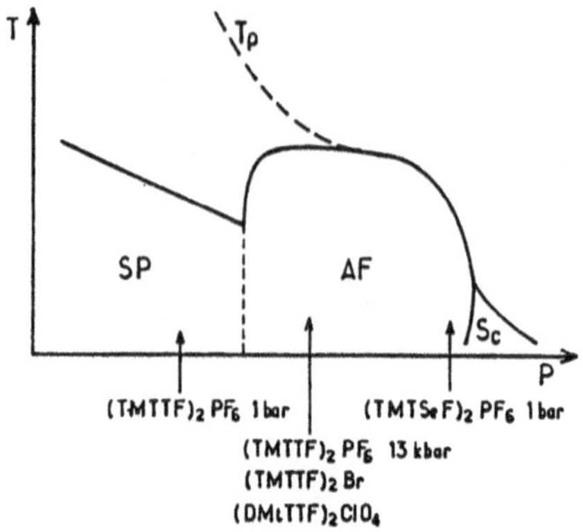

Fig. 9. Theoretical phase diagram from Bourbonnais et al (ref 51,54). P is a parameter which decreases with localization.

Experimentally, P is for example the pressure (since T_ρ decreases with pressure).

Many experimental results support this theory. Weakly localized salts like $(TMTTF)_2Br$ or $(DMtTTF)_2ClO_4$ are AF. In this description, the TMTSeF salts are on the right part of the diagram (not discussed in this paper) where the AF ground state competes with superconductivity. When pressure is increased, T_ρ decreases for all the TMTTF salts and simultaneously a cross-over from SP to AF ground states is observed for $(TMTTF)_2PF_6$,[36].

A few compounds seem to contradict this classification. For example, $(TMTTF)_2SCN$ and $(TMTTF)_2SbF_6$ present an AF ground state although the corresponding localization suggests a SP ground state. In the same way, $(DMtTTF)_2SCN$ remains paramagnetic down to 2.5 K although one would predict a SP ground state. These exceptions are not really surprising since several parameters are kept constant to predict the phase diagram mentioned above. In particular, the elasticity of the lattice and the electron-phonon coupling are assumed to be constant. These parameters may indeed be modified by the 160 K phase transition in $(TMTTF)_2SCN$. Another assumption in the deduction of the phase diagram shown Figure 9 is to keep constant the interchain coupling, for both the AF and SP ground states. However, we have seen that interstack interactions are rather complex and change among the series.

As far as SP ground state is concerned, one possible origin of the interchain coupling is the Coulomb interaction since the order parameter is for example the periodic lattice distortion appearing on the organic chains below Tc. The only case for which the periodicity of the distortion is known is $(TMTTF)_2PF_6$ for which a $(a^*/2, b^*/2, c^*/2)$ wave vector has been found[61]. This commensurate superstructure is quite natural to take profit of the $4k_f$ localization and as shown in Figure 10, corresponds to very simple displacements of the TMTTF molecules. Whatever the actual displacement may be, the coupling between periodic lattice distortions on neighboring chains is proportional to $\cos(\varphi)$ where φ is their difference of phase.

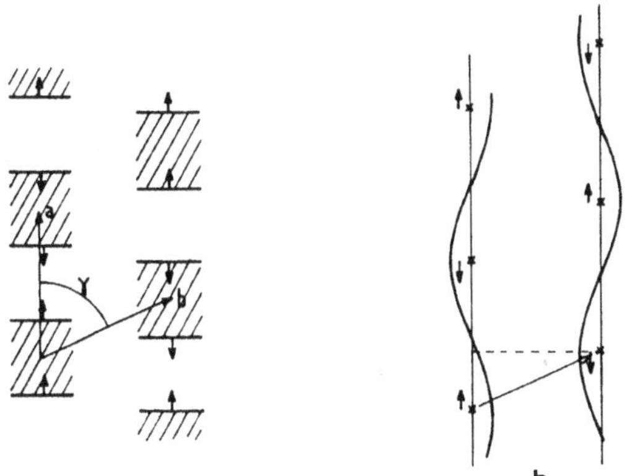

Fig. 10: a) Periodic lattice distortion at the Peierls phase transition. The actual direction of the displacement symbolized by the arrows is not known. The $4k_f$ localization is also shown.
b) Schematization in terms of a $2k_f$ periodic lattice distortion to show the interchain coupling.

Due to the triclinic symmetry, φ is given by: $\varphi = \pi + \pi b/a \cos(\gamma)$. Then, there is a critical situation corresponding to $\varphi = \varphi_c = \pi/2$ for which the Coulomb interchain coupling is zero. A close examination of cristallographic data indicates that some TMTTF salts are not far from this situation and small variations of φ can induce large variations of the transverse coupling and consequently of the temperature of the phase transition. For example, using the low temperature unit cell parameters[63] one obtains $\cos(\varphi) = 0.44$ and $\cos(\varphi) = 0.42$ for $(TMTTF)_2PF_6$ and AsF_6 respectively. On the other hand, $\cos(\varphi) = 0.95$ for $(DIMET)_2AsF_6$ at 300 K and the ratio between the mean field temperature (To = 30 K) and the actual transition temperature (Tc = 17 K) is smaller than for other compounds. In the same way, an unfavorable φ may explain the AF ground state in $(TMTTF)_2SbF_6$ together with the occurence of 1D fluctuations. Unfortunately, cristallographic data below the 155 K anomaly are still missing to support this idea.

Finally, the distinctive behavior of the strongly localized salts should also be mentioned since they present an AF ground state despites a very strong localization. One possible explanation for this behavior is a lack of adiabaticity between the spins and the phonons, this peculiar situation being obtained because of comparable values of the intrachain exchange and Debye temperature[64,65]. A more complete discussion would require the comparison with the magnetic properties of some TCNQ salts. For these compounds, the same order of the exchange is found together with a SP ground state but however with unconventional 3D precursor effects and with sometimes the presence of a soft mode at temperatures well above Tc[66].

CONCLUSION

Coming to the end of this paper, it is certainly useful to emphasize some of the experimental results that we have described. This paper has been devoted to a large number of compounds which seem at first very different. They are made with various organic molecules derived from TSeF or TTF and with anions of various shape and symmetry. Comparing their transport properties, one gets the impression of very different behaviors, going from semiconductors to very good metals. However, an obvious link appears between these salts considering their structural organization since all of them have the same stoichoimetry and closely related crystal structures. Analogies also exist in their magnetic properties with in particular a relatively large number of samples presenting an antiferromagnetic ground state at low temperature.

From a physical point of view, one important parameter is the commensurate band filling (one hole per diad) which brings a natural explanation for the electronic localization if correlation effects are introduced. In intermediate coupling ($U \approx 4t$), the localization can be due to Umklapp terms and transport properties among the series are nicely explained within this description. Following the same theory, the low temperature magnetic properties can also be classified.

One important point is the relevance of 1D physics in this theory. The argument for the electronic localization is essentially 1D and more generally there are theoretical arguments and experimental evidences that a 1D-2D crossover is only present at low temperature[51,54]. Consequently, the description of the paramagnetic phase is essentially dependent on intrachain interactions. On the contrary, the description of phase transitions requires a realistic description of interchain couplings. Thanks to numerical studies, transfer integrals between organic molecules and interchain interactions are now better understood[52,67]. Either for the SP or AF instabilities we have shown how the study of the phase transition requires a detailed knowledge of the structure or transfer integrals.

Finally, taking into account all the relevant parameters, it is clear that a coherent and unified understanding of the series is possible. This is an important asset for the discussion of new materials.

ACKNOWLEDGMENTS

It is a pleasure to acknowledge illuminating discussions with V. Emery, C. Bourbonnais, J. P. Pouget, A. Abderraba, L. Ducasse and F. Creuzet. Many experimental results presented in this paper were obtained in collaboration with R. Laversanne. I also thank P. Vaca for critical reading of the manuscript.

REFERENCES

1. For a recent review on Organic conductors of the TMTSeF series, see for example: D. Jerome and A. J. Schulz, Adv. Phys. 31:299 (1982).
2. D. Jerome, A. Mazaud, M. Ribault, K. Bechgaard, J. Physique 41:L95 (1980).
3. K. Bechgaard, K. Carneiro, M. Olsen, F. S. Rasmussen, C. S. Jacobsen, Phys. Rev. Lett. 46:852 (1981).

4. S. S. P. Parkin, M. Ribault, D. Jerome and K. Bechgaard, J. Phys. C 4:5305 (1981).
5. J. C. Scott, J. Appl. Phys. 53:1845 (1982).
6. R. Moret, J. P. Pouget, R. Comes and K. Bechgaard, J. Physique Coll. C3 44:957 (1983).
7. S. S. P. Parkin, E. M. Engler, R. R. Schumaker, R. Lagier, V. Y. Lee, J. Voiron, K. Carneiro, J. C. Scott, R. L. Greene, J. Physique 44:791 (1983).
8. For a recent review see for example: Proceedings of the ICSM'86, Synth. Met. 19 (1987).
9. The highest Tc is obtained for $(BEDT)_2I_3$ at ambient pressure,
   (a) after a special pressure cycle: F. Creuzet, C. Bourbonnais, G. Creuzet, D. Jerome, D. Schweitzer and H. J. Keller, Synth. Met. 19:157 (1987)
   (b) after a thermal treatment of the crystals: D. Schweitzer and H. J. Keller, these proceedings.
10. J. B. Torrance, J. Physique Coll. C3 44:799 (1983).
11. C. Coulon, R. Laversanne and J. Amiell, Physica 143B:425 (1986).
12. V. J. Emery, these proceedings.
13. N. Thorup, G. Rindorf, K. Soling, K. Bechgaard, Acta Cryst. B37:1236 (1981).
14. J. L. Galigne, B. Liautard, S. Peytavin, G. Brun, M. Maurin, J. M. Fabre, E. Toreilles and L. Giral, Acta Cryst. B35:2609 (1979).
15. B. Gallois, J. Gaultier, F. Bechtel, D. Chasseau, C. Hauw, L. Ducasse, Synt. Met. 19:419 (1987).
16. D. Chasseau, Nouv. Journ. Chem. 9:347 (1985).
17. T. J. Emge, H. H. Wang, P. C. W. Leung, P. R. Rust, J. D. Cook, P. L. Jackson, K. D. Carlson, J. M. Williams, M. H. Whangbo, E. L. Venturini, J. E. Shirber, L. J. Azevedo and J. R. Ferraro, J. Amer. Chem. Soc. 108:695 (1986).
18. C. Coulon, J. Amiell, D. Chasseau, E. Manhal, J. M. Fabre, J. Physique 47:157 (1986).
19. A. Mortaki, J. Gaultier, private communication.
20. K. Bechgaard, C. S. Jacobsen, K. Mórtensen, H. J. Pedersen, N. Thorup, Solid State Commun. 33:1119 (1980).
21. C. Coulon, P. Delhaes, S. Flandrois, R. Lagnier, E. Bonjour, J. M. Fabre, J. Physique 43:37 (1982).
22. F. Creuzet, S. S. P. Parkin, D. Jerome, K. Bechgaard and J. M. Fabre, J. Physique Coll. C3 44:1099 (1983).
23. C. Coulon, P. Delhaes, J. Amiell, J. P. Manceau, J. M. Fabre, L. Giral, J. Physique 43:1721 (1982).
24. K. Mortensen, E. M. Engler, Phys. Rev. B29:842 (1984).
25. C. Coulon, R. Laversanne, J. Amiell, E. Dupart, E. Manhal, J. M. Fabre, Synth. Met. 19:367 (1987).
26. R. Laversanne, C. Coulon, J. Amiell, E. Dupart, P. Delhaes, J. P. Morand and C. Manigand, Solid State Commun. 58:765 (1986).
27. C. Coulon, A. Maaroufi, J. Amiell, E. Dupart, S. Flandrois, P. Delhaes, R. Moret, J. P. Pouget and J. P. Morand, Phys. Rev. B26:6322 (1982).
28. R. Laversanne, C. Coulon, B. Gallois, J. P. Pouget, and R. Moret, J. Physique Lett. 45:L393 (1985).
29. C. Coulon, S. S. P. Parkin and R. Laversanne, Phys. Rev. B31:3583 (1985).
30. K. Mortensen, Solid State Commun. 33:1119 (1980).
31. S. P. P. Parkin, C. Coulon, R. Moret, J. P. Pouget, Phys. Rev. B, in press.
32. S. Barisic, S. Brazovskii, in: "Recent developments in Condensed Matter Physics", 1, J. T. Devreese, ed., Plenum, New York (1981).
33. V. J. Emery, R. Bruisma, S. Barisic, Phys. Rev. Lett. 48:1039 (1982).

34. C. Coulon, PhD Thesis, Bordeaux (1982).
35. We thank V. J. Emery for bringing this argument to our attention.
36. F. Creuzet, PhD Thesis, Orsay (1987).
37. J. C. Scott, H. J. Pedersen, K. Bechgaard, Phys. Rev. Lett. 45:2125 (1980).
38. R. Laversanne, J. Amiell, C. Coulon, C. Garrigou-Lagrange, P. Delhaes, Mol. Cryst. Liq. Cryst. 119:317 (1985).
39. J. B. Torrance, Phys. Rev. B17:3099 (1978).
40. J. B. Torrance, H. J. Pedersen, K. Bechgaard, Phys. Rev. Lett. 49:881 (1982).
41. W. M. Walsh, F. Wudl, E. Aharon-Shalom, L. W. Rupp, J. M. Vandenberg, K. Andres and J. B. Torrance, Phys. Rev. Lett. 49:885 (1982).
42. K. Mortensen, Y. Tomkiewicz, T. D. Schultz and E. M. Engler, Phys. Rev. Lett. 46:1238 (1981).
43. K. Mortensen, Y. Tomkiewicz, K. Bechgaard, Phys. Rev. B25:3319 (1982).
44. F. Creuzet, T. Takahashi, D. Jerome, J. M. Fabre, J. Physique Lett. 43:755 (1982).
45. S. S. P. Parkin, J. C. Scott, J. B. Torrance and E. M. Engler, Phys. Rev. B26:6319 (1982).
46. Experimental results on high quality samples which show the pseudo gap can be found in Ref. 36.
47. C. Coulon, J. C. Scott, R. Laversanne, Phys. Rev. B33:6235 (1986).
48. M. Tokumoto, H. Ankai, T. Ishiguro, G. Saito, H. Kobayashi, R. Kato and A. Kobayashi, Synth. Met. 19:215 (1987).
49. R. Laversanne, PhD Thesis, Bordeaux (1987).
50. D. J. Scalapino, Y. Imry, P. Pincus, Phys. Rev. B11:2042 (1975).
51. C. Bourbonnais, Proceedings of the NATO Advanced Study Institute on Low Dimensional Conductors and Superconductors, Sherbrooke, Canada (1986), in press.
52. A. Abderraba, PhD Thesis, Bordeaux (1987).
53. R. Laversanne, C. Coulon, J. Amiell, Europhys. Lett., 2:401 (1986).
54. F. Creuzet, C. Bourbonnais, L. G. Caron, D. Jerome, K. Bechgaard, Synth. Met. 19:289 (1987).
55. J. M. Delrieu, M. Roger, Z. Toffano, A. Moradpour, K. Bechgaard, J. Physique 47:839 (1986).
56. T. Takahashi, Y. Maniwa, H. Kawamura, G. Saito, J. Phys. Soc. Japan 51:2837 (1986).
57. H. J. M. De Groot, L. J. De Jongh, Physica 141B:1 (1986).
58. A. Maaroufi, PhD Thesis, Bordeaux (1986).
59. C. Coulon, R. Laversanne, J. Amiell, P. Delhaes, J. Phys. C 19:L753 (1986).
60. M. Roger, J. M. Delrieu, E. W. Mbougue, Phys. Rev. B34:4952 (1986).
61. (a) J. P. Pouget, Chemica Scripta 17:85 (1981).
(b) J. P. Pouget, R. Moret, R. Comes, K. Bechgaard, J. M. Fabre, L. Giral, Mol. Cryst. Liq. Cryst. 79:129 (1982).
62. R. Laversanne, J. Amiell, C. Coulon, Mol. Cryst. Liq. Cryst. 137:169 (1986).
63. T. Granier, private communication.
64. A. J. Schulz, private communication.
65. L. G. Caron, C. Bourbonnais, Phys. Rev. B29:4230 (1984).
66. For a review see: J. W. Bray, L. V. Interrante, I. S. Jacobs, J. C. Tonner, in: "Extended Linear Chain Compounds" 3:353, J. S. Miller, ed, Plenum Press, New York (1983).
67. L. Ducasse, M. Abderraba, B. Gallois, J. Phys. C 18:L947 (1985).

MOLECULAR METALS AND SUPERCONDUCTORS: BEDT-TTF RADICAL SALTS

Dieter Schweitzer and Heimo J. Keller*
Max-Planck-Institut für Medizinische Forschung
Abt. Molekulare Physik
Jahnstr. 29, 69 Heidelberg, and
*Anorganisch Chemisches Institut der Universität
Im Neuenheimer Feld, 6900 Heidelberg
Fed. Rep. Germany

INTRODUCTION

The first discovery of superconductivity in an organic metal - in the radical salts of TMTSF (tetramethyltetraselenafulvalene) - under pressure /1/ and ambient pressure /2/ has demonstrated that in addition to the usual intrastack contacts between the donor molecules in quasi one dimensional metals intermolecular contacts between molecules in neighbouring stacks are important. These interstack contacts result in a less pronounced one dimensional electronic behaviour leading to a stabilization of the metallic character down to low temperatures.

In 1982, SAITO et al. /3/ showed that due to the somewhat nonplanar structure of BEDT-TTF (bis-ethylenedithiolotetrathiafulvalene) the electrocrystallisation of radical salts of this donor results in even stronger interstack contacts. Electrochemically prepared crystals of $(BEDT\text{-}TTF)_2 \cdot ClO_4(TCE)_{0.5}$ showed in fact the typical electronic behaviour of a quasi two-dimensional organic metal /3, 4/ down to low temperatures, but did not become superconducting. In a similar salt $(BEDT\text{-}TTF)_4(ReO_4)_2$ PARKIN et al. /5, 6/ did find a superconducting state near 2 K under an isotropic pressure of about 4 kbar, whereby the isotropic pressure was necessary to suppress a metal-insulator transition at 80 K and ambient pressure. For the first time in an organic sulphur donor system superconductivity was observed. Nevertheless, the electrochemical preparation of this radical salt is not easy due to the fact that a large number of radical salts with different stoichiometries and crystal structures might grow simultaneously. For that reason radical salts of BEDT-TTF with I_3^- counterions in THF solutions were prepared /7-9/ leading to two radical salts with identical stoichiometries but different structures (see fig. 1), the so-called α-and β-$(BEDT\text{-}TTF)_2I_3$ phases. Both types of crystals can be easily distinguished by eye because α-phase crystals have usually a plate-like shape while β-phase crystals are canted rhombohedrons. In addition the room temperature ESR linewidth of the conduction electrons can be used to discriminate unequivocal between both salts (70 to 110 Gauss for α-phase crystals and 20 to 25 Gauss for β-phase crystals depending on the direction of the crystals with respect to the magnetic field /10, 11/.

α-$(BEDT\text{-}TTF)_2I_3$ crystals are two-dimensional organic metals with a nearly

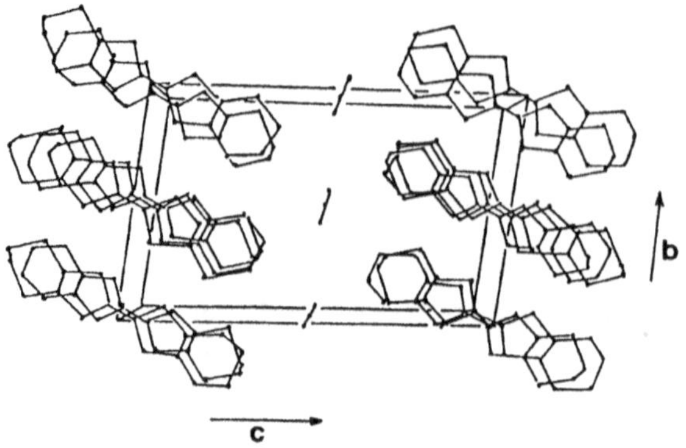

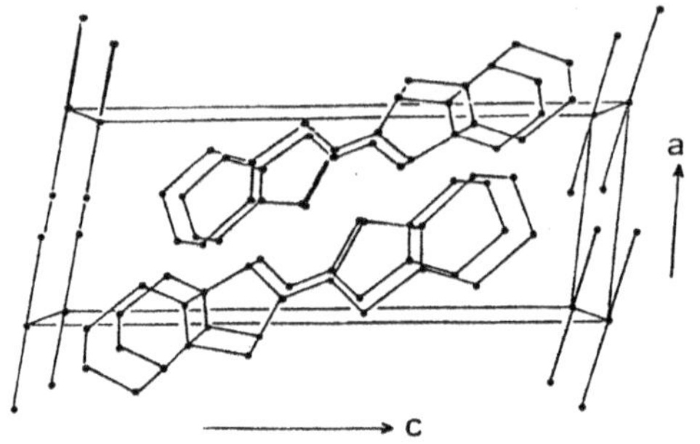

Figure 1. Stereoscopic projections of the structure of α-(BEDT-TTF)$_2$I$_3$ (top) and β-(BEDT-TTF)$_2$I$_3$ (bottom) along the stacking axes.

isotropic electrical conductivity ($\sigma_{300} \approx$ 60-250 S/cm) within the ab-plane, a 1000-times smaller conductivity in the c*-direction and a metal insulator transition at 135 K /8, 9/. β-(BEDT-TTF)$_2$I$_3$ crystals are two-dimensional metals as well ($\sigma_{300} \approx$ 10-50 S/cm) but stay metallic down to low temperature and - as was shown first by the group of SHEGOLEV /12/ - become superconducting at 1.3 K and ambient pressure.

In the following a review of the structural and physical properties of α- and β-(BEDT-TTF)$_2$I$_3$ shall be given and the conditions will be described under which superconductivity at ambient pressure and 8 K in both types of crystals is observed.

β-(BEDT-TTF)$_2$I$_3$

The discovery of superconductivity in β-(BEDT-TTF)$_2$I$_3$ at T_c = 1.3 K and ambient pressure by YAGUBSKII et al. /12/ was confirmed by WILLIAMS et al. /13/ and Meissner-effect measurements /14/ demonstrated the bulk property of the superconductivity in these crystals. A short time later it was found that under a pressure of 1.3 kbar the superconducting transition in β-phase crystals can be raised to 7.5 K /15, 16/. After a particular pressure-temperature cycling procedure -pressurization up to 1.5 kbar at room temperature and a release of the helium gas pressure at temperatures below 125 K - superconductivity at T_c = 8 K and ambient pressure was found /17-19/. The observed sharp superconducting transition - as measured by the resistivity of the crystal - as well as the suppression of the superconducting state by a magnetic field along the c*-axis of about 6 T at various temperatures is shown in fig. 2. The confirmation of bulk superconductivity at 8 K as well was obtained by Meissner-effect /20/ and ac-susceptibility measurements /18/. However, this specially prepared superconducting state at 8 K and ambient

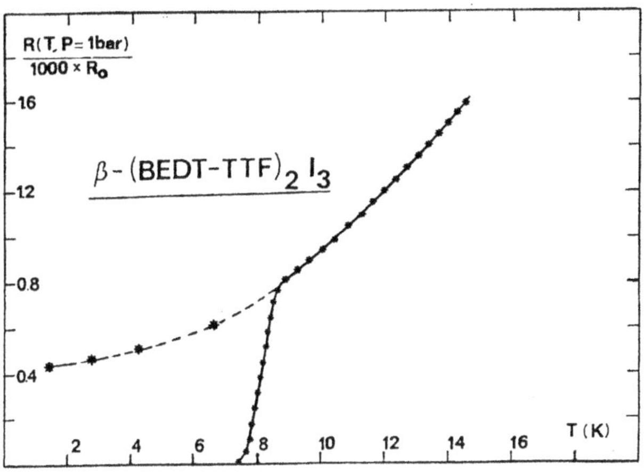

Figure 2. A superconducting transition at 8 K and ambient pressure in β-(BEDT-TTF)$_2$I$_3$ after a special pressure-temperature cycling procedure /17/ as well as the suppression by a magnetic field along the c*-axis at various temperatures.

pressure is meta-stable because superconductivity at 8 K can only be obtained as long as the crystal temperature did not exceed 125 K. At this temperature a metal metal phase transition was observed by thermopower measurements /10/. Recently it was shown that this meta stable 8 K superconducting state can be obtained as well by an electronic excitation with laser light at temperatures below 125 K /21/, but here, due to the fact that the light is absorbed within a few microns only in a thin surface layer.

In order to understand these somewhat surprising facts, the structural properties of the β-(BEDT-TTF)$_2$I$_3$ crystals under the different conditions have to be discussed. At room temperature β-crystals are triclinic /22, 23/ (a = 15.243, b = 9.070, c = 6.597 A; α = 109.73, β = 95.56; γ = 94.33°; V = 848.9 A^3) and the structure is very similar to the Bechgaard-salts. Only one type of crystallographic equivalent stacks exists. Below 195 K at ambient pressure an incommensurate structural modulation exists /24-26/ and the origin of this modulation is connected with an anion cation interaction. Below 125 K a commensurate superstructure with

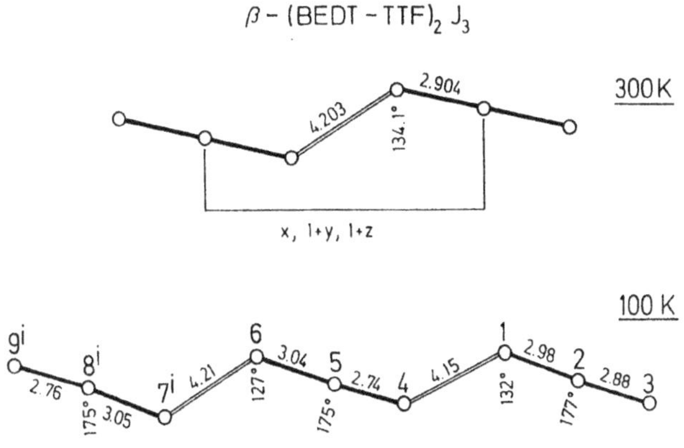

Figure 3. Comparison of the geometry of the triiodide chains in β-(BEDT-TTF)$_2$I$_3$ at room temperature (top) and 100 K (bottom). The unit cell at 100 K contains the three triiodids as shown.

a unit cell of about three times as large as at room temperature was observed /27/ (a = 18.269, b = 21.04, c = 6.543 A; α = 93.56, β = 94.84, γ = 99.86°; V = 2461 A^3). The basic structural change below 125 K with respect to the room temperature structure lies in a pronounced distortion of the triodide chains and especially in a change of the linear and symmetric I_3^- anions at room temperature into non-linear and asymmetric anions (see fig. 3). This finding was confirmed by resonance Raman-investigations /21/. Therefore, it was assumed /27/ that the symmetric linear structure of the I_3^- anions at room temperature is stabilized down to low temperatures by the special temperature cycling procedure /17, 18/ and this more symmetric sructure results in the high T_c-superconducting transition at 8 K. In fact, this assumption was confirmed recently by neutron diffraction experiments /28, 29/ and it was shown that no incommensurate modulated structure at 4.5 K exists, but furthermore, that the terminal ethylene-groups of the BEDT-TTF-molecules in the stacks are all ordered in contrast to the room temperature structure where only on one side the ethylene groups in the stacks are ordered while on the other side they occupy the two possible conformations statistically /28, 29/.

An electronic excitation of the I_3^- anions in β-phase crystals at temperatures below 125 K by laser light can induce - at least in a thin layer on the surface - the structural transformation from the low temperature T_c = 1.3 K structure into the more ordered and symmetric high temperature T_c = 8 K structure /21/. This transformation is observed in the resonance Raman-spectrum by a disappearance of the splitting of the symmetric stretching mode of the I_3-anions with time at constant laser power (see fig. 4) or immediately at high light intensity (≈50 mW) /21/. Therefore, at temperatures between 1.5 and 8 K at least the surface of the crystal - in the moment it is not clear whether IR-radiation might even switch the whole volume of the crystal - can be switched optically from the normal conducting into the superconducting state. But only those parts of the surface are switched which were irridiated by light (optical storage!). However, this transformation is again only stable as long as the crystal temperature does not exceed 125 K /21/.

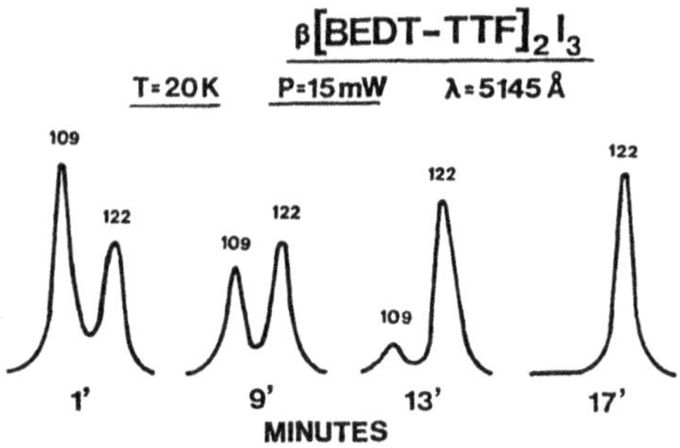

Figure 4. Structural transformation from the low into the high temperatue superconducting phase in β-(BEDT-TTF)$_2$I$_3$ by irridiation with light /21/ as observed by the change of the resonance Raman spectrum of the symmetric stretching mode (109 and 122 cm^{-1}) of the I_3-anions with time at constant laser power (15 mW, at 5145 A).

α- AND α_t-(BEDT-TTF)$_2$I$_3$

The unit cell of α-(BEDT-TTF)$_2$I$_3$ crystals (triclinic: a = 9.211, b = 10.85, c = 17.488 A; α = 96.95, β = 97.97, γ = 90.75°, V = 1717 A^3 /9/) at room temperature is twice as large as the unit cell of the β-phase. In contrast to the β-phase crystals in α-(BEDT-TTF)$_2$I$_3$ two crystallographically different stacks occur and there exist large dihedral angles between the molecular planes (59.4 and 70.4°) of neighbouring donor molecules. The I_3^- anions are linear and all the terminal ethylene groups of the BEDT-TTF-donor molecules in both crystallographically non-equivalent stacks are ordered. Under ambient pressure the α-(BEDT-TTF)$_2$I$_3$ undergoes a metal-insulator phase transition at 135 K /7-9/, which can be suppressed by an isotropic pressure of >12 kbar, but no sign of superconductivity could be observed down to 100 mK /30/. It was claimed /31/ that doping the α-phase crystals with iodine at room temperature results in a metallic state below the insulator phase transition at 135 K followed by a transition to superconductivity at around 3.2 K. Similar results

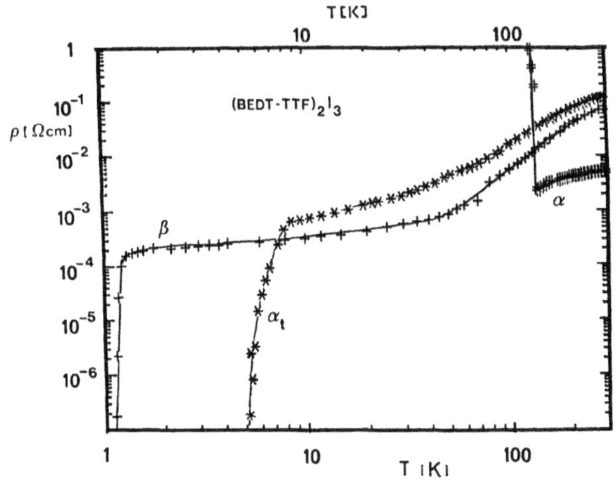

Figure 5. Resistivity versus temperature (logarithmic scales) for α-, α_t- and β-(BEDT-TTF)$_2$I$_3$

as in ref. /31/ for the resisitivity for such iodine-doped α-crystals could be reproduced /32/, but microwave conductivity as well as ac-susceptibility measurements /32, 33/ showed that this metallic state is not a bulk effect in these crystals.

Recently, BARAM et al. /34/ reported a structural transformation of α-(BEDT-TTF)$_2$I$_3$ into β-phase crystals by tempering the α-phase at a temperature of 70-100°C for about 10 to 20 hours. This is a quite surprising result because of the essential differences in the structures of both phases but the structural transformation was confirmed by Weissenberg-pictures /34/. The most surprising fact after the structural transformation was that these crystals show a sharp decrease in resistivity at around 8 K indicating a superconducting transition at this temperature, but prove by a simultaneously applied magnetic field was not reported. Further it was not clear whether or not the eventual superconducting transition is a bulk effect in these tempered crystals.

Another open question is what the differences in the structures between the normal grown β-(BEDT-TTF)$_2$I$_3$ and the tempered α_t-(BEDT-TTF)$_2$I$_3$ crystals are. (In the following tempered α-phase crystals are called α_t-(BEDT-TTF)$_2$I$_3$). This is an important question since β-(BEDT-TTF)$_2$I$_3$ becomes superconducting under normal ambient pressure conditions at 1.3 K while in α_t-(BEDT-TTF)$_2$I$_3$ this transition seems to be at around 8 K, whereas the β-(BEDT-TTF)$_2$I$_3$ shows superconductivity at 8 K only after special treatment (temperature pressure cycling).

A systematic study of the temperature-dependence of the resistivity (fig. 5, 6) and ac-susceptibility (fig. 7) both with and without applying a magnetic field /35/ as well as ESR (fig. 8), NMR (fig. 9), Resonance Raman (fig. 10) and thermopower investigations on α_t-(BEDT-TTF)$_2$I$_3$ /35/ have shown that bulk superconductivity at 8 K and ambient pressure exist in such α_t-crystals. In contrast to the specially prepared 8 K meta-stable superoncucting state in β-crystals here in α_t-(BEDT-TTF)$_2$I$_3$ the superconducting state is stable and entirely reproducable for several temperature cycles up to 380 K /35/.

^{13}C-NMR solid state investigations /35/ (fig. 9) have shown that by tempering α-crystals above 70°C for several days in air all crystals used in the experiment (≈0.3 g) were totally converted into the new α_t-(BEDT-TTF)$_2$I$_3$-phase.

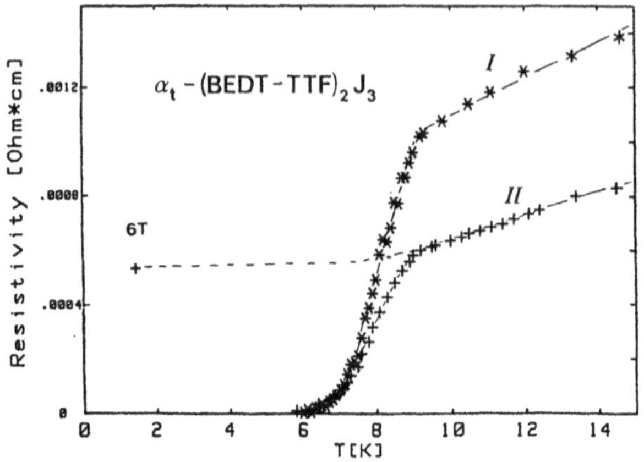

Figure 6. Resistivity of two α_t-(BEDT-TTF)$_2$I$_3$ crystals (I and II) in the temperature region between 0 and 15 K /35/. The suppression of superconductivity in sample II by applying a magnetic field of 6 T at 1.3 K parallel to the c*-axis is shown as well.

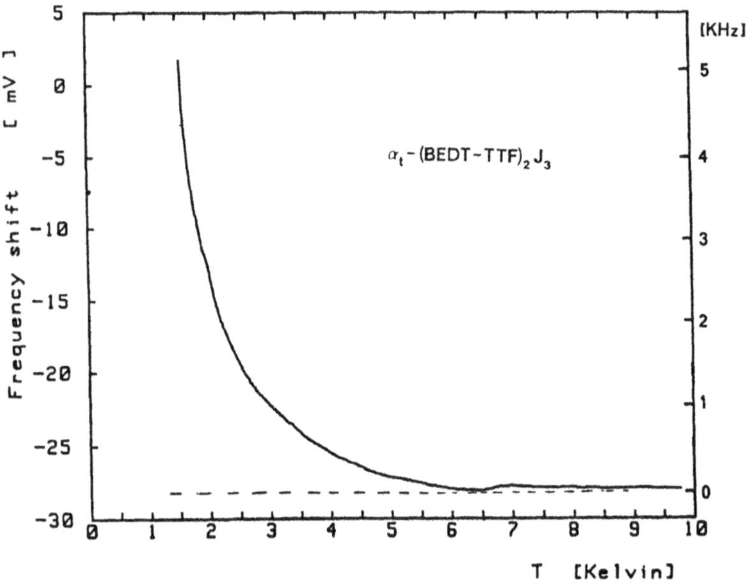

Figure 7. Increase in resonance frequency of a LC circuit due to exclusion of the rf-field by diamagnetic shielding currents (ac susceptibility) in α_t(BEDT-TTF)$_2$I$_3$ (sample II of fig. 5) by lowering the temperature /35/.

The ESR linewidth measurements (fig. 8) as well as the resonance Raman spectra (fig. 10) indicate the phase transition after tempering as well /35/. The upper critical fields H_{c2} (fig. 11) /35/ are anisotropic but with 2.5 to 11 T (depending on the direction of the magnetic field with respect to the crystal axes) relatively high. All the results from ref. /34, 35/ indicate that the α_t-crystal structure at room temperature is identical with the one of the high T_c-superconducting phase of the β-crystals. This is probably due to the fact that in the

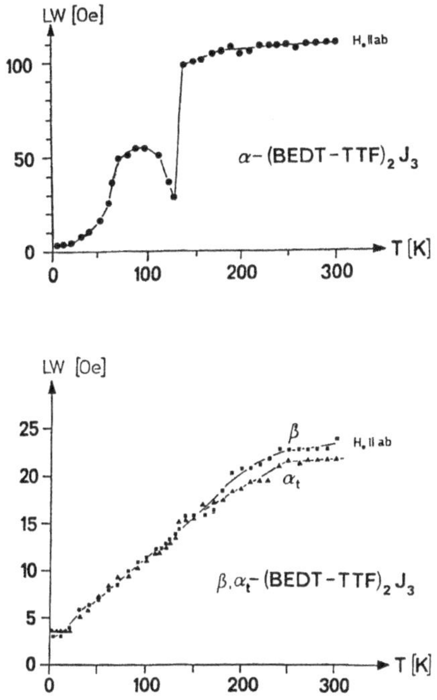

Figure 8. Temperature dependence of the ESR-linewidth for α-(BEDT-TTF)$_2I_3$ (top) as well as for β-(■) and α_t-(BEDT-TTF)$_2I_3$ (▲) (bottom) /35/.

α-phase crystals at room temperature (and probably at 75°C the tempering temperature as well) all terminal ethylene-groups of the BEDT-TTF donor molecules are ordered similar as in the high T_c-superconducting β-phase and therefore the phase transition at 75°C results into an ordered structure. To clarify the situation further X-ray structure investigations are necessary.

An important question is whether or not the usual electron phonon coupling is responsible for the superconductivity in the radical salts of BEDT-TTF. Tunneling experiments in the normal metallic /36/ as well as in the superconducting state /37, 38/ were carried out. Tunneling experiments of HAWLEY et al. /37/ on β-(BEDT-TTF)$_2$·[AuI$_2$] crystals (superconducting transition at ambient pressure at 4.5 K /39, 40/) did show a superconducting gap Δ_T in the ab-plane about 5 times larger than the expected BCS-value. More recent tunneling experiments on β-(BEDT-TTF)$_2$I$_3$ and on β-(BEDT-TTF)$_2$ [AuI$_2$] in the ab-plane as well /38/ show values of Δ_T only about 15% larger than the BCS weak coupling value and are in good accordance

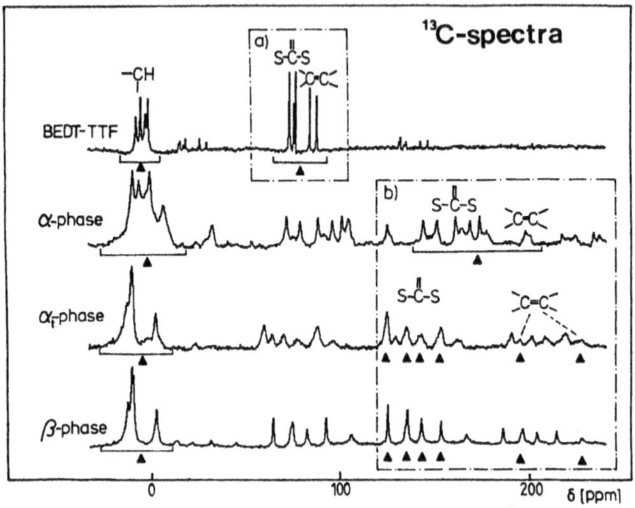

Figure 9. Magic angle spinning (MAS) ^{13}C-NMR spectra at 300 K of BEDT-TTF and of the organic metals α-, β- and α_t-(BEDT-TTF)$_2$I$_3$. The center bands are marked by ▲ in the framed parts as well as in the region of the ethylene groups (ν = 68 MHz, spinning frequency between 4-5 KHz) /35/.

with the conventional electron phonon theory of superconductivity. However, the superconducting gap might be quite anisotropic and in fact preliminary band calculations /41/ indicate that the Fermi-surface has serveral separate sheets, which obviously may possess different superconducting gaps. In such a situation the average gap may correspond to the BCS-value and the maximum value of the gap may be considerably higher /38/. Therefore more experimental data for tunneling contacts with different orientations in the ab-plane are needed.

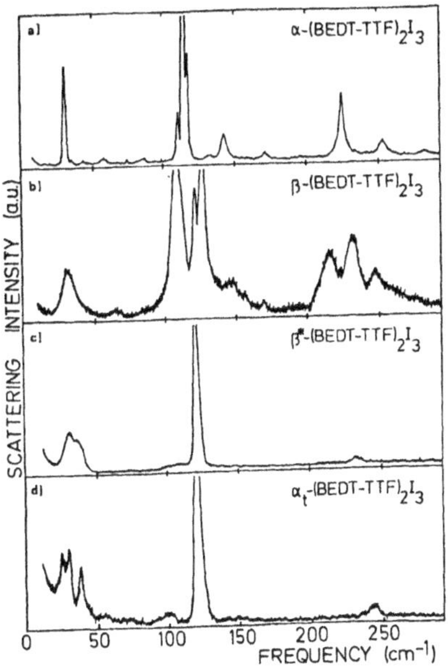

<u>Figure 10.</u> Low energetic parts of the resonance Raman-spectra at 2 K of
a) α -(BEDT-TTF)$_2$I$_3$
b) β -(BEDT-TTF)$_2$I$_3$ (low temperature superconducting (1.3 K) phase
c) β -(BEDT-TTF)$_2$I$_3$ (high temperature superconducting (8 K) phase, as prepared by optical excitation /21/).
d) α_t-(BEDT-TTF)$_2$I$_3$ (excitation wave length 4880 A, 10 mW) /35/.

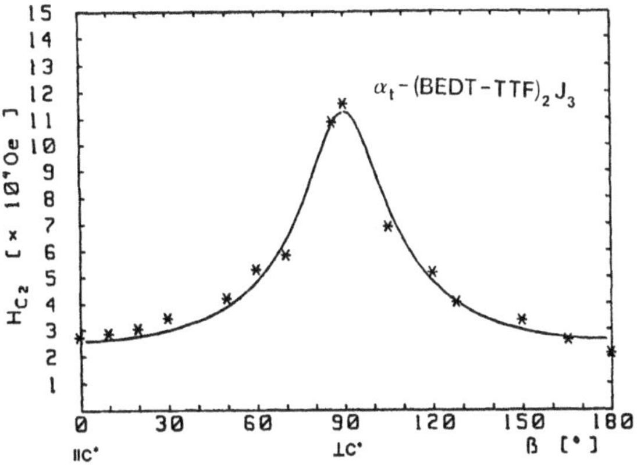

Figure 11. Upper critical fields H_{c2} at a temperature of 1.3 K for α_t-(BEDT-TTF)$_2$I$_3$ as evaluated by the mid-transition of the resistivity curve for several magnetic field directions /35/.

CONCLUSIONS

Bulk superconductivity at 8 K and ambient pressure exists in β-(BEDT-TTF)$_2$I$_3$ and α_t-(BEDT-TTF)$_2$I$_3$ crystals. While in the β-crystals the superconducting state has to be prepared under special conditions (temperature pressure cycling procedure) and is only meta stable, in α_t-(BEDT-TTF)$_2$I$_3$ this superconducting state at 8 K is stable and entirely reproducable for many temperature cycles up to 380 K.

ACKNOWLEDGEMENT

We would like to thank all our colleagues and co-workers (see references) for the very effective cooperation during all the investigations in the last years. Our special thanks to Professors H. Endres, M. Weger and Dr. Swietlik for many helpful discussions and efficient teamwork in Heidelberg.

REFERENCES

1. D. Jérome, A. Mazaud, M. Ribault and K. Bechgaard:, J. Phys. Lett. 4: L 95 (1980).
2. K. Bechgaard, K. Caneiro, M. Olsen, F.B. Rasmussen and C.S. Jacobsen, Phys. Rev. Lett, 46: 852 (1981).
3. G. Saito, T. Enoki, K. Toriumi and H. Inokuchi, Solid State Comm. 42: 557 (1982).
4. G. Saito, T. Enoki, H. Inokuchi and H. Kobayashi, Journal de Physique, 44: C 3, 1215 (1983).
5. S.S.P. Parkin, E.M. Engler, R.R. Schumaker, R. Lagier, V.Y. Lee, J.C. Scott and R.L. Greene, Phys. Rev. Lett. 50: 270 (1983).
6. S.S.P. Parkin, E.M. Engler, R.R. Schumaker, R. Lagier, V.Y. Lee, J. Voiron, K. Carneiro, J.C. Scott and R.L. Greene, Journal de Physique, 44: C 3, 791 (1983).
7. D. Schweitzer, I. Hennig, K. Bender, K. Dietz, H. Endres and H.J. Keller: in: Verhandlungen d. Deutschen Physikalischen Gesellschaft, Physikertagung Münster, März 1984, MO 32, p. 584 (1984).
8. K. Bender, K. Dietz, H. Endres, H.W. Helberg, I. Hennig, H.J. Keller, H.W. Schäfer and D. Schweitzer, Mol. Cryst. Liq. Cryst. 107: 45 (1984).
9. K. Bender, I. Hennig, D. Schweitzer, K. Dietz, H. Endres and H.J. Keller, Mol. Cryst. Liq. Cryst. 108: 359 (1984).
10. I. Hennig, K. Bender, D. Schweitzer, K. Dietz, H. Endres, H.J. Keller, A. Gleitz and H.J. Helberg, Mol. Cryst. Liq. Cryst. 119: 337 (1985).
11. R. Rothaemel, L. Forró, J.R. Cooper, J.S. Schilling, M. Weger, P. Bele, H. Brunner, D. Schweitzer and H.J. Keller, Phys. Rev. B 34: 704 (1986).
12. E.B. Yagubskii, I.F. Shegolev, V.N. Laukhin, P.A. Kononovich, M.V. Kartsovnik, A.V. Zvarykina and L.I. Buravov, Sov. Phys. JETP Lett. 39: 12 (1984).
13. J.M. Williams, T.J. Emge, H.H. Wang, M.A. Beno, P.T. Copps, L.N. Hall, K.D. Carlson and G.W. Crabtree, Inorg. Chem. 23: 2558 (1984).
14. H. Schwenk, C.P. Heidmann, F. Gross, E. Hess, K. Andres, D. Schweitzer and H.J. Keller, Phys. Rev. B 31: 3138 (1985).
15. K. Murata, M. Tokumoto, H. Anzai, H. Bando, G. Saito, K. Jajimura and T. Ishiguro, J. Phys. Soc. Japan 54: 1236 (1985).
16. V.N. Laukhin, E.E. Kostyuchenko, Yu.V. Sushko, I.F. Shegolev and E.B. Yagubskii, Soviet Physics JETP Lett. 41: 81 (1985).
17. F. Creuzet, G. Creuzet, D. Jérome, D. Schweitzer and H.J. Keller, J. Physique Lett. 46: L-1079 (1985).
18. F. Creuzet, D. Jérome, D. Schweitzer and H.J. Keller, Europhys. Lettr. 1: 461 (1986).
19. F. Creuzet, C. Bourbonnais, D. Jérome, D. Schweitzer and H.J. Keller, Europhys. Lett. 1: 467 (1986).

20. H. Veith, C.P. Heidmann, F. Gross, A. Lerf, K. Andres and D. Schweitzer, Solid State Commun. 56: 1015 (1985).
21. R. Swietlik, D. Schweitzer and H.J. Keller, Phys. Rev. B. in print.
22. T. Mori, A. Kobayashi, Y. Sasaki, H. Kobayashi, G. Saito and H. Inokuchi, Chem. Lett. 957 (1984).
23. V.F. Kaminskii, T.G. Prokhorova, R.P. Shibaeva and E.B. Yagubskii, JETP Lett. 39: 17 (1984).
24. P.C.W. Leung, T.J. Emge, M.A. Beno, H.H. Wang and J.M. Williams, J. Am. Chem. Soc. 106: 7644 (1984).
25. T.J. Emge, P.C.W. Leung, M.A. Beno, A.J. Schultz, H.H. Wang, L.M. Sowa and J.M. Williams, Phys. Rev. B. 30: 6780 (1984).
26. P.C.W. Leung, T.J. Emge, M.A. Beno, H.H. Wang, J.M. Williams, V. Petricek and P. Coppens, J. Am. Chem. Soc. 107: 6184 (1985).
27. H. Endres, H.J. Keller, R. Swietlik, D. Schweitzer, K. Angermund and C. Krüger, Z. Naturforsch. 41a: 1319 (1986).
28. A.J. Schultz, M.A. Beno, H.H. Wang and J.M. Williams, Phys. Rev. B 33: 7823 (1986).
29. A.J. Schultz, H.H. Wang, J.M. Williams and A. Filhol, J. Am. Chem. Soc. 108: 7853 (1986).
30. H. Schwenk, F. Gross, C.P. Heidmann, K. Andres, D. Schweitzer and H.J. Keller, Mol. Cryst. Liq. Cryst. 119: 329 (1985).
31. E.B. Yagubskii, I.F. Shegolev, V.N. Laukhin, R.R. Shibaeva, E.E. Kostyuchenko, A.G. Khomenko, Yu.V. Sushko, A.V. Zvarykina, Pis'ma Zh. Eksp. Theor. Fiz. 40: 387 (1984).
32. W. Kremer, H.W. Helberg, E. Gogu, D. Schweitzer and H.J. Keller, Ber. d. Bunsengesellschaft, in print.
33. M. Przybylski, H.W. Helberg, D. Schweitzer and H.J. Keller, Synthetic Metals, 19: 191 (1987).
34. G.O. Baram, L.I. Buravov, L.C. Degtariev, M.E. Kozlov, V.N. Laukhin, E.E. Laukhina, V.G. Orischenko, K.I. Pokhodnia, M.K. Scheinkmann, R.P. Shibaeva and E.B. Yagubskii, JETP Lett. 44: 293 (1986).
35. D. Schweitzer, P. Bele, H. Brunner, E. Gogu, U. Haeberlen, I. Hennig, T. Klutz, R. Swietlik and H.J. Keller, Zeitschr. f. Physik B, condensed Matter 67: in print.
36. A. Nowack, M. Weger, D. Schweitzer, H.J. Keller, Solid State Comm. 60: 199 (1986).
37. M.E. Hawley, K.E. Gray, B.D. Terris, H.H. Wang, K.D. Carlson, J.M. Williams, Phys. Rev. Lett. 57: 629 (1986).
38. A. Nowack, U. Poppe, M. Weger, D. Schweitzer and H. Schwenk, Zeitschr. f. Physik B. condensed Matter 67: in print.
39. H.H. Wang, M.W. Beno, U. Geiser, M.A. Firestone, K.S. Webb, L. Nunez, G.W. Crabtree, K.D. Carlson, J.M. Williams, L.J. Azevedo, J.F. Kwak, J.E. Schirber, Inorg. Chem. 24: 2466 (1985).
40. E. Amberger, H. Fuchs and K. Polborn, Angew. Chem. 97: 968 (1985).
41. J. Kübler, M. Weger, C.B. Sommers, Sol. State Comm. in print.

POLYNUCLEAR METAL CLUSTER COMPOUNDS: A NEW CHEMICAL SUBMICRON STRUCTURE

L.J. de Jongh

Kamerlingh Onnes Laboratorium der Rijksuniversiteit Leiden
Nieuwsteeg 18, P.O. Box 9506
2300 RA Leiden, The Netherlands

"CHEMICAL" VERSUS "PHYSICAL" SUBMICRON STRUCTURES

One of the most important advances in the study of condensed matter of the last few years is that men has started to manipulate matter on the atomic and molecular scales. In line with the spectacular development of ultra-high vacuum techniques, a great many novel materials are being made by chemical or physical vapour deposition, materials which could never have been obtained by means of conventional chemical synthesis. The advent of the scanning tunneling microscope and the enormous increase in resolution of the electron microscopes enable us to actually observe matter "in real space" on the atomic scale. Since the artificially obtained materials often have a layered structure, with a repetition rate that can extend from a single monolayer up to hundreds of layers, the study of surfaces and interfaces has become of prime importance.

We may conveniently denote these artificial structures by the term "physical submicron structures", and juxtapose them to the "chemical submicron structures", which are the main theme of this NATO workshop. For we should not forget that the methods of chemical synthesis likewise lead to an overwhelmingly rich amount of layered or chainlike structures, or to even more intricate systems, in many cases with surprising new physical properties. The term "lattice engineering" or "molecular engineering" has already been in use for many years for this kind of synthesis [1]. To arrive at such materials the chemist may use to his advantage the latest results in solid state chemistry, coordination chemistry, metallorganic chemistry, and last but not least organic chemistry. Compared with the physical methods to obtain submicron structures, the chemical route has a number of obvious advantages of which we name the following:

- The cost of synthesizing chemical compounds is in most cases extremely low compared to the investments needed to construct the UHV apparatus needed for many of the "physical" submicron structures.
- Since the chemical submicron structure are compounds, its constituting elements quite often possess a degree of perfection that is difficult to reach with deposition techniques. For example, in a quasi 2-d magnetic compound, consisting of magnetic monolayers separated by nonmagnetic material, each monolayer in principle is perfect, the only limit being set by lattice defects. Also, one is not bothered by the strong interaction with the substrate that is inherent to UHV deposited layers.

- The yield of chemical synthesis is in most cases quite large compared to UHV deposition techniques or condensation in atomic beams or in matrices. This means that for the chemical submicron structures samples up to a few grams may often be obtained, in some cases even single crystals. Accordingly, the range of physical properties that can be studied is much larger.
- Lastly, with the cehmical route one starts right at the atomic or molecular level, building up the submicron structure by suitable choices and combinations of the building blocks. To make a chemical submicron structure with a repetition length of 1-10 nm is therefore quite easy, whereas this lies at the very end of the scale that is strived for in e.g. submicron lithography.

In conclusion, in our view the chemical and physical submicron structures in many ways complement one another. It should be remarked that the developments of both fields have only just started. This is illustrated for instance by the recent surprising discovery of the high-T_c superconductivity in layered perovskite compounds. These materials may rightfully be viewed as a chemical submicron structure consisting of a repetition of superconducting layers of Cu and O, these sheets being separated by layers of nonconducting material.

In this respect we would like to repeat our plea to the chemists to extend their efforts to synthesize chemical "sandwich"- and "fibre" structures. So far, chemical quasi 1-d or 2-d materials in most cases consisted of assemblies of mutually separated single **chains** or monolayers. It seems of prime importance to extend this to structures in which fibres composed of bundles of parallel chains, or sheets composed of a few or even many layers appear. These fibres and thin films may again consist of either magnetic, semiconducting, conducting or even superconducting material.

POLYNUCLEAR METAL CLUSTER COMPOUNDS

In this section we shall show that polynuclear metal cluster compounds [2-4] form an interesting new class of chemical submicron structures that so far have been scarcely studied by the physical community. In general a cluster can be defined as a unit of a limited number of atoms that may be of the same kind or also of different types. Thus metal clusters or Van der Waals clusters containing a few up to hundreds of atoms can be obtained by condensation in atomic beams or in noble gas matrices. For catalytic purposes, small metal clusters can be formed in zeolites or on substrates. In synthetic chemistry a molecular unit that is used as a building block to generate structures of a 1-d or 2-d polymeric nature is also often called a cluster. Examples are the MoS_6 and $MoSe_6$ units in the Chevrel phases. (These phases are indeed called cluster compounds by some chemists.)

The polynuclear metal cluster compounds that we shall consider, are rather special in that they consist of macromolecules, each macromolecule being composed of a "core" of a certain number (n) of metal atoms, surrounded by a "shell" of ligands! A few examples are given in fig. 1. In many cases the macromolecule is an ion, which, together with a suitable counterion builds up a regular lattice. In other cases it is a neutral molecule, so that the solid formed by it should be regarded as a dense packing of spheres with only short-range crystalline order, similar as in a glass. The important point is that since we are dealing with chemical compounds, the macromolecules in a given compound are identical. Consequently, the solid formed can be described as a macroscopically large array of identical metal particles, embedded in a dielectric matrix. The latter is formed by the ligand shells surrounding the metal particles, plus the counter ions, when these are present. This chemical submicron structure of mutually separated identical metal clusters is quite unique,

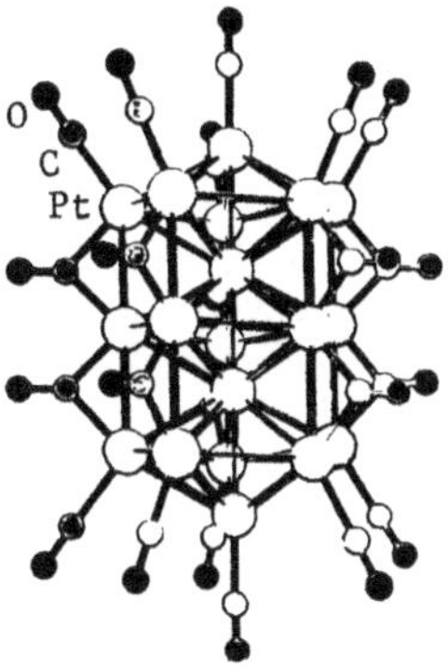

$[Pt_{19}(CO)_{22}]^{4-}$

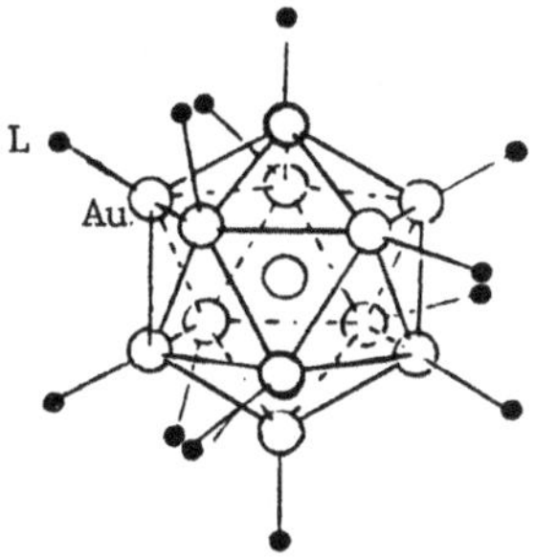

$[Au_{13}L_6]^{5+}$

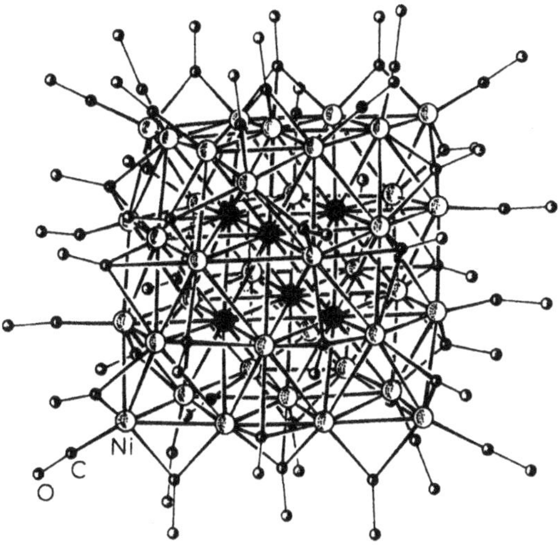

$[Ni_{38}Pt_6(CO)_{48}H]^{5-}$

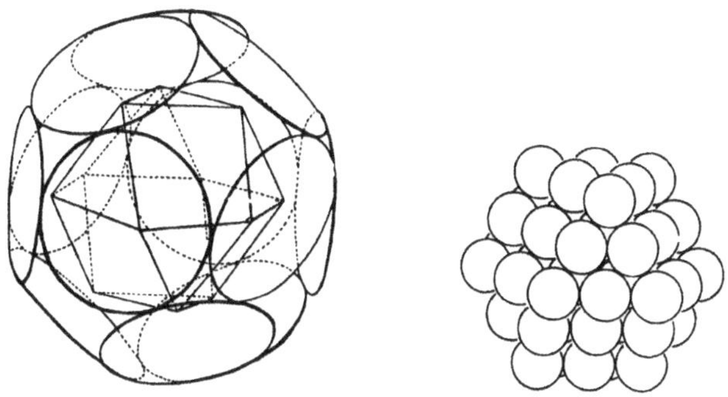

$Au_{55}[P(C_6H_5)_3]_{12}Cl_6$

Fig. 1. Some examples of metal cluster compounds containing 19 Pt atoms, 13 Au atoms, 38 Ni and 6 Pt atoms, and 55 metal atoms where the metal can be Au, Pt, Ru, Rh or Co.

since it cannot be realised by the physical methods employed so far, e.g. by condensation in atomic beams. The problem is that with those physical methods, one always obtains a large distribution in size of the particles so-formed. The FWHM of the size distribution may even equal the size corresponding to the maximum value of the distribution. A quite similar problem arises in metal colloid chemistry. Although in principle the metal clusters formed in atomic beams can be mass-selected, leading to a collection of uniform size, there remain two formidable problems to be solved. Firstly, the yield of such mass-selection machines is so low that it would take a time of the order of months or even years to obtain a sample large enough for, say, a conventional specific heat experiment (for this a sample of at least 1 μg- 1 mg is needed). Secondly, even if macroscopic quantities of mass-selected metal particles could be obtained, it will be extremely difficult, if not impossible to keep them apart in a 3-d (or even in a 2-d) array.

These considerations underline the great advantages of the polynuclear metal cluster compounds for the study of ultra small metal particles as a function of cluster size. In each compound the metal particles are completely identical. In going from one compound to the other the type of metal atom can be varied. At present there exist already a few hundred different metal cluster compounds, with metal atoms of almost every transition element in the periodic table. For the same metal the number n of metal atoms can also be varied by considering different compounds. As mentioned this is a point of great interest since it enables to study the physical properties for different cluster sizes. In the last few years the maximum number n_{max} of atoms in the metal core has been greatly increased. Actually we have n_{max} = 55 for Au, Rh, Ru and Pt, which is realized in the "Schmid-clusters" shown in fig. 1. About two years ago a Russian group [5] published the synthesis of a giant Pd cluster molecule, containing a metal core of as much as 561 Pd atoms! Recently, Schmid and coworkers have been able to repeat this effort, and have indeed obtained similar Pd_{561} cluster compounds [6]. These clusters are in fact so large that in the X-ray the structure of the metal core itself can already separately be seen.

In view of these developments the conclusion appears to be justified that we are only seeing the beginning of an evolution towards larger and larger metal cluster compounds in the near future.

FROM METAL CLUSTERS TO CLUSTER-METALS

One of the latest results of Schmid and coworkers [6,7] that deserves to be mentioned here is the discovery of a way to generate 3-dimensional metallic lattices of a surprising new structure. The method is illustrated in fig. 2 and can be summarized as follows. After dissolving the M_{55} "Schmid-clusters" in dichloormethane a dc voltage is put over two platinum electrodes. This leads to a decomposition of the cluster molecules that is not just a simple electrolysis, since the process occurs at both the anode and the kathode. Probably what happens is that upon contact with the electrode the ligand-shell of the cluster molecule becomes unstable, the ligands come off taking away the peripheral metal atoms of the cluster to which they were attached. Surprisingly, the inner M_{13} cuboctahedral metal survives this fragmentation process as a stable unit. Besides the deposition of free normal M metal at the electrodes, micro-crystals of a new metal structure are found at both electrodes. This unconventional metal lattice can be interpreted as built up from the M_{13} fragments in the way sketched in fig. 2. Basically, thirteen of these M_{13}-cuboctahedra form a $(M_{13})_{13}$ supercluster according to a dense cubic packing of spheres. In the same way, thirteen of the so-formed $(M_{13})_{13}$ superclusters come together to form a $[(M_{13})_{13}]_{13}$ unit. This superlattice structure is detected by additional characteristic X-ray reflections with f.c.c. symmetry, which can clearly be distinguished in the Deleije-Scherrer powder diagrams [7]. The

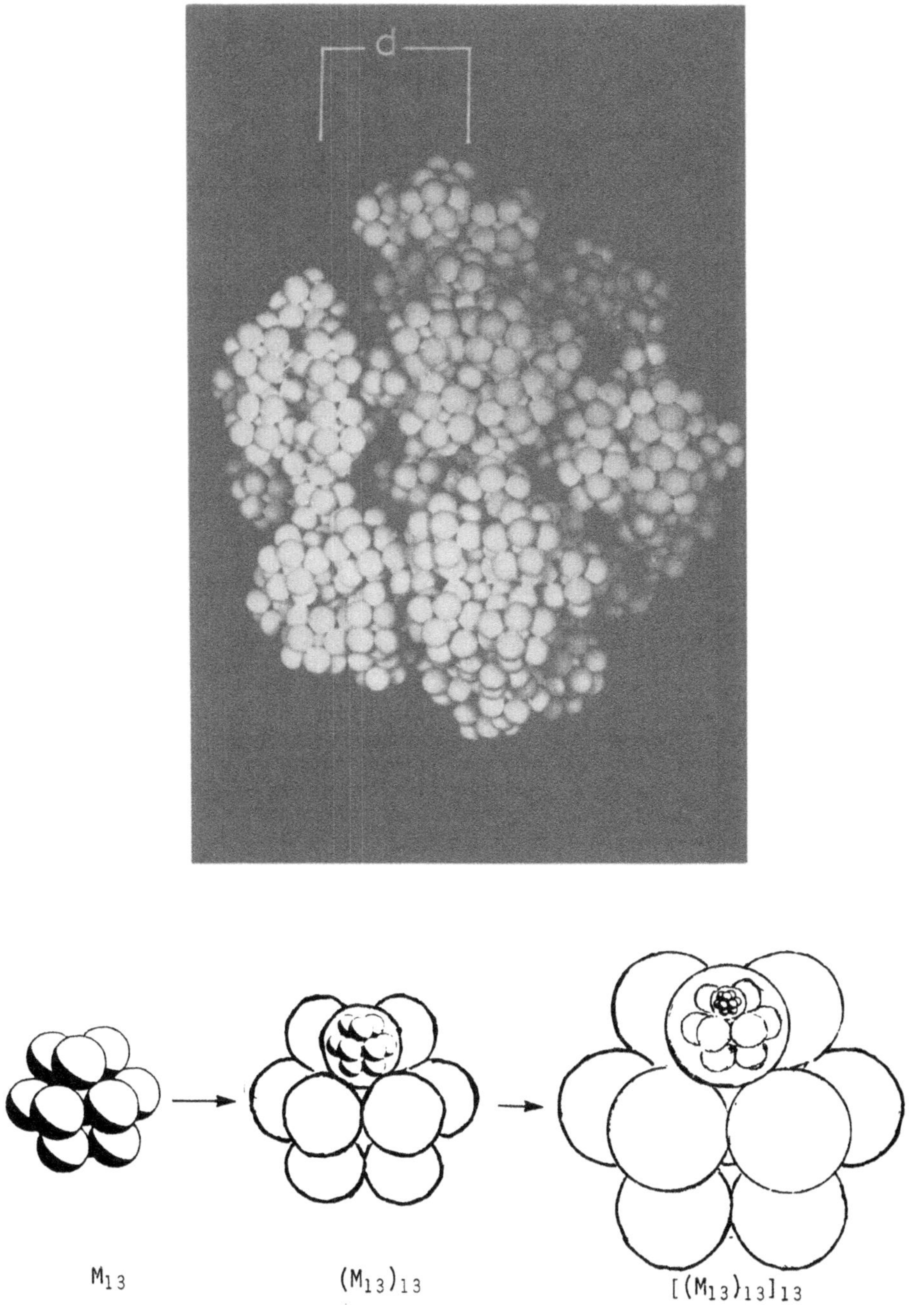

Fig. 2. Lower part shows the way in which the "polymeric" new metal lattices with fractal structure are built up by combining larger and larger metal clusters in the same cuboctahedral packing arrangement. Upper part gives an impression of the structure that results after repeating three times this procédé. The distance d is about 16 Å.

aggregation of similar cuboctahedral building blocks in increasingly larger units according to the same cubic close packing suggests an n-fold polymeric repetition of this process. This has still to be verified, but, if true, it would be quite interesting since the metal lattice generated in this way would correspond to a 3-dimensional fractal, the structure being self-similar down to a length-scale corresponding to the diameter of the basic M_{13} building block. The fractality should have important consequences for physical properties such as electronic transport and lattice vibrations.

Clearly, the so-obtained metal lattice can be viewed as a very special submicron structure. It seems highly improbable that it could ever by obtained by other methods than by chemical synthesis. The novel fractal metal structure, which has been named "cluster-metal" by Schmid [6], appears to have a surprisingly high stability. For the case of gold it was found that only by heating it to temperatures exceeding 400 °C, a stepwise transformation to the structure of normal bulk gold metal occurs. This should be compared to the melting temperature of 1064 °C for Au metal.

Experiments on these cluster-metals have just started. Besides their physical properties, it is obviously of great interest to study their chemical properties, in particular as regards catalysis.

SOME EXPERIMENTAL RESULTS

Our approach to unravel the electronic properties of the polynuclear metal clusters by physical measurements is to combine where possible the measurement of macroscopic quantities like magnetization, susceptibility, electric conduction, specific heat, with microscopic probes such as NMR, ESR and Mössbauer effect. Although a substantial amount of data have already been obtained, the study of the physics of these materials has really only just begun. Here we briefly summarize a number of results.

Starting with the magnetic susceptibility, we recall that in a previous note [8] we have reported the temperature dependence of the magnetic susceptibility of polynuclear carbonyl clusters, namely $[Pt_{38}(Co)_{44}H_2](PPN)_2$ and a number of Ni clusters, both large (Ni_{34}, Ni_{38}) and small (Ni_9, Ni_{12}). Independently, Teo et al. [9] have published susceptibility data on a large series of Pt carbonyl clusters, ranging from Pt_6 up to Pt_{38}. In all cases the temperature dependence of the susceptibility can be described by the formula:

$$\chi(T) = \chi_{dia} + \chi_0 + C_0/(T-\theta)$$

Here χ_{dia} is the negative diamagnetic susceptibility, which is temperature independent, χ_0 is a positive temperature independent contribution that is obtained after correction for χ_{dia}, and the last term is a contribution in the form of a Curie-Weiss law that dominates the behaviour at low temperatures ($\lesssim$ 100 K). Values found for the Curie-Weiss temperatures θ are rather small ($\theta \lesssim 10$ K). The effective moments deduced from the experimental Curie constants C_0 (assuming a spin value $S=\frac{1}{2}$) are quite small for the Pt clusters (0.1 - 0.5 μ_B per cluster), are only slightly larger for the small Ni clusters ($\simeq 1$ μ_B/cluster), and increase to 4 - 9 μ_B/cluster for the large Ni clusters. For comparison, we recall that in bulk Ni metal the magnetic moment is about 0.6 μ_B/atom, which would correspond to 20 - 23 μ_B/cluster for Ni_{34} and Ni_{38}. This evidences a strong reduction of the magnetic moment with respect to the bulk.

Similar results have now been obtained for the Schmid-clusters [10]. For the clusters of the nonmagnetic metals Au_{55}, Rh_{55}, Ru_{55} and Pt_{55} the susceptibility is low-valued, although in all cases a weak contribution of the Curie-Weiss form can be clearly distinguished at low temperatures. As an example fig. 3 shows the results for the Ru_{55} cluster. For the Co_{55}

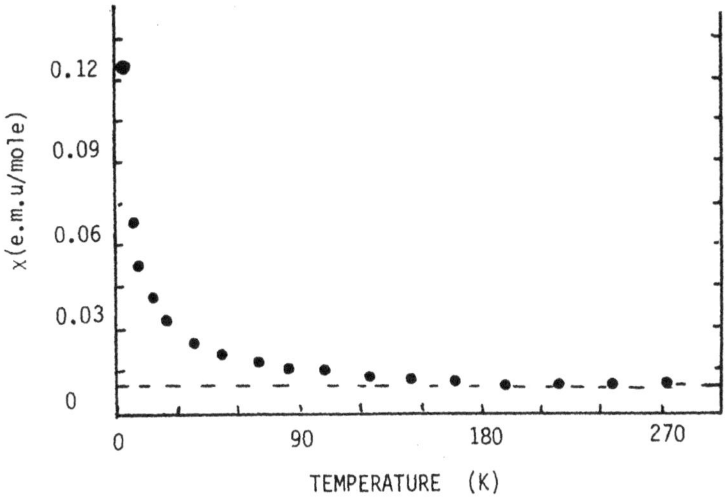

g. 3. Magnetic susceptibility versus temperature for the Ru_{55} Schmid-cluster.

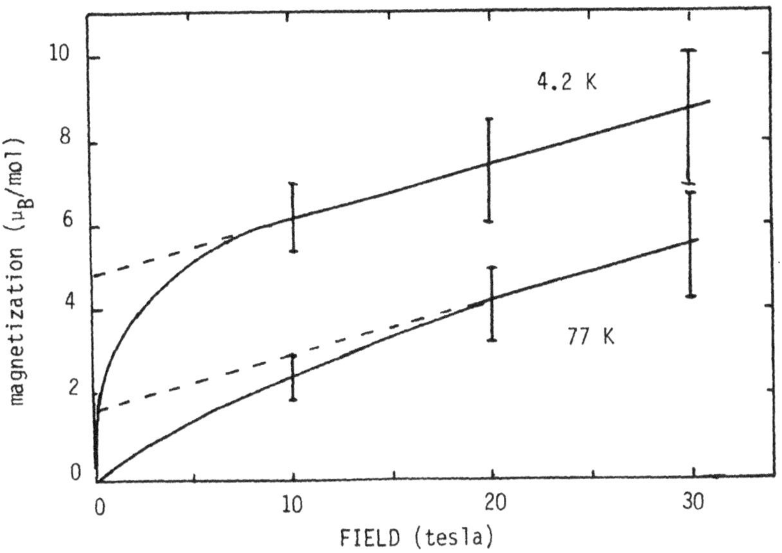

ig. 4. High-field magnetization curves at T = 4.2 K and T = 77 K for the Co_{55} Schmid-cluster.

cluster the susceptibility is considerably higher, although the value of about 5 μ_B/cluster is still very small as compared to the bulk Co metal (in Co metal the moment is 1.7 μ_B/atom which would give 95 μ_B/cluster for the Co_{55} cluster).

It thus appears that in all these metal cluster compounds there is a certain fraction of unpaired spin density residing on the cluster, the relative amount of which depends on whether the metal atom would be magnetic or nonmagnetic in the bulk, on the cluster size, etc.. Also the value of the temperature independent term χ_0 depends on the type of metal atom. Also for a series of small ruthenium carbonyl clusters (Ru_3, Ru_4, Ru_5, Ru_6, ...) that we are currently investigating in collaboration with B.F.G. Johnson and P.P. Edwards (Univ. of Cambridge), our susceptibility data show the typical diverging contribution at low temperatures in almost all cases.

Obviously we have been worried about the possible presence of magnetic impurities in the samples. Repeating the measurements on different samples of the same material, however, yielded essentially the same results. Only in some cases where the samples were air-sensitive differences have been observed that could be brought down to partial decomposition of the material, with the associated formation of free metal particles. These effects have been avoided by keeping the materials and sample holders constantly in nitrogen atmosphere.

Strong evidence that the magnetic behaviour cannot be due to paramagnetic impurities of the usual kind is obtained by measurement of the magnetization in very high magnetic fields. In fig. 4 the results for the Co_{55} cluster are shown up to 30 Tesla at T = 4.2 K. At this low temperature a paramagnetic impurity magnetization should saturate in fields of 2-4 Tesla already. By contrast after an initial more rapid increase the magnetization of Co_{55} keeps increasing up to the highest fields in a gradual, almost linear fashion. Previously we have reported similar behaviour for the large Ni clusters. Interestingly, magnetization data on the small Ru clusters at T = 1.6 K up to 5 Tesla also show no sign of saturation [10].

Detailed theoretical calculations for the observed magnetic behaviour are still lacking at present. Qualitative explanations run along the following lines. For **bare** (naked) small metal particles experimental evidence as well as Molecular Orbital calculations indicate that the magnetic moment per cluster should roughly correspond to what is estimated on basis of the bulk metal per atom value. At the same time, experimental results for the effects on the magnetic moment of metal surface atoms due to adsorption of molecules (CO, H_2) as well as M.O. calculations for such effects on magnetic clusters, clearly indicate a severe reduction of the magnetic moment. The extremely strong reductions of the magnetic moments of the large Ni and Co clusters with respect to the bulk values may thus be understood from the fact that the majority of the metal atoms are at the surface, where they are ligated to the ligand molecules. According to the M.O. calculations the effect of the ligands is to destabilize one or more levels that are occupied for the bare cluster [11]. The electrons of the destabilized orbital(s) pair-off the originally unpaired spins near the Fermi-level, thereby reducing the total magnetic moment.

It thus appears that the effect of ligand coordination to the metal core has strong analogies with the crystal-field effects on single metal atoms due to their surrounding ligands. These effects form an extensively studied part of the theory of transition metal compounds. Depending on the symmetry and strength of the ligand field the single magnetic metal atom can be in a high-spin or in a low-spin state. We are planning to extend these theories to clusters of more than one metal atom.

The appearance of unpaired spin density on the clusters of the nonmagnetic metals may be similarly understood. By the effect of the ligand field the energy level structure will profoundly be changed. The ensuing redistribution of the available electrons may then very well result in some small amount of unpaired spin density. The noninteger values found for the magnetic moments per cluster may be explained in terms of s-p hybridization; similar as in the band picture for magnetic metals. The behaviour of the high-field magnetization has strong analogy with the Stoner model for a magnetic metal, where the sub-bands of spin-up and spin-down levels are shifted with respect to one another, and full saturation is thus only reached for field values equal to the band width. In the metal cluster the energy level structure is discrete due to the finite size, but one can nevertheless distinguish "sub-bands" of spin-up and spin-down levels, as e.g. in the M.O.-calculations. The initial more rapid increase of the magnetization can be ascribed to saturation of the same unpaired spins that are responsible for the Curie-Weiss terms in the magnetic susceptibility. The unpaired spin density can also be seen in ESR experiments on e.g. the Pt carbonyl clusters. The ESR signal has a g-value slightly larger than the free-electron one and a temperature dependence that mirrors that of the magnetic susceptibility [12].

The electrical conductivity of some of these materials also shows highly interesting behaviour. For the Au_{55} cluster we have reported [13] the dc conductivity in the range 70 K - 350 K, where we found that the resistance of a pressed powder pellet was excellently described by the exponential law $R \propto \exp(T_0/T)^{\frac{1}{2}}$. The same behaviour has recently been measured for the Pd_{561} cluster compound over an even wider temperature range. The resistivity shows strong nonohmic effects at low temperatures.

It is quite surprising that these two features, viz. the square root in the exponential and the nonohmic behaviour are seen in a rather large variety of different physical systems, such as granular metals (cermets) and low-dimensional conductors. A consisent theoretical explanation of all these phenomena so far is lacking. The existing theoretical models are mainly extensions or modifications of Mott's original variable range hopping model for thermally activated conduction processes in disordered systems, and in particular on the analysis in terms of critical percolation paths by Ambegaokar et al. [14]. However, in view of the susceptibility results we have to conclude that the energy level structure in the metal clusters studied by us is still far from continuous, so that models for hopping charge motion between **metallic** particles may not be applicable. In this respect it is of interest to link the conductivity with models for molecular hopping conduction, as have been proposed for small-polaron conduction [15], and for intervalence charge transfer in mixed-valence compounds [16]. Indeed, it is well-known that the electron-phonon coupling in metal cluster compounds is very strong, i.e. even the smallest changes in the geometry of the cluster molecule have a profound effect on the electronic energy level distribution; in some cases also the valency of the molecule is changed. Thus the cluster molecule may easily accept an extra charge by slightly distorting, the charge transfer between neighbouring molecules being aided by vibronic couplings.

We finally discuss some of the NMR and Mössbauer measurements performed so far. The ^{195}Au Mössbauer spectrum of $Au_{55}(PPh_3)_{12}Cl_6$ had been studied earlier at T = 4.2 K [17]. We have recently extended these results [18] by measuring the spectra as a function of temperature in the range 0.5 K - 30 K. The Mössbauer effect provides a microscopic probe by which one may study the chemical environment of the metal atoms via their nuclear spins. Indeed, the spectra for the Au_{55} cluster can be decomposed into four different contributions related to four different types of Au atoms. These are the 13 Au atoms that form the central core of the cuboctahedron, the 24 uncoordinated peripheral, the 12 peripheral atoms coordinated to the PPh groups, and the 6 peripheral atoms coordinated to the Cl atoms. Thus one distinguishes three different "surface" sites and one "volume" site, where the 13 inner Au atoms are completely surrounded by other metal atoms. Indeed the Mössbauer parameters for the volume site are found to be close to those of bulk Au metal, whereas those for the surface sites lie in the range of parameters appropriate for non-conducting Au compounds.

The temperature dependence of the spectra is much more pronounced than for Au metal, for which the spectrum hardly changes with temperature for $T < 80$ K. On the other hand it is quite similar to what has been observed for small Au particles in colloidal Au [19]. The temperature dependences of the Mössbauer spectra reflect the Debije-Waller factors associated with the different metal sites. Thus the Mössbauer effect provides a microscopic method to obtain information on the phonon excitations within the metal particle itself, i.e. apart from its environment. In addition it is found that to explain the observed temperature dependence, also a vibration of the metal particle as a whole (within the surrounding matrix) has to be taken into account [18].

With the NMR technique one may also use the nuclear spin of the metal atom, e.g. Pt, to probe their physical and chemical properties. The NMR resonance line in insulating Pt compounds is determined by the chemical shifts, whereas in bulk Pt metal the conduction electrons produce an additional, much larger shift, the well-known Knight-shift. In principle this provides one of the most direct ways to decide whether a particle is metallic or not. A second criterion is provided by the temperature dependence of the nuclear-spin-lattice relaxation time (T_1), which in a metal should follow the Korringa relation (T_1T = constant), and which in a dielectric is very much longer. Measurements of solid state pulse NMR on $[Pt_{38}(CO)_{44}H_2](PPN)_2$ and $[Pt_{26}(CO)_{32}](PPh_4)_2$ by means of the spin-echo technique yield resonance positions that are in the range of known chemical shifts and T_1's which do not follow the Korringa law [12]. These findings could be explained by the fact that in these materials all but a few Pt atoms are still coordinated by the ligands (CO). On the other hand theoretical calculations of the Knight shift for very small **bare** metal particles also predict an extreme reduction of the overall Knight shift. Furthermore, the NMR lineshapes predicted from these calculations, as well as those observed experimentally in bare Pt-particles of comparable sizes, are both quite similar to what is observed in these two Pt carbonyl cluster compounds [12]. Therefore it is still difficult to decide whether the finite particle size or the ligand coordination (or a combination of both) are responsible.

ACKNOWLEDGEMENTS

The research on metal cluster compounds is the joint effort of a large number of people, physicists as well as chemists. I am greatly indebted to all the colleagues and coworkers who have contributed, each in their own way, to obtain the results described.

The work in Leiden is part of the research program of the "Stichting voor Fundamenteel Onderzoek der Materie" (Foundation for Fundamental Research on Matter) and was made possible by financial support from the "Nederlandse Organisatie voor Zuiver-Wetenschappelijk Onderzoek" (Netherlands Organization for the Advancement of Pure Research). The investigations are sponsored by the Leiden Materials Science Group ("Werkgroep Fundamenteel Materialen Onderzoek"). The research program is also sponsored by the Commission of the European Communities.

REFERENCES

1. See e.g. L.J. de Jongh in "Magneto Structural Correlations in Exchange Coupled Systems, ed. R.D. Willett, D. Gatteschi and O. Kahn, Reidel, 1983.
2. P. Chini, Gazetta Chimica Italiana 109:225-240 (1979).
3. "Transition metal clusters", ed. B.F.G. Johnson, Wiley, 1980.

4. G. Schmid in "Structure and Bonding", 62:51-85 (1985).
5. M.N. Vargaftik, V.P. Zagorodnikov, I.P. Stolyarov, I.I.Moiseev, V.A. Likholobov, D.I. Kochubey, A.L. Chuvilin, V.I. Zaikovsky, K.I. Zamaraev, and G.I. Timofeeva, J. Chem. Soc., Chem. Commun., 937 (1985).
6. G. Schmid, Nachr. Chem. Tech. Lab. 34: nr. 3, p.249 (1987).
7. G. Schmid and N. Klein, Angew. Chem., 98:910 (1986).
8. B.J. Pronk, H.B. Brom, L.J. de Jongh, G. Longoni and A. Ceriotti, Solid State Commun. 59:349 (1986).
9. B.K. Teo, F.J. Disalvo, J.W. Waszczak, G. Longoni and A. Ceriotti, Inorg. Chem. 25:2262 (1986).
10. A. Mees, J. Darriet, Albino O. de Aguiar, M.J.G.M. Jurgens, to be published.
11. D.R. Salahub and F. Raatz, Int. J. Quantum Chem. S18:173 (1984); Surface Science 146:L609 (1984); G.F. Holland, D.E. Ellis, and W.C. Trogler, J. Chem. Phys. 83:3507 (1985).
12. B.J. Pronk, H.B. Brom, G. Longoni and A. Ceriotti, submitted to Solid State Commun.
13. M.P.J. van Staveren, H.B. Brom, L.J. de Jongh and G. Schmid, Solid State Commun. 60:319 (1986); and to be published.
14. V. Ambegaokar, B.I. Halperin, and J.S. Langer, Phys. Rev. B4:2612 (1971).
15. T. Holstein, Ann. Phys. 8:325 (1959); J. Appel. Solid State Phys. 21:193 (1968); I.G. Auster and N.F. Mott, Adv. Phys. 18:41 (1969).
16. K.Y. Wong and P.N. Schatz, Progr. Inorg. Chem. 28:369 (1981); K. Nasu and Y. Toyazawa, J. Phys. Soc. Japan 51:2098 (1982); P. Day, contribution to this volume.
17. G. Schmid, R. Pfeil, R. Boese, F. Bandermann, S. Meyer, G.H.M. Calis and J.W.A. van der Velden, Chem. Ber. 114:3634 (1981).
18. H.H.A. Smit, R.C. Thiel, L.J. de Jongh and G. Schmid, to be published.
19. M.P.A. Viegers and J.M. Trooster, Phys. Rev. B15:72 (1977).

METAL CLUSTER ACTIVE SITES IN PROTEINS

Edward I. Solomon*, James E. Pate, T. David Westmoreland,
Lung-Shan Kau, Mark D. Allendorf, and Darlene J. Spira-Solomon

Department of Chemistry, Stanford University
Stanford, California 94305

INTRODUCTION

Binuclear and multimetal cluster centers are found in a wide variety of enzymes involved in electron transfer and small molecule activation. It is important to define the contributions of interactions between the metal ions in an active site to its structure and function. Attention is focussed on the coupled binuclear copper sites which occur in hemocyanin (reversible O_2 binding), tyrosinase (O_2 activation and hydroxylation) and laccase (O_2 reduction to water). Excited state spectral studies on active site derivatives are used to define exogenous ligand bridging between the two coppers in the active sites of hemocyanin and tyrosinase. A chemical and spectroscopic comparison of the coupled site in hemocyanin to the binuclear center (T3) in laccase indicates that the sites are similar with respect to an endogenous bridge responsible for antiferromagnetic coupling but differ in that exogenous ligands bind to only one copper at the T3 site. In addition to the T3 coupled binuclear copper site, laccase contains two other coppers, the Type 1 and Type 2 centers. X-ray absorption edge spectral studies of the 1s->4p transition at 8984 eV are used to demonstrate that in the absence of the T2 copper the T3 site is reduced and will not react with O_2. The role of the T2 and T3 coppers in exogenous ligand binding has been probed through low temperature MCD which allows a correlation between the excited state and ground state spectral features. Through these low temperature MCD studies it is demonstrated that one azide produces charge transfer intensity to both the paramagnetic T2 and the antiferromagnetic T3 center defining a new trinuclear copper active site which appears to be important in the irreversible multi-electron reduction of dioxygen to water.

Many metalloenzymes contain more than one metal atom or a metal cluster in their active sites. These enzymes catalyze a wide variety of specialized functions which include electron transfer and small molecule activation. The main classes of these multicenter enzyme active sites are summarized in Figure 1. The largest class is based on the

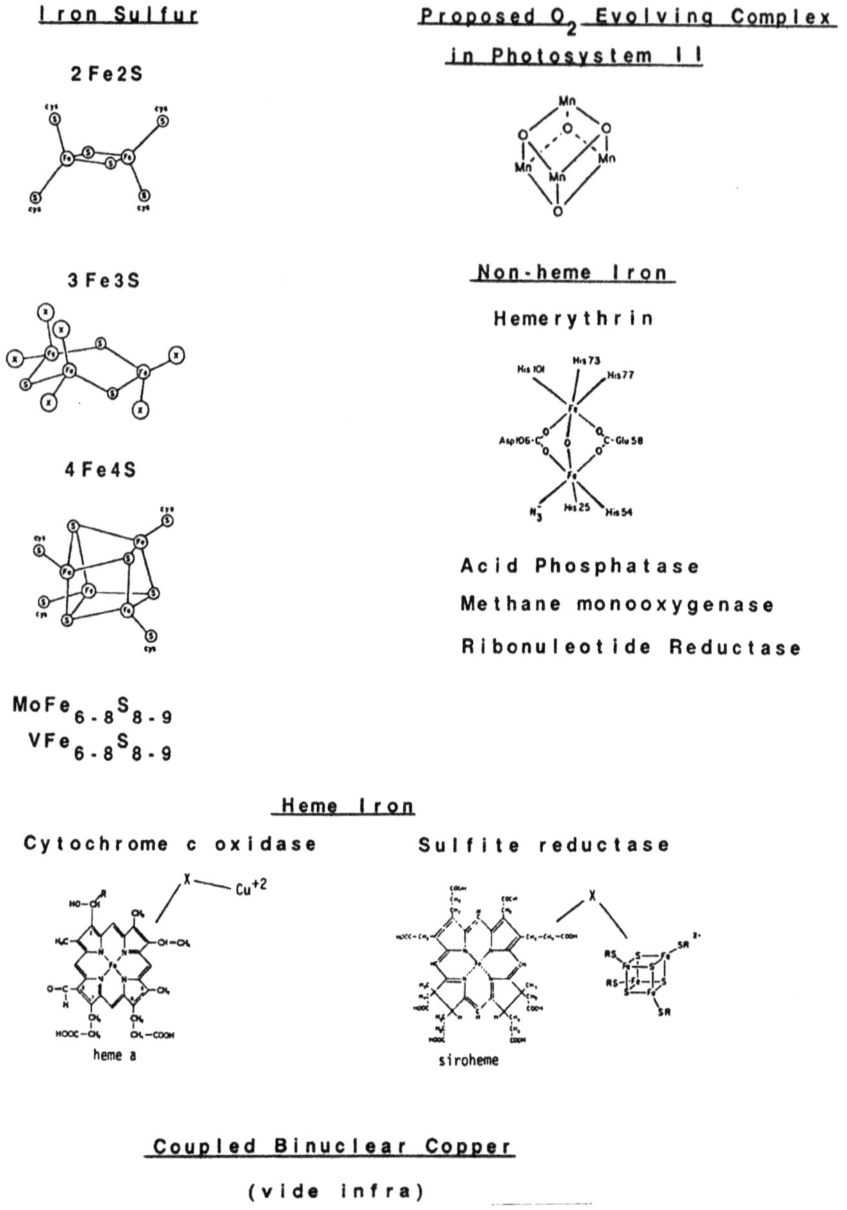

Figure 1. Summary of the classes of coupled metal clusters found in protein active sites.

iron-sulfur clusters which contain two, three, or four iron atoms in association with a number of bridging sulfide and terminal thiolate ligands. These sites are generally involved in electron transfer reactions.[1] In addition, the six electron reduction of dinitrogen to ammonia by nitrogenase takes place at a $\sim$six-iron cluster which also contains molybdenum[2] or vanadium.[3] These iron sulfur clusters exist in a variety of oxidation states, each with distinctive electronic structural features. Of particular note is the two-iron center (2Fe2S) which in its [Fe(III)Fe(II)] oxidation state shows properties of a class II mixed-valent complex with an antiferromagnetic S=1/2 ground state.[4] Alternatively, the four-iron cluster is described as two antiferromagnetically coupled 2Fe2S clusters in which each two-iron center is class III mixed-valent with a ferromagnetic S=9/2 ground state.[5]

A similar Mn_4O_4 cluster has recently been proposed for the oxygen evolving complex in Photosystem II based on EPR data.[6] An energy level ordering of S=3/2 (ground state) < 1/2 < 5/2 results when the observed data is fit to a spin hamiltonian which contains both ferromagnetic and antiferromagnetic exchange interactions. Thus, the manganese cluster appears to display electronic interactions similar to those in the 4Fe4S clusters.

A class of binuclear non-heme non-sulfur iron centers also exists. In general, these enzymes in their ferrous oxidation states react with dioxygen either reversibly (hemerythrin) or irreversibly (ribonucleotide reductase,etc.). Low temperature magnetic circular dichroism (MCD) spectroscopy of deoxyhemerythrin has shown that the [Fe(II)Fe(II)] form is antiferromagnetically coupled through a hydroxide bridge.[7] The binding of exogenous[8] ligands causes the bridge to protonate and an EPR observable ferromagnetic S=4, ground state results. Alternatively, reaction with dioxygen produces a [Fe(III)Fe(III)] dimer which is strongly antiferromagnetically coupled through an oxo bridge. Since exogenous ligands can coordinate to only one iron, it is important to define the role of the superexchange pathway associated with the endogenous bridge ($H_2O \longleftrightarrow OH^- \longleftrightarrow O^{2-}$) in multi-electron reductions at the binuclear non-heme iron site.

There are two active site clusters involving heme iron. Cytochrome c oxidase contains a high spin ferriheme a antiferromagnetically coupled to a Cu(II) center ($-2J > 200$ cm^{-1}) yielding an S=2 ground state.[9] The sulfite and nitrite reductases contain high spin ferric siroheme coupled to a 4Fe4S cluster. Both types of coupled sites are involved in multielectron interactions with small molecules.[10]

A major focus of multicenter metalloenzyme research has been the coupled binuclear copper active site. This site has been found in a wide variety of proteins and enzymes involved in various dioxygen utilizing reactions as summarized in Table I. Particularly, hemocyanin and tyrosinase in the deoxy [Cu(I)Cu(I)] state react with dioxygen to produce stable oxygenated species with similar spectroscopic features (Figure 2). In contrast to hemocyanin, which binds dioxygen reversibly, tyrosinase oxygenates monophenols to ortho-diphenols. The simplest of the multicopper enzymes is laccase[11] which contains in addition to a coupled binuclear site, two other copper centers that together catalyze the irreversible four-electron reduction of dioxygen to water at the binuclear site. These proteins provide an important series of active sites where one can ultimately hope to correlate geometric and electronic structure with biological function.

Hemocyanin	Arthropod	deoxy + $O_2 \rightleftarrows$ oxy
	Mollusc	deoxy + $O_2 \rightleftarrows$ oxy $2H_2O_2 \rightleftarrows O_2 + 2H_2O$
Tyrosinase		deoxy + $O_2 \rightleftarrows$ oxy $2H_2O_2 \rightleftarrows O_2 + 2H_2O$ phenol + $2e^- + O_2 + 2H^+ \rightleftarrows$ <u>o</u>-diphenol + H_2O 2<u>o</u>-diphenol + $O_2 \rightleftarrows$ 2<u>o</u>-quinone + $2H_2O$
Laccase Ceruloplasmin Ascorbic Acid Oxidase		$4AH + O_2 \rightarrow 4A + 2H_2O$

Table I. Summary of proteins containing coupled binuclear copper active sites and their functions.

Resonance Raman spectroscopy has shown that in oxyhemocyanin and oxytyrosinase bound dioxygen has been formally reduced to peroxide and thus the binuclear site exists in the [Cu(II)Cu(II)] oxidation state.[12] However, the spectroscopic features of oxyhemocyanin shown in Figure 2 are quite unusual when compared to those of normal, monomeric, tetragonal cupric complexes. Instead of the normal weak d-d transitions at ~ 600 nm, oxyhemocyanin exhibits two intense absorptions bands in the visible and near-UV. Further, in contrast to the usual tetragonal Cu(II) EPR spectra, oxyhemocyanin is EPR nondetectable. Since both coppers in the active site are known to be d^9 cupric from X-ray absorption edge studies,[13,14] they must be strongly antiferromagnetically coupled. SQUID magnetic susceptibility studies have only been able to put a lower limit on the magnitude of this coupling of $-2J > 600\ cm^{-1}$ ($H_{ex} = -2JS_i \cdot S_j$).[15]

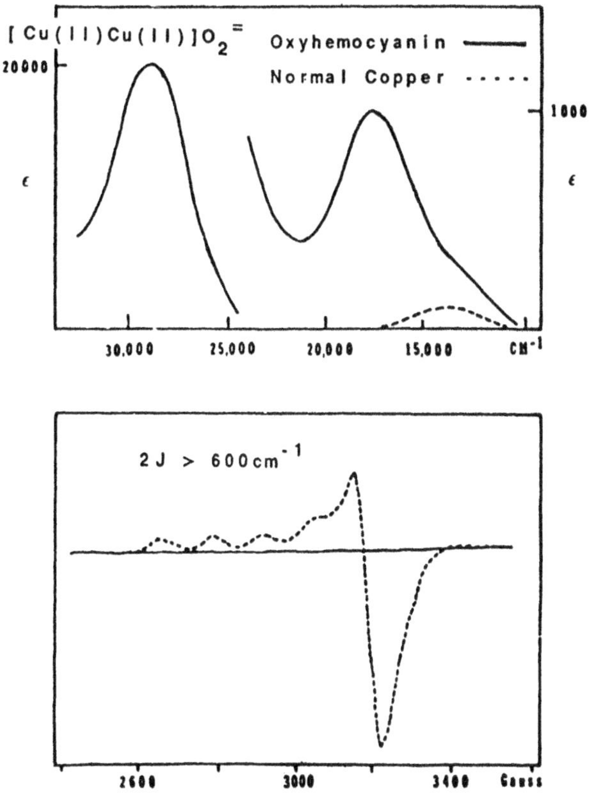

Figure 2. Comparison of absorption (top) and EPR (bottom) spectra of normal tetragonal copper(II) and oxyhemocyanin.

Since the ground state is spectroscopically inaccessible, the excited state features must be employed to probe the interactions of dioxygen with the coppers in the oxyhemocyanin active site. Figure 3 presents the spectroscopic changes observed by displacement of peroxide from the oxyhemocyanin site, producing the met derivative [Cu(II)Cu(II)].[16] The two intense bands in the oxyhemocyanin absorption spectrum at 350 nm (ε = 20,000 $M^{-1}\ cm^{-1}$) and 600 nm (ε = 1000

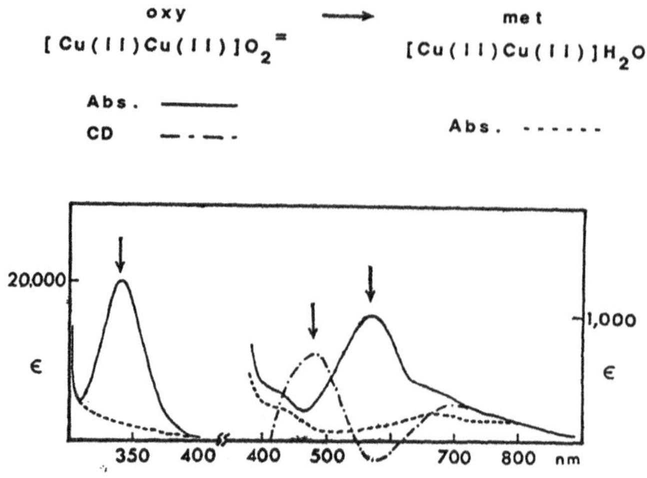

Figure 3. Absorption and circular dichroism spectra of oxy-hemocyanin compared with absorption of methemocyanin.

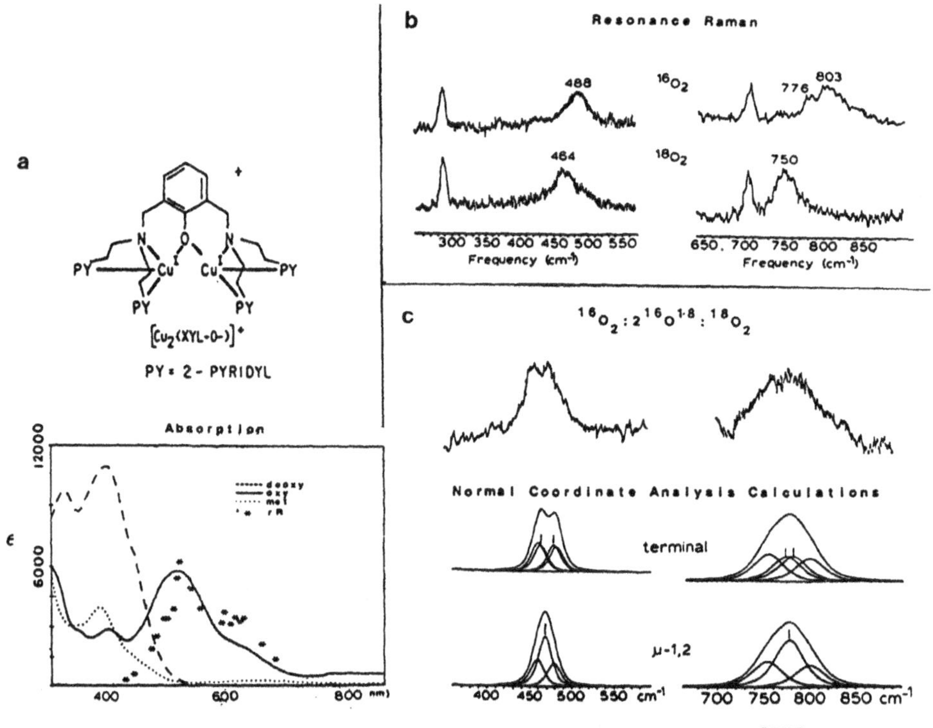

Figure 4. (a) Molecular structure of O_2 binding model complex[17] and absorption spectra of deoxy, oxy and met forms. Also included is the resonance Raman profile for the O-O stretch in the oxy complex. (b) Resonance Raman spectra in $^{16}O_2$ and $^{18}O_2$ of the oxy complex. The unlabelled peaks are due to the CH_2Cl_2 solvent. (c) Comparison of observed resonance Raman spectra for isotope mixture with predicted spectra for two coordination geometries.

M^{-1} cm^{-1}) are eliminated as is a feature in the circular dichroism (CD) spectrum at 486 nm ($\Delta\varepsilon$ = +2.53 M^{-1} cm^{-1}). These three features are therefore assigned as O_2^{2-} --> Cu(II) charge transfer transitions. Weak d-d transitions which are sensitive to the average ligand field at the coppers remain at ∿630 nm (ε ∿ 250 M^{-1} cm^{-1}) and reflect an approximately tetragonal coordination geometry.

It is thus important to define the charge transfer spectral features associated with a O_2^{2-}-Cu(II) bond and to use these features to obtain insight into the charge transfer spectrum of oxyhemocyanin. Karlin and coworkers have synthesized a binuclear copper(I) model complex which reacts with dioxygen to give the absorption spectrum changes illustrated in Figure 4a.[17] Resonance Raman excitation into these features in the presence of $^{16}O_2$ reveals a peak at 803 cm^{-1} which shifts to 750 cm^{-1} in $^{18}O_2$ indicating that dioxygen is bound as peroxide (Figure 4b). The mode of peroxide coordination was probed through an isotopic perturbation of the resonance Raman spectrum combined with a normal coordinate analysis. In Figure 4c is presented the resonance Raman spectrum of the oxygenated complex in the Cu-O_2 and O-O stretch regions with a statistical mixture of oxygen isotopes ($^{16}O_2$: 2$^{16}O^{18}O$: $^{18}O_2$). Also shown is the vibrational spectrum calculated using experimental lineshapes for both terminal and bridging peroxide. The metal-ligand stretch clearly allows these possibilities to be distinguished and indicates that peroxide is bound asymmetrically and thus to only one copper. Therefore, from the resonance Raman enhancement profile of the O-O stretch (Figure 4a) a terminally bound peroxide-cupric complex exhibits two charge transfer transitions, a band at 503 nm (ε = 6300 M^{-1} cm^{-1}) and a less intense lower energy shoulder at 625 nm (ε = 1100 M^{-1} cm^{-1}).

These O_2^{2-} --> Cu(II) charge transfer bands can be assigned using the energy level diagram shown in Figure 5. The highest energy occupied orbital of peroxide is a doubly degenerate π^* level. Upon coordination to Cu this degeneracy is lifted and results in a π_σ^* level which is oriented along the Cu-O bond and a π_v^* which is perpendicular to the Cu-O bond. The π_σ^* is more strongly stabilized by the bonding interaction with the copper. Further, the highest energy, half-occupied orbital of tetragonal Cu(II) is $d_{x^2-y^2}$ which has a lobe directed at the peroxide ligand in the equatorial plane. Since the intensity of charge transfer transitions is proportional to overlap of the donor and acceptor orbitals, this analysis predicts a relatively weak low energy π_v^* --> Cu(II) transition and a more intense π_σ^* --> Cu(II) transition.

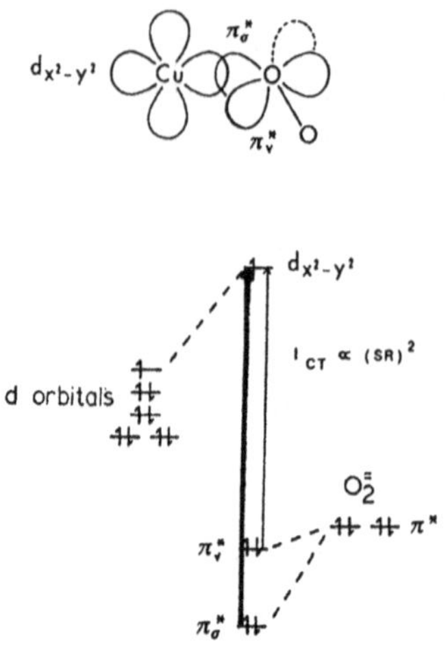

Figure 5. Molecular orbital diagram for terminally bound peroxide. The two possible charge transfer transitions are indicated with relative intensity.

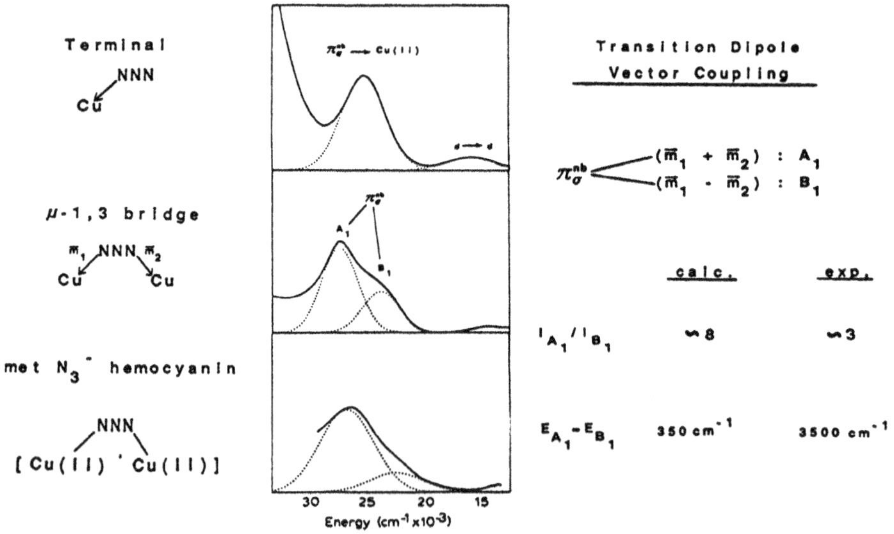

Figure 6. Absorption spectra with Gaussian resolution for terminally bound and bridging azide model complexes and for met-azide hemocyanin. Also shown are results of TDVC calculations for the bridging model complex.

Oxyhemocyanin, however, exhibits three charge transfer transitions. Since the above analysis shows that only two bands are possible for peroxide bound to a single copper, the three band pattern must have another origin.

This analysis must then be extended to consider the effects of interactions with two coppers on the charge transfer transitions of a bridging ligand. Structurally defined monomers and dimers of Cu(II) with azide exist and are suitable models for probing these bridging effects.[18] As shown in Figure 6, azide bound to a single copper gives rise to one relatively intense absorption band. In analogy to peroxide, this transition originates from the π_{σ}^{nb} level; the corresponding π_{v}^{nb} --> Cu(II) transition is too weak to be observed. In the dimer, the π_{σ}^{nb} absorption band is observed to split into two bands with the one at higher energy having more intensity. A Transition Dipole Vector Coupling (TDVC) Model has been developed to analyze these data.[16] In the dimer, the transition dipoles of the N_3^- --> Cu(II) charge transfer transitions to each copper will couple to form symmetric and antisymmetric combinations. From group theory, these coupled transition dipoles have A_1 and B_1 symmetry (in C_{2v}), respectively. The intensity of each dimer transition is given by the square of the vector sum or difference of the individual transition moments. For the dimer, the intensity ratio I_{A1}/I_{B1} is predicted to be ~ 8 indicating that the higher energy band is $\pi_{A_1}^{nb}$ --> Cu(II) and the band at 625 nm is $\pi_{B_1}^{nb}$ --> Cu(II). Further, the coulomb interaction between the two transition moments can be estimated for the μ-1,3 geometry to obtain an approximate value for the energy difference between the A_1 and B_1 dimer states. The calculated value gives the correct energy ordering with the A_1 component at higher energy but is much smaller than the observed splitting, which indicates that exchange terms contribute significantly to the splitting. Finally, addition of azide to methemocyanin yields a charge transfer spectrum which shows an A_1, B_1 pattern very similar to the spectrum of the μ-1,3 azide model complex. Thus azide binds in an analogous μ-1,3 bridging geometry to the binuclear site in methemocyanin.

This approach can now be applied to interpret the charge transfer spectrum of oxyhemocyanin with peroxide bridging in a μ-1,2 fashion.[16] In the dimer, the π_{σ}^* and π_{v}^* charge transfer transitions of the monomer each split into two bands. The symmetry types, selection rules, intensity ratios, and coulomb splittings have been calculated and are

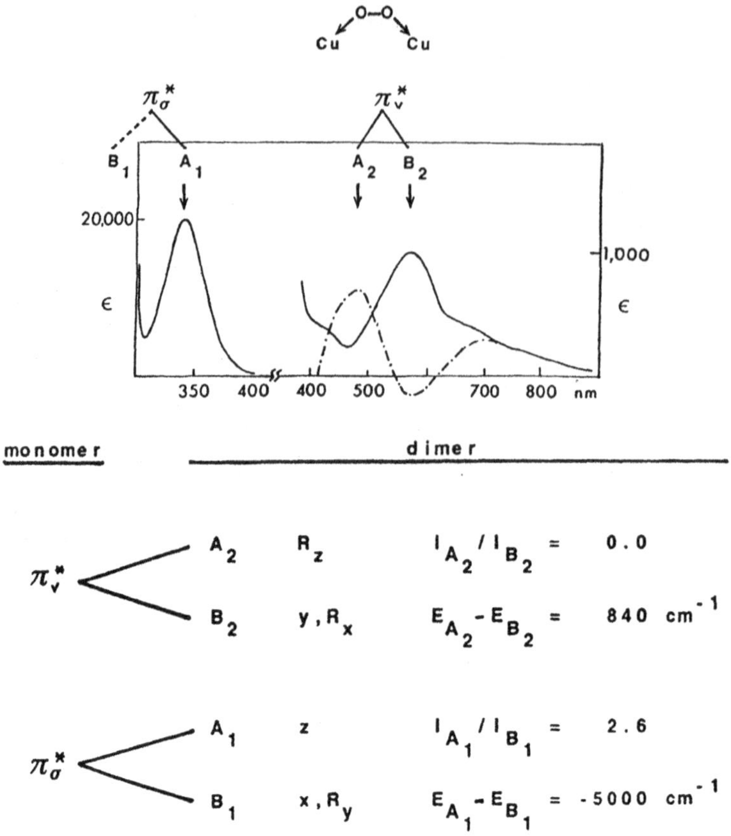

Figure 7. Assignment of the $O_2^{2-} \rightarrow Cu(II)$ charge transfer spectrum of oxyhemocyanin with TDVC calculations.

given in Figure 7. The π_v^* --> Cu(II) transition will split into two components, a lower energy B_2 band which should contain most of the absorption intensity and a higher energy A_2 transition having little absorption intensity. This higher energy transition is, however, magnetic dipole allowed and is therefore predicted to be relatively intense in the CD spectrum. The predicted relative energies and absorption and CD intensities correlate with the bands at 600 nm and 486 nm, allowing their assignment as the B_2 and A_2 components, respectively. The π_σ^* --> Cu(II) transition should be analogously split into a low energy A_1 component with most of the intensity and a B_1 component at higher energy. The A_1 band is associated with the very intense 350 nm band while the B_1 component is obscured by intense protein absorption at higher energy.

From the preceding interpretation of the unique spectroscopic features of oxyhemocyanin and oxytyrosinase a spectroscopically effective model of the active site may be generated as illustrated in Figure 8.[16] In the site, two tetragonal cupric ions are bridged by both an endogenous ligand (OR^-) and the exogenous μ-1,2 peroxide. The endogenous bridge is responsible for the strong antiferromagnetic coupling between the coppers and the bridging peroxide gives rise to the intense electronic absorption features.

This model for the coupled binuclear copper site in hemocyanin and tyrosinase may now be compared to the parallel site in laccase which contains a blue copper (denoted Type 1 or T1), a normal copper (Type 2, T2), and a coupled binuclear copper (Type 3, T3) center. Initial studies have focussed on the simplified Type 2 depleted (T2D) derivative[19] in which the T2 center has been reversibly removed. Native laccase exhibits an absorption band at 600 nm ($\varepsilon = 5700\ M^{-1}\ cm^{-1}$) which is associated with the T1 copper and a peak at 330 nm originally assigned to the T3 center (Figure 9).[20] Upon removal of the T2 copper the 330 nm absorption disappears, raising doubts regarding its assignment as a Type 3 spectral feature. As shown in Figure 9, both the T1 and T2 sites contribute to the EPR spectrum of native laccase (the T3 cupric ions are coupled and hence EPR nondetectable as in hemocyanin) while the T2 contributions are absent in the T2D derivative.

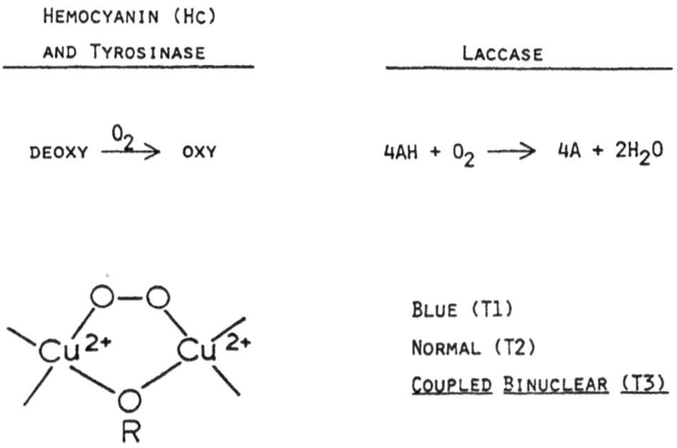

Figure 8. The spectroscopically effective active site of hemocyanin and tyrosinase. Also indicated is a structure-function comparison with laccase.

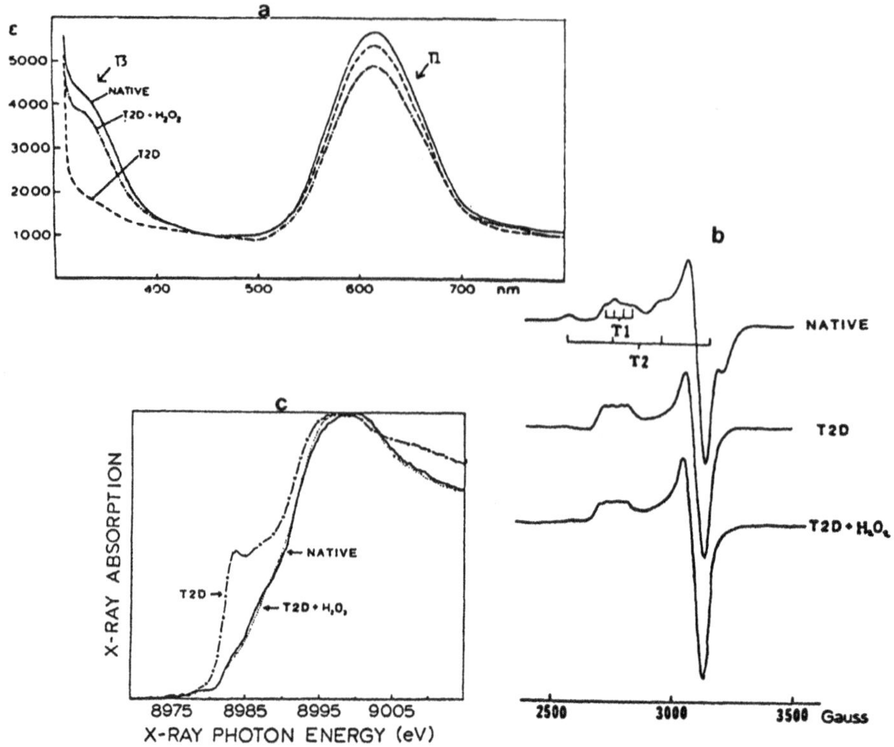

Figure 9. (a) Optical absorption, (b) EPR, and (c) X-ray absorption edge spectra for native and T2D laccase and after reaction of T2D with peroxide.

When T2D laccase is reacted with peroxide the 330 nm band reappears.[21] This observation suggested the possibility that the T3 site might be reduced in the T2D derivative but could be oxidized by H_2O_2. Alternatively, other researchers assigned the 330 nm band as an O_2^{2-} --> Cu(II) charge transfer transition in analogy to oxyhemocyanin and therefore associated this spectral feature with peroxide binding to the T3 site.[22] An absorption feature at 330 nm, however, cannot directly distinguish between these possibilites. Also, the EPR spectrum is not diagnostic of the T3 oxidation state since both a reduced site and an antiferromagnetically coupled oxidized binuclear cupric site would be EPR nondetectable. A direct probe of oxidation state is, however, available through X-ray absorption edge spectroscopy. The T2D form exhibits a peak at 8984 eV which is eliminated by reaction with peroxide, qualitatively indicating that Cu(I) is being oxidized to Cu(II).[21] Such conclusions must be drawn cautiously however, since covalent cupric complexes can also exhibit X-ray absorption edge intensity at low energy.[23]

The oxidation state dependence of the X-ray absorption edge feature was defined and quantitated by a systematic study of a series of nineteen Cu(I) and forty Cu(II) complexes.[14,24] The Cu K edge band energies and shapes depend on oxidation state, ligation, and coordination geometry. Figure 10 (top) presents representative edges for two, three, and four coordinate Cu(I), and tetragonal Cu(II) complexes with a variety of ligand sets. Two or three coordinate Cu(I) complexes always exhibit a peak at energies below 8984 eV. In addition, the three coordinate complexes display a double peak with lower intensity. No Cu(II) complex studied shows a peak below 8986 eV. There is however a low energy tail in the 8984 eV region which is associated with a peak at higher energy.

These spectral differences can be understood through ligand field theory.[14] For two coordinate Cu(I) with an approximately linear geometry, the electric dipole allowed 1s --> 4p transition is predicted to split into 1s --> $4p_z$ and 1s --> $4p_{x,y}$ components. Further, the 1s --> $4p_z$ transition is expected to be highest in energy due to repulsive interactions with the axial ligands. Polarized single crystal X-ray edge spectroscopy confirms the assignment of the lowest energy peak as 1s --> $4p_{x,y}$. As shown at the bottom of Figure 10, increasing the ligand field

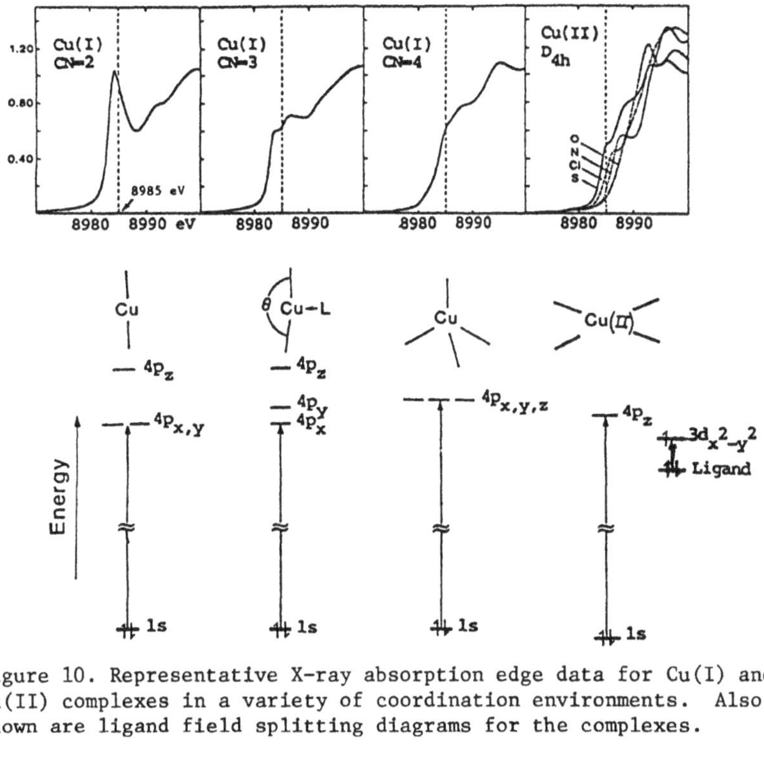

Figure 10. Representative X-ray absorption edge data for Cu(I) and Cu(II) complexes in a variety of coordination environments. Also shown are ligand field splitting diagrams for the complexes.

strength by adding a third ligand along the y axis leads to an additional repulsive interaction which raises the energy of the $4p_y$ level relative to the $4p_x$. This accounts for the experimentally observed edge splitting in the three coordinate Cu(I) complexes. For a tetrahedral geometry, all p orbitals are equally destabilized by the ligand field and the 1s --> 4p transitions are all above 8985 eV.

For the tetragonal Cu(II) complexes, both the low energy tail and its associated higher energy peak shift to lower energy as the covalent interaction with the equatorial ligand set increases. Since this band is z polarized with an energy shift dependent on equatorial ligand ionization energy, this transition can be assigned as the 1s --> $4p_z$ combined with a ligand --> Cu(II) charge transfer shake-up, which is also supported by final state calculations.[25]

Turning to the X-ray absorption edge spectrum of the T3 site in T2D laccase (Figure 11), a peak is observed below 8984 eV. This energy is characteristic of Cu(I) while the shape of the peak suggests that it is due to three coordinate copper. The amount of reduced copper present can then be quantitated from the normalized edge intensities of copper model

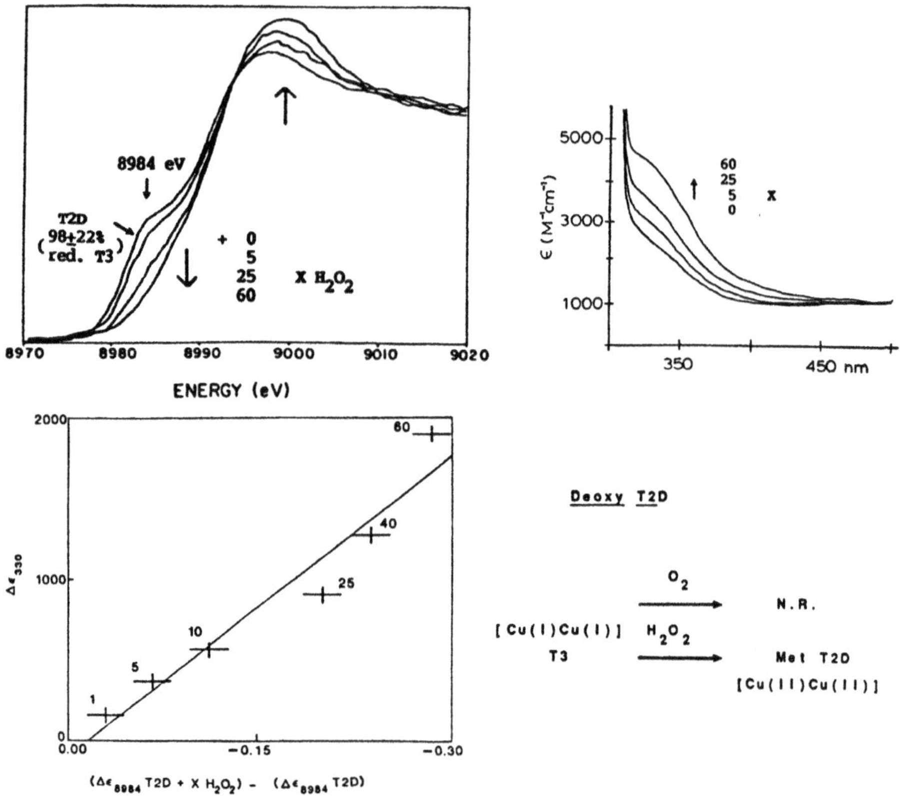

Figure 11. Correlation of X-ray edge data and optical absorption for the reaction of T2D laccase with peroxide.

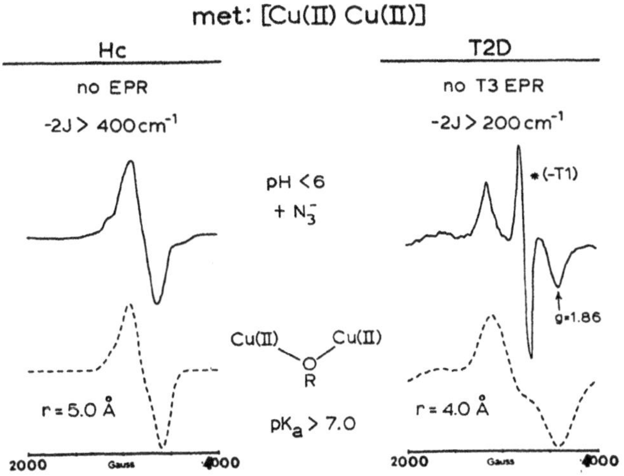

Figure 12. Comparison of EPR spectra of the uncoupled met derivatives of hemocyanin and T2D laccase with simulations for dipolar interacting copper(II)'s at distances indicated. The peak indicated by the asterisk results from subtraction of the T1 signal.

complexes with the appropriate coordination number. For T2D laccase, the 8984 eV peak indicates that 98 ± 22% of the T3 sites are reduced. Peroxide reacted T2D laccase shows no 8984 eV feature and corresponds quantitatively to a fully oxidized T3 site.

The possibility of the 330 nm absorption band being O_2^{2-} --> Cu(II) charge transfer transition was investigated through a correlation of the 330 nm absorption intensity increase with the loss of the 8984 eV Cu(I) peak as T2D was titrated with H_2O_2. The linear correlation in Figure 11 demonstrates that all change at 330 nm is associated with oxidation of the T3 site with no evidence for additional peroxide binding to the oxidized Cu(II). Thus, the T3 site in T2D laccase is a deoxy [Cu(I)Cu(I)] form which, in strong contrast to deoxyhemocyanin and deoxytyrosinase, does not react with dioxygen. It is however oxidized by peroxide to give the binuclear cupric met derivative, [Cu(II)Cu(II)].[26]

In comparing the spectroscopic properties of the coupled binuclear copper site in met T2D laccase with that of methemocyanin, it is noted that both have no EPR signal. Since both sites contain two Cu(II)'s, the sites must be antiferromagnetically coupled. SQUID magnetic susceptibility measurements place a lower limit on the singlet-triplet splitting of $-2J > 400\ cm^{-1}$ for methemocyanin and $-2J > 200\ cm^{-1}$ for met T2D laccase.[27] By lowering the pH and adding azide a new broad EPR signal is produced in both met derivatives (Figure 12).[26,28] This signal accounts for ∿15% of the T3 sites and can be simulated based on two dipolar interacting cupric ions. In met T2D laccase this signal clearly shows a high field component at g = 1.86 associated with a larger zero field splitting and thus shorter Cu-Cu distance relative to methemocyanin. These results are interpreted in terms of an endogenous bridge being present in both sites, providing the superexchange pathway for strong antiferromagnetic coupling and the lack of an EPR signal. Protonation of this bridge uncouples the sites and a quantitative study of the pH dependence of the g = 1.86 signal in laccase indicates an intrinsic pK_a of > 7.0 for the endogenous bridge.[26,28] Thus, met T2D laccase and methemocyanin are similar with respect to exchange coupling through an endogenous bridging ligand.

Reduction of the met derivatives by one electron produces the mixed-valent 1/2-met [Cu(II)Cu(I)] derivatives. A comparison of the chemical and spectroscopic properties of 1/2-met hemocyanin and 1/2-met T2D laccase reveals a number of important differences (Figure 13). In particular, the 1/2-met T3 site in T2D has EPR spectra and equilibrium binding constants for exogenous ligands which are similar to those observed for normal tetragonal Cu(II) complexes. In 1/2-met hemocyanin, however, binding constants of exogenous ligands are, in general, two orders of magnitude greater and the spectroscopic features are very unusual relative to those of normal cupric complexes.

The unusual chemical and spectroscopic features of 1/2-met hemocyanin have been studied in some detail. Specifically, a met apo

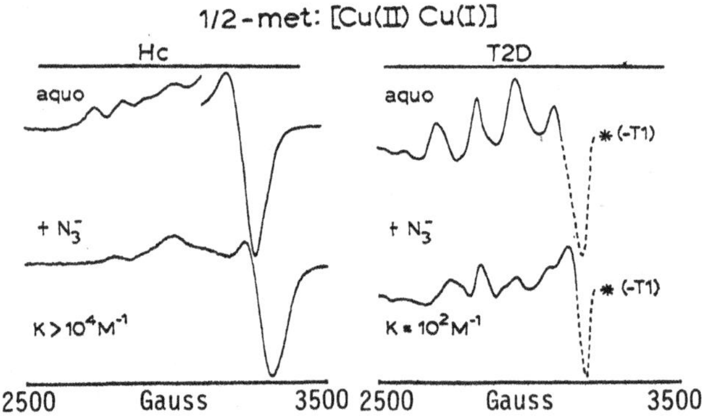

Figure 13. Comparison of EPR spectra of 1/2-met hemocyanin and T2D laccase in the absence and presence of azide. The T1 copper signal has been subtracted from the T2D laccase data.

[Cu(II) -] derivative of hemocyanin has been prepared which contains a single Cu(II) in the active site.[29] The strength of exogenous ligand binding in this derivative very closely parallels aqueous Cu(II) chemistry. For example, as shown in Figure 14a, addition of azide causes changes in the Cu(II) EPR spectrum which indicates azide binding. Dialysis against azide-free buffer rapidly restores the original met apo EPR signal. In contrast, addition of azide to the 1/2-met derivative followed by extensive dialysis does not produce the original EPR signal, but instead the unusual spectrum at the bottom of Figure 14b which corresponds to one extremely tightly bound azide at the active site. Further, reversible coordination of CO to the Cu(I) changes this Cu(II) EPR signal and the azide is now rapidly removed from the active site by dialysis. Thus, the presence of the Cu(I) greatly increases the affinity of the 1/2-met site for azide while coordination of CO to the Cu(I) greatly reduces this affinity. Both observations indicate that in addition to the Cu(II), Cu(I) must also bind the exogenous ligand at the active site. This bridging mode is supported by the unusual spectroscopic features of the 1/2-met hemocyanins.

In Figure 15 are shown the absorption and EPR spectra of a series of 1/2-met hemocyanin derivatives.[29] In contrast to the four parallel hyperfine lines normally observed for tetragonal Cu(II) complexes, the EPR spectra for the 1/2-met derivatives of hemocyanin are much more complex. Also, in addition to the d-d transitions for tetragonal Cu(II), a new absorption band appears at low energy which increases in intensity with variations in exogenous ligand ($NO_2^- < Cl^- < Br^- < I^- < N_3^-$). These

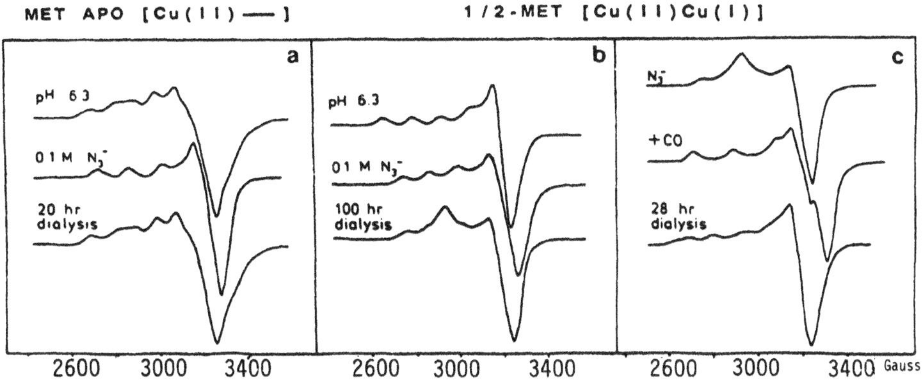

Figure 14. (a) EPR data indicating reversible binding of azide to met apo hemocyanin. (b) EPR data indicating irreversible azide binding to 1/2-met hemocyanin. (c) Effect of CO binding on high affinity azide binding in 1/2-met hemocyanin.

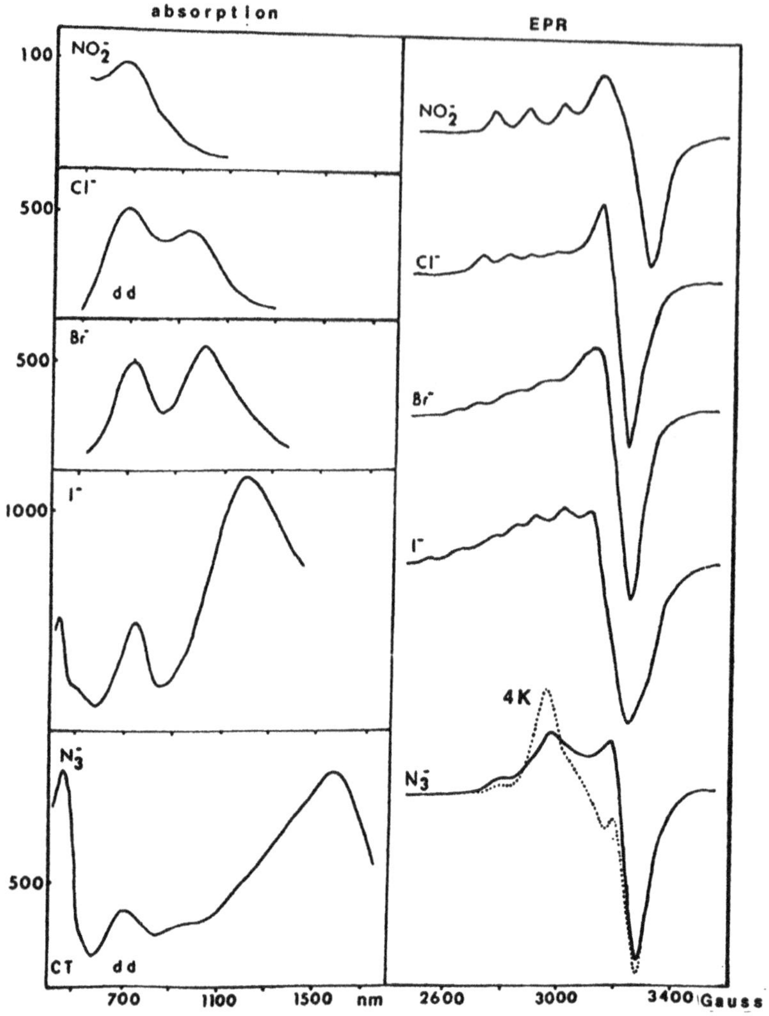

Figure 15. Low temperature (7K) absorption spectra and EPR spectra (77K) for a series of 1/2-met hemocyanin derivatives.

new spectral features are probed in detail in Figure 16. S, X, and Q-band EPR spectra have been obtained and simulated. In particular, the Q-band data show that only one paramagnetic specie is present and has axial g values. The resolution of the additional hyperfine splittings is improved in the S-band data which demonstrate that the unpaired electron interacts with both copper nuclei. The fitted g values and parallel hyperfine coupling constants are listed in the table in Figure 16. Further correlation of the low temperature MCD spectra with the absorption data indicates that the new low energy absorption band has no significant C term intensity associated with it. From the general expression for orientationally averaged C terms ($I_c = g_z m_x m_y + g_x m_y m_z +$

$g_y m_x m_z)$[30] it is clear that this must result from the lack of two perpendicular non-zero electric dipole transition moments for this absorption band. The unidirectional nature of this transition allows it to be assigned as an intervalent transfer (IT) transition corresponding to optical excitation of an electron from the Cu(I) to the Cu(II) and thus polarized only along the Cu-Cu vector. The experimentally observed oscillator strengths for the IT transitions are also given in the table.

From the table in Figure 16, both the relative magnitude of the hyperfine coupling to each copper and the IT absorpton intensity

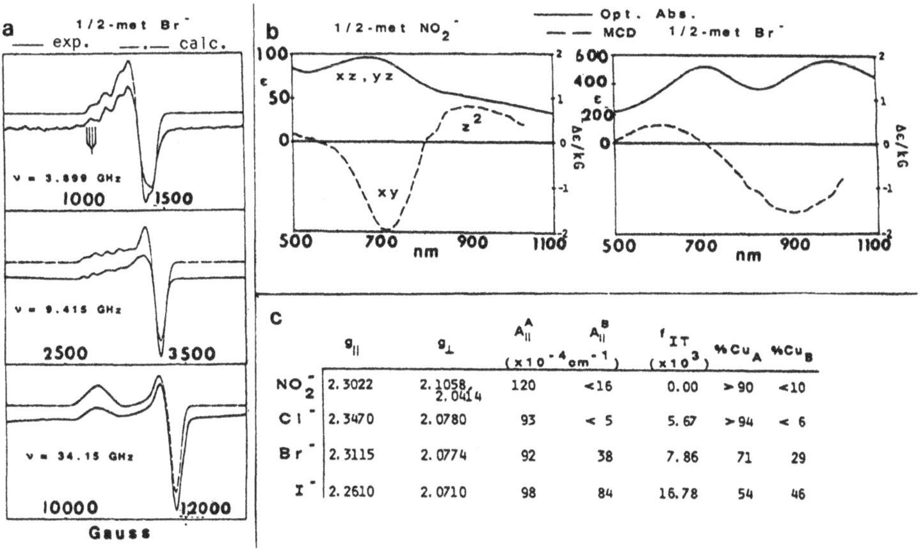

	$g_{\parallel}$	$g_{\perp}$	$A_{\parallel}^{A}$ ($\times 10^{-4}$ cm^{-1})	$A_{\parallel}^{B}$	f_{IT} ($\times 10^{3}$)	%Cu_A	%Cu_B
NO_2^-	2.3022	2.1058, 2.0414	120	<16	0.00	>90	<10
Cl^-	2.3470	2.0780	93	< 5	5.67	>94	< 6
Br^-	2.3115	2.0774	92	38	7.86	71	29
I^-	2.2610	2.0710	98	84	16.78	54	46

Figure 16. (a) S-band, X-band, and Q-band spectra for 1/2-met Br^- hemocyanin with calculated spectra using the parameters given in the table. (b) Comparison of the low temperature MCD and absorption spectra of 1/2-met NO_2^- and 1/2-met-Br^- hemocyanin. (c) Spectroscopic fitting parameters for the series.

correlate with the covalent nature of the tightly bound exogenous ligand. Thus, this exogenous ligand must provide the pathway for electron delocalization and therefore bridge between the two coppers, as shown in Figure 17 (left). In contrast, the lack of both high affinity binding

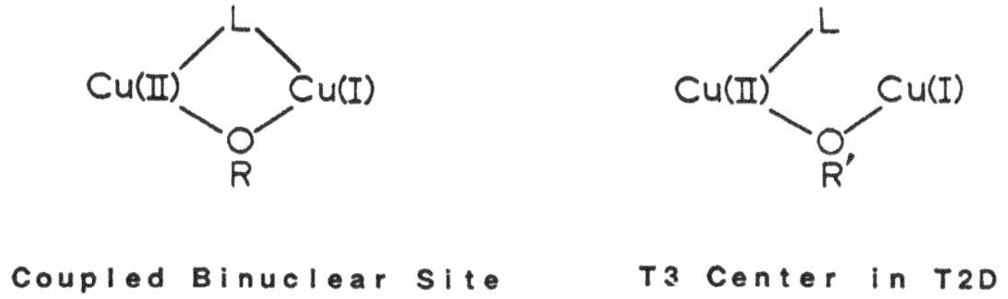

Figure 17. Comparison of exogenous ligand binding modes in 1/2-met hemocyanin and 1/2-met T2D laccase.

and mixed-valent spectral features in the 1/2-met T2D laccase strongly suggests that exogenous ligands bind only to one copper at the T3 site. Native laccase reduces dioxygen to water but the reduced T3 site in T2D laccase does not react with dioxygen. Thus, the T2 copper must play an important role in exogenous ligand interactions with the T3 site. Low temperature MCD spectroscopy was employed to probe exogenous ligand interactions with the T2 and T3 centers since the optical features associated with the paramagnetic T2 site can be distinguished from those associated with the antiferromagnetic T3 centers.[31] As illustrated in Figure 18 (left) application of a magnetic field to the T2 (S=1/2) center

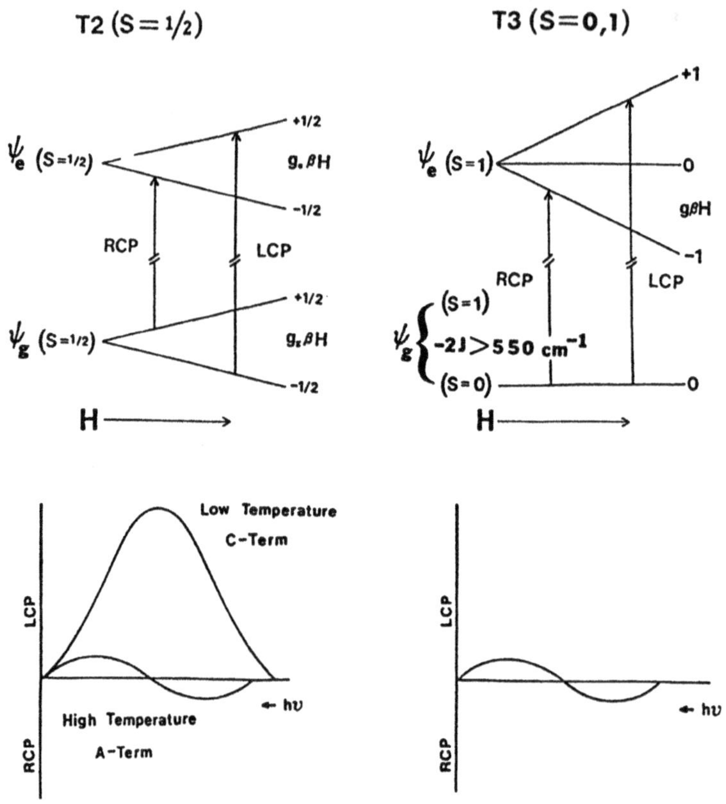

Figure 18. Origin of low temperature MCD features for the T2 and T3 centers in native laccase.

splits both the ground and the excited state energies by $g\beta H$. The selection rules for MCD transitions are $\Delta m = +1$ for left circularly polarized (LCP) light and $\Delta m = -1$ for right circularly polarized (RCP) light. Thus, one predicts for the T2 center two transitions with equal intensity and opposite sign. Since the Zeeman splitting is on the order of 10 cm^{-1} and the absorption bands are several thousand cm^{-1} wide, the transitions will cancel except for a broad, weak derivative-shaped signal at high temperatures (an A term). However, as the temperature is lowered the Boltzmann population of the m_s=+1/2 level is reduced relative to that of the m_s=-1/2 level. This leads to a large increase in absorption of LCP light and results in a temperature dependent C term. At very low temperatures ($\sim$4K), these C terms will be two to three orders of magnitude more intense than the A terms. In contrast, the strong antiferromagnetic coupling in the T3 center leads to an S=0 ground state and a thermally inaccessible S=1 excited state (-2J > 550 cm^{-1}). Since the S=0 level is nondegenerate, it cannot be split by a magnetic field and there are no changes in ground state population with temperature. Thus, there are no C terms associated with the T3 center and the low temperature MCD spectrum of native laccase should be dominated by the contributions from the paramagnetic T2 and T1 centers where only the T2 center is believed to be capable of binding exogenous ligands.[32]

The binding of azide to the T3 and T2 sites in native laccase has been probed via absorption and low temperature MCD spectroscopic studies of the N_3^- --> Cu(II) charge transfer transitions in the 400-500 nm

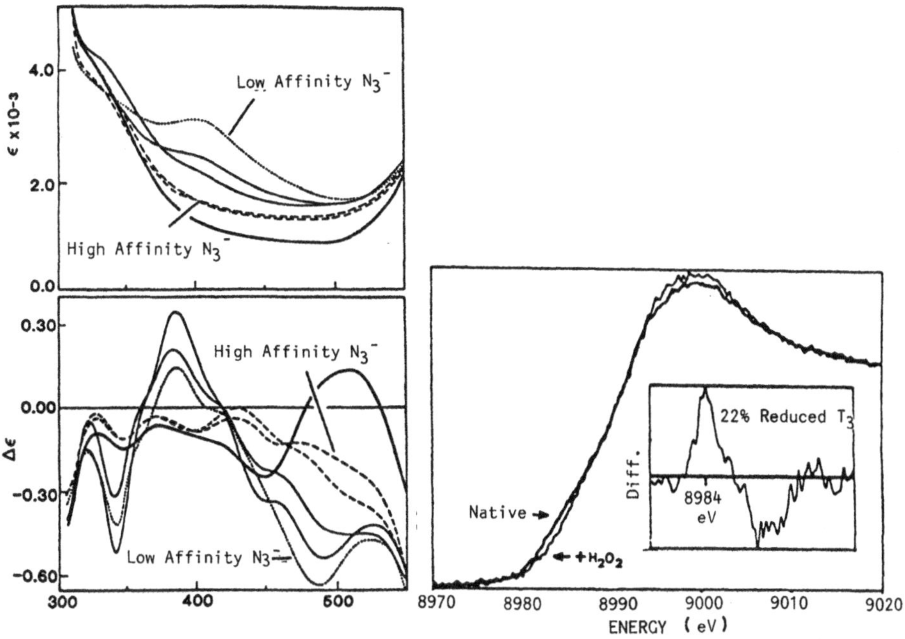

Figure 19. Absorption and low temperature MCD spectra of azide binding to native laccase. Also included are X-ray absorption edge data for reaction of native laccase with peroxide.

region (Figure 19). The low temperature MCD and absorption behavior in this region is quite complex and consists of two contributions. The spectral changes associated with high affinity (HA) binding are completed with the addition of less than one equivalent of azide ($K_{HA} \geq 10^4\ M^{-1}$). Prior addition of H_2O_2 eliminates these features, suggesting that they may be associated with ligand binding to a fraction of native laccase molecules in which the T3 site is reduced. This interpretation has been confirmed through X-ray absorption edge studies, shown in Figure 19, in which the edge features of H_2O_2 treated laccase have been subtracted from those of native enzyme.[14a] A peak is present at 8984 eV with an intensity that corresponds to ∿ 22% reduced T3 sites in the native enzyme.

The features associated with low affinity azide binding to fully oxidized laccase sites can then be obtained by subtraction of the high affinity azide spectral contributions. The low affinity absorption and low temperature MCD spectra are shown in Figure 20 for increasing [N_3^-]. Two N_3^- -> Cu(II) charge tranfer bands are observed in absorption, one at 500 nm and a more intense band at 400 nm. Both increase with azide

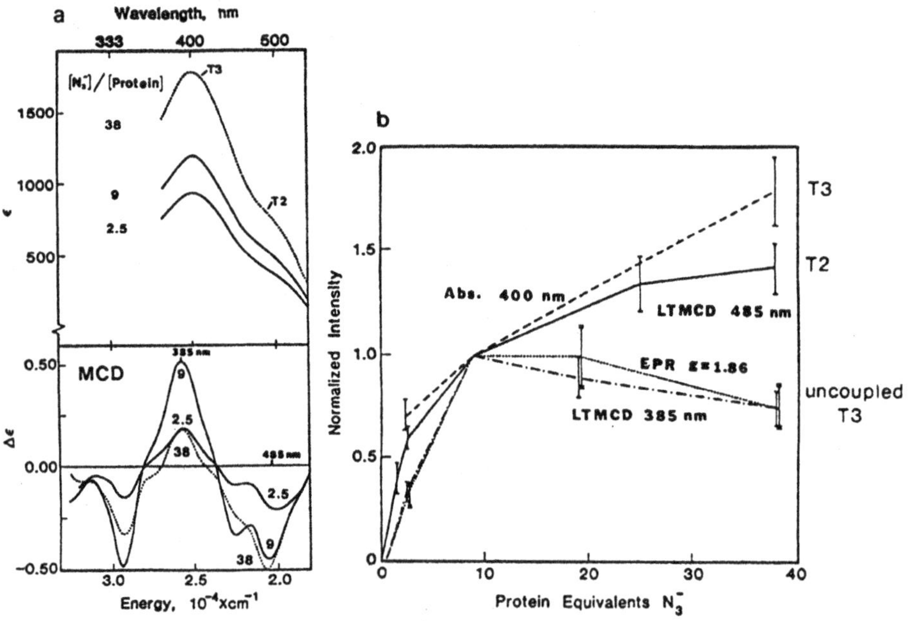

Figure 20. (a) Absorption and low temperature MCD spectra for binding of low affinity azide to native laccase. (b) Correlation of the LT MCD signals at 485 and 385 nm to the 400 nm absorption and the g = 1.86 EPR signal with increasing [N_3^-].

concentration; the 400 nm absorption intensity is plotted as a function of $[N_3^-]$ in Figure 20b. In the low temperature MCD spectrum a negative peak is observed at 485 nm which increases in magnitude with increasing azide concentration. It is thus clear that the corresponding ∿ 500 nm absorption band is associated with a paramagnetic ground state and must correspond to azide binding to the T2 center. In the region of the intense 400 nm absorption, a positive feature at 385 nm is observed in the low temperature MCD spectrum. The intensity of this feature first increases, then decreases with increasing azide concentration. As shown in Figure 20b, the intensity of this low temperature MCD feature does not correlate with the 400 nm absorption intensity, but in fact parallels the appearance of the g=1.86 EPR signal which is associated with the small fraction of uncoupled T3 sites (Figure 12, right). Thus, no low temperature MCD intensity is associated with the intense 400 nm absorption band and it must be associated with azide binding to the antiferromagnetically coupled T3 site.

From the low temperature MCD data it has been found that azide binds to both the T2 and T3 centers with similar binding constants. The stoichiometry of this binding has been determined through a series of ligand competition experiments, one of which is summarized in Figure 21. The appearance of fluorine superhyperfine splitting in the EPR spectrum

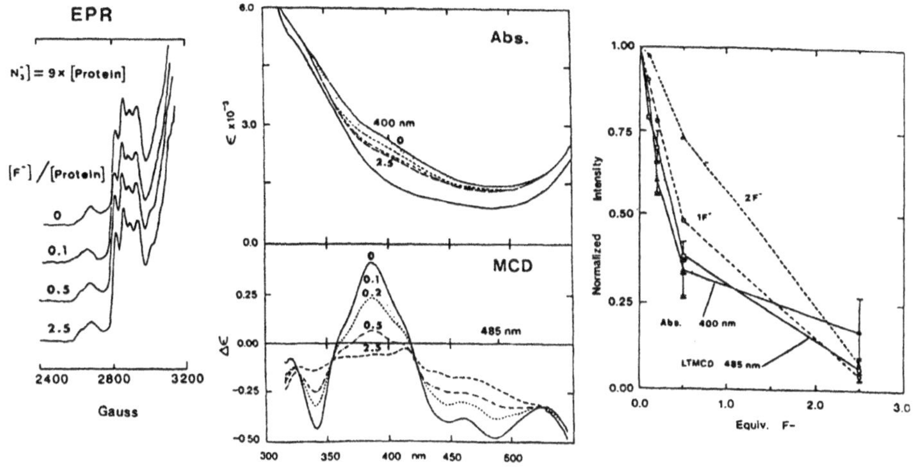

Figure 21. Results of N_3^-/F^- binding competition experiments for native laccase.

(I_n=1/2 for ^{19}F) indicates that at 2.5-fold excess, one fluoride binds to the T2 copper center. In the absorption and low temperature MCD spectra, the low affinity N_3^- -> Cu(II) charge transfer transitions associated with the T2 (485 nm MCD) and the T3 (400 nm abs.) centers are eliminated with the addition of increasing concentrations of fluoride. The quantitative loss of these features parallel each other as shown by the solid line in Figure 21 (right). Fitting these data to models[33] for competitive binding of one or two fluorides clearly demonstrates that only one fluoride binds to the T2 center and is involved in displacing azide from both the T2 and T3 centers. Therefore one exogenous azide ligand must bridge between the T2 and T3 centers in native laccase.

Figure 22 summarizes the differences between exogenous azide binding to the coupled binuclear copper site in hemocyanin (and tyrosinase) as

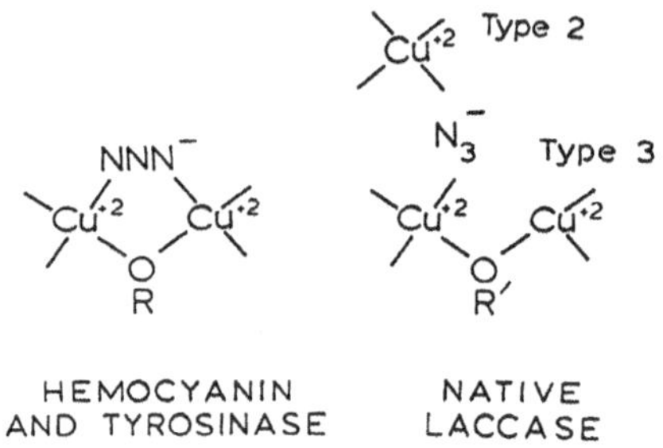

Figure 22. Comparison of the spectroscopically effective models for azide binding at the binuclear copper active site in hemocyanin and the trinuclear copper cluster in laccase.

compared to laccase. For hemocyanin and tyrosinase exogenous ligands bridge the two coppers at the active site while in laccase azide binds to only one copper of the T3 center and bridges to the T2 center forming a trinuclear copper cluster. It is the focus of current research to define in detail the role of the μ-1,2 peroxy bridge in reversible O_2 binding and activation and the T2-T3 trinuclear copper cluster with respect to its role in the multi-electron reduction of dioxygen to water.

Acknowledgements: The authors wish to thank the students and postdoctoral associates who have contributed to this research: Dr. Richard S. Himmelwright, Dr. Nancy C. Eickman, Dr. Cynthia D. LuBien, Dr. Yeong T. Hwang, Dr. Marjorie E. Winkler, Dr. Thomas J. Thamann, Dr. Arturo G. Porras, and Dr. Dean E. Wilcox. EIS also acknowledges the contributions of the following collaborators: Prof. K. Lerch, Prof. K.O. Hodgson, Prof. T.G. Spiro, Prof. K.D. Karlin, Prof. C.A. Reed, Prof. T.N. Sorrell, Dr. W.B. Mims, and Dr. M.S. Crowder. NIH grant #DK31450 is gratefully acknowledged for supporting this work.

References

1. Note that the enzyme aconitase contains a 3Fe3S cluster which can be converted to a 4Fe4s cluster where substrate is found to bind to the additional iron atom. (a) T.A. Kent, B.H. Huynh, and E. Munck, Proc. Natl. Acad. Sci. USA 77, 6574 (1980). (b) H. Beinert, M.H. Emptage, J.-L. Dreyer, R.A. Scott, J.E. Hahn, K.O. Hodgson, and A.J. Thomson, Proc. Natl. Acad. Sci. USA 80, 393 (1983).

2. (a) B.M. Hoffman, J.E. Roberts, and W.H. Orme-Johnson, J. Am. Chem. Soc. 104, 860 (1982). (b) B.M. Hoffman, R.A. Venters, J.E. Roberts, M. Nelson, and W.H. Orme-Johnson, J. Am. Chem. Soc. 104, 4711 (1982).

3. R.L. Robson, R.R. Eady, T.H. Richardson, R.W. Miller, M. Hawkins, J.R. Postgate, Nature 322, 388 (1986).

4. (a) G. Palmer, W.R. Dunham, J.A. Fee, R.H. Sands, T. Iizuka, and T. Yonetani, Biochim. Biophys. Acta 245, 201 (1971). (b) H. Blum, F. Adar, J.C. Salerno, and J.S. Leigh, Biochem. Biophys. Res. Commun. 77, 650 (1977).

5. (a) B.C. Antanaitis and T.H. Moss, Biochim. Biophys. Acta 405, 262 (1975). (b) A.J. Thomson, J. Chem. Soc. Dalton 1180, (1981). (c) L. Noodleman, J.G. Norman, Jr., J.H. Osborne, A. Aizman, and D.A. Case, J. Am. Chem. Soc. 107, 3418 (1985). (d) L. Noodleman, E.J. Baerends, J. Am. Chem. Soc. 106, 2316 (1984).

6. S.C. dePaula, W.F. Beck, and G.W. Brudvig, J. Am. Chem. Soc. 108, 4002 (1986).

7. (a) R.C. Reem and E.I. Solomon, J. Am. Chem. Soc. 106, 8323 (1984). (b) R.C. Reem and E.I. Solomon, J. Am. Chem. Soc. 109, 1216 (1987).

8. An endogenous ligand is one which is intrinsic to the protein system, generally either a protein residue or a solvent molecule, while an exogenous ligand is extrinsic to the protein system.

9. (a) M.F. Tweedle, L.J. Wilson, L. Garcia-Iniguez, G.T. Babcock, and G. Palmer, J. Biol. Chem. 253, 8065 (1978). (b) T.A. Kent, E. Munck, W.R. Dunham, W.F. Filter, K.L. Findling, T. Yoshida, and J.A. Fee, J. Biol. Chem. 257, 12489 (1982). (c) T.A. Kent, L.J. Young, G. Palmer, J.A. Fee, and E. Munck, J. Biol. Chem. 258, 8543 (1983). (d) G.W. Brudvig, T.H. Stevens, R.H. Morse, and S.I. Chan, Biochemistry 20, 3912 (1981).

10. J.A. Christner, E. Munck, P.A. Janick, and L.M. Siegel, J. Biol. Chem. 258, 11147 (1983).

11. (a) R. Malkin and B.G. Malmstrom, Adv. Enzymol. 33, 177 (1970). (b) B.G. Malmstrom, L.-E. Andreasson, and B. Reinhammar, in The Enzymes: Boyer, P.D., ed., Academic: New York 1975, Vol. XII. (c) J.A. Fee, Struct. Bonding 23, 1 (1973).

12. T.B. Freedman, J.S. Loehr, and T.M. Loehr, J. Am. Chem. Soc. 98, 2809 (1976).

13. J.M. Brown, L. Powers, B. Kincaid, J.A. Larrabee, and T.G. Spiro, J. Am. Chem. Soc. 102, 4210 (1980).

14. (a) L.S. Kau, D.J. Spira, J.E. Penner-Hahn, K.O. Hodgson, and E.I. Solomon, J. Am. Chem. Soc. in press, (1987). (b) L.S. Kau, E.I. Solomon, and K.O. Hodgson, J. Physique 47, 289 (1986).

15. (a) E.I. Solomon, D.M. Dooley, R.H. Wang, H.B. Gray, M. Cerdonio, F. Mogno, and G.L. Romani, J. Am. Chem. Soc. 98, 1029 (1976). (b) D. Dooley, R.A. Scott, J. Ellinghaus, E.I. Solomon, and H.B. Gray, Proc. Nat. Acad. Sci. USA 75, 3019 (1978).

16. N.C. Eickman, R.S. Himmelwright, and E.I. Solomon, Proc. Natl. Acad. Sci. USA 76, 2094 (1979).

17. K.D. Karlin, R.W. Cruse, Y. Gultneh, J.C. Hayes, and J. Zubieta, J. Am. Chem. Soc. 106, 3372 (1984).

18. (a) V. McKee, J.V. Dagdigian, R. Bau, and C.A. Reed, J. Am. Chem. Soc. 103, 7000 (1981). (b) V. McKee, M. Zvagulis, J.V. Dagdigian, M.G. Patch, and C.A. Reed, J. Am. Chem. Soc. 106, 4765 (1984). (c) K.D. Karlin, B.I. Cohen, J.C. Hayes, A. Farooq, and J. Zubieta, Inorg. Chem. 26, 147 (1987).

19. M.T. Graziani, L. Morpurgo, G. Rotilio, and B. Mondovi, FEBS Lett. 70, 87 (1976).

20. B. Reinhammar, Biochim. Biophys. Acta 275, 245 (1972).

21. C.D. LuBien, M.E. Winkler, T.J. Thamann, R.A. Scott, M.S. Co, K.O. Hodgson, and E.I. Solomon, J. Am. Chem. Soc. 103, 7014 (1981).

22. (a) O. Farver, P. Frank, and I. Pecht, Biochem. Biophys. Res. Comm. 108, 273 (1982). (b) P. Frank, O. Farver, and I. Pecht, J. Biol. Chem. 258, 11112 (1983). (c) P. Frank, O. Farver, and I. Pecht, Inorg. Chim. Acta 91, 81 (1984).

23. L. Powers, W.E. Blumberg, B. Chance, C. Barlow, J.S. Leight, Jr., J.C. Smith, T. Yonetani, S. Vik, and J. Peisach, Biochim. Biophys. Acta 546, 520 (1979).

24. T.A.G. Smith, Ph.D. Thesis, Stanford University, 1985.

25. R.A. Blair, W.A. Goddard, Phys. Rev. B. 22, 2767 (1980). (b) N. Kosugi, T. Yokoyama, K. Asakuna, and H. Kuroda, Springer Proc. Phys. 2, 55 (1984). (c) N. Kosugi, T. Yokoyama, K. Asakuna, and H. Kuroda, Chem. Phys. 91, 249 (1984).

26. D.J. Spira and E.I. Solomon, submitted to J. Am. Chem. Soc. (1987).

27. D.E. Wilcox, T.D. Westmoreland, P.O. Sandusky, and E.I. Solomon, unpublished results.

28. D.E. Wilcox, J.R. Long, and E.I. Solomon, J. Am. Chem. Soc. 106, 2186 (1984).

29. R.S. Himmelwright, N.C. Eickman, and E.I. Solomon, J. Am. Chem. Soc. 101, 1576 (1979).

30. (a) P.J. Stephens, Ann. Rev. Phys. Chem. 25, 201 (1974). (b) S.B. Piepho and P.N. Schatz, Group Theory in Spectroscopy, Wiley-Interscience: New York, 1983.

31. (a) M.D. Allendorf, D.J. Spira, and E.I. Solomon, Proc. Natl. Acad. Sci. USA 82, 3063 (1985). (b) D.J. Spira, M.D. Allendorf, and E.I. Solomon, J. Am. Chem. Soc. 108, 5318 (1986).

32. The oxidized T1 site in laccase is spectroscopically similar to that of plastocyanin and azurin, suggesting that exogenous ligands can coordinate only to the T2 and T3 coppers at the native active site.

33. (a) J. Steinhardt, and J.A. Reynolds, in Multiple Equilibria in Proteins, Academic: New York, 1969. (b) I.M. Klotz, H. Triwush, F.M. Walker, J. Am. Chem. Soc. 70, 2935 (1948).

LANGMUIR-BLODGETT FILMS OF DONOR-SIGMA-ACCEPTOR MOLECULES AND PROSPECTS FOR ORGANIC RECTIFIERS §

Robert M. Metzger* and Charles A.Panetta

*Department of Chemistry - University of Alabama
Tuscaloosa AL 35487-9671 USA
Department of Chemistry - University of Mississippi
University MS 38677 USA

INTRODUCTION

Our goal is to assemble and test a unimolecular rectifier of electrical current, which could be part of a very thin (5 nm thick) electronic device. This idea, originated by Aviram in 1973, depends on the asymmetry of molecules D-σ-A, where D is a good one-electron donor (but poor acceptor), A is a good one-electron acceptor (but poor donor), and σ is a covalent bridge that keeps the molecular orbitals of D separate from those of A. We have found five molecules which self-assemble as monolayers; three contain the TCNQ moiety; three contain "greasy" dodecyl groups on the donor end (which helps in monolayer formation), but one contains only hexyl groups. All of them can be transferred to a glass or Al substrate as Langmuir-Blodgett films. Recent FTIR data for a single monolayer are presented.

THE ORGANIC RECTIFIER - THEORETICAL, SYNTHETIC, AND ASSEMBLY CRITERIA

This is a progress report on the Organic Rectifier Project, i.e. on our efforts to synthesize molecules of the type D-σ-A, which may be potential rectifiers of electrical current. Here D = strong one-electron organic donor, such as TTF (tetrathiafulvalene, **1**), or TMPD (N,N,N',N'-tetramethyl-para-phenylenediamine, **2**), σ = covalent sigma bridge, A = strong one-electron acceptor, such as TCNQ (7,7,8,8-tetracyanoquinodimethan, **3**) or chloranil (2,3,5,6-tetrachloro-para-benzoquinone, **4**). Aviram and co-workers predicted in 1973 [1-3] that D-σ-A molecules, such as the proposed molecule **5**, sandwiched as an oriented monolayer between ordinary metallic thin films M and N, as in **6**, would act as a rectifier of electrical current. The obvious advantage of such a device is its small thickness: the molecule **5** should be only 2 nm thick. If very thin conventional films M, N (1.5 nm each) can be used, then a 5 nm thick device becomes possible, much thinner than the working direction of conventional Si or GaAs devices (1 to 3μm).

The incremental results of the Organic Rectifier Project have been reported elsewhere [4-15]. Here we summarize the design criteria that have evolved, and discuss the implications of our most recent results.

§ Supported in part by NSF-DMR Grant 84-17563

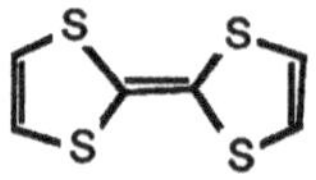

1

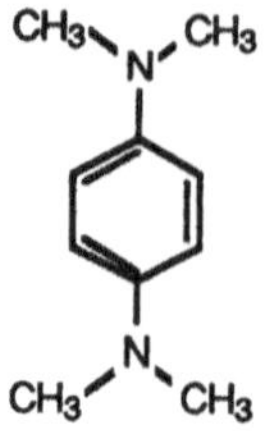

2

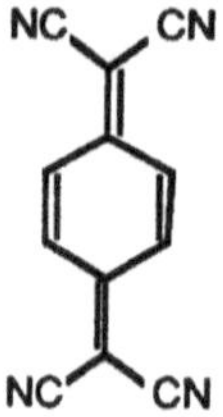

3

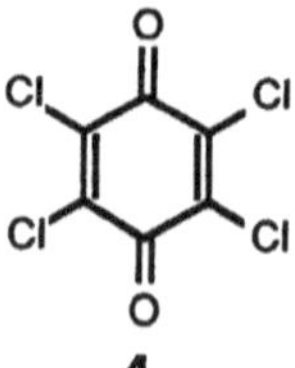

4

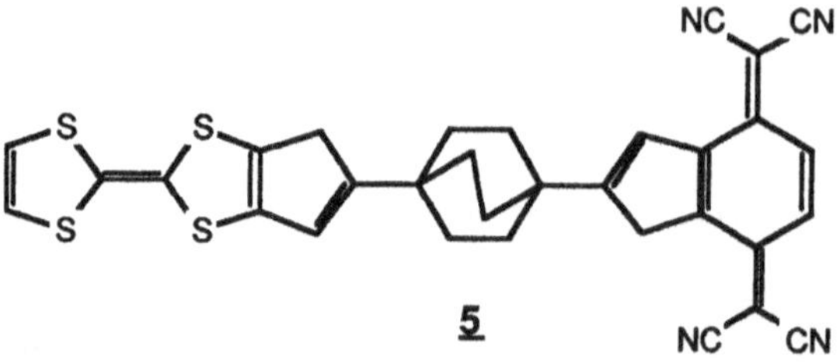

5

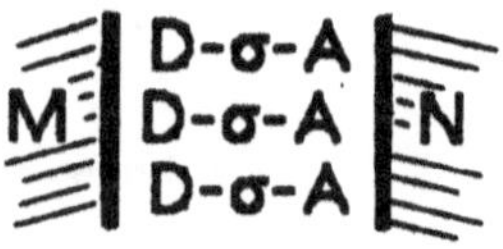

6

Simplified Mechanism

If device **6** can be assembled, then the mechanism for rectification is given by [2-5,10,16]:

FORWARD BIAS: GO: $M \mid D\text{-}\sigma\text{-}A \mid N \text{----}> M^- \mid D^+\text{-}\sigma\text{-}A^- \mid N^+ \text{----}> M^- \mid D\text{-}\sigma\text{-}A \mid N^+$ **7**
7a **7b (ITF)**

REVERSE BIAS: NO GO: $M \mid D\text{-}\sigma\text{-}A \mid N \text{----xxx--}> M^+ \mid D^-\text{-}\sigma\text{-}A^+ \mid N^-$ **8**

REVERSE BIAS: GO: $M \mid D\text{-}\sigma\text{-}A \mid N \text{-----}> M \mid D^+\text{-}\sigma\text{-}A^- \mid N \text{-----}> M^+ \mid D\text{-}\sigma\text{-}A \mid N^-$ **9**
9a (ITR) **9b**

Under zero or moderate forward bias the electron transfer (ET) can occur by elastic, through-space (TS) [17] tunneling though a chemisorptive barrier from D to M, and from N to A (**7a**); thus is formed the zwitterionic molecular state $D^+\text{-}\sigma\text{-}A^-$; this is followed by through-bond (TB)[17] inelastic tunneling (call it **ITF, 7b,** or forward inelastic tunneling) from A^- to D^+: this produces charge separation. Under reverse bias, the formation of zwitterion $D^-\text{-}\sigma\text{-}A^+$ (**8**) is very unlikely, because good organic one-electron donors are miserable acceptors, and good acceptors are poor donors. Indeed, in the gas phase, the energy required to form the ions $TTF^+(g)$, $TCNQ^-(g)$ at infinite mutual separation is known to be 4.0 eV[18,19], whereas the energy required to form the ion pair $TTF^-(g)$, $TCNQ^+(g)$ is estimated at over 9 eV [20,21]:

$$TTF(g) + TCNQ(g) = TTF^+(g) + TCNQ^-(g) \quad \Delta H = 6.83 - 2.8 \text{ eV} = 4.0 \text{ eV} \qquad \textbf{10}$$

$$TTF(g) + TCNQ(g) = TTF^-(g) + TCNQ^+(g) \quad \Delta H = 9.6 \text{ eV (est)} \qquad \textbf{11}$$

However, under reverse bias another mechanism is possible (**9a**): a zwitterion $D^+\text{-}\sigma\text{-}A^-$ forms first, then charge separation from M to N occurs: this is reverse "uphill" tunneling **ITR**, and is much less likely than ITF. Of course, tunneling is here between states of different energies, so one sees these inelastic processes as elastic processes between virtual states of equal energy, followed by inelastic "through- bond" processes.

Langmuir-Blodgett Films

So far, our efforts have been centered upon molecules D-σ-A which self-assemble at the air-water interface as "Pockels-Langmuir" monolayers [our term for such monolayers, honoring Irving Langmuir (1881-1957) and Agnes Pockels].[22]. Such monolayers can, usually, be transferred to glass or metal or other substrates by the Langmuir-Blodgett (LB) technique pioneered by Langmuir and by Katharine B. Blodgett (1898-1979) [23-26]. The typical molecule that forms excellent PL and LB films is cadmium arachidate, $Cd(n\text{-}C_{19}H_{39}COO)_2$, **12**.

12

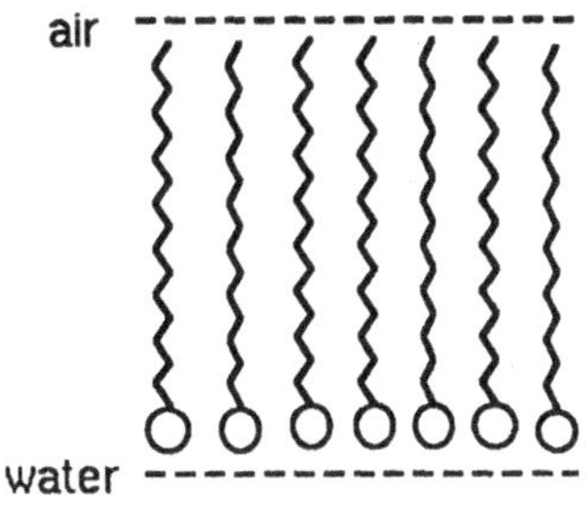

13

At the air-water interface a PL monolayer of cations **12** (shaped like tadpoles) points with the carboxylate ends toward the water subphase, as in **13**: the cadmium counterions are in the subphase, coordinated rather closely with the carboxylate head group. Usually, such work is done today by using an automated Langmuir trough or film balance.

The LB technique consists of slowly inserting a macroscopic metal, glass, silicon, or other substrate through the monolayer into the subphase; either upon insertion into the water, or upon slow withdrawal from the water, the PL monolayers transfer quantitatively to the substrate with little or no distortion: thus are formed the LB monolayers. If the substrate is dipped several times through the monolayer, LB multilayers can be transferred to the substrate [23-28].

Other techniques for monolayer coverage of metals

Another method of transferring self-assembling monolayers to substrates without using a Langmuir trough is the Bigelow oleophobic (BO) film-casting technique [29-32]. The forces binding an LB or BO monolayer to the substrate are usually weak physisorptive or chemisorptive forces. A better adhesion is by direct covalent bonding to a surface. Bonding to silanized metal surfaces rarely achieves monolayer coverage [33]. Also, attaching silanized molecules to oxide-bearing metal surfaces produces compact monolayers [34], as does attaching molecules bearing disulfide bonds to gold surfaces [35]. Another method of providing "strong" films is to polymerize LB films containing diacetylene linkages in situ using ultraviolet radiation [36].

Despite the existence of these other potential techniques for assembling an organic M | D-σ-A | N rectifier, we concentrate below on D-σ-A molecules that will self-assemble as PL and LB monolayers.

Synthetic and device assembly criteria

The criteria for successful synthesis of candidate D-σ-A molecules and for their assembly into the organic rectifier **6** have been stated before [4,5,8,10,16] and can be summarized as follows:

(1) The donor D must have a low ionization potential I_D (<7.5 eV) and be fairly flat.

(2) The acceptor A must have a high electron affinity A_A (>2 eV) and must be fairly flat.

(3) The σ bridge must be saturated (not conjugated). It must be long enough to prevent extensive ground-state mixing of the donor molecular orbitals with the acceptor molecular orbitals (> 3 carbon-like atoms), yet short enough (< 9 carbon-like atoms) or rigid enough to prevent the curling of the D end over the A end of the molecule. Molecular modeling shows that 5 or 6 carbon atoms seems to be the optimal length. The σ bridge must be flat enough to provide good lateral LB film packing.

(4) The D-σ-A molecule must pack well in monolayers. For this, one end must be hydrophobic, the other hydrophilic. There should be no overlap of D-σ-A over A-σ-D (which would cancel dipole moments and destroy the directionality of the device **6**). The lateral π-π attractive interactions, and maybe a mixed-valence ground state, as exists in the quasi-one-dimensional salt TTF TCNQ, may reduce the cost of ionization of D-σ-A to D^+-σ-A^- from 4 eV to maybe 1 or 2 eV.

(5) The I_D and A_A values must match as closely as possible the work functions of the metal layers M and N (see Table 1). As can be seen, the match is far from perfect., even for the best donors and acceptors known to date.

Table 1. Ionization potentials I_D for the donors D, electron affinities A_A for the acceptors A, and work functions ϕ for the metals M, N

I_D (eV)		A_A(eV)		ϕ (eV)	
TMPD	6.25 [37]	DDQ	3.13 [39]	Al	3.74 [41]
TTF	6.83 [18]	TCNQ	2.8 [19]	Au	4.58 [41]
pyrene	7.41 [38]	Chloranil	2.76 [40]	Pt	5.29 [41]

(6) The bridge-building organic reaction **14** leading to D-σ-A must be more likely, or more efficient, than the competing reaction of charge transfer (CT) salt formation **15**:

D-X + Y-A ---------------> D-σ-A **14**

D-X + Y-A ---------------> X-D^+ Y-A^- **15**

(7) The Franck-Condon reorganization of the molecular geometry of D to D^+ and of A to A^- [42] must be small, or fast, so that the overall ET within D-σ-A does not become slow [42-44]. Under the right conditions, ET through the molecule could be faster than 1 ns [43-46].

(8) The monolayers must be close-packed and defect-free for at least a few μm in the lateral direction (to prevent electrical shorting of metal layer M to metal layer N). It is known that over hydrophobic, oxide-free metals (Pt, Au, Ag) cadmium arachidate films have larger defects (probably at domain boundaries) [47], and have maybe a disordered, fan-like structure [48]. Over hydrophilic, oxide-covered metals (Al, Sn) or glass or quartz, the defects seem smaller, allowing for a percolative curent j across the film and through disclinations [49] (the voltage dependence of j is $\log j = aV^{1/4}$ [50] and not $\log j = b\ V^{1/2}$ [51]). One way of achieving freedom from defects on the scale of a few microns would be to use a thin interdigitated layer M, where the film perfection is sampled on several small disks of radius 3 μm radius (these disks would be connected to macroscopic electrodes).

(9) At the M to N distance of, typically, 2 nm, the directed TB tunneling must be much more likely than the undirected TS tunneling. Sofar, in inelastic electron tunneling spectroscopy (IETS) [52] for random films of insulators placed between superconducting Pb and normal Al electrodes at 4.2 K, the ratio TB/(TB+TS) is only 0.01. For the organic rectifier to succeed, this ratio (for compact monolayers) must be much closer to 1.0.

(10) The process of laying down the second metal layer N atop M | D-σ-A as a sputtered or evaporated film, and also the exothermicity of the TB tunneling in the device **6**, must both avoid heating the monolayer above 100°C. Similarly, high voltages that may cause dielectric breakdown must be avoided.

RESULTS

Coupling reactions

In 1976 Hertler prepared carbamate, or urethane, polymers of TTF with TCNQ if one reacts the 2,2'-bisisocyanate of TTF with a 2,5-dihydroxyethoxyTCNQ; the resulting insoluble black polymer was semiconducting, not metallic [53]. However, the utility of Hertler's contribution was to show that carbamates of strong donors with strong acceptors could be prepared, and that reaction **14** was preferred over reaction **15** for carbamates. Baghdadchi prepared the monoisocyanate of TTF, **16**, and coupled it with 2-bromo-hydroxyethoxyTCNQ, **17** [4-6, 54] to yield the **carbamate 18** of TTF with TCNQ. However, two forms were found, of which one seemed zwitterionic, the other neutral (as evidenced by the infrared CN stretching frequencies). Neither product could be isolated in acceptable purity.

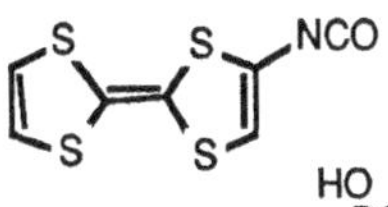

16

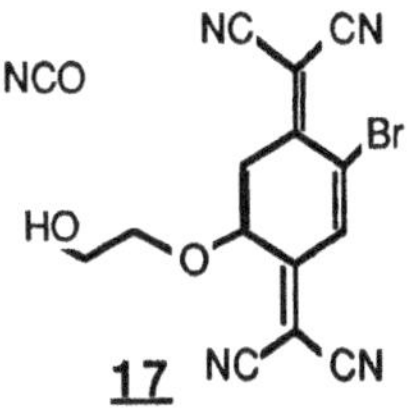

17

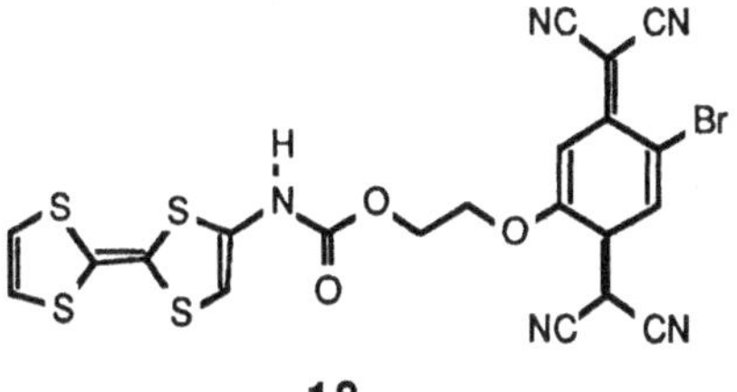

18

Esters of TTF with TCNQ **17** were also prepared, using the 2-acyl chloride of TTF, but, again, two products were obtained [6]. The neutral form of the TTF carbamate with TCNQ did form LB monolayers, but with the wrong geometry (probably with the molecule lying almost flat on the water subphase, rather than normal to it).

Since that time, most of our efforts have been concentrated on the carbamate coupling, although one may conceive of other possible coupling reactions.

Acceptors

Most of the work has been done with Hertler's TCNQ alcohol **17** (BHTCNQ), whose crystal structure was recently determined (Fig. 1) [13]. From cyclic voltammetric studies we estimate that the electron affinity of **17** is 2.9 ± 0.2 eV [13]. The inefficient synthesis of **17** has caused us to seek better, or more accessible acceptors. The anthraquinone analog **19** has been prepared in excellent yield [9], but the molecule is severely bent, and **19** is a weak two-electron acceptor, rather than a strong one-electron acceptor. The synthesis of **20** has been accomplished [55], and the crystal structure of its methyl ester, **21**, has been determined [14] (Fig. 2).

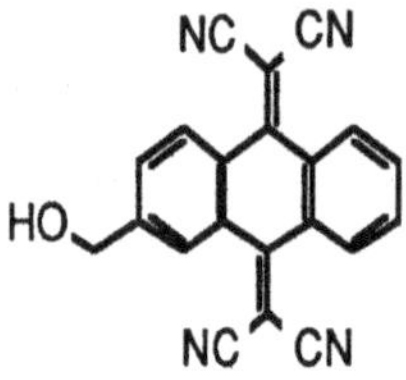

19

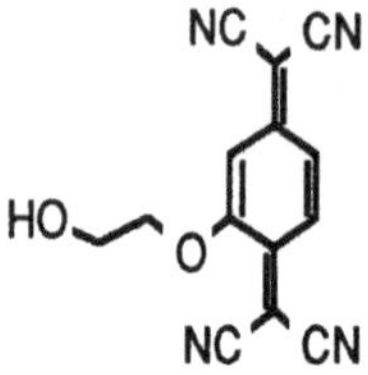

20

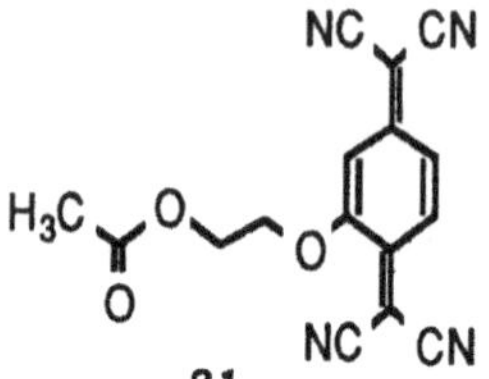

21

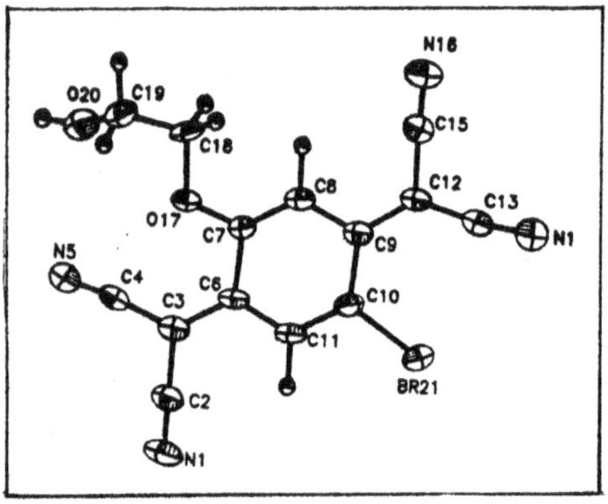

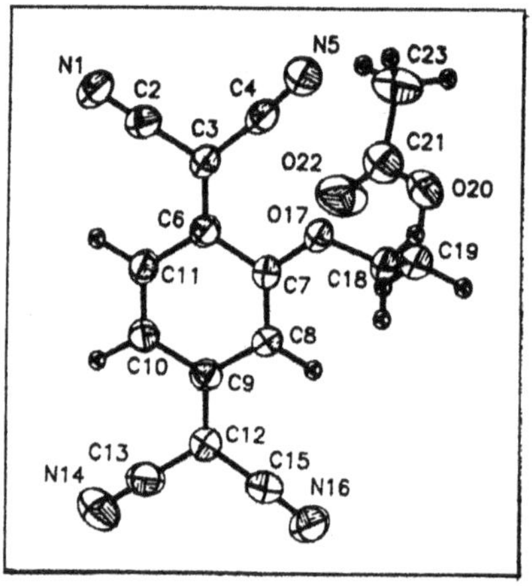

Fig. 1. ORTEP plot of BHTCNQ, **17** (monoclinic, space group $P2_1/n$, **a**= 9.258 Å, **b**= 13.618 Å, **c**= 10.947 Å, β = 92.14°, Z = 4, R = 3.9% [13])

Fig. 2. ORTEP plot of AETCNQ, **21** (triclinic, space group P1bar, **a**= 7.165 Å, **b**= 9.058 Å, **c**= 13.244Å, α= 70.06°, β= 87.14°, γ= 68.22°, Z =2 R = 3.4% [14]).

Donors

After the early work on TTF isocyanate (a strong donor) [4-6], our attention turned to the isocyanate of pyrene, **22** (a medium donor), and to substituents of phenyl isocyanate (a weak donor); adding a dodecyloxy group made weak donor **24**; adding the dimethylamino group made the medium donor DMAP-NCO, **25.** The crystal structure of its methyl carbamate, DMAPCMe, **26,** was determined (Fig. 3) [11]; its calculated (AM1) ionization potential (7.17 eV) and its cyclic voltammetry showed that **26** is a donor weaker than TMPD or TTF but stronger than pyrene [12]. There is extensive hydrogen bonding in the crystal of DMAPCMe [11], in solution

[12], and also in solution of its electrochemically generated radical anion [12]. We have also prepared the bis-dodecyl and the bis-hexyl derivatives of **25**.

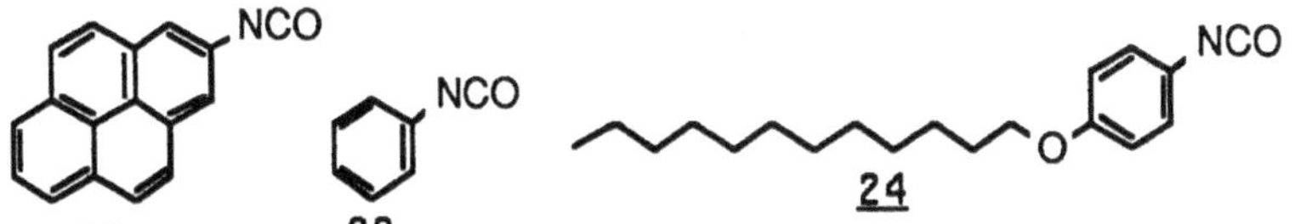

22 23 24

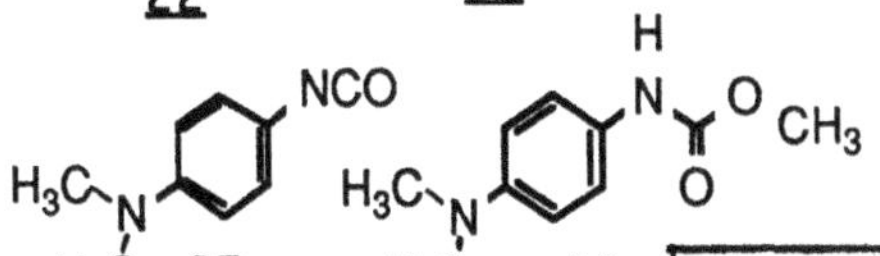

25 26

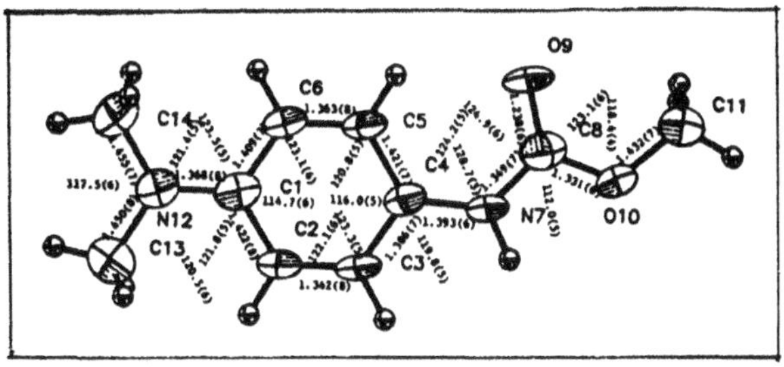

Fig. 3. ORTEP plot of DMAPCMe, **26** (orthorhombic, space group Pbca, **a** = 13.926 Å, **b** = 9.999 Å, **c** = 14.845 Å, Z = 8, R = 5.9 % [11])

Crystal structures of Donor-sigma-Acceptor Molecules

To test the proposition that the σ bridge allows for an extended structure of

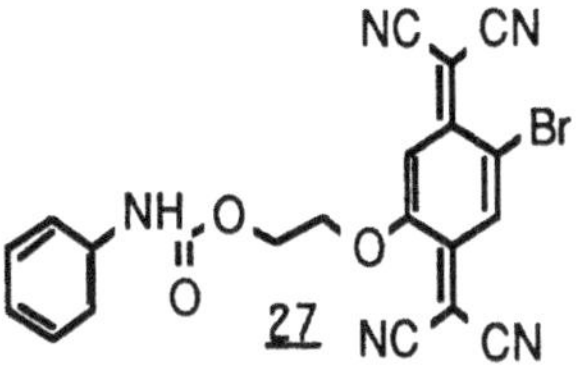

27

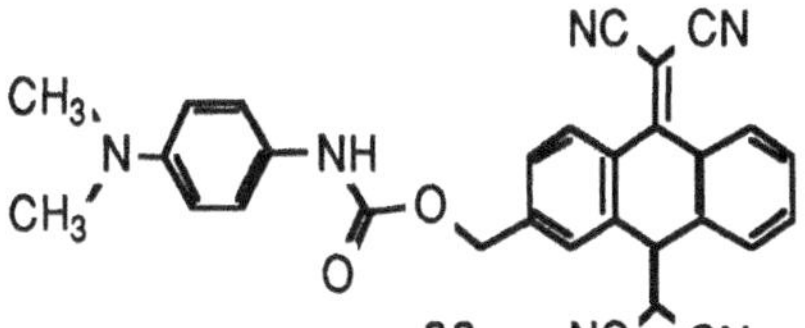

28

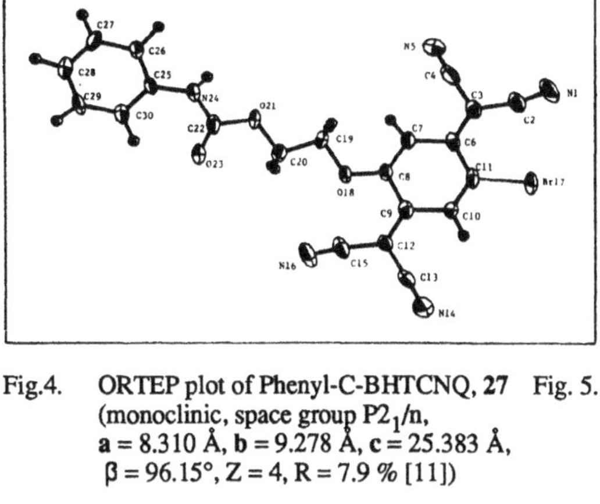

Fig.4. ORTEP plot of Phenyl-C-BHTCNQ, **27** (monoclinic, space group $P2_1/n$, **a** = 8.310 Å, **b** = 9.278 Å, **c** = 25.383 Å, β = 96.15°, Z = 4, R = 7.9 % [11])

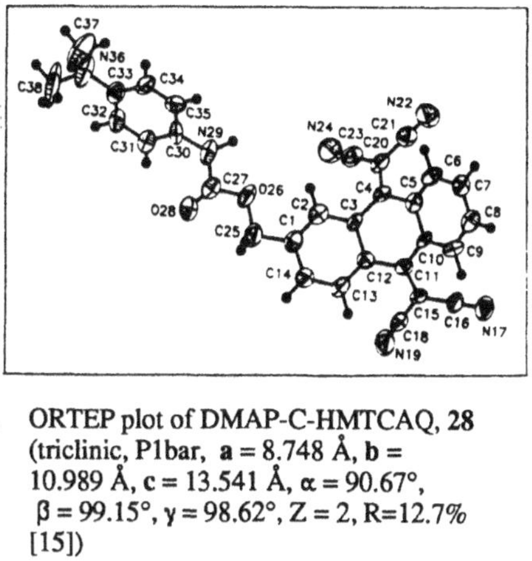

Fig. 5. ORTEP plot of DMAP-C-HMTCAQ, **28** (triclinic, P1bar, **a** = 8.748 Å, **b** = 10.989 Å, **c** = 13.541 Å, α = 90.67°, β = 99.15°, γ = 98.62°, Z = 2, R=12.7% [15])

the D-σ-A adducts, crystal structures have been determined for phenyl-C-BHTCNQ, **27** (Fig. 4) [11] and for DMAP-C-HMTCAQ, **28** (Fig. 5) [15]. In both cases the molecules do not form LB films, but the crystal structure reveals an extended conformation; the crystals belong to the centrosymmetric space group P1 bar; the molecules pack in the crystal so that D-σ-A packs nearly over A-σ-D.

Pockels-Langmuir and Langmuir-Blodgett Films

It was found that only certain D-σ-A molecules **29-33** form monolayers at the air-water interface; the data are given in Table 2, and the relevant pressure-area isotherms (and molecular structures) are shown in Figs. 6-10.

The collapse pressure Π_c is defined to be the differential surface tension, which, if exceeded, leads to the collapse of the film, and to the irreversible "riding of ice floes over each other". The area per molecule can be' defined in several ways: the area at the collapse point is A_c, and is the smallest; the area per molecule at zero pressure, A_0, is the point in the Π-A isotherm obtained when the steepest part of the curve is extrapolated linearly to zero pressure; A_m is defined as the mid-point between A_c and A_0. Of all these points, A_c gives what probably is the best estimate of the lateral dimensions of the molecule in a well-packed film.

The shape of the Π-A isotherm is indicative of the fluidity or rigidity within the film. For cadmium arachidate, the slope is almost vertical, because the film is almost crystalline in its two-dimensional packing. For the molecules described in Figs. 6-10, the slope indicates a more fluid-like environment, in which the molecules probably do not pack as rigidly as they would in a crystal. Of the molecules given above, only Py-C-BHTCNQ lacks an alkyl "greasy chain"; DDOP-C-BHTCNQ, BDDAP-C-BHTCNQ and BDDAP-C-HMTCAQ all have dodecyl groups that help provide a hydrophobic region in the molecule (if the dodecyl groups are replaced by methyl groups, no monolayers are formed). Since, however, the presence of long alkyl chains may retard the TB ET, it was important to find a shorter alkyl chain which would help in providing monolayer formation. Thus we are pleased with the recent result that the bis-hexyl chain in **33** was "sufficient" to provide good PL monolayers. Work on similar molecules is continuing.

Langmuir-Blodgett films and tests on rectification

Efforts were made to find rectification behavior in sandwiches of molecules **18** [4,5], **29** [7], and **31** [7]. For **18**, sandwiches Pt|**18**|Hg and SnO|**18**|Hg were electrical shorts [4,5]. This was not surprising, since the large molecular A_c for **18** indicated that it was "sitting" with the longest axis almost parallel to the water subphase, and therefore almost parallel to the Pt of SnO substrate; if this is true, then the fim thickness, probably only 6 Å, was so small that TS tunneling, or tunneling through defects could not be avoided. For **29** and **31**, electrical shorts were observed in similar sandwiches [7]. However, in the same laboratory, sandwiches Al|**12**|Hg using monolayers of cadmium arachidate were not insulating[7], even though they had been prepared previously as high-resistance sandwiches in other laboratories [51]. Therefore it can be stated that a proper test of rectification of **29** or **31** had not been made.

Important structural information is provided by Fourier Transform Infrared (FTIR) spectroscopy [32,56-60]. Figs. 11-13 were obtained on a Nicolet 60SXR spectrometer, operated at 4 cm^{-1} resolution, using grazing angle specular reflectance and a Brewster angle polarizer. In all cases the samples were transferred as LB films onto aluminium-coated glass slides. The molecule studied was BDDAP-C-HMTCAQ, **32**; in Figs. 11,12 a single monolayer is present, yet one can resolve the aliphatic CH stretch bands at 2951, 2910, and 2849 cm^{-1} and the CN stretch at 2221 cm^{-1}. In Fig. 13 the LB films consists of 16 monolayers.

Table 2. Molecular areas and collapse pressures for PL monolayers

Molecule	No.	T K	Π_c mN/m	A_s Å^2	A_m Å^2	A_o Å^2	Ref.
TTF-C-BHTCNQ	**18**	292	12.7	134±50	--	--	[5]
DDOP-C-BHTCNQ	**29**	292	20.2	50	55	60	[5,16]
BDDAP-C-BHTCNQ	**30**	293	47.3	57	69	82	[16]
		303	45.9	54	70	82	
Py-C-BHTCNQ	**31**	283	28.2	53	60	66	[16]
BDDAP-C-HMTCAQ	**32**	293	22.3	58	71	83	[16]
BHAP-C-HMTCAQ	**33**	293	35.8	42	47	53	

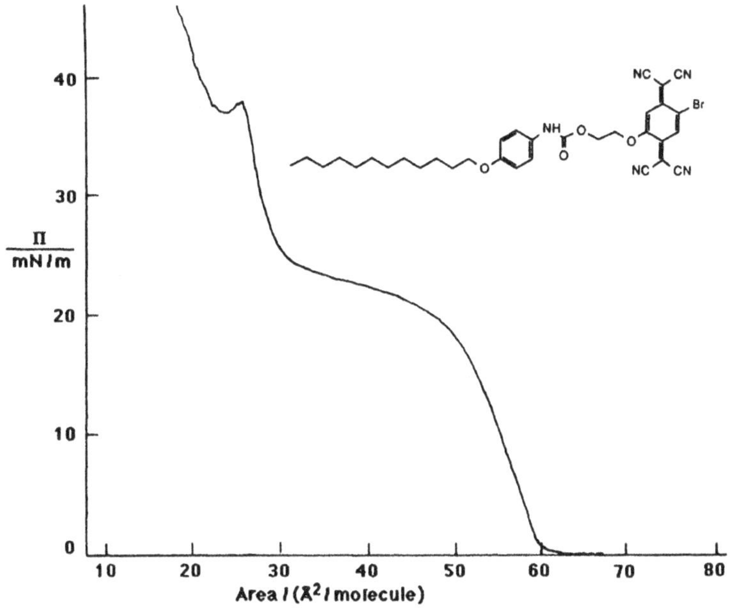

Fig. 6. Pressure-area isotherm for DDOP-C-BHTCNQ, **29** [5,16].

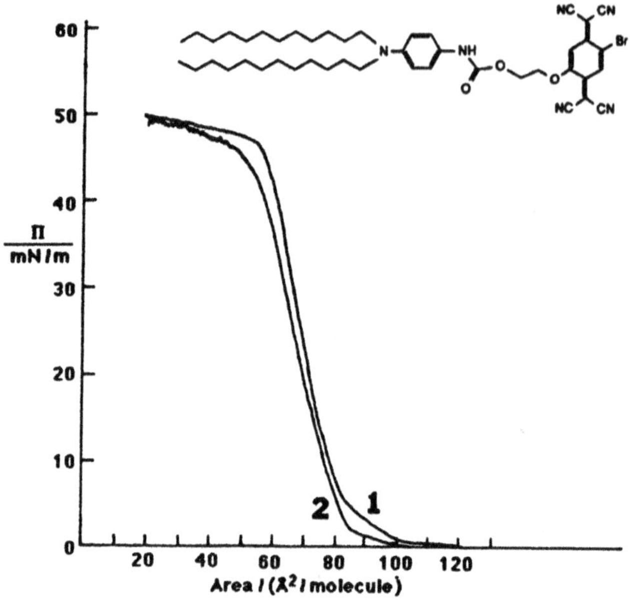

Fig. 7. Pressure-area isotherm for BDDAP-C-BHTCNQ, 30, at 293 K (curve 1) and at 303 K (curve 2) [16].

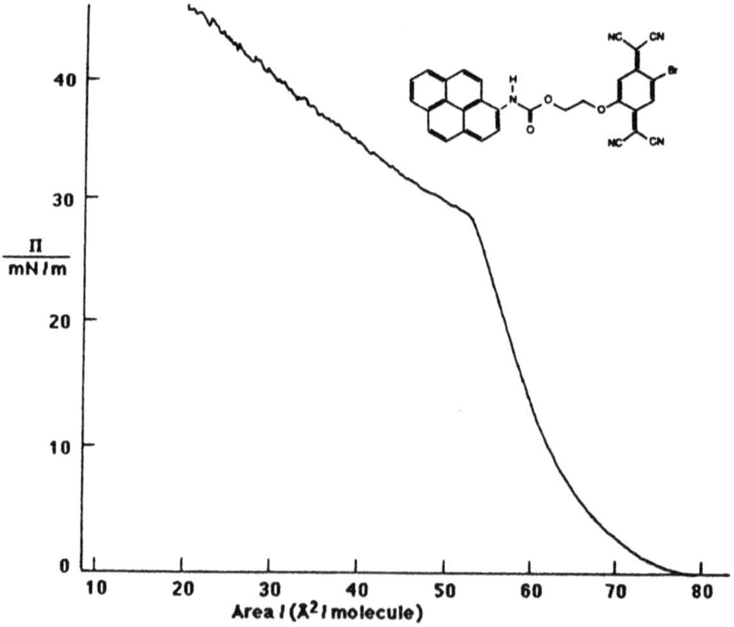

Fig. 8. Pressure-area isotherm for Py-C-BHTCNQ, 31 [16].

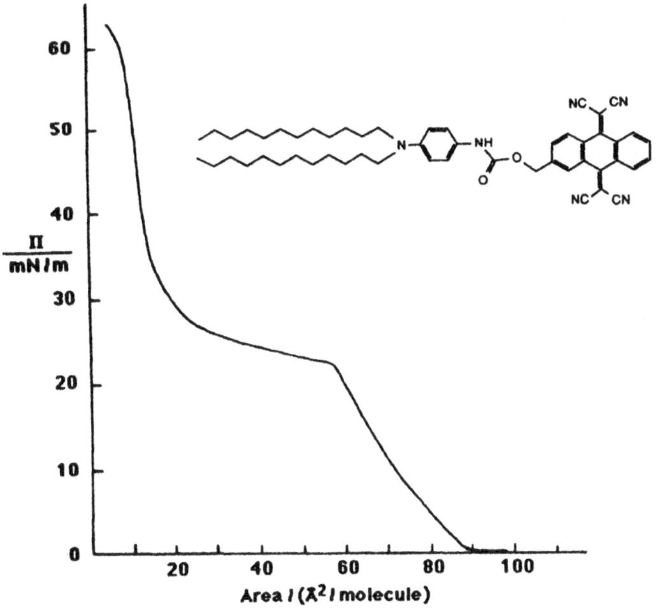

Fig. 9. Pressure-area isotherm for BDDAP-C-HMTCAQ, **32** [16].

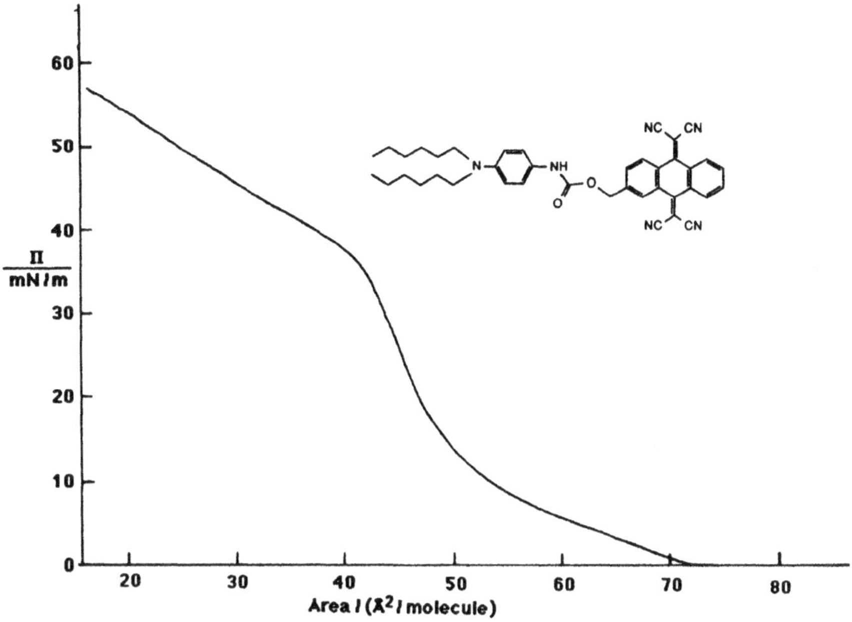

Fig. 10. Pressure-area isotherm for BHAP-C-HMTCAQ, **33** (this study).

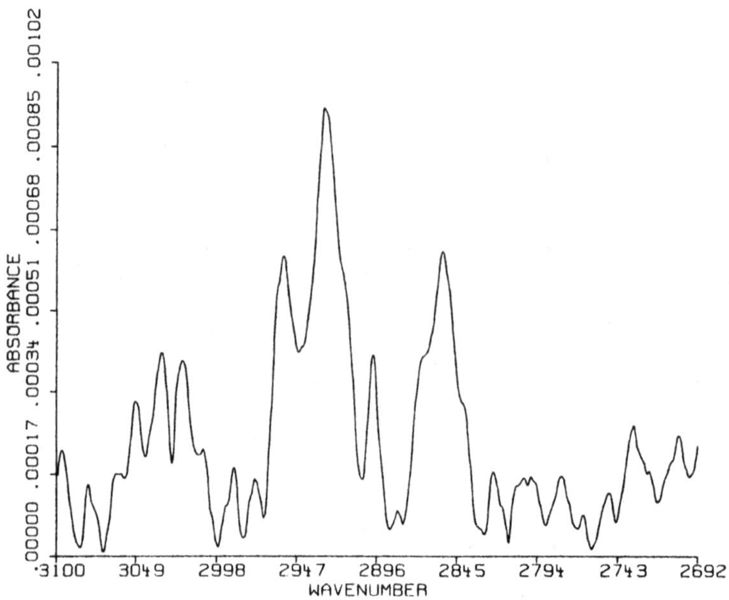

Fig.11. FTIR of single LB monolayer of BDDAP-C-HMTCAQ, **32**, on aluminum, showing aliphatic CH stretch bands.

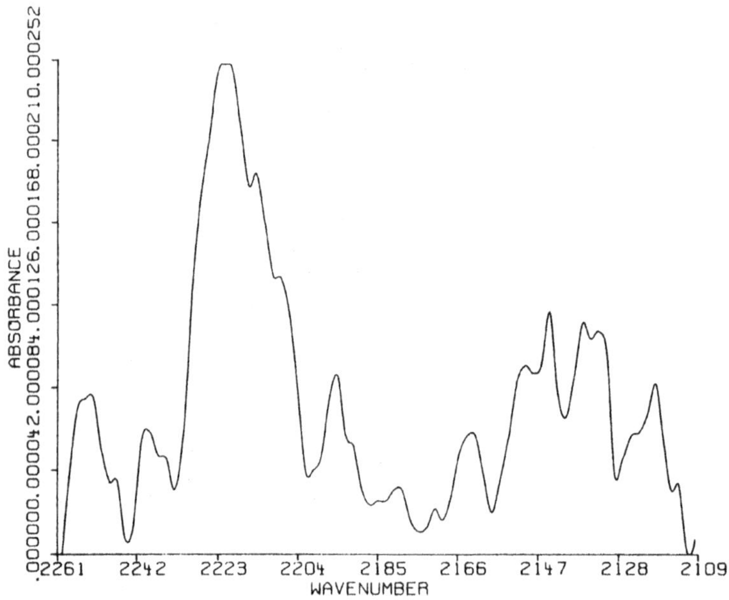

Fig.12. FTIR of single LB monolayer of BDDAP-C-HMTCAQ, **32**, on aluminum, showing CN stretch bands.

Conceptually, our synthetic program is related to efforts by many groups to model the photosynthetic reaction center: linked D-σ-A systems (D = porphyrin, A = quinones) are under intense study in several laboratories [61-66].

It should be mentioned that Kuhn et al. [67] and later Sugi et al. [68] have prepared rectifiers based on seven or more monolayers of cadmium arachidate, three of them randomly doped (5:1) with an electron donor dye, one a pure cadmium arachidate layer, the other three doped (5:1) with an electron acceptor dye. However, the device does not work if there are fewer than seven monolayers.

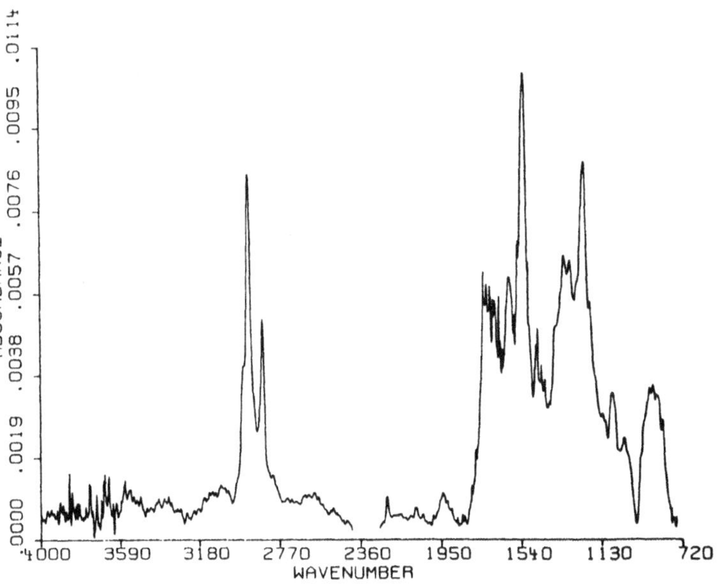

Fig. 13. FTIR of 16 LB monolayers of BDDAP-C-HMTCAQ , **32**, on aluminum.

Recently Fujihira prepared a LB photodiode [68] using a single monolayer of molecule **34**, which consists of a donor D (ferrocene), a sensitizer S (pyrene), and an acceptor A (viologen), deposited as an LB film onto a semitransparent gold electrode (with the viologen on the Au surface), and placed in a 0.1 molar potassium chloride solution, with a platinum counter electrode. The electron transfer is from solution to D, then to the ground state of S; upon irradiation of the film at 330 nm, the electron is pushed to the excited state of S, S*, which is above the LUMO of A, which finally gets the electron and transfers it to the Au layer. 2 nA of current at 0.0 V versus SCE was measured when light was on [69].

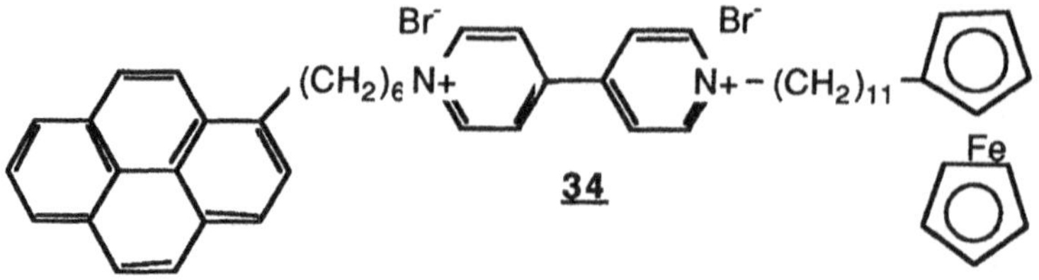

34

CONCLUSIONS

We have demonstrated that LB films of D-σ-A molecules can be synthesized; we are on our way to fabricating metal I organic I metal sandwiches which should give us a realistic test for the organic rectifier.

ACKNOWLEDGMENTS

It is a pleasure to thank all of our collaborators past and present: Dr. Jamil Baghdadchi (now at BASF Buffalo), Dr. A. M. Bhatti (now at Univ. of Southern Mississippi), Prof. Sukant Tripathy (GTEL, now Univ. of Lowell), Prof. Yozo Miura (Osaka City University), Profs. Norman E. Heimer and Daniell L. Mattern (Mississippi), Profs. M. P. Cava, J. L. Atwood, and L. D. Kispert (Alabama), and Drs. R. Schumaker, J. L. Grant and R. K. Laidlaw (Alabama), and Mr. Epifanio Torres (Mississippi). We are grateful to Mr. Ken Kempfert and Dr. Robert Rosenthal of Nicolet Analytical Instruments for measuring the FTIR spectra. The work would not have been possible without the generous support of the National Science Foundation and the Office of Naval Research.

REFERENCES

[1] A. Aviram, M. J. Freiser, P. E. Seiden, and W. R. Young, U.S.Patent **US-3,953,874** (27 April 1976).
[2] A. Aviram and M. A. Ratner, Chem. Phys. Lett. **29**, 277-283 (1974).
[3] A. Aviram, P. E. Seiden, and M. A. Ratner, in "Molecular Electronic Devices", F. L. Carter, ed. (Dekker, New York, 1982) page 5.
[4] R. M. Metzger and C. A. Panetta, J. Phys. (Les Ulis, Fr.) Colloque **44**, C3-1605 to C3-1611 (1983).
[5] R. M. Metzger and C. A. Panetta, in "Molecular Electronic Devices, Vol. II", F. L. Carter, ed. (Dekker, New York, 1987) page 1.
[6] C. A. Panetta, J. Baghdadchi, and R. M. Metzger, Mol. Cryst. Liq. Cryst. **107**, 103-113 (1984).
[7] R. M. Metzger, C. A. Panetta, N. E. Heimer, A. M. Bhatti, E. Torres, G. F. Blackburn, S. K. Tripathy, and L. A. Samuelson, J. Molec. Electronics, **2**, 119-124 (1986).
[8] R. M. Metzger, C. A. Panetta, Y. Miura, and E. Torres, Synth. Metals **18**, 797-802 (1987).
[9] E. Torres, C. A. Panetta, and R. M. Metzger, J. Org. Chem. **52**, 2944-2945 (1987).
[10] R. M. Metzger and C. A. Panetta, Proc. of the Eighth Winter Conference on Low-Temperature Physics, Cuernavaca, Mexico, January 1987, in press.
[11] R. K. Laidlaw, Y. Miura, C. A. Panetta, and R. M. Metzger, Acta Cryst., submitted.
[12] R. K. Laidlaw, Y. Miura, J. L. Grant, L. Cooray, M. Clark, L. D. Kispert, and R. M. Metzger, J. Chem. Phys., in press.
[13] R. K. Laidlaw, J. Baghdadchi, C. A. Panetta, Y. Miura, E. Torres, and R. M. Metzger, Acta Cryst., submitted.
[14] Y. Miura, R. K. Laidlaw, C. A. Panetta, and R. M. Metzger, Acta Cryst., submitted.
[15] R. M. Metzger, R. K. Laidlaw, E. Torres, and C. A. Panetta, J. Cryst.Spectr.Res., submitted.

[16] R. M. Metzger, R. R. Schumaker, M. P. Cava, R. K. Laidlaw, C. A. Panetta, and E.Torres, Langmuir, in press.
[17] R.Hoffman, Acc. Chem. Res. **4**, 1-9 (1971).
[18] R. Gleiter, E. Schmidt, D. O. Cowan, and J. P. Ferraris, J. Electron Spectrosc. **2**, 207-210 (1973).
[19] R. N. Compton and C. D. Cooper, J. Chem. Phys. **66**, 4325-4329 (1977).
[20] Z. G. Soos, Chem. Phys. Lett. **63**, 179-183 (1979).
[21] F. Herman and I. P. Batra, Phys. Rev. Lett. **33**, 94 -97 (1974).
[22] See e.g. G. L. Gaines, Jr. "Insoluble Monolayers at Liquid - Gas Interfaces" (Interscience, New York, 1966).
[23] K. B. Blodgett, J. Am. Chem. Soc. **57**, 1007-1022 (1935).
[24] K. B. Blodgett and I. Langmuir, Phys. Rev. **51**, 964-982 (1937).
[25] H. Kuhn, D.Moebius, and H. Buecher in "Techniques of Chemistry, Vol. I - Physical Methods of Chemistry - Part V - Determination of Thermodynamic and Surface Properties" A. Weissberger and B. W. Rossiter, eds. (Wiley - Interscience, New York, 1972) pages 577 - 702.
[26] H. Kuhn, Pure Appl. Chem. **51**, 341-352 (1979).
[27] H. Kuhn, Pure Appl. Chem. **53**, 2105-2122 (1981).
[28] See e. g. Thin Solid Films **68** (1980), **99** (1983), **132-134** (1985).
[29] W. C. Bigelow, D. L. Pickett, and W. A. Zisman, J. Colloid Sci. **1**, 513-538 (1946).
[30] L. Netzer, R. Iscovici, and J. Sagiv, Thin Solid Films **99**, 235-241 (1983).
[31] L. Netzer, R. Iscovici, and J. Sagiv, Thin Solid Films **100**, 67-76 (1983).
[32] J. Gun, R. Iscovici, and J. Sagiv, J. Colloid Interf. Sci. **101**, 201-213 (1984).
[33] R. W. Murray, Acc. Chem. Res. **13**, 135-141 (1980).
[34] R. Maoz and J. Sagiv, Thin Solid Films **132**, 135-151 (1985).
[35] R. G. Nuzzo, F. A. Fusco, and D. A. Allara, J. Am. Chem. Soc. **109**, 2358-2368 (1987).
[36] G. Wegner, Z. Naturforschung **24b**, 829 (1969).
[37] M. Batley and L. E. Lyons, Mol. Cryst. **3**, 357 -374 (1968).
[38] E. Clar, J. M. Robertson, R. Schloegl, and W. Schmidt, J. Am. Chem. Soc. **103**, 1320-1328 (1981).
[39] E. C. M. Chen and W. E. Wentworth, J. Chem. Phys. **63**, 3183-3191 (1975).
[40] C. D. Cooper, W. F. Frey, and R. N. Compton, J. Chem. Phys. **69**, 2367-2374 (1978).
[41] D. R. Gray, Ed., "American Institute of Physics Handbook, II Edition" (McGraw-Hill, New York, 1963) pages 9-147 to 9-149.
[42] R. A. Marcus, Disc. Faraday Soc. **29**, 21-31 (1960).
[43] L. T. Calcaterra, G. L. Closs, and J. R. Miller, J. Am. Chem. Soc. **105**, 670-671 (1983).
[44] J. R. Miller, L. T. Calcaterra, and G. L. Closs, J. Am. Chem. Soc. **106**, 3047-3049 (1984).
[45] N. S. Hush, M. N. Paddon-Row, E. Cotsaris, H. Oevering, J. W. Verhoeven, and M. Heppener, Chem. Phys. Lett. **117**, 8-11 (1985).
[46] P. L. Dutton, R. C. Prince, D. M. Tiede, K. M. Petty, K. J. Kaufmann, K. J. Netzel, and P. M. Rentzepis, Brookhaven Symp. Biol. **28**, 213-237 (1977).
[47] I. R. Peterson and G. J. Russell, Thin Solid Films **134**, 143-152 (1985).
[48] S. Garoff, H. W. Deckman, J. H. Dunsmuir, M. S. Alvarez, and J. M. Bloch, J. Phys. (Les Ulis, Fr.) **47**, 701-709 (1986).
[49] I. R. Peterson, J. Mol. Electronics, submitted.
[50] G. G. Roberts, P. S. Vincett, and W. A. Barlow, J. Phys. C **11**, 2077-2085 (1978).
[51] B. Mann and H. Kuhn, J. Appl. Phys. **42**, 4398-4405 (1971).
[52] P. K. Hansma, Ed. "Tunneling Spectroscopy: Capabilities, Applications, and New Techniques" (Plenum, New York, 1982).
[53] W. R. Hertler, J. Org. Chem. 41, 1412-1416 (1976).
[54] J. Baghdadchi, Ph. D. dissertation, Univ. of Mississippi, Dec. 1982.
[55] Y. Miura, C. A. Panetta, and R. M. Metzger, J. Org. Chem., in press.
[56] N. E. Schlotter, M. D. Porter, T. B. Bright, and D. L. Allara, Chem. Phys. Letters **132**, 93-98 (1986).
[57] W. G. Golden, D. D. Saperstein, M. W. Severson, and J. Overend, J. Phys. Chem. **88**, 574-579 (1984).

[58] J. F. Rabolt, M. Jurich, and J. D. Swalen, Applied Spectrosc. **39**, 269-272 (1985).
[59] J. F. Rabolt, F. C. Burns, N. E. Schlotter, and J. D. Swalen, J. Chem. Phys. **78**, 946-952 (1983).
[60] J. D. Swalen and J. F. Rabolt, in "Fourier Transform Infrared Spectroscopy, Vol. 4 (Academic, 1985) pages 283-314.
[61] J. Lindsay, D. Mauzerall,and H. Linschitz, J. Am. Chem. Soc. **105**, 6528-6529 (1983).
[62] A. R. McIntosh, J. R. Bolton, J. S. Connolly, K. L. Marsh, D. R. Cook, T.-F. Ho, and A. C.Weedon, J. Phys. Chem. **90**, 5640-5646 (1986).
[63] A. D. Joran, B. A. Leland, G. G. Geller, J. J. Hopfield, and P. B. Dervan, J. Am. Chem. Soc. **106**, 6090-6092 (1984).
[64] S. Nishitani, N. Kurata, Y. Sakata, S. Misumi, A. Karen, T. Okada, and N. Mataga, J. Am. Chem. Soc. **105**, 7771-7772 (1983).
[65] T. A. Moore, D. Gust, P. Mathis, J.-C.Mialocq, C. Chachaty, R. V. Bensasson, E. J. Land, D. Doizi, P. A. Liddell, W. R. Lehman, G. A. Nemweth, and A. L. Moore, Nature **307**, 630-632 (1984).
[66] C. Krieger, J. Weiser, and H. A. Staab, Tetrahedron Lett. **26**, 60550-6058 (1985).
[67] E. E. Polymeropoulos, D. Moebius, and H. Kuhn, Thin Solid Films, **68**, 173-190 (1980).
[68] M. Sugi, K. Sakai, M. Saito, Y. Kawabata, and S. Iizima, Thin Solid Films **132**, 69-76 (1985).
[69] M. Fujihira, K. Nishiyama, and H. Yamada, Thin Solid Films **132**, 77-82 (1985).

MIXED VALENCE COMPOUNDS : POTENTIAL APPLICATIONS IN MATERIALS AND MOLECULAR ELECTRONICS

Jean-Pierre Launay

Laboratoire de Chimie des Métaux de Transition, Université Pierre et Marie Curie, 4 place Jussieu, 75252 Paris Cedex O5 France

INTRODUCTION

Mixed valence compounds have been known for a long time, but their rediscovery twenty years ago was motivated by their original physicochemical properties[1] . As a matter of fact, their electronic structure shows the presence of quasi degenerate ground states differing by the localization of one or more electrons. Thus thermal as well as optical electron transfer is a fundamental property of these compounds.

The concept of "mixed valence" was first formulated in Inorganic Chemistry where oxidation states can be relatively well defined. Subsequently it has been extended to encompass organic compounds where oxidized and reduced forms of a given molecule are simultaneously present, like in Bechgaard's salts. In fact, there is no reason to deny the mixed valence nature to all systems where several localized electronic structures with nearly equal energies can be written, differing one from the other only by a simple electron transfer. This is what happens for instance in multisite organic radicals[2] or in one-electron reduced forms of $Ru(bipy)_3^{2+}$ type complexes [3] and also in several conducting polymers.

Here we shall focus on the applications (present or potential) of mixed valence compounds. However, for obvious technological reasons, most of them do not concern crystalline materials but rather amorphous or polymeric systems. As we shall see, the applications in materials science are beginning to develop at the industrial level. But owing to the progress in chemical synthesis and theoretical descriptions it is now possible to conceive applications in Molecular Electronics where the desired properties would depend ultimately on the behaviour of a single molecule.

POTENTIAL APPLICATIONS IN MATERIALS

Electrical properties

Conduction and semi-conduction According to their electronic structures which can be more or less delocalized (class II versus class III in Robin-Day's classification[1]), extended mixed valence compounds can exhibit a wide variety of behaviours, i.e. semi conducting, conducting and superconducting behaviours. For the first two types, a large number of compounds have already been described[4] . However conducting and semi conducting compounds are interesting for application only if some other property is present (ease of processing, sensitivity to a physical or chemical perturbation, special physical properties,...).

Thus gels of non stoichiometric V_2O_5 exibit semi-conduction due to electron hopping between V^V and V^{IV} states[5] .These gels can be easily spread on a substrate and are presently used as anti-static coatings in the photographic industry.

In the field of non-stoichiometric oxides, the interest has been driven recently on substances like TiO_2 which can be used as oxygen sensors. The equilibrium with oxygen can be written as :

$$O_{2(gas)} + 4\,Ti^{3+} + 2\,\square \rightleftarrows 4\,Ti^{4+} + 2\,O^{2-}$$

which leads to a $p_{O2}^{-1/6}$ dependence of the Ti^{3+}concentration[6] .The equilibrium with an oxygen-containing atmosphere changes the Ti^{3+} concentration and thus the conductivity. These oxide semiconductor type oxygen sensors have some advantages over electrochemical zirconia oxygen sensors. They allow the real time control of air to fuel ratio in engines and begin to appear in automobile industry in Japan[7] .

Coordination complexes can also lead to conducting mixed-valence systems, especially when a one-dimensional structure is achieved. This is what happens in KCP[8] and related materials. However, no application seem to have been studied at the present time, probably because of the difficulties of single crystal growing and the use of precious metals.

Conducting polymers are also relevant to the field of mixed valence chemistry. Thus polypyrrole and polythiophene become conductors after they have been partially oxidized[9,10]. More interesting, their conductivity depends upon the presence of certain chemicals such as NH_3, but the exact mechanism of this action is unknown. However their use as sensors is actively investigated[11] .

One of the most fascinating polymeric systems is polyaniline. It is obtained as a deep blue powder by persulfate oxidation of aniline in acid medium[12] . The conductivity is strongly dependent upon the oxidation state and is maximum when a "mixed" state is present[13,14] (Fig. 1). Although the exact mechanism of conductivity is not completely understood, some points are well established : thus the conductivity is electronic in nature[12] and seems to involve variable range hopping[15] . Electron spin resonance and magnetic properties show that the electrons are associated by pairs in bipolarons[16] , as usual in organic molecules. But the real interest lies in the spectacular influence of H^+ on the conductivity, which can vary from ca $10^{-10}\ \Omega^{-1}\ cm^{-1}$ in basic medium, to ca $5\ \Omega^{-1}\ cm^{-1}$ in acidic medium. This huge variation is an electronic process leading to an insulator/metal transition. Thus sensor applications are conceivable.

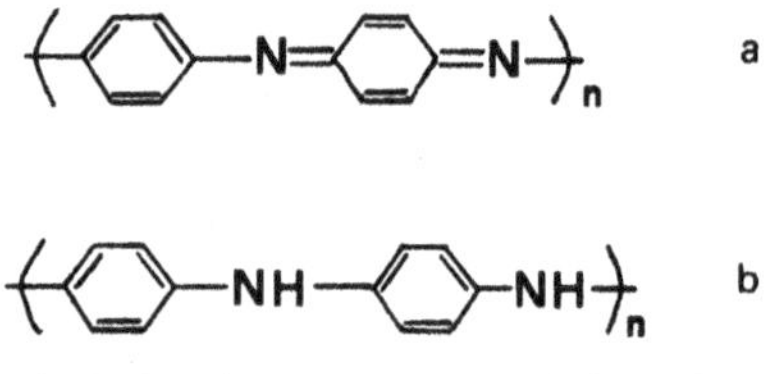

Fig. 1. Polyaniline a : Fully oxidized form
b : Fully reduced and protonated form

Switching effects and other non-linear responses Switching effects have been observed in Copper and Silver salts of tetracyanoquinodimethane (TCNQ) radical anions[17] . These compounds can be grown as thin films sandwiched between two electrodes. Threshold and memory effects are observed on the current voltage characteristics with two possible states : a low impedance ("ON") and a high impedance ("OFF") one . A likely equation for switching is

$$\underset{\text{(OFF)}}{[Cu^+(TCNQ^-)]_n} \rightleftharpoons \underset{\text{(ON)}}{Cu^{\circ}_x + [Cu^+(TCNQ^-)]_{n-x} + (TCNQ)_x}$$

i.e. one starts from a homovalent Cu^+TCNQ^- radical ion salt which is insulating as a result of its simple stoichiometry with one unpaired electron per site. The electronic interaction between TCNQ sites being small, the material is a Mott insulator. In the conducting state, a mixed valence compound is formed containing TCNQ and $TCNQ^-$ sites. The transition between these two states is believed to be triggered by the electric field.

Non linear behaviours are also observed when thin films of conducting polymers are connected to an electrode and studied by electrochemistry. The basic process has been very clearly explained by R.W. Murray[18] . The best examples are thin films of coordination polymers such as polyvinyl ferrocene or polyvinylbipyridine containing Ruthenium. Their "conductivity" is due to the self exchange of electrons between oxidized and reduced forms and is thus maximum when the electrode potential is near the redox potential of the couple. For other values, the film is almost insulating. This gives rise to non linear intensity-potential characteristics for which the term "redox conduction" has been coined, to distinguish it from the usual conduction process in which the composition of the material is not modified by the current flow. Moreover, the use of bilayer devices gives rise to rectification and memory effects, with the possibility to trap charges in an outer film[18] . Finally molecule-based transistors have been described[19] in which a film of conducting polymer is used to bridge two electrodes (source and drain), while a third electrode (gate) provides the control of redox conduction via the potential of the solution. Such systems have been shown to be able to amplify currents, hence the term "transistor" is fully justified[20] .

Superconductivity Superconductivity is now well established in mixed valence organic metals such as Bechgaard's salts[21] , which are reviewed in this Workshop. It is also found in the so-called Chevrel phases, which are formed by the association of mixed-valence clusters[22] . More recently superconductivity has also been observed in 1D coordination complexes[23].However until the last months, the highest critical temperatures were still the fact of metallic alloys, and did not exceed 23 K.

Then suddenly came the last months breakthrough of high critical temperatures for mixed oxides of the type $La_{2-x}\,Ba_x\,Cu\,O_{4-y}$ [24] . At the time this paper is written, the subject is still in rapid evolution and critical temperatures above the liquid nitrogen temperature have been obtained for Y Ba Cu O phases [25] . Although the exact nature of the process is not fully understood, two general comments can be made : First, these materials are one more time mixed-valence systems, here containing Copper in oxidation states II and III. The advantage provided by the mixed valence nature is the possibility to change the electronic filling of the bands thus adjusting the position of the Fermi level. Second, the basic process responsible of superconductivity, i.e. the BCS pairing of electrons, is usually explained by the electron phonon coupling, which plays a crucial role in the behaviour of mixed valence compounds. According to B.K.Chakraverty[26] , there is a close relation between bipolarons and itinerant Cooper pairs, so that superconductivity should be searched in the conditions where strong electron-phonon coupling occurs, in particular near insulator-metal transitions. In the case of the present oxides, it is the existence of a strongly Jahn-Teller distorted oxidation state of Copper, i.e. Copper(II), which would provide the necessary strong electron-phonon interaction[27] .

Optical Properties

Mixed-valence compounds usually exhibit a peculiar electronic transition, the intervalence band, corresponding to an optical electron transfer between the different redox sites. Since the intensity of this band depends upon the ratio of the concentrations of the two oxidation states, it is possible to change the optical properties of the material by a redox reaction. This can be performed by an electrochemical process (or less frequently by a photochemical process). The first one will be called here Electrochemichromism, to emphasize its relevance to Electrochemistry, and distinguish it from Electrochromism, in which it is the direct influence of an electric field which gives rise to a change in optical properties.

Electrochemichromism Usually, one starts from a colorless compound, such as WO_3 (i.e. W^{VI}, electronic configuration d^0) coated on a transparent electrode. The electrochemical reduction yields a deep blue mixed-valence compound containing W^{VI} and W^{V} [28]. The process is reversible but is relatively slow (ca 100 msec) since it is based on a reaction with an electrode. However, applications are actively studied for static display[29] and solar attenuation (smart windows)[30] .

Besides metal oxides, other mixed-valence compounds are studied, such as Prussian Blue[31] and conducting polymers[32] . For the latter, the optical absorption is usually not depicted as an intervalence transition but rather as transitions involving the valence band, the conduction band and polarons or bipolarons bands[33] . However, it is sensitive to the oxidation state of the material, the reduced form being usually semiconducting and colorless, while the partly oxidized form is conductive and strongly coloured. An interesting exception is polyisothianaphthene, for which, owing to a very small bandgap (ca 1 eV), it is the oxidized form which is colorless[34] , giving a transparent conducting polymer.

Photochromism In the same way, a coloured mixed-valence compound can be generated by a photochemical reaction. Thus colloidal Tungsten oxides as well as several easily reducible Tungsten and Molybdenum polyoxometallates are photochemically reduced to mixed-valence "blues" in the presence of electron donors[35] . However the process is very slow, especially in the fading direction, and no application has been considered yet.

POTENTIAL APPLICATIONS IN MOLECULAR ELECTRONICS

If now we turn from the field of material science to the domain of molecular electronics, two new features appear : first, the size of the system under consideration decreases dramatically since it is assumed to work ultimately at the molecular level; thus addressing and setting of the molecular components become crucial problems. Second, it is necessary to devise active molecules, which could be able to process a signal. These two points will now be considered.

The addressing problem

In principle any structure leading to a one-dimensional behaviour could be used to connect the active molecules by "molecular wires" to the outside. However it is necessary that it exhibits a modular structure with the possibility to replace a given element. An interesting possibility is provided by the so called "shish-kebab" systems in which metallated macrocycles are connected in apical positions by bridging ligands[36] (Fig.2). These systems exhibit a semi-conducting behaviour when the metal ions are present in a mixed oxidation state. Furthermore oligomers with two to four bridging units can be obtained[37]. So these compounds can be used to bridge the gap between binuclear model systems (e.g. the Creutz Taube ion[38]) and extended systems allowing the connexion to the outside world.

Recently, mixed valence conductors have been incorporated in Langmuir-Blodgett films[39] .This has been realized with an amphiphilic charge transfer salt N-docosylpyridinium$^+$ $TCNQ^-$. After transfer on a solid substrate, the film was partially oxidized by Iodine vapours to generate a mixed valence $TCNQ/TCNQ^-$ system, with a resulting conductivity near $10^{-1}\ \Omega^{-1}\ cm^{-1}$. It is thus possible to prepare conducting planes sandwiched between the insulating hydrophobic chains of the N-docosylpyridinium ions. This represents a significant progress for the use of these compounds because the

Langmuir-Blodgett technique appears more and more as a method of choice to deposit small quantities of active molecules with a controlled orientation, and the TCNQ-based conductor could be used for in-plane wiring. In the meantime, these conducting films are interesting for sensors applications[39].

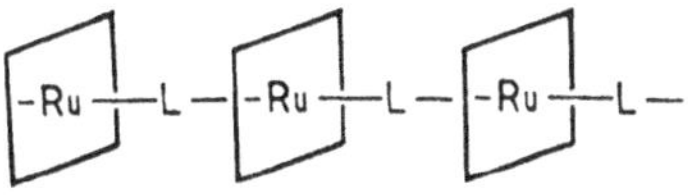

Fig. 2. Linear chain conductors obtained from Ru-macrocycle systems[36]

Finally, the addressing problem could also be tackled by optical communication. Elegant experiments have been described showing the possibility to trigger an intramolecular electron transfer by flash photolysis[45].

Control of intramolecular electron transfer

Assuming that the addressing problem can be solved, we now consider the possibility to control the intramolecular electron transfer in a mixed valence molecule, in order to realize an active device. In principle, the basic process of electron transfer could be able to carry a signal since a mixed valence molecule is a two-state system and if the two ends of the molecule are different, these two states must exhibit different spectroscopic and chemical properties. Thus for instance, in the analogous case of bichromophoric molecules, it is possible to detect a long distance electron transfer by the specific luminescence of the charge separated state[40]. But in order that this transferred electron can be used, a rather long ligand will be necessary so that the two ends of the molecule can display different interactions with their environment. Unfortunately, most of the time, one of the two states is much more stable than the other and contributes almost exclusively to the ground state. But recently, it has been shown by T.J. Meyer and coworkers that it is possible to control the oxidation state distribution by changing the environment. A first possibility is to use a mixed valence Ruthenium-Osmium dimer in which one of the metals (Ru) is coordinated to a water molecule or a OH^- ligand depending on the pH. Thus a change in pH triggers a change in redox potential of the $Ru^{II/III}$ couple with a concomitant electron transfer[41] (Fig.3). The other possibility is to use the difference in solvent sensitivity of two couples. Thus a mixed valence Ru-Os system can be driven from the Os^{III} - Ru^{II} state in CH_3NO_2 to Os^{II} - Ru^{III} in DMF, with spectacular changes in the absorption spectrum[42]. This is an extreme case of solvatochromism, and it suggests that analogous changes could perhaps be obtained by an electric field. This would require an oriented disposition of the mixed valence system in a thin film and the application of an electric field of the order of 10^6 to 10^7 V/cm, which is not unrealistic by the Langmuir-Blodgett technique. Such materials would present extremely strong and fast electrochromic effects with applications in the field of display devices and transducers.

Following another line, we have recently undertaken the study of model binuclear mixed valence systems in which the two metal atoms are linked by a bridging ligand exhibiting either a "structural mobility" or a chemical sensivity[43]. The metal sites are $Ru(NH_3)_5$ moieties which exhibit acceptable inertness to substitution, and good π donor properties to ensure enough electronic interaction through the bridge. The degree of electronic interaction is monitored through the intensity of the intervalence band which appears in the near infra red upon progressive oxidation of the Ru^{II} - Ru^{II} dimer.

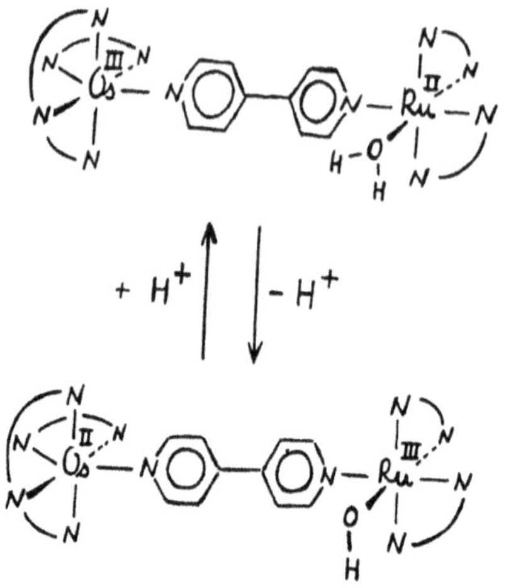

Fig. 3. p_H induced electron transfer[41]

An interesting class of ligands is provided by α–ω bipyridyl polyenes. Thus we have prepared $[(NH_3)_5 Ru^{II} (EE\ bipybu)\ Ru^{II} (NH_3)_5]^{4+}$, where EE bipybu = *all trans* bipyridylbutadiene. Its cyclic voltammogram shows a single anodic wave at +O.16 volts vs SCE in acetone, corresponding to the simultaneous oxidation of the two Ruthenium atoms. However, due to the existence of a comproportionation equilibrium, the progressive oxidation by Br_2 in heavy water shows the appearence of the mixed valence compound which is identified by its intervalence band near 9OO nm. The extinction coefficient, when corrected from the comproportionation equilibrium, shows no tendency to decrease with the length of the conjugated system. This is confirmed by the same experiment with "all trans" bipyridyl hexatriene[46]. So these ligands provide a convenient way to couple metal atoms 16 to 2O Å apart.

In order to test the possibility of molecular switching, we have prepared the ZE isomer of bipyridylbutadiene. This is easily performed by irradiation of the EE isomer at 34O nm, followed by selective dissolution of the ZE isomer in water. Due to steric crowding, the ZE isomer cannot be planar and several theoretical calculations predict an important twist angle between one of the pyridine rings and the central butadiene skeleton[44] (Fig.4). Thus the degree of electronic coupling through the ligand should be markedly weakened.

We have prepared the binuclear Ruthenium complex of the ZE isomer and studied its oxidation. Preliminary results show indeed a marked decrease in the electronic interation. Thus in the homovalent Ruthenium II state, the intensity of the metal-to-ligand charge transfer band decrease and its energy increases with respect to the EE isomer. For the mixed valence species, the energy and the extinction coefficient both decrease.

Although these results are encouraging, considerable work remains to be done in order to find a true molecular switch, with a reversible behaviour and a geometry suitable for incorporation in an extended structure.

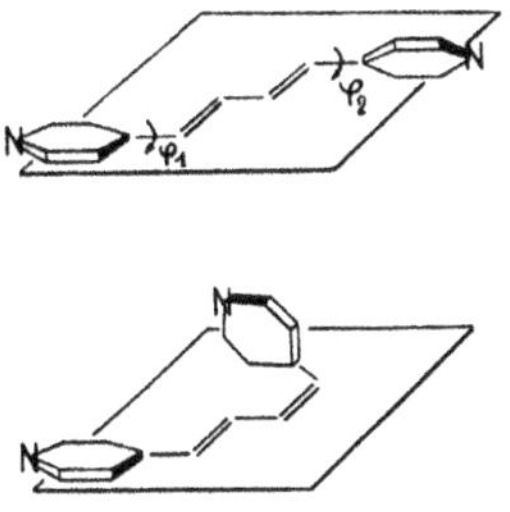

Fig. 4. Equilibrium geometries for the EE (above) and ZE (below) isomers of bipyridylbutadiene

Another interesting ligand for control of the intramolecular electron transfer is 44' Azopyridine (44' azpy), since the diazo group provides a protonation site. We have thus prepared $[(NH_3)_5Ru(44'\ azpy)Ru(NH_3)_5]^{4+}$. It exhibits a metal to ligand charge transfer band at a particularly low energy (λ_{max} = 700 nm in acetonitrile) which complicates the observation of the intervalence band. However in nitrobenzene we have been able to observe the characteristic increase and decrease of the optical density upon bromine oxidation. Apparently, the low energy of the π^* orbitals of 44' azopyridine ensures a rather strong interaction with Ruthenium orbitals and this ligand seems to have good electron mediating properties.

However the most spectacular effect arises upon protonation of the II-II dimer. The blue species turn to red (λ_{max} = 475 nm), while the protonation of the diazo group was expected to move the charge transfer band towards higher wave lengths. A study of the solvatochromic effect on both forms suggests that an internal redox reaction has occured, with the reduction of the diazo group to an hydrazo group:

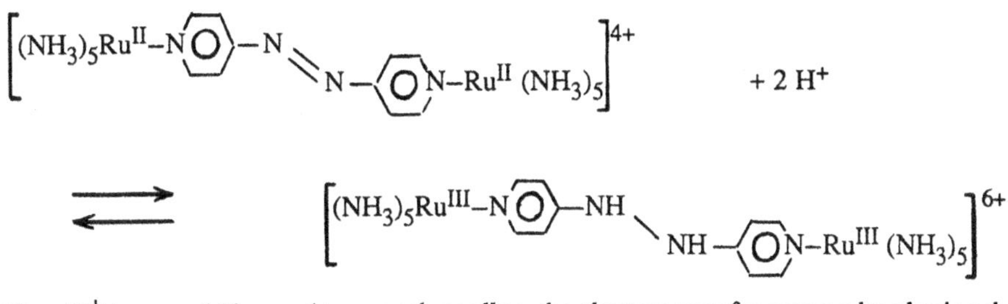

Thus H^+ ions could be used to quench or allow the electron transfer process by altering the conjugated character of the ligand.

CONCLUSION

After being the subject of much fundamental studies, mixed valence compounds are more and more involved in the field of applications. Industrial realizations can now be found in materials science, as a result of their electrical, optical, and particularly their *combined* optical and electrical properties. But their evolution potentialities are great because they lie at the intersection between Solid State Chemistry, Organic Chemistry, and Coordination Chemistry. Thus following a general trend, it can be predicted that they will be progressively involved in more subtle devices than bulk materials, and they certainly will play a role as active elements in Molecular Electronics.

REFERENCES

1. M.B. Robin and P. Day, Adv. Inorg. Chem.Radiochem. 1O,247(1967)
   N.S. Hush, Progr. Inorg. Chem. 8,391(1967).
2. J.E. Harriman and A.H. Maki, J. Chem. Phys. 39,778(1963).
3. C.D. Tait, D.B. Mc Queen, R.J. Donohoe, M.K. De Armond, K.W. Hanck and D.W. Wertz, J. Phys. Chem. 9O,1766(1986).
4. P. Day, Int. Rev. in Phys. Chem. 1,149(1981).
   See also "Mixed Valence Compounds", ed by D.B. Brown, D. Reidel Dordrecht, 198O.
5. J.Bullot, P.Cordier, O.Gallais, M.Gauthier and J.Livage , J.Non Cryst. Solids 68,123 (1984)
6. J.B. Goodenough, Les Oxydes des Métaux de Transition, Gauthier-Villars, Paris, 1973,p 282.
7. T. Takeuchi, Proc. 2nd Int. Meet. Chem. Sensors, Bordeaux 1986,PL-6.
8. A.E. Underhill, Phil. Trans. R. Soc. London, Ser A, 314,125(1985).
9. G.B. Street, T.C. Clarke, R.H. Geiss, V.Y. Lee, A. Nazzal, P. Pfluger and J.C. Scott, J. Physique 44,C3,599(1983).
10. M. Gazard, J.C. Dubois, M. Champagne, F. Garnier and G. Tourillon, J. Physique 44,C3,537(1983).
    G. Tourillon and F. Garnier, J. Electroanal. Chem. 135,173(1982).
11. G. Gustafsson and I. Lundström, Proc. 2nd Int. Meet. on Chem. Sensors, Bordeaux 1986,7-14.
12. M. Doriomedoff, F. Hautière-Cristofini, R. de Surville, M.Jozefowicz, L.T. Yu and R. Buvet, J. Chim. Phys. 68,1O55(1971).
13. M. Nechtschein, private Communication.
14. L.A. Burke and M. Kertesz, to be published.
15. J.P. Travers, J. Chroboczek, F. Devreux, F. Genoud, M. Nechtschein,A. Syed, E.M. Genies and C. Tsintavis, Mol. Cryst. Liq. Cryst. 121,195(1985).
16. F. Devreux, F. Genoud, H. Lecavelier, M. Nechtschein, C. Santier, J.P. Travers, B. Villeret, P. Audebert, G. Bidan and A. Syed, Colloque National Matériaux II, Paris 1986.
17. J.O. Poehler and R.S. Potember, Mol. Cryst. Liq. Cryst.1O7,91(1984).
18. P. Denisevich, K.W. Willman and R.W. Murray, J. Am. Chem. Soc. 1O3,4727(1981).
19. H.S. White, G.P. Kittlesen and M.S. Wrighton, J. Am. Chem. Soc. 1O6,5375(1984).
20. E.P. Lofton, J.W. Thackeray and M.S. Wrighton, J. Phys. Chem. 9O,6O8O(1986).
21. K. Bechgaard and D. Jerome, Sci. Am. 247,5O(1982).
22. J.C.Armici, M.Decroux, O.Fischer, M.Potel, R.Chevrel, and M.Sergent Solid State Comm. 33,607 (1980)
23. L. Brossard and P.Cassoux, Compt. Rend. Acad.Sc.Ser 2, 3O2, 2O5 (1986)
24. J.G. Bednorz and K.A. Müller, Z. Phys. B.64,189(1986).
25. M.K. Wu, J.R.Ashburn, C.J.Torng, P.H.Hor, R.L.Meng, L.Gao, Z.J.Huang, Y.Q.Wang, and C.W.Chu Phys. Rev. Lett. 58,908 (1987)
26. B.K. Chakraverty. J. Physique 42,1351(1981).
27. L.F. Mattheiss, Phys. Rev. Lett. 58,1O28(1987).

28. I.F. Chang, B.L. Gilbert and T.I. Sun., J. Electrochem. Soc. 122, 955(1975). S.K. Mohapatra, J. Electrochem. Soc. 125,284(1978).
29. B. Reichman and A.J. Bard, J. Electrochem. Soc. 126,583(1979). C. Sanchez, F. Babonneau, R. Morineau, J. Livage and J. Bullot, Phil. Mag. B,47,279(1983).
3O. S.F.Cogan, T.D.Plante,M.A.Parker,and R.D.Rauh Sol.En.Mater.14,185 (1986). R.B.Goldner, R.L.Chapman, G.Foley, E.L.Goldner, T.Haas, P.Norton, G.Seward, and K.K.Wong, Sol.En. Mater. 14,195(1986)
31. R.J. Mortimer and D.R. Rosseinsky, J. Chem. Soc. Dalton Trans.1984,2O59.
32. R.J. Waltman, A.F. Diaz and J. Bargon, J. Electrochem. Soc. 131,74O(1984).
33. J.L. Brédas, J.C. Scott, K. Yakushi and G.B. Street, Phys. Rev B, 3O,1O23(1984).
34. M. Kobayashi, N. Colaneri, M. Boysel, F. Wudl and A.J. Heeger, J. Chem. Phys. 82,5717(1985).
35. A. Chemseddine, C. Sanchez, J. Livage, J.P. Launay and M. Fournier, Inorg. Chem. 23,26O9(1984). A.Léaustic, F.Babonneau and J. Livage, J. Phys. Chem. 9O,4193(1986).
36. A. Datz, J. Metz, O. Schneider and M. Hanack, Synth.Metals 9,31(1984). J.P. Collman, Euchem Conference on Molecular Materials, Les Arcs, France, Jan 1986.
37. C. Marzin, G. Tarrago, I. Zidane, E. Bienvenue, P. Seta, C. Andrieux, H. Gampp and J.M. Savéant, Inorg. Chem. 25,1778(1986).
38. C. Creutz and H. Taube, J. Am. Chem. Soc. 91,3988(1969).
39. A. Barraud, P.Lesieur, A.Ruaudel-Teixier, and M.Vandevyver Thin Solid Films 134,195(1985) L.Henrion, G.Derost, A.Ruaudel-Teixier and A.Barraud Proc.2nd Int.Meet.Chem.Sensors Bordeaux, 1986 p. 7-06
4O. P.Pasman, G.F.Mes, N.W.Koper, and J.W.Verhoeven J.Am.Chem.Soc. 107,5839 (1985)
41. G.A. Neyhart and T.J. Meyer, Inorg. Chem. 25,48O8(1986).
42. J.T. Hupp, G.A. Neyhart and T.J. Meyer, J. Am. Chem. Soc.1O8,5349(1986).
43. J.P. Launay, S. Woitellier, M. Sowinska, M. Tourrel and C. Joachim, Proceedings 3rd Int. Symp. on Molecular Electronic Devices, North Holland, in Press.
44. I. Baraldi, F. Momicchioli and M.C. Bruni, J. Chem. Soc. Farad. Trans 2 68,1571(1972). C. Joachim and J.P. Launay, Chem. Phys. 109,93(1986).
45. C.Creutz, P.Kroger, T.Matsubara, T.L.Netzel, and N.Sutin J.Am.Chem.Soc. 101, 5442 (1979) P.Pasman, G.F.Mes, N.W.Koper, and J.W.Verhoeven J.Am.Chem.Soc. 107, 5839 (1985)
46 C.W.Spangler, S.Woitellier and J.P.Launay Work in progress.

DIMETHYL-DICYANOBENZOCHINODIIMINE WITH Li^+, Ag^+ AND Cu^+ AS COUNTER-IONS: NEW RADICAL-ANION SALTS WITH EXTREMELY HIGH CONDUCTIVITY

J.U.v. Schültz, M. Bair, H.P. Werner and H.C. Wolf

3 Phys. Inst. der Universität Stuttgart
Pfaffenwaldring 57, D-7000 Stuttgart, W. Germany

A. Aumüller, P. Erk and S. Hünig

Inst. für Org. Chemie der Universität Würzburg, Am Hubland
D-8700 Würzburg, W. Germany

INTRODUCTION

The title compound 2,5-DM-DCNQI, a strong organic acceptor, is in its properties very similar to TCNQ, but is much easier accessible. By conventional chemistry only one step is necessary to obtain DCNQI from the corresponding chinone (1).

In its reduced state it forms complexes with organic or metallic counterions, mostly crystalizing in a segregated stacks arrangement of donators and acceptors. This configuration leads preferentially to low dimensional band structures and electronic conduction. In case of $(DM\text{-}DCNQI)_2Cu$ an extremely high conductivity was observed (2). With ca. 500.000 Scm^{-1} for T<10K, this represents the highest value of all "organic metals" so far in the non-superconducting state. In this paper we present additionally the data of the Li and Ag salt which also form 2:1 complexes. Both salts are accessible by the methods of conductivity and of ESR which help to reveal the properties of charge and spin-transportation.

CRYSTAL STRUCTURE

The three investigated salts of 2,5-DM-DCNQI are isomorphological in structure (space group $I4_1$). Each metal-ion stack is surrounded by four anion stacks. Consequently, the metal-ions are 4-fold coordinated in a tetrahedral fashion.

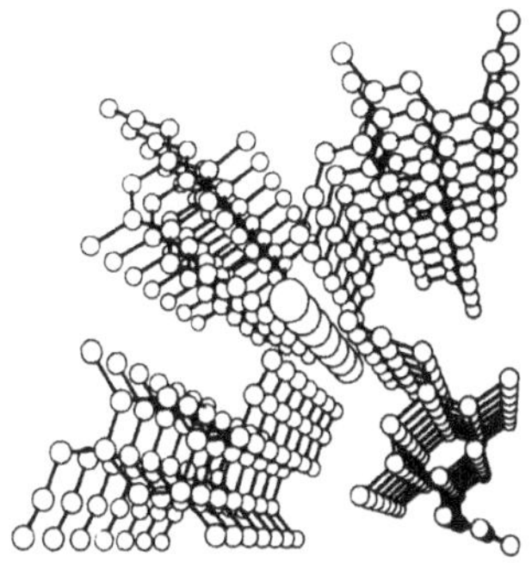

Fig.1: Perspective view of the crystal structure along the c-axis of $(2,5\text{-}DM\text{-}DCNQI)_2Cu$ from Ref.(2)

EXPERIMENTAL
The conductivity measurements were performed by conventional 4-probe technique via evaporated metal strips further contacted. The cooling was achieved via an Oxford ESR-9 cryostat, a Helium bath cryostat and a He^3-system (3). The ESR equipment was a Varian Century Lines model E-109 with a home-build goniometer.

RESULTS
(DM-DCNQI)$_2$Li,Ag: The conductivity, ESR and susceptibility data of the 3 salts are given in Fig.2 and 3.

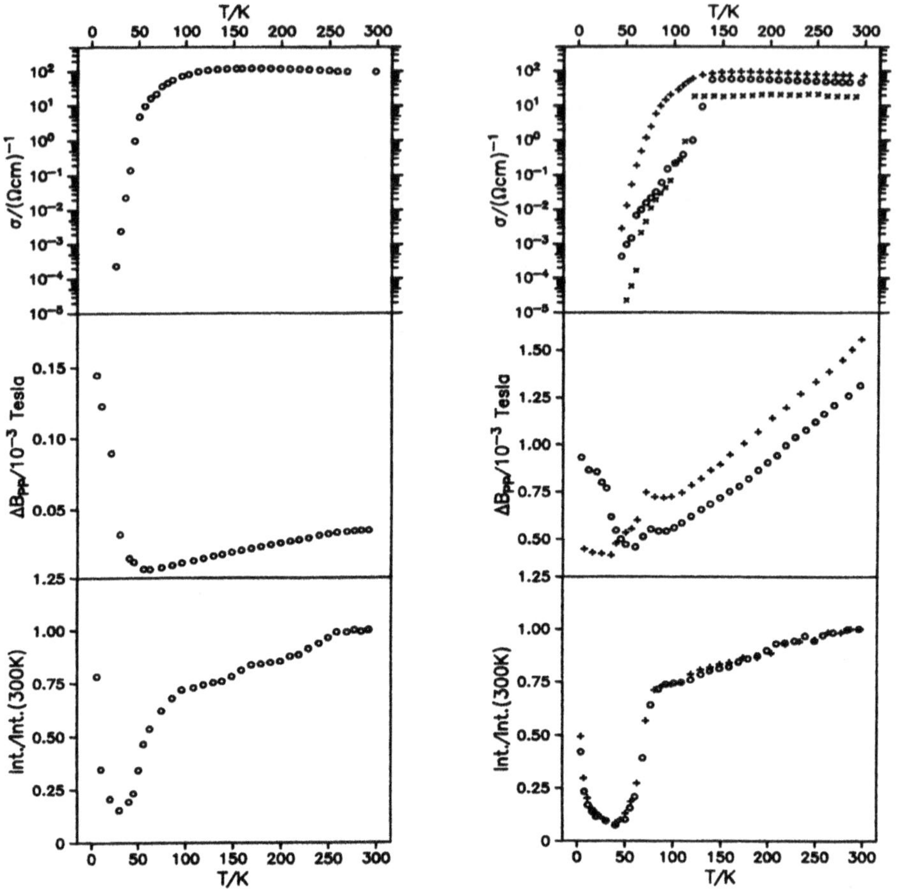

Fig.2: Conductivity, ESR linewidth and ESR intensity of (DM-DCNQI)$_2$Li (left) and of (DM-DCNQI)$_2$Ag (right) as function of T.
The different σ-data of the Ag-salt correspond to different samples, the two ESR-linewidth curves to different orientations of the crystal in respect to the external field.

Both salts exhibit, at least qualitatively, similar behavier: σ is about 100 Scm^{-1} at RT and stays constant till about 100K (T_σ) to drop below. Concomitantly, in the ESR intensities a phase transition (T_χ) is seen as well. It should be mentioned that in the Li and the Ag-salt the angular-dependence of the ESR-linewidth corresponds at high temperatures to that of Δg (5,6), ($\Delta g = g_{exp} - g_e$), but is opposite below T_χ.

$(DM\text{-}DCNQI)_2Cu$:

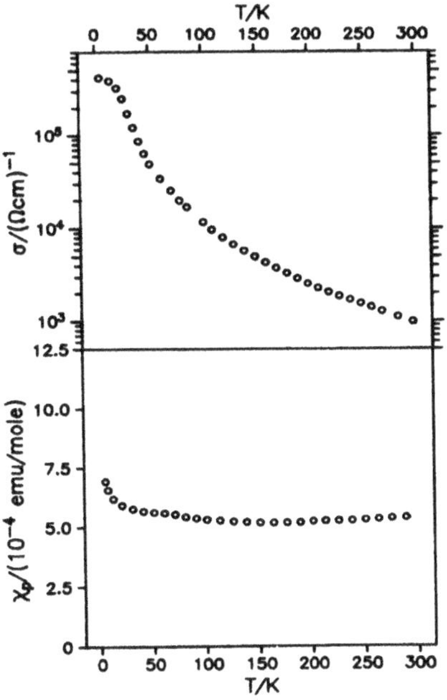

Fig.3: The temperature dependence of conductivity and susceptibility of $(DM\text{-}DCNQI)_2Cu$

In contrast to the other compounds, σ increases from 300K till the lowest temperature reached (450mK), and there is no ESR signal detectable although the susceptibility is very high. It is further worth mentioning, that about 25% of the Cu's are Cu^{2+} (4). There is no sign for a transition to a superconducting state. Applying pressure leads to an insulating state (8). There are hints that by pressure we change the band-filling towards a Mott-Hubbard insulator, presumably via a transition Cu(I) to Cu(II) as demonstrated by ESR on a pressed pellet (6), or by a disproportion $2\ Cu(I) \rightarrow Cu(II) + Cu(0)$.

DISCUSSION

$(DCNQI)_2Li,Ag$: The temperature range above T_σ is called "metallic", the one below "insulating" or "semiconductive". The minor temperature dependence in the first range points towards a conductivity strongly limited by scattering on defects in the chain, as well known for low dimensional systems. In this case electron-phonon interaction only governs the interstack scattering rate, which does not enter into the conductivity but in the ESR linewidth via the modified Elliot relation $T_2^{-1}=\Delta g^2/\tau_\perp$. In the 1d-case intrastack scattering ($\tau_\parallel$)does not mix Bloch- and Zeeman states (7). Therefore the linear decrease of ΔB_{pp} in T (Li:350-70mG,Ag:13-5G) is explained by a one photon process activating the interstack scattering rate. The difference in the absolute value is due to a weaker heavy atom effect in case of the Li-salt. Below T_χ, $\Delta B_{pp}(\theta)$ does not coincide any more with $\Delta g(\theta)$, $\Delta B_{pp}(\theta)$ follows roughly an angular-dependence as expected for dipolar interaction between localized spins (6). We therefore assume that at T_χ a dimerisation occurs (Peierls-transition) which-in combination with a strong electron-electron-repulsion (big U)-as deduced from the high susceptibility, leads to a full band and localized charges. The spins can, but don't must as seen in other systems, undergo a pairing. The fact that $T_\chi \leq T_\sigma$ and is not sample dependent, indicates a charge pinning on defects above the Peierls transition temperature.

$(DM\text{-}DCNQI)_2Cu$: The strong increase of σ lowering the temperature is clearly due to a slowing down of the electron-phonon scattering. In comparison to the Li and Ag salt, the conductivity therefore can not be pure 1d. Interstack interaction may screen the defects within the stacks. A considerable enhancement of the interstack scattering rate $(\tau_\perp Cu)^{-1}$ is expected by to reasons:

The N-Cu-distance with 1.98 Å (N-Ag: 2.3 Å) is in the dimension of chemical bonds, additionally the Cu^{2+} admixture opens a conduction path from stack to stack via the N-Cu-N bridge.

$(\tau_\perp Cu)^{-1} \gg (\tau_\perp Ag)^{-1}$ also explains the lack of an ESR-line although the paramagnetic susceptibility is high. A lower value for the linewidth of about 0.1 Tesla can be estimated.

CONCLUSION

All DCNQI-salts (1), except those with Cu^+ as counterions have their conductivity in the anion stack solely. Defects limit the absolute value of σ to about 100 Scm^{-1}. Such defects are "screened" in the Cu-salt by more effective interstack scattering due to the narrow N-Cu-distances and a considerable Cu^{2+}-percentage. Higher conductivities are reachable and phase transitions are suppressed, therefore.

REFERENCES

(1) A.Aumüller, S.Hünig: Liebigs Ann. Chem. 142, (1986) 165.
(2) A.Aumüller, P.Erk, S.Hünig, G.Klebe, J.U.v.Schütz, H.P.Werner: Angewandte Chemie 98, (1986) 159.
(3) These measurements were performed at the 1.Phys.Inst. of Stuttgart with the help of H.Schöffler which is greatly acknowledged.
(4) E.Schmeißer et. al.: to be published.
(5) M.Bair, Diplomarbeit Universität Stuttgart,1987.
(6) H.P.Werner, Doktorarbeit Universität Stuttgart,1987.
(7) H.J.Pedersen, J.C.Scott and K.Bechgaard: Phys. Rev. B18, (1981) 5014.
(8) D.Jerome, S.Tomics et. al.: to be published.

CONDUCTIVE SOLIDS BASED ON SOME NEW MOLECULES WITH ISOTHIAZOLO-RINGS

G.C. Papavassiliou, G.A. Mousdis, V. Gionis, J.S. Zambounis and S.Y. Yiannopoulos

Theoretical and Physical Chemistry Institute,
National Hellenic Research Foundation,
48,Vass.Constantinou Ave., Athens 116/35, Greece

INTRODUCTION

During the last three years, a number of conductive solids based on substituted derivatives of the tetrathiafulvalene with at least one pyrazino-ring [1]or pyridino-ring [2] have been reported. The corresponding metal 1,2-dithiolenes with pyrazino-ring [1], [3] or pyridino-ring [4] do not give conductive solids.

In this paper we report preliminary results on the preparation and investigation of a number of conductive solids based on some derivatives of tetrathiafulvalene with at least one isothiazolo-ring (π-donors) and on some metal 1,2-dithiolenes with two isothiazolo-rings (π-acceptors). The new π-donors and π-acceptors were prepared from disodium 5-cyanoisothiazoledithiolate [5], as starting material, according to the Scheme 1. The conductive solids were prepared from the compounds (3), (5), (8) and (10) by electrocrystallization in presence of Bu_4NX(where $X=BF_4$, Br_3 etc) and from the compounds (6) by electrocrystallization in presence of a π-donor.

EXPERIMENTAL

Details on the preparation of the compounds (3), (5) and (6) from disodium 5-cyanoisothiazoledithiolate (1) [5] are described in [6]. The compound (8) was prepared as follows. To a solution of 14g of KOH in 600ml of water, 1.4g of (3) was added and the mixture was heated at reflux temperature for 4 hours. After cooling the mixture was filtered . To the filtrate 30ml of aq.HCl (37%) was added dropwise. The precipitate, which contains (7) was filtered, washed with water and dried. Same product (7) was obtained by heating (3) in aq. H_2SO_4(70%v) at reflux temperature. The crude product (1.6g) was diluted in DMF (60ml) and heated at reflux temperature for 1 hour. After cooling the blue precipitate was filtered off and then the DMF was removed by evaporation. After extraction with CH_2Cl_2 and separation on silica gel column chromatography 250mg of (8) were obtained. Compound (8) is a pale-

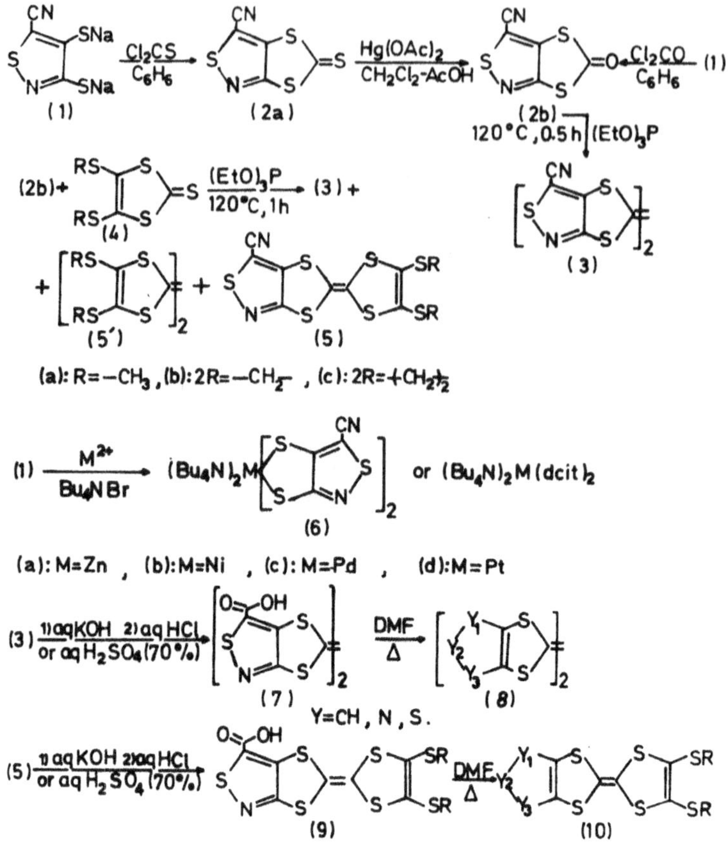

Scheme 1

-yellow solid; mp=276°C; mw=318 (:mass spectrum);UV(CH_3CN) (λ/nm, absorbance/a.u.)342(0.65), 310(0.485sh), 279(0.57),260 (0.49sh);IR(KBr) (cm^{-1}) 3071(m), 1380(m), 1293(s), 826(m), 783(m), 655(m). UV-spectrum, in comparison to those of (3) and (7), is an evidence that the structure of (8) is that with Y_1=CH, Y_2=N and Y_3=S. Compound (10c) (mp=197°C)was prepared by similar method. The preparation of (8) and metal 1,2-dithiolene analogous from 8-cyanoisothiazolopentathiepin [7] is in progress.

Compound $(5'c)_x Ni(dcit)_2$ (x∿2) was prepared by electrocrystallization at 0.4mm platinum electrode in a stardard H-cell containing a solution of 20mg of (5´c) and 23mg of (6b) [8] in 65ml of CH_2Cl_2 and using a current density of 6μA/cm^2. After 4 days of growth black flakes, green in the reflected light, were obtained at the anode. Analysis:

Cald(for x=2)	C28.71, H2.88, N4.78, S60.13
Found	C30.54, H1.36, N5.32, S60.05

Compound $(8)_xBF_4$ was prepaped by similar method using a solution-suspension of 20mg of (8) and 120mg of Bu_4NBF_4 in 60 ml of CH_2Cl_2 and a current density of $1.5\mu A/cm^2$. After 10 days of growth black needles, green or violet in the reflected light were obtained at the anode.

RESULTS AND DISCUSSION

Fig.1 shows the reflectance spectra of polycrystalline compactions* of $(5'c)_xNi(dcit)_2$ and $(8)_xBF_4$ and the absorption spectra of these compounds rubbed on CaF_2-plates. The spectra indicate that the compounds are semiconductors with activation energy ca 185 and 300meV, respectively. The dc-conductivity on polycrystalline pellets* at room temperature was

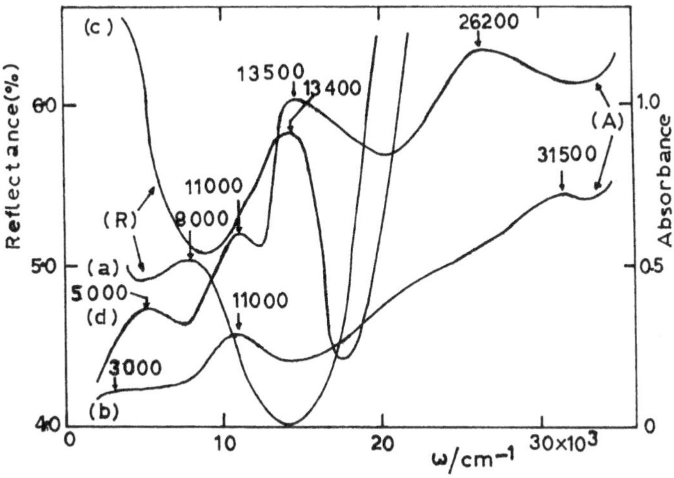

Fig.1. Reflectance (R) and absorption(A) spectra of $(5'c)_x$ $Ni(dcit)_2$ (a), (b) and $(8)_xBF_4$(c),(d).

found to be of the order of 10 and $0.1\Omega^{-1}cm^{-1}$, respectively. These results are similar to those obtained from $Z_xM(dmit)_2$ [9] and $(BPTTF)_2BF_4$ [10] respectively. The preparation and investigation of new materials having at least one isothiazolo-ring are in progress.

REFERENCES

1. G.C.Papavassiliou, S.Y.Yiannopoulos and J.S.Zambounis, "Bis(diazino)tetrathiafulvalenes and Similar π-Donors" Mol.Cryst., 120, 333(1985); "Pyrazinoethylenedithiotetrathiafulvalene: a New Unsymmetrical π-Donor" J.Chem. Soc., Chem.Commun.,820(1986); "Bis(pyrazino)tetrathiafulvalene and Similar π-Donors", Chem.Scripta, 27,..., (1987).

* For instrumentation see ref.2

2. G.C.Papavassiliou, "Bis[4,5-b]pyridino-1,1′,3,3′-tetrathiafulvalene: Synthesis and Charge Transfer Complexes" Chim. Chron. (New Series) 15, 161 (1986); G.C.Papavassiliou, S.Y. Yiannopoulos and J.S.Zambounis, "Pyrazino-tetraheterofulvalenes and a Few of Their Salts" this volume.
3. J.Becher, C.E.Stidsen, H.Toftund, and F.M.Assad, "Sodium 2-Methyl-2-propylthiolate, a Versatile Sulfur Reagent. Preparation of Protected Polysulfur Ligands and Their Nickel (II) Complexes" Inorg.Chim.Acta, 121, 23 (1986).
4. M.N.Edinberry, G.E.Gymer, and S.Jevons, "Antifungal Pyridine-2-thiones", Brit.UK Pat.Appl.2, 053,189, 4 Febr.1981; Chem.Abstr. 95, 8074z, (1981).
5. S.A.Vladuchick, T.Fukunaga, H.E.Simmons and O.W.Webster, "Thiacyanocarbons.6.1,4-Dithiino-[2,3-c;6,5-c′]diisothiazole-3,7-dicarbonitrile, Isothiazole[3,4-f][1,2,3,4,5] pentathiepine-8-carbonitrile, and Disodium 5-cyanoisothiazoledithiolate", J.Org.Chem. 45, 5122 (1980).
6. G.C.Papavassiliou, G.A.Mousdis, V.Gionis, J.S.Zambounis, and S.Y.Yiannopoulos, "Disodium 5-Cyanoisothiazoledithiolate as a Starting Material for Preparation of New Conductive Solids", Z.Naturforsch.b, in press.
7. B.L.Chenard, R.L.Harlow, A.L.Johnson, and S.A.Vlanduchick, "Synthesis, Structure and Properties of Pentathiepins", J.Am.Chem.Soc., 107, 3871 (1985).
8. A.Terzis, "Crystal Structure of $(Bu_4N)_2$[trans-Ni(dcit)$_2$]" Acta Cryst., to be published.
9. G.C.Papavassiliou, A.M.Cotsilios and C.S.Jacobsen, "Spectroscopic Investigation of Metal Dimercaptodithiolenes and Selenium Analogs", J.Mol.Structure, 115, 41 (1984); G.C.Papavassiliou, J.S.Zambounis, A.E.Underhill, B.Kaye, and H.P.Geserich, "Electrical Properties of 3,4,5-Tris (alkylthio)-1,2-dithiolium Charge Transfer Complexes". Mol.Cryst., Liq.Cryst., 134, 53 (1986); A.Kobayashi, R.Kato and H.Kobayashi, "New Aspects of Molecular Conductors of Metal Complexes With Multisulfur π-Donor and π-Acceptor Molecules", Synth.Metals., 19, 635 (1987).
10. G.C.Papavassiliou, H.P.Geserich, S.Y.Yiannopoulos and J.S.Zambounis, "Optical Properties of Bis(pyrazino)tetrathiafulvalene Salts", J.Mol.Structure, 143, 215 (1986); A.E.Underhill, B.Kaye, G.C.Papavassiliou, and S.Y.Yiannopoulos "Three Dimensional Conducting Solids: $(BPTTF)_2BF_4$, $(BPTTF)_3(ReO_4)_2$, $(BPTTF)ClO_4$ [where BPTTF=bis(pyrazino) tetrathiafulvalene] and Similar Compounds", Mol.Cryst. Liq.Cryst., 134, 59 (1986).

PYRIDINO-TETRAHETEROFULVALENES AND A FEW OF THEIR SALTS

G.C. Papavassiliou, S.Y. Yiannopoulos and
J.S. Zambounis

Theoretical and Physical Chemistry Institute
National Hellenic Research Foundation
48, Vass. Constantinou Ave., Athens 116/35, Greece

INTRODUCTION

Recently, the unsymmetrical π-donor DMET (dimethylethylenedithio-dithiadiselenafulvalene) has attracted much interest since the discovery of superconductivity in the cation radical salt $(DMET)_2Au(CN)_2$ [1]. In this paper we report the preparation of some new unsymmetrical tetraheterofulvalenes having a pyridino-ring and a few of their salts.

EXPERIMENTAL

The new π-donors (9) and (11) were prepared by cross coupling of 1,3-pyridino[4,5-b]-dithiole-2-one [2]-[4]* and 4,5-dimethyl-1,3-dithiole-2-thione [5], 1,3-benzodithiole-2-thione [6], 4,5-bis(alkylthio)-1,3-dithiole-2-thiones [7] and 4,5-dimethyl-1,3-diselenole-2-selone [8], via triethyl phosphite, $(EtO)_3P$, according to the Scheme 1.** 3-amino-2-chloropyridine (1)[3]*** was used as starting material. The preparation of (9) and (11) by using 4-amino-3-bromopyridine [4],[9] as starting material was unsuccessful because of the low yield of the intermediate products [4]. The preparation of these donors by an alternative method starting from 4-nitro-3-bromopyridine is in progress [4]. Some charge transfer complexes and some cation radical salts were prepared by direct reaction (DR) of the π-donors (9) and (11) with TCNQ, Br_2 and Bu_4NI_3 in CH_2Cl_2. Also some cation radical salts were prepared by electrocrystallization (EL) of some π-donors in presence of Bu_4NX (where $X=I_3$, IBr_2 etc) in CH_2Cl_2. Preparative data are given in Table 1. Electrical conductivity measurements were performed with a four-probe method. Samples were mounted with four fine gold wires and electrical conductivity measurements were supplied by a Keithley Model 220 programmable current source; voltage was measured on a Keithley Model 602 electrometer.

---

*Compound (6) is a white solid, mp=110°C [4].
**Compound (7) was obtained as a yellow solid, mp>300°C [4].
***Commercial (1) was used without further purification.

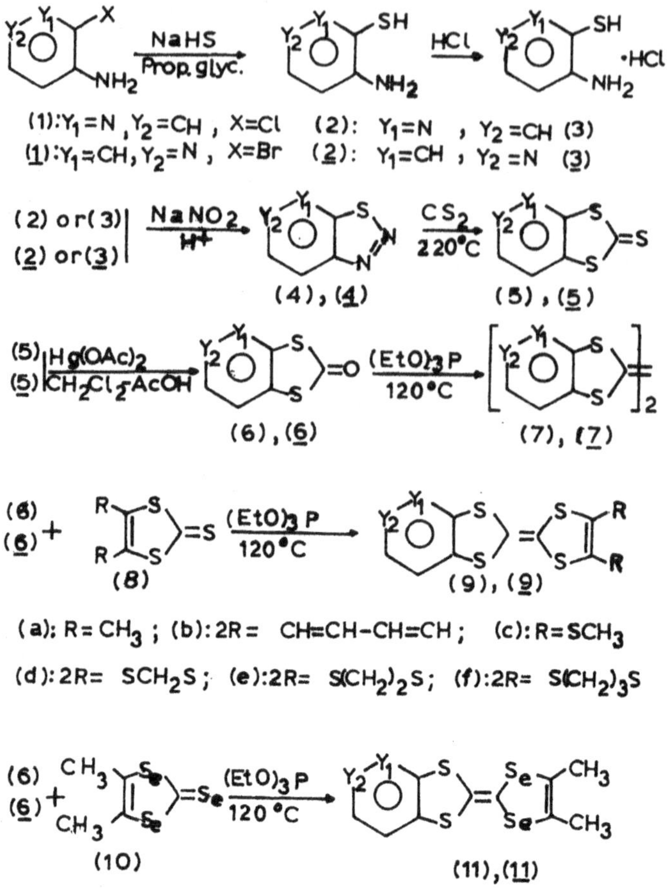

Scheme 1

RESULTS AND DISCUSSION

Conductivity measurements on polycrystalline compactions of (9b)TCNQ, β-(9e)$_x$Br$_3$ and (11)TCNQ showed that the compounds are neutral complexes similar to (DBTTF)TCNQ [10]. All the rest salts of the Table 1 were found to be conductive. Conductivity measurements on single crystals of α-(9e)$_2$IBr$_2$ along the needle axis (which is the a-axis [11])showed a semiconducting behaviour [12] with activation energy 230meV and σ_{RT}= 3 - 7x10^{-3} $\Omega^{-1}cm^{-1}$. Conductivity measurements on single crystals of the new salts are now underway.

ACKNOWLEDGEMENT

We would like to thank Dr.D.Rigas and Prof.N.Alexandrou for recording mass spectra and Dr.E.I.Kamitsos and M.A.Karakassides for optical and electrical measurements.

Table 1. Preparative data

Compound	Method	Yield(%)	Appearance	mp/^{o}C	UV(λ/nm)$^{+}$
(9a)		9	yellow	199	390
(9b)		4	yellow	248	350
(9c)		7	orange-yellow	117	358
(9d)		6	yellow	206	450
(9e)		10	orange-yellow	238	358
(9f)		3	yellow	208	354
(11)		15	orange	217	360
(9a)TCNQ*	DR		black powder		
β-$(9a)_xI_3$	DR		small black needles		
(9b)TCNQ*	DR		small brown needles		
β-$(9d)_xI_3$	DR		small black-golden needles		
$(9e)_x$TCNQ	DR		small black needles		
α-$(9e)_xBr_3$	EL		small black - bronze crystals		
β-$(9e)_xBr_3$	DR		brown needles		
α-$(9e)_xI_3$	EL		bronze needles or plates		
β-$(9e)_xI_3$	DR		brown-bronze powder		
α-$(9e)_2IBr_2$**	EL		black needles		
$(11)_x$TCNQ	DR		small orange-brown plates		
α-$(11)_xI_3$++	EL		bronze needles		
α′-$(11)_xI_3$++	EL		bronze-golden spears		
β-$(11)_xI_3$	DR		small black-bronze needles		
α-$(11)_xIBr_2$	EL		small black needles or plates		
β-$(11)_xIBr_2$	DR		black powder		
α-$(11)_xPF_6$	EL		black needles		

+ Peak position of the longest wavelength band (CH_3CN)

* From elemental analysis

**From x-ray crystal structure solution

++The resonance Raman spectra of both salts (α-,α′-) showed bands at 107, 214, 320...cm^{-1}, which are characteristic of I_3 (linear, symmetric).

REFERENCES

1. K.Kikuchi, M.Kikuchi, T.Namiki, K.Saito, I.Ikemoto, K.Murata, I.Ishiguro, and K.Kobayashi, "Superconductivity in the Cation Radical Salt $(DMET)_2Au(CN)_2$", Chem.Lett., in press.
2. K.Kowichi, "3-Mercaptothiopyridone-2", Pol., J.Chem.,52, 2039(1978)
3. G.C.Papavassiliou, "Bis[4,5-b]pyridino-1,1′,3,3′-tetrathiafulvalene: Synthesis and Charge Transfer Complexes" Chim.Chron.(New Series), 15, 161(1986).
4. G.C.Papavassiliou and S.Y.Yiannopoulos, unpublished work.
5. J.P.Ferraris, T.O.Poehler, A.N.Bloch and D.O.Cowan "Synthesis of the Highly Conducting Organic Salts: Tetramethyltetrathiofulvalenium-tetracyano-p-qui nodimethanide", Tetrahedron Lett., 2553(1973).
6. S.Hunig and E.Fleckenstein, "Eine Einfache Synthese von o-Phenylendithiol" Leibigs Ann.Chem., 738, 192(1970).
7. G.C.Papavassiliou, J.S.Zambounis, and S.Y.Yiannopoulos, "Some New π-Donors: Bis(propylenedithio)-tetrathiafulvalene, 4,5-Bis(alkylthio)-1,2-dithiolium Tetrafluoroborates and Similar Compounds" Chem.Scripta 27,...(1987).
8. A.Moradpour, Y.Peyrussan,I.Johansen and K.Bechgaard, "High-Yield Synthesis of Tetramethyltetraselenafulvalene Avoiding the Use of Gaseous H_2Se",J.Org.Chem.48, 388(1983); G.Bates, to be published.
9. O.S.Tee, and M.Paventi, "Kinetic and Mechanism of the Bromination of 4-Pyridone and Related Derivatives in Aqueous Solution", Can.J.Chem.61, 2256(1983).
10.A.Girlando, C.Pecile, A.Brillante, and K.Syassen, "High Pressure Optical Studies of Neutral-Ionic Phase Transitions in Organic C.T.Crystals", Synth.Metals, 19, 503 (1987), and refs cited therein.
11.A.Terzis, V.Psycharis, A.Hountas, and G.C.Papavassiliou, "Structures of the conducting Solids of Pyrazinoethylenedithiotetrathifulvalene (PEDTTTF) and [4,5-b]Pyridinoethylenedithiotetrathiafulvalene ([4,5-b]PEDTTTF): α-$(PEDTTTF)_2IBr_2$ and α-[4,5-b]$PEDTTTF)_2IBr_2$" Acta Cryst. 43...(1987).
12.A.E.Underhill, B.Kaye, and G.C.Papavassiliou, unpublished work.

SYSTEMATIC VARIATIONS OF THE BEDT-TTF RING SYSTEMS:

SYNTHESIS AND PHYSICAL PROPERTIES OF NEW RADICAL CATION SALTS

R.R. Schumaker, E. Dupart, R. Laversanne, C. Coulon and P. Delhaès

Centre de Recherche Paul Pascal - Domaine Universitaire de Bordeaux I
33405 Talence Cedex, France

INTRODUCTION

Since the observance of superconductivity for a perrhenate salt of bis(ethylenedithio) tetrathiafulvalene (BEDT-TTF) [1] a large number of investigations have concentrated on the electrochemical crystal growth of various BEDT salts. Another pathway for exploration proceeding from the BEDT molecule is to modify systematically the periphery of the ring system and to prepare new radical-cation salts.

CHEMICAL PREPARATION

A series of symmetrical and unsymmetrical molecules have been obtained from thiones by classical cross-coupling techniques. The details of the chemical synthesis will be published elsewhere [2] ; we must notice that the chemical preparations are quite similar to those recently described by another group [3]. The BEDT-TTF family of prepared compounds (3 to 7) is presented in the following scheme.

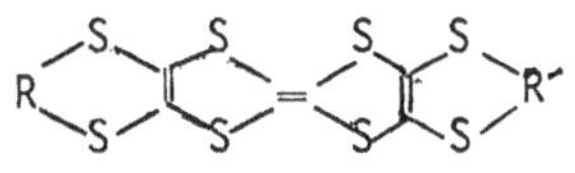

1 R = R' = $-CH_2-CH_2-$; BEDT - TTF

2 R = R' = $-CH_2-$; BMDT - TTF

3 R = R' = $-CH=CH-$; BVDT - TTF

4 R = R' = $-C(CH_3)=C(CH_3)$; BDMVDT - TTF

5 R = $-CH_2-$, R' = $-CH_2-CH_2-$; METT - TTF

6 R = $-CH=CH-$, R' = $-CH_2-CH_2-$; VETT - TTF

7 R = $-C(CH_3)=C(CH_3)-$; R' = $-CH_2-CH_2-$; DMVETT - TTF

From these molecules we have carried out the crystal growth of different salts with several usual monovalent anions. We have used a classical electro-oxidation technique playing with the main parameters: constant current density, solvent and temperature. Our positive attempts are gathered in Table 1.

Table 1. Radical cation salts of BEDT derivatives

Compounds	Electrocrystallisation conditions (at constant current i = 1 - 5 μA) (Solvent)	(Temperature)	Salts	Crystalline morphology	a-c electrical conductivity $\sigma_{290\,K}$ $(\Omega^{-1}cm^{-1})$
1/ Symmetrical					
3 BVDT	DCM (dichloromethane)	50°C	$(BVDT)_xAsF_6$	plates	50 - 70
	DCM	50°C	$(BVDT)_xSbF_6$	plates	80 - 100
4 BDMVDT	1,1,1 - TCE (trichloromethane)	50°C	$(BDMVDT)_xAsF_6$	powder	-
	1,1,1 - TCE	50°C	$(BDMVDT)_xSbF_6$	needles	150 - 200
	THF (tetrahydrofurane)	20°C	$(BDMVDT)_xICl_2$	bundle of fibers	50
2/ Unsymmetrical					
5 METT	1,1,2 - TCE	20°C	$(METT)_2AsF_6$*	plates	15
	1,1,2 - TCE	20°C	$(METT)_1ClO_4$*	plates	10^{-2} - 10^{-1}
	1,1,2 - TCE	20°C	$(METT)_1SbF_6$*	plates	3 - 5 10^{-2}
6 VETT	1,1,1 - TCE	50°C	$(VETT)_xAsF_6$	powder	-
7 DMVETT	1,1,2 - TCE	50°C	$(DMVETT)_xSbF_6$	micro-plates	-
	1,1,2 - TCE	20°C	$(DMVETT)_xAsF_6$	micro-plates	-

* Estimated from electronic microprobe method

PHYSICAL PROPERTIES

We have performed low frequency a.c. electrical conductivity measurements and EPR X band spectroscopy on every single crystal presented on Table 1. The conductivity temperature dependences measured down to liquid helium temperature are in agreement with the room temperature absolute values (Figure 1). The unsymmetrical (METT) salts are semiconductors whatever their stoichiometry whereas the symmetrical ones (BVDT and BDMVDT) are rather valuable conductors. For the two BVDT salts (X = AsF_6, SbF_6) we observe a sharp metal-insulator around 150 K as proved by the plot of the logarithmic derivative versus temperature which is given in insert of the Figure 1.

EPR experiments were done to determine the principal axes and the corresponding eigen values of the g tensor at room temperature. Experiments at low temperature were performed with the static magnetic field approximatively perpendicular to the plane of the crystals. This direction is generally close to the long axis of the molecule (i.e. close to the eigen direction of g_{max}), however a misorientation is usually observed in these compounds and temperature dependences of EPR data cannot easily be measured along simple g directions.

Two significant examples are given in Figure 2 where data for $(BVDT)_x$ SbF_6 and $(BDMVDT)_xSbF_6$ are given. In both cases standard g eigen values are found ($g_{min} \sim 2.002$, $g_{int} \sim 2.007$, $g_{max} \sim 2.009$). The misorientation is weak for $(BVDT)_xSbF_6$ and the data given Figure 2 is taken close to g_{max}. Due to a larger misorientation, the result obtained for $(BDMVDT)_xAsF_6$ corresponds to a direction at about 25° from g_{max} and the measured g factor is smaller.

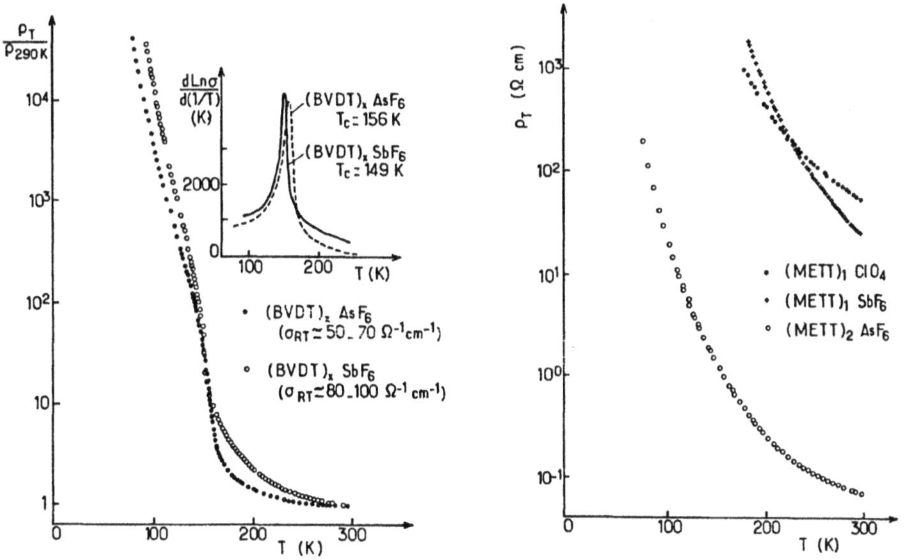

Fig. 1. Electrical conductivity temperature dependences

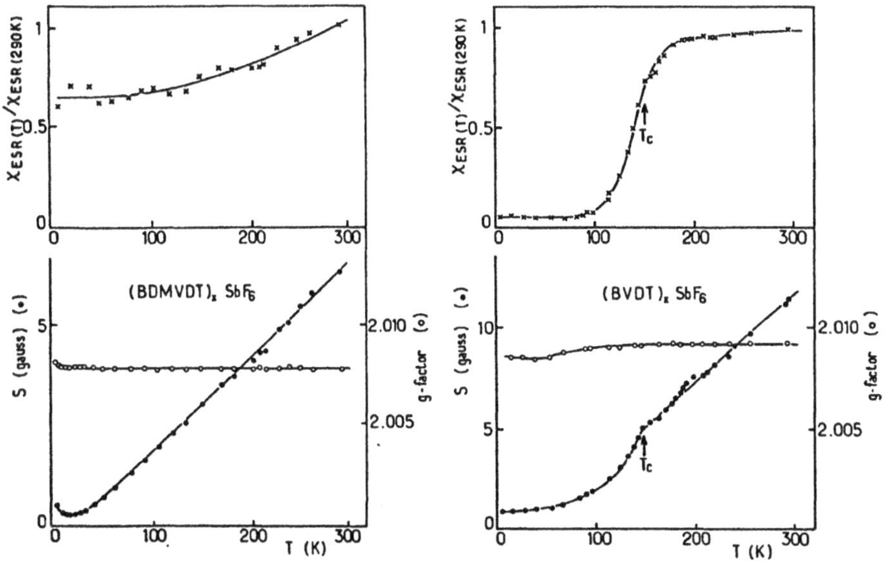

Fig. 2. E.S.R. experiments: spin susceptibility, linewidth and g-factor temperature dependences.

On the BDMVDT salt we observe a weak T variation of spin susceptibility without any particular behavior on linewidth and g-factor values excepted at very low temperatures where these physical quantities are increasing. This new feature may be due to the occurence of antiferromagnetic fluctuations. On the BVDT salts we observe a sharp transition on the linewidth as well as on the spin susceptibility at $T_C \simeq 150\,K$. These results confirm the presence of a phase transition which induces a gap in both the electronic and magnetic excitation spectra.

CONCLUSION

The most promising results are obtained with the salts issued from the symmetrical molecules. The BVDT salts ($X = AsF_6$, SbF_6) present a sharp metal-insulator around 150 K which appears very similar to the structural phase transition observed just below room temperature on identical BEDT salts [4]. The $(BDMVDT)_xSbF_6$ salt probably shows antiferromagnetic fluctuations which may lead to an ordered magnetic state at very low temperature. This new series of compounds appears attractive because a competition between different cooperative ground states is observed as in BEDT salts.

REFERENCES

1. S.S. Parkin, E.M. Engler, R.R. Schumaker, R. Lagier, V.Y. Lee, J.C. Scott, R.L. Greene, Phys. Rev. Letters, 50, 270 (1983).
2. R.R. Schumaker, E. Dupart, J.P. Morand, C. Coulon, P. Delhaès, to be published.
3. T. Nogami, K. Inoue, T. Nakamura, S.I. Iwasaka, H. Nakano and H. Mikawa, Synthetic Metals, 19, 539 (1987).
4. R. Laversanne, J. Amiell, P. Delhaès, D. Chasseau, C. Hauw, Sol. Stat. Commun., 62, 177 (1985).

TRANSFER INTEGRALS IN $(DIMET)_2MF_6$:

COMPARISON OF EHT AND CNDO CALCULATIONS

L. Ducasse[1], M. Abderrabba[1], A. Fritsch[1] and B. Gallois[2]

[1]Laboratoire de Physicochimie Théorique, UA 503
[2]Laboratoire de Cristallographie, UA 144

Université de Bordeaux I, 33405 Talence Cedex, France

INTRODUCTION

The salts based on dimethylethylenedithio-tetrathiafulvalene (or DIMET) and centrosymmetrical anions MF_6^-(M=As,P,Sb) exhibit rather striking differences. The P and As salts crystallize as needles (as do their TMTTF homologs) but the Sb compound, as rods. Room-temperature electrical conductivities are very different ($\sigma(P) \simeq \sigma(As) \simeq 10\Omega^{-1}cm^{-1}$; $\sigma(Sb) \simeq 10^{-3}\Omega^{-1}cm^{-1}$),as well are their low-temperature ground-states[1] (spin-Peierls below 17 K for As constrating with an antiferromagnetic ordering under 10 K for Sb). An usual way to characterize organic compounds is to determine their transfer integrals wchich allow to calculate their band structure. We compare in this paper the transfer integrals of $(DIMET)_2AsF_6$ and $(DIMET)_2SbF_6$.

TRANSFER INTEGRALS CALCULATION

The transfer integrals are evaluated using the dimer splitting approximation

$$t = \langle HOMO_1 | H | HOMO_2 \rangle$$

where H is the (effective) one-electron hamiltonian coupling the HOMO's of two interacting DIMET molecules. Considering an interaction which makes the combination $HOMO_1$-$HOMO_2$ of a lower energy than the $HOMO_1$+$HOMO_2$ combination, the corresponding t is positive[2].

- EHT results

The crystallographic structures of $(DIMET)_2AsF_6$ and $(DIMET)_2SbF_6$ are given on Fig. 1[3]. The AsF_6 configuration is very similar to those already observed in TMTTF salts. The SbF_6 stacking along the chain appears to be disrupted between the molecules I and $\bar{I}$+c : the crystal seems to involve rather isolated dimers. The calculated transfer integrals are also reported on Fig. 1.

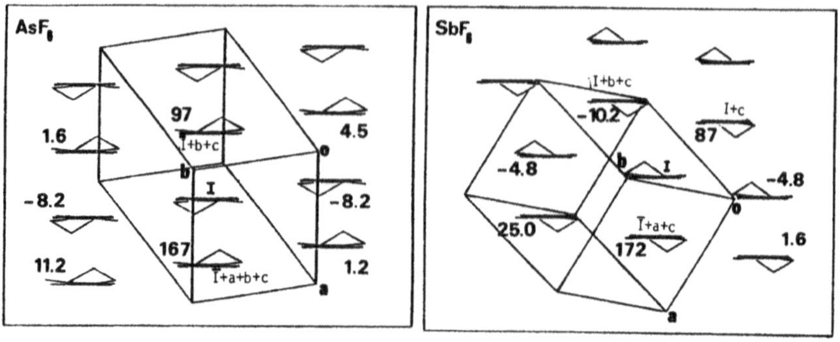

Figure 1. Crystal structure and transfer integrals (meV) in $(DIMET)_2MF_6$.

We shall only discuss in this paper about the intrastack transfer integrals, t_{S1}($\bar{I}$+a+b+c for As ; $\bar{I}$+a+c for Sb) and t_{S2}($\bar{I}$+b+c for As ; Ī+c for Sb) with regard to the room-temperature behaviour and the electronic localization in these salts. In this case, the transverse integrals which are somewhat larger in the Sb salt, are not pertinent. From Fig. 1, one notes that the dimerization, evaluated by t_{S1}/t_{S2}, is roughly identical in the As and Sb salts. This similarity contrasts with the 4 orders for magnitude difference in their conductivities. We discuss below about two effects which are not included in the present EHT calculations : first the possible role of the anions and second the introduction of a self consistent process in the calculation of the DIMET molecular orbitals.

- Role of the anions

Some authors have emphasized the possible role of the anions because the $4k_F$-component of the anion potential may contribute to the Umklapp processes (g_3 parameter) which are pertinent to describe the localization of the electrons. In order to evaluate the cation-anion interaction, we use a local approach of the electrostatic potential within a very simple point-charge approximation. The potential between at point M induced by the net charges q_H of the atoms A at a distance R_{AM} from M is :

$$V(M) = + \frac{1}{4\pi\varepsilon_o} \sum_A \frac{q_A}{R_{AM}}$$

The A charges were evaluated by the CNDO[4] method which is known to give more reliable values than the EHT method. The isopotential maps were calculated for 4 DIMET molecules along the same stack interacting with 8 anions MF_6^-, and they are drawn on Fig. 2.. For Sb, two figures were drawn to take account of the shift of Ī+c molecule with respect to the I and Ī+b+c stacking.(In a first step, we have checked that these results were not different from the ones obtained by a neutral system of 4 molecules and 2 anions). Within this very crude approach, we note from Fig. 2 that each intermolecular space is qualitatively perturbed in a similarity by the anion potential.

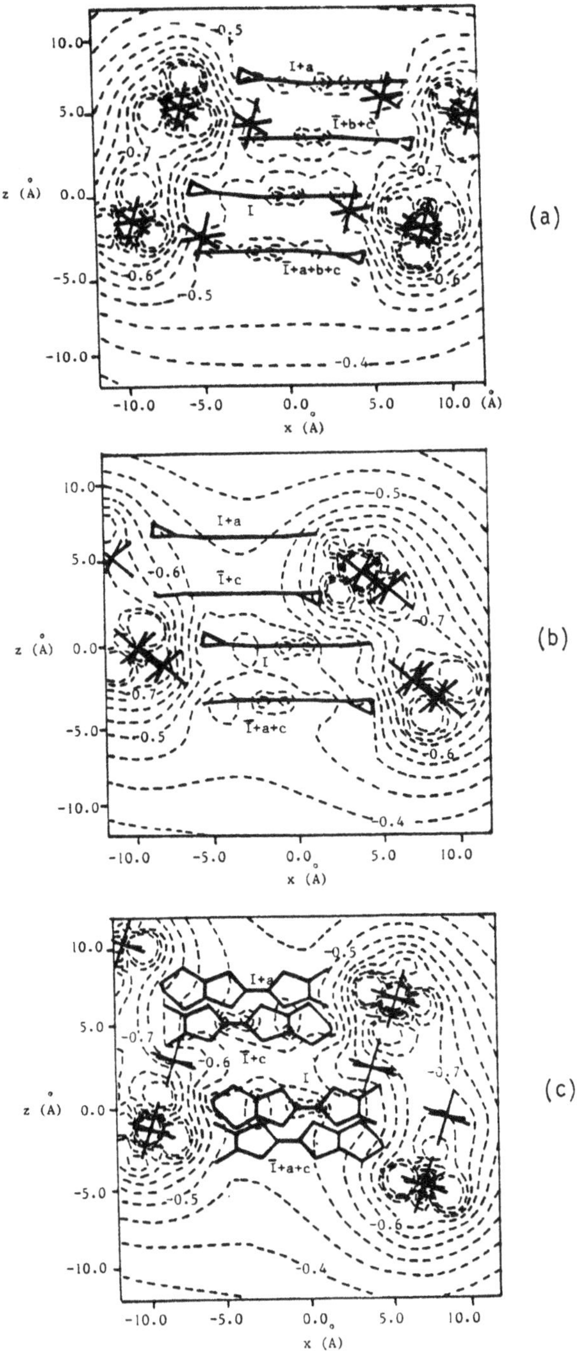

Figure 2. Isopotential maps drawn with $1/4\pi\varepsilon_0=1$, in the xOz plane (x//I long axis and z⊥ I mean plane): (a)AsF_6, (b)SbF_6, (c)SbF_6 with z along the line joining the centers of the I and $\bar{I}$+c molecules.

From that point of view, no difference appears between the S_1 and S_2 interactions. Thus the anion-cation interaction cannot be involved to recalibrate the calculated integrals. This picture does not corroborate the potential model of Laversanne[5] which leads to a different renormalization of t_{S1} and t_{S2} for the As and Sb salts.

- Role of self-consistency in the calculation of the transfer integrals: CNDO approach

CNDO calculations were performed to evaluate the molecular orbitals of a dimer $(DIMET)_2^{+q}$ as a function of q=0 or 1. The transfer integrals (in meV.) are given below :

	q=0		q=1	
	AsF_6	SbF_6	AsF_6	SbF_6
t_{S1}	297.9	311.6	1004.0	1021.7
t_{S2}	191.8	123.8	874.8	no convergence
t_{S1}/t_{S2}	1.55	2.52	1.15	-

Restricted Hartree-Fock formalism together with the half-electron correction were used for q=+1. Apart from the absolute magnitude of the t's, the CNDO results lead to similar dimerizations t_{S1}/t_{S2}, compared to the EHT integrals, in the case q=0 for both compounds (with a somewhat larger value for Sb) and for As only when q=1. The striking result is that the SCF convergence cannot be obtained for the S_2 interaction in Sb. Although all the SCF cycles begin in every case (iteration 0) but the two highest occupied orbitals of symmetry $\pi_1 \pm \pi_2$, the iteration process fails to converge, after some iterations, only for this S_2 interaction. These CNDO calculations indicate that the electronic energy of two monomers with the S_2 geometry in SbF_6 correspond to frontier orbitals which are not of the usual π symmetry. One may consider that an electron moving along the stack in the $(DIMET)_2SbF_6$ by delocalization on the π orbital monomer is prevented from going to one dimer to another dimer if the interaction between them is of the S_2 type : this corresponds to an increased localization in SbF_6 compared to the AsF_6 compound. More detailed work is in progress in order to better analyse this problem.

REFERENCES

1. R. Laversanne, C. Coulon, J. Amiell, E. Dupart, P. Delhaès, J.P. Morand and C. Manigand, Physical properties of radical cation salts of the dimethylethylenedithioletrathiafulvalene, Solid State Commun. 58 : 765 (1986).
2. L. Ducasse, M. Abderrabba, J. Hoarau, M. Pesquer, B. Gallois and J. Gaultier, Temperature dependence of the transfer integrals in the $(TMTSF)_2X$ and $(TMTTF)_2X$ families, J. Phys. C. : Solid State Phys. 19 : 3805 (1986).
3. B. Gallois, J. Gaultier, F. Bechtell, D. Chasseau, C. Hauw and L. Ducasse, $(DMEDT-TTF)_2X$: structure of two radical cation salts with centrosymmetrical anions $X=SbF_6^-$ and $X=AsF_6^-$, Synth. Metals, 19 : 419 (1987).
4. J.A. Pople and D.L. Beveridge in "Approximate Molecular Orbital theory", J.L. Smith and A. Stryker-Rodda, Mc Graw Hill, New-York (1970).
5. R. Laversanne, in "Localisation électronique et mise en ordre des spins dans les conducteurs organiques", Thesis, Bordeaux (1987).

CRYSTAL STRUCTURE AND PHYSICAL PROPERTIES OF A NEW CONDUCTING CHARGE-TRANSFER SALT : $(BEDT\text{-}TTF)_3Cl_2.2H_2O$.

D. Chasseau[a†], D. Watkin[a], M. Rosseinsky[b], M. Kurmoo[b], and P.Day[b]

[a]Chemical Crystallography Laboratory
9Parks Road
Oxford, OX1 3PD

[b]Inorganic Chemistry Laboratory
South Parks Road
Oxford, OX1 3QR

† Permanent address: Laboratoire de Cristallographie,UA 144
351 Cours de la Libératior
33405 Talence (France)

INTRODUCTION

The family of BEDT-TTF* charge transfer salts have been shown to encompass a wide range of cation : anion ratios with several crystal morphologies and structures[1]. These differences have resulted in a variety of electrical and magnetic properties[2]. In the past few years, following the discovery of superconductivity in the salts, $(BEDT\text{-}TTF)_2ReO_4$ [3] and $(BEDT\text{-}TTF)_2I_3$ [4], much effort has been devoted to studying a range of new BEDT-TTF salts with other inorganic tetrahedral and linear triatomic anions[5]. In an effort to understand the effect of changing anions on the structure and properties and to check whether incorporating sulphur containing anions will increase the dimensionality, we have attempted to prepare materials with alkyl and aryl-sulphonate, -dithiocarbamate and -thiophosphate anions.

In the course of synthesising the dibenzyl-dithiocarbamate salt in CH_2Cl_2 the title compound was isolated. The present communication details the crystal structure and physical properties of this salt, which is not only the first BEDT-TTF chloride salt to be isolated but also the first BEDT-TTF salt containing H_2O.

EXPERIMENTAL

Crystals were prepared electrochemically in a three compartment cell, using platinum electrodes, $(C_4H_9)_4N[(C_7H_7)_2NCS_2]$ as electrolyte and CH_2Cl_2 as solvent for ca. 10 days at a constant voltage of 1.4V.

*BEDT-TTF=3,4,3′,4′-bis (ethylenedithio) - 2,2′,5,5′ tetrathiafulvalene.

Conductivity was measured by both AC (110Hz) and DC using a current of 10 μA. Contacts were made with silver paint on evaporated gold pads (ca. 20 ohms per pair). X-band EPR spectra were recorded by use of a reflection spectrometer. Samples were mounted on quartz rods using silicone grease and helium-cooled in an Oxford Instruments continuous flow cryostat.

Intensities of 12614 reflections were collected with a Nonius CAD4 diffractometer on an elongated black plate-like crystal (0.4 x 0.15 x $0.03mm^3$) using Cu Kα radiation monochromatised with a graphite plate. 9467 were independent with 6013 observed ($I > 3\sigma(I)$) and corrected for absorption and Lorentz - polarisation. The structure was solved by direct methods and refined by large block matrix least squares to final R=0.045.

RESULTS AND DISCUSSION

Although the attempt to synthesise the dibenzyl-dithiocarbamate salt was unsuccessful, it is interesting that the material reported here is formed by anion extraction from the solvent. To our knowledge, this is the first observation of such a reaction. The existence of chlorine was confirmed by X-ray fluorescence and we have been unable to grow any crystal in THF as solvent: subsequent addition of CH_2Cl_2 induces growth.

The crystal belongs to the triclinic system, space group $P\bar{1}$, a=11.214(2), b=13.894(2), c=15.924(2) Å, α=94.74(1), β=109.27(1), γ=97.03(1)°, Z=2, V=2304.3 $Å^3$ and D_c=1.817 g cm^{-3}. The crystal structure is shown in Figure 1. The asymmetric unit contains three independent cations, two chloride anions and two water molecules. The BEDT-TTF cations are arranged face to face in crystallographically equivalent stacks parallel to the b-axis. .

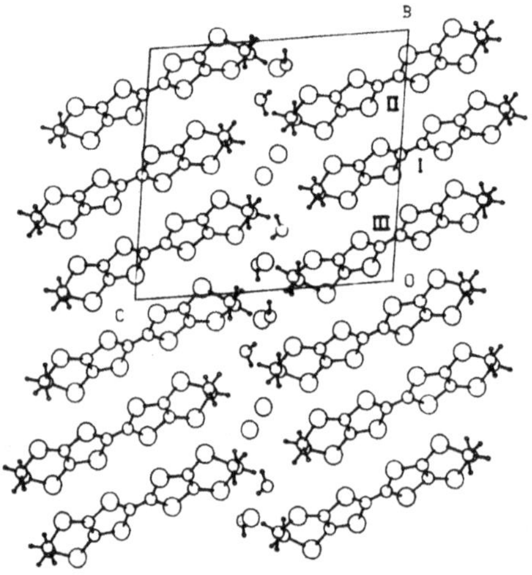

FIGURE 1: View of the unit cell along the a axis.

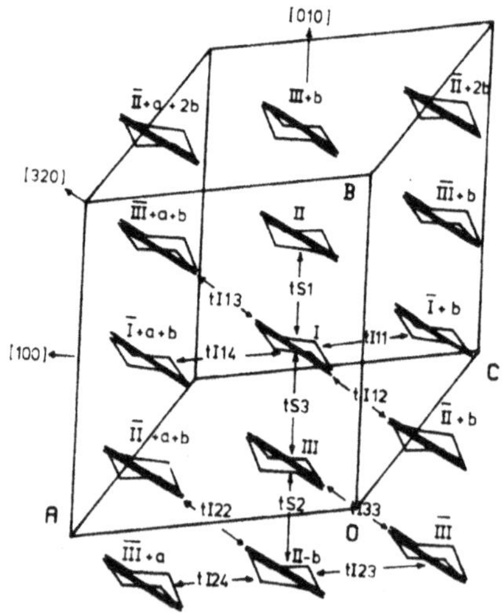

FIGURE 2: View of the BEDT-TTF molecules along their long axis.

The cationic stacks are close to one another in the a-direction and separated by the anions and water molecules in the c-direction. This results in a 2D-framework parallel to the ab-plane (Figure 2).

This structure represents a new one in this class of materials, showing different packing motifs and a greater number of cation-cation interactions (Figure 2). These interactions are best described by considering three directions, namely [010], [110] and [320]. Along [010], the planes of the BEDT-TTF molecules are parallel to one another and make an angle of 63° to the b-axis; the overlapping modes, ts_1 and ts_2 are similar but the interplanar distances are different (3.59 and 3.67 Å respectively). The third overlapping mode, ts_3, is different to the other two and is characterised by only a transverse shift. The interplanar distance is 3.68 Å.

The interactions tI_{11}, tI_{14}, tI_{23} and tI_{24} along [110] are all similar; there is no real overlap between the BEDT-TTF molecules. The S-S distances are greater than 3.74 Å. The angle (ϕ) between the BEDT-TTF planes and the direction of the centres is ca. 35°.

Along [320] there are four different interactions. The BEDT-TTF molecules are almost in one plane (ϕ=5°). Due to the fact that the interaction, tI_{22}, between two centrosymmetric molecules is similar to the interactions tI_{12} and tI_{13} (S-S distances between external sulphur atoms ranging from 3.33 to 3.46 Å), the repeat unit is in fact six BEDT-TTF molecules. The fourth interaction between units, tI_{33} is expected to be much weaker than the others (shortest S-S = 3.66 Å). This mode of packing and the different types of interaction are unique to this material as compared to the whole family of BEDT-TTF salts.

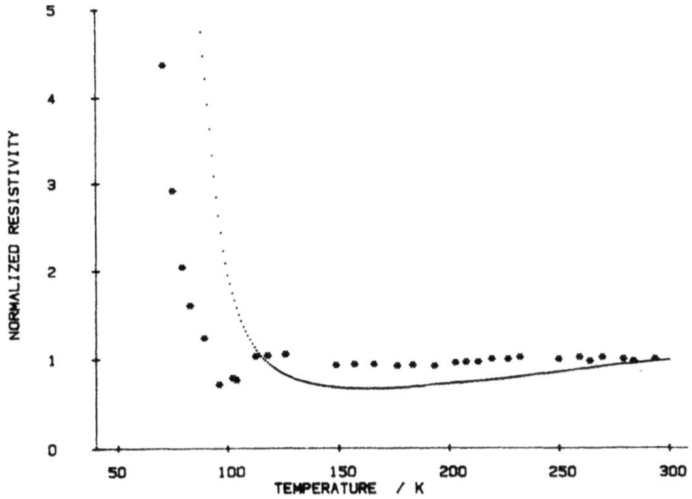

FIGURE 3:Resistance (electrical ...and from EPR ***) versus temperature.

Another interesting feature of this material, and observed for the first time, is the inclusion of water of crystallisation in a BEDT-TTF charge transfer salt. The structure of the anion is a complex hydrogen bonded unit comprising four chlorine atoms and four water molecules.

Four probe electrical conductivity measurements, σ (300K) = 0.6S cm^{-1}, (Figure 3) indicate a metal-insulator transition at ca. 100K which resembles those of other 3:2 salts with ClO_4^- and ReO_4^- as anions[6], and has been associated with Peierls instability accompanied by $4k_F$ distortion. EPR measurements in the temperature range 4-300K have been recorded with the static magnetic field oriented parallel to three orthogonal axes. The spectra consist of a single peak at g=2.009(4). The lineshape is Lorentzian and symmetric in two orientations and asymmetric (Dysonian) in the third. The peak to peak width decreases gradually with lowering of temperature until ca. 100K where it is temperature independent. The conductivity obtained by deconvoluting the asymmetric peak into its real and imaginary components is consistent, within experimental error, to the electrical measurements (Figure 3). The spin susceptibility has similar behaviour to the peak to peak width with temperature. Rotation experiments reveals periodicity of 180° for g value, H_{pp} and spin susceptibility

CONCLUSION

Although obtained in an indirect way, $(BEDT\text{-}TTF)_3Cl_2.2H_2O$ is the first chloride salt to be synthesised and also the first to contain water of crystallisation. The crystal structure indicates 11 modes of interactions. The material has a metal-insulator transition at ca. 100K as observed by transport and EPR measurements.

ACKNOWLEDGMENTS

The authors are grateful to Oxford University and the United Kingdom S.E.R.C. for funds. D.C. is indebted to the "Centre National de la Recherche Scientifique (France)" and the Royal Society of London for the award of a European Science Exchange Fellowship, during his stay in Oxford where this work was carried out.

1. S.S.P. Parkin, E.M. Engler, V.Y. Lee, R.R. Schumaker; Mol.Cryst.Liq.Cryst., 119: 375. (1985).
2. Proceedings of the International Conference on Synthetic Metals, Kyoto, Japan; Synthetic Metals, 19 (1987).
3. S.S.P. Parkin, E.M. Engler, R.R. Schumaker, R. Lagier, V.Y. Lee, J.C. Scott, andR.L. Greene; Phys. Rev. ,50: 270 (1983).
4. E.B. Yagubskii, I.F. Schlegolev, V.N. Laukhin, P.A. Kononovich, M.V. Kartsovnik,. A.V. Zvarykina and L.I. Buravov; JETP, 39: 12 (1984).
5. I.D. Parker, S.D. Obertelli, R.H. Friend, D.R.Talham, M. Kurmoo, P. Day, J.A.K. Howard, and A.M. Stringer; Synthetic Metals, 19: 185 (1987).
6. T. Enoki, K.Imaeda, M. Kobayashi and H. Inokuchi; Phys. Rev., B33: 1553 (1986).

ELECTRONIC EXCITATIONS IN (BEDT-TTF)-SALTS

Hans W. Helberg

Drittes Physikalisches Institut
University of Goettingen
D-3400 Goettingen, West Germany

INTRODUCTION

The purpose of the present work was to investigate the transition from intra- to intermolecular excitations in the several compounds of the BEDT-TTF* salts. Measurements start on crystals of α-(BEDT-TTF)$_3$(NO$_3$)$_2$ and α-(BEDT-TTF)$_2$I$_3$.

The used method enables to measure the dispersion of the polarizability tensor orientation and the absorption tensor in transmitted light. At short waves in the visible spectrum (VIS) the polarizability tensor is orientated like the molecule axes of the planar stack forming molecules due to the intramolecular excitations. In the near infrared range (NIR) the orientation is governed by intermolecular transitions, particularly the transitions between adjacent stacks or between stack molecules and counterions.[1]

RESULTS AND DISCUSSION

α-(BEDT-TTF)$_3$(NO$_3$)$_2$ crystallizes (010) faced in the monoclinic system. Thus one principal axis of the indicatrix must coincide with the $\vec{b}$ axis. In α-(BEDT-TTF)$_3$(NO$_3$)$_2$ the optical direction Z parallels $\vec{b}$ (X, Y, Z principal or optical directions of the indicatrix corresponding to the principal refractive indices n_α, n_β, n_γ, respectively, with the conventional relation $n_\alpha < n_\beta < n_\gamma$). X and Y ly in the plane (010). A strong bisectrix dispersion was observed.[2] The indicatrix rotates by about 50° around the $\vec{b}$ axis in the whole spectrum range. This dispersion indicates intramolecular BEDT-TTF excitations in the VIS and intermolecular transitions of the S...O contacts and the hydrogen bridged CH...O contacts in the NIR between the BEDT-TTF$^+$ and the NO$_3^-$ anions, respectively.[2]

α-(BEDT-TTF)$_2$I$_3$ crystallizes (001) faced in the triclinic system. Contrary to α-(BEDT-TTF)$_3$(NO$_3$)$_2$, no bisectrix dispersion was found.[1] In the observed frequency range Y coincides with the stack direction $\vec{a}$ and Z with the longitudinal axes L of the BEDT-TTF molecules, respectively.

Fig. 1 shows the relative birefringence of both materials versus wavenumber. Besides a pronounced increase in the NIR there is only a broad maximum at $2.1\cdot10^4$ cm^{-1} for the α-(BEDT-TTF)$_3$(NO$_3$)$_2$. The above mentioned intermolecular contact excitations give rise to the increase in the NIR.

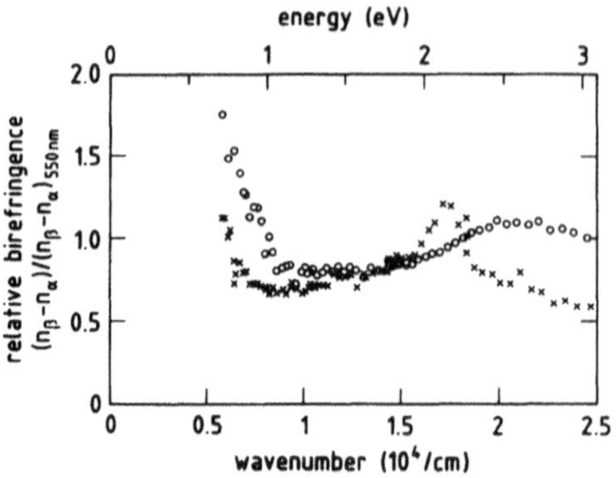

Fig. 1. Relative birefringence ($n_\beta-n_\alpha$) (referred to the birefringence at 550 nm) of α-(BEDT-TTF)$_2$(NO$_3$)$_2$ (o)[2] and α-(BEDT-TTF)$_2$I$_3$ (X) versus wavenumber.

The birefringence of α-(BEDT-TTF)$_2$I$_3$ shows a distinct peak indicating a dispersion of n_β centered at $1.85 \cdot 10^4$ cm^{-1}.

In both materials the principal axes of the absorption tensor coincide with principal axes of the indicatrix. The absorption in each case measured in two principal directions are plotted in Figs. 2 and 3 for α-(BEDT-TTF)$_3$(NO$_3$)$_2$ and α-(BEDT-TTF)$_2$I$_3$, respectively. In both materials there is a strong absorption region around $2 \cdot 10^4$ cm^{-1} obviously due to intramolecular transitions in the BEDT-TTF molecules. The absorption polarized parallel Y is much more pronounced. The absorption peaks for the both polarization directions are characterized by Davydov splitting. Besides this main absorption range there is another much smaller absorption peak at $1.15 \cdot 10^4$ cm^{-1}.

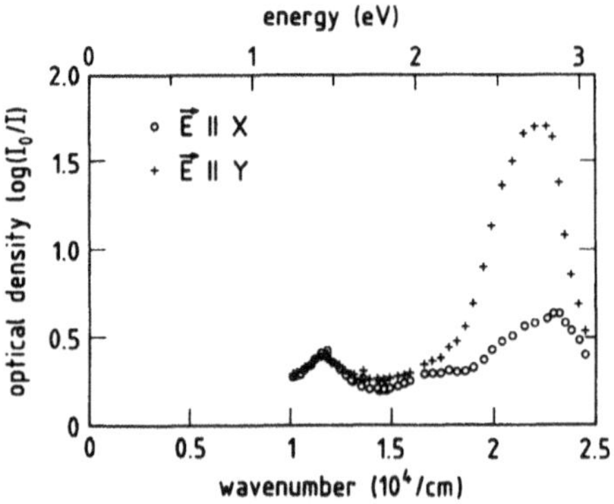

Fig. 2. Absorption of transmitted light polarized along the optic direction X and Y, respectively, in α-(BEDT-TTF)$_3$(NO$_3$)$_2$ versus wavenumber. X and Y ly in the plane (010) varying their directions with wavenumber. The polarization is always parallel to the principal directions of the absorption tensor.

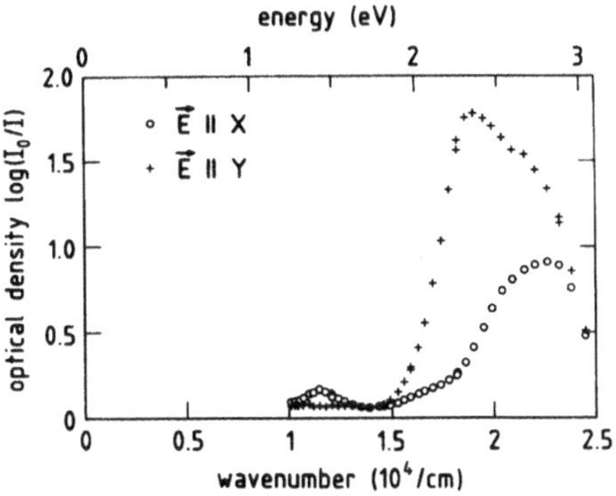

Fig. 3. Absorption of transmitted light polarized along the optical directions X and Y, respectively, in α-(BEDT-TTF)$_2$I$_3$ versus wavenumber. Y coincides with the stack direction $\vec{a}$. X is perpendicular to the longitudinal axes of the (BEDT-TTF) molecules in the ($\vec{b}'$, $\vec{c}^*$) plane ($\vec{b}'$ is perpendiculary directed to $\vec{a}$ and $\vec{c}^*$ in (001)). The polarization is always parallel to the principal directions of the absorption tensor.

The main absorption peaks in α-(BEDT-TTF)$_3$(NO$_3$)$_2$ are located at $2.23 \cdot 10^4$ cm^{-1} (∥Y) and $2.30 \cdot 10^4$ cm^{-1} (∥X) resulting a splitting of about 700 cm^{-1} (Fig. 2). The ratio of maximum absorption is 1.7 (∥Y)/0.6 (∥X) ≈ 2.8.

In α-(BEDT-TTF)$_2$I$_3$ (Fig. 3) the main absorption peaks are lying at $1.90 \cdot 10^4$ cm^{-1} (∥Y) and $2.26 \cdot 10^4$ cm^{-1} (∥X). A splitting of about 3600 cm^{-1} results. The ratio of maximum absorption is 1.8 (∥Y)/0.9 (∥X) ≈ 2.0. The pronounced absorption peak (∥Y) at $1.9 \cdot 10^4$ cm^{-1} corresponds to the dispersion of n_β in Fig. 1. A broad shoulder follows to higher wavenumbers.

In a first simplified step the point dipole-point dipole interactions were calculated taking into account only the nearest-neighbour interactions. Table 1 shows the induced field strength in nearest-neighbour BEDT-TTF molecules parallel to the in-plane longitudinal axes L and transversal axes T of these molecules if one molecule in the center is excited.

In the case of α-(BEDT-TTF)$_3$(NO$_3$)$_2$ the longitudinal molecule axes L are inclined to each of the three optimal directions allowing a direct excitation along L proportionally to the component for that direction. Besides this direct excitations Table 1 gives the excitations via the dipole interactions. The negative sign means an opposite direction of the induced field compared to the direction of the excitation field. In the case of a negative sign a blue shift results. Therefore, the peak ∥X is more located to higher energies than the peak ∥Y (Fig. 2).

In the case of α-(BEDT-TTF)$_2$I$_3$ the optical directions X and Y are perpendicular to L. The excitation along L is only via the dipole interactions (Table 1). The component parallel L (negative sign) is more excited from X than from Y. Thus the splitting direction is the same like Fig. 2. The peak ∥X is more located to higher energies than the peak ∥Y (Fig. 3).

These calculations give only a general view. More detailed investigations are prepared, also measurements on other BEDT-TTF salts.

Table 1. Electrical field strength induced in nearest-neighbour molecules parallel to the molecule axes L and T referred to the excitation dipole moment p.

substance	optic direction field	direction of excitation	relative induced field $E\,4\pi\epsilon_0/\|p\|$ parallel to axes L nm^{-3}	T nm^{-3}
α-$(BEDT\text{-}TTF)_3(NO_3)_2$	X	⊥[$\bar{2}$01] in (010)	-12.3	5.4
	Y	‖ [$\bar{2}$01]	10.3	21.1
	Z	‖ $\vec{b}$	-37.9	13.6
α-$(BEDT\text{-}TTF)_2I_3$	X	⊥L, ⊥$\vec{a}$	41.1	22.7
	Y	‖ $\vec{a}$, stack	26.8	15.0
	Z	‖ L	-35.2	11.7

I thank D. Schweitzer and H.J. Keller for the crystals and the Deutsche Forschungsgemeinschaft for supporting this work.

REFERENCES

1. H.W. Helberg, "Dispersion of the polarizability tensor by inter- and intrastack excitations in organic conductors," Mol. Cryst. Liq. Cryst. 119:179 (1985).
2. H.W. Helberg, "Electronic excitations in α-$(BEDT\text{-}TTF)_3(NO_3)_2$," Physica 143B:488 (1986).

EXISTENCE OF TWO PHASES IN β-$(BEDT\text{-}TTF)_2I_3$:

PROOF BY RESONANCE RAMAN SPECTROSCOPY

*R. Świetlik, D. Schweitzer and H.J. Keller[1]

Max-Planck-Institut für Medizinische Forschung, Abteilung für Molekulare Physik and [1]Anorganisch-Chemisches Institut der Universität, 6900 Heidelberg, Fed. Rep. Germany

Among various organic metals and superconductors based on the BEDT-TTF molecule [bis(ethylenedithiolo)tetrathiafulvalene] and different polyhalide anions the most intensively studied is β-$(BEDT\text{-}TTF)_2I_3$. Partially it is caused by the fact that these crystals exhibit the superconductivity with the highest critical temperature reported for organic materials, but the main reason of this interest is the existence of two superconducting states in these crystals. Under ambient pressure the superconductivity in β-$(BEDT\text{-}TTF)_2I_3$ is observed below T_c≈1.3K[1-4] (low-T_c phase) but after a particular pressure-temperature cycling procedure the superconductivity can be stabilized under ambient pressure at T_c≈8.1K[5,6] (high-T_c phase). α-$(BEDT\text{-}TTF)_2I_3$ is another modification of the salt formed between BEDT-TTF and I_3^-; at ambient pressure α-phase crystals undergo a metal-insulator phase transition at T=135K[4,7]. Recently, it was discovered that by tempering α-phase crystals at about 75°C for several days crystals can be obtained with similar properties as the high-T_c state of the β-crystals[8,9] (these crystals are further denoted as α_t-$(BEDT\text{-}TTF)_2I_3$).

The existence of two superconducting states in β-$(BEDT\text{-}TTF)_2I_3$ can be connected with a commensurate superstructure[10] developed below 125K which can be suppressed by an applied pressure. The most important feature of this superstructure is a pronounced distortion of the I_3^- anions (low-T_c phase). However, the pressure-temperature cycling procedure suppresses the development of the superstructure and a completely ordered high-T_c state can be obtained[11]. In other superconductors based on BEDT-TTF the structural situation is simpler in comparison to β-$(BEDT\text{-}TTF)_2I_3$. For example, in the salt β-$(BEDT\text{-}TTF)_2IAuI$ only one superconducting state is observed with T_c≈4K and the structure is ordered in the whole temperature range below 300K; i.e. the $IAuI^-$ anions are linear and symmetric[12].

The studies of the resonance Raman (RR) spectra of the I_3^- anions in the β-$(BEDT\text{-}TTF)_2I_3$[13] have shown that below 125K a splitting of the band related to the symmetrical stretching mode ν_1 of the linear and symmetric I_3^- anions into three or two lines (109, 120 and 126cm^{-1} at T=2K) exists (Fig.1). For some crystals the band at 109cm^{-1} disappeared irreversible during the illumination by intense laser light and only one line at about 122cm^{-1} was left (Fig.2). For other crystals the band at 109cm^{-1} did not completely disappear even for long time of irradiation and relatively high laser power but the intensity ratio between the fundamental and split bands $I(122cm^{-1})/I(109cm^{-1})$ increased. Above 125K the spectra were the same for both kinds of β-$(BEDT\text{-}TTF)_2I_3$ crystals

*On leave from: Institute of Molecular Physics, Polish Academy of Sciences, 60-179 Poznań, Poland

and only minor changes due to the temperature were observed.

The splitting of the ν_1 band in β-(BEDT-TTF)$_2$I$_3$ can be related to the commensurate superstructure developed below 125K; i.e. to the distortion of the I_3^- anions and the fact that three different anions exist in the unit cell. The irreversible change induced by the laser light can be connected with the disappearence of the distortion of the I_3^- anions[13]. Neutron diffraction studies have shown that the high-T_c phase of β-(BEDT-TTF)$_2$I$_3$ crystals is completely ordered[11]. Therefore, we conclude that the excitation by laser light can induce a transition from the low-T_c to the high-T_c state of β-(BEDT-TTF)$_2$I$_3$, at the surface at least. In the RR experiment the I_3^- and possibly also the donor cations are excited into the first excited singlet state. During the radiationless desactivation of this excited state by internal convertion and vibronic relaxation the I_3^- anion may fall down to the vibronic levels of the ground state S_o in such a way that it reaches either the symmetric or the asymmetric configuration. The efficiency of this process should be proportional to both the time of illumination and the power of the exciting laser light. However, for some crystals the irreversible change by light could not be performed totally (this is not understood in the moment). Nevertheless, the ratio I(122cm^{-1})/I(109cm^{-1}) increases with the power giving evidence that a transition takes place in these crystals as well, although it is not complete and is not fully irreversible. It should be emphasized that the splitting could be observed again after temperature cycling when the upper temperature exceeded 125K; i.e. the temperature below which the superstructure is developed.

The investigations of α_t-(BEDT-TTF)$_2$I$_3$ crystals have shown that the split band at about 109cm^{-1} is absent. The α_t-(BEDT-TTF)$_2$I$_3$ is in the high-T_c state for the whole temperature range (i.e. the structure is completely ordered), therefore, only one line at 120cm^{-1} was observed[9] (Fig.3). Similary, only one band assigned to the ν_1 mode of IAuI$^-$ anions was observed for β-(BEDT-TTF)$_2$IAuI crystals (Fig.3), in agreement with the structural studies which have shown that the IAuI$^-$ anions are not distorted and that no structural change takes place in this salt[14]. On the other hand, the ν_1 mode of the I_3^- anions in the RR spectra of the α-(BEDT-TTF)$_2$I$_3$ crystals was split into three lines as well (115, 118 and 121cm^{-1} at T=55K), despite the I_3^- anions are linear and symmetric (Fig. 1). However, this splitting was qualitatively different in comparison with β-(BEDT-TTF)$_2$I$_3$ as was shown by RR measurements as a function of temperature, frequency and

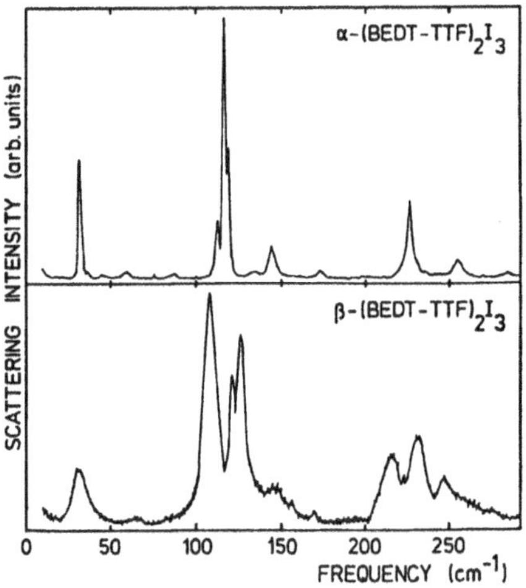

Fig.1. The resonance Raman spectra of α-(BEDT-TTF)$_2$I$_3$ and β-(BEDT-TTF)$_2$I$_3$ (T=2K, λ=4880 Å).

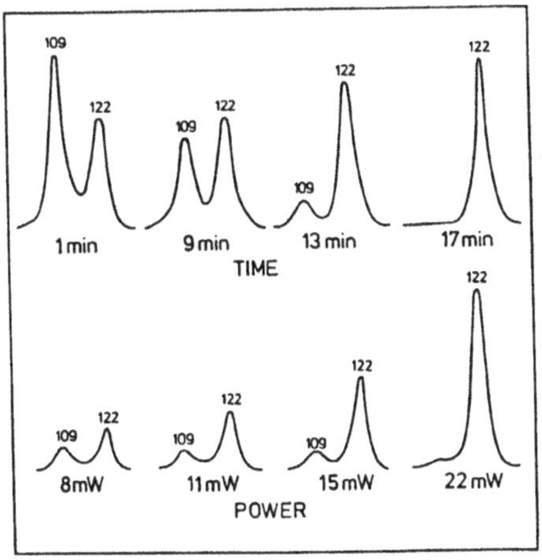

Fig.2. The disappearence of the band at 109cm^{-1} in β-(BEDT-TTF)$_2$I$_3$ crystals, λ=5145Å (top: with time by an illumination with P=15mW, T=20K; bottom: with increasing power of the laser light, T=35K)

power of the exciting laser light. The experimental evidence for α-(BEDT-TTF)$_2$I$_3$ shows that the splitting cannot be related to the presence of different phases in the crystal but it should be related to the influence of crystal field effects on I_3^- anions[13].

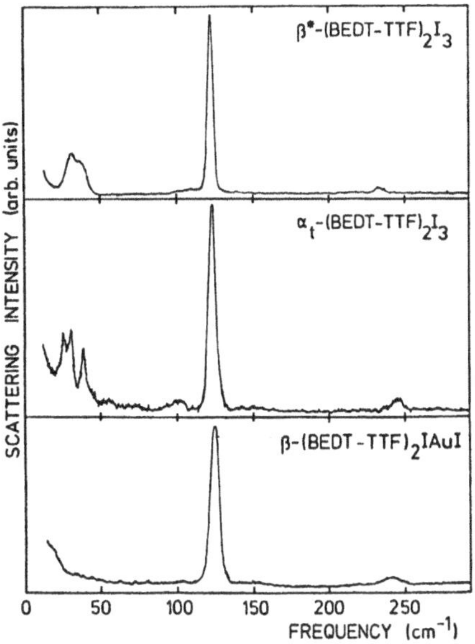

Fig.3. The comparison of the RR spectra of β*-(BEDT-TTF)$_2$I$_3$ (for which the irreversible change by light has been performed) α$_t$-(BEDT-TTF)$_2$I$_3$ and β-(BEDT-TTF)$_2$IAuI (T=2K, λ=4880Å)

In conclusion, in β-(BEDT-TTF)$_2$I$_3$ two superconducting states exist at ambient pressure: the low-T_c state with $T_c \approx 1.3K$ and the high-T_c state with $T_c \approx 8K$. The formation of the high-T_c state is connected with the suppression (by pressure-temperature procedure) of the development of a commensurate superstructure at 125K; without this procedure the crystal stays in its low-T_c state. The existence of the two different states in β-(BEDT-TTF)$_2$I$_3$ crystals was shown by the resonance Raman studies on the I_3^- anions as well. Moreover, on the basis of the observed changes of the RR spectra during illumination by the exciting laser light, one can draw a conclusion that intense light induces the transition between these states as well. The suitable electrical conductivity measurements are under progress.

REFERENCES

1. E.B. Yagubskii, I.F. Schegolev, V.N. Laukhin, P.A. Kononovich, M.V. Kartsovnik, A.V. Zvarykina and L.I. Buravov, Sov. Phys. JETP Lett. **39**, 12 (1984)
2. J.M. Williams, T.J. Emge, H.H. Wang, M.A. Beno. P.T. Copps, L.N. Hall, K.D. Carlson and G.W. Crabtree, Inorg. Chem. **23**, 2558 (1984)
3. H. Schwenk, C.P. Heidmann, F. Gross, E. Hess, K. Andres, D. Schweitzer and H.J. Keller, Phys. Rev. **B31**, 3138 (1985)
4. H. Schwenk, F. Gross, C.P. Heidmann, K. Andres, D. Schweitzer and H.J. Keller, Mol. Cryst. Liq. Cryst. **119**, 329 (1985)
5. F. Creuzet, G. Creuzet, D. Jérome, D. Schweitzer and H.J.Keller, J. Physique Lett. **46**, L-1079 (1985)
6. F. Creuzet, D. Jérome, D. Schweitzer and H.J. Keller, Europhys. Lett. **1**, 462 (1986)
7. K. Bender, I. Hennig, D. Schweitzer, K. Dietz, H. Endres and H.J. Keller Mol. Cryst. Liq. Cryst. **108**, 359 (1984)
8. G.O. Baram, L.I. Buravov, L.C. Degtariev, M.E. Kozlov, V.N. Laukhin, E.E. Laukhina, V.G. Orischenko, K.I. Pokhodnia, M.K. Scheinkman, R.P. Shibaeva and E.B. Yagubskii, JETP Lett. **44**, 293 (1986)
9. D. Schweitzer, P. Bele, H. Brunner, E. Gogu, U. Haeberlen, I. Hennig, T.Klutz R. Świetlik and H.J.Keller, Z. Phys. B (in press)
10. H. Endres, H.J. Keller, R. Świetlik, D. Schweitzer, K. Angermund and C. Krüger, Z. Naturforsch, **41a**, 1319 (1986)
11. A.J. Schultz, H.H. Wang, J.M. Willliams and A. Filhol, J. Am. Chem. Soc. **108**, 7853 (1986)
12. H.H. Wang, M.A. Beno, U. Geiser, M.A. Firestone, K.S. Webb, L. Nuñez, G.W. Crabtree, K.D.Carlson, J.M.Williams, L.J. Azevedo, J.F. Kwak and J.E. Schirber, Inorg. Chem. **24**, 2465 (1985)
13. R. Świetlik, D. Schweitzer and H.J. Keller, Phys. Rev. B (in press)
14. R. Świetlik, H. Grimm, D. Schweitzer and H.J. Keller, Z. Naturforsch. **42a** (in press)

RESONANCE RAMAN SPECTRA OF I_3^- IN TEDA(TRIETHYLENEDIAMINE)$I_{2.5}$TCNQ AND β-(BEDT-TTF)$_2I_3$

André D. Bandrauk, Kim D. Truong, and Cosmo Carlone

Groupe de Recherche sur les Semiconducteurs et Diélectriques
Faculté des Sciences, Université de Sherbrooke
Sherbrooke, Qué, J1K 2R1, Canada

INTRODUCTION

Resonance Raman Spectra are presented for single crystals of TEDA$I_{2.5}$TCNQ and β-(BEDT-TTF)$_2I_3$. For both conductors, excitation at 647.1 nm proves to be the optimum resonant frequency for monitoring phase transformations. It is concluded that in TEDA$I_{2.5}$TCNQ, the I_3^- ions bend as the temperature is lowered in order to accomodate a contracting lattice. In the case of the β-(BEDT-TTF)$_2I_3$, a splitting of the I_3^- symmetric mode is observed, but no deformation of the anion is evident as one goes into the incommensurate phase at low temperatures.

The I_3^- ion is an ion which occurs readily as a counterion for organic donnors which have the tendency to form conducting segregated stacks[1,2] and also nonconducting salts[3]. We have shown in earlier work that temperature dependent measurements of the intensity and width of the ν_1 symmetric mode of I_3^- at $\sim$ 115 cm^{-1} is an excellent probe for phase transitions in linear chain compounds containing such ions. We have thus been able to infer the participation of the I_3^- anion in the phase transition of the ternary compound TMA(trimethylammonium)ITCNQ[4]. In the course of research on hydrogen bonding effects in organic conductors, we have found that the donor triethylenediamine, TEDA, forms hydrogen bonded linear chains parallel to TCNQ chains in $TEDA_2TCNQ_3$[5,6]. Instead of the form $M_3^+(TCNQ_3)^{-2}I_3^-$[7,8], we have obtained a new ternary compound with TEDA containing both I^- and I_3^- anions, TEDA$I_{2.5}$TCNQ[9]. The crystal structure of this new compound is illustrated in figure 1 where one sees clearly <u>nonlinear</u> chains of I_3^-, contrary to the previous ternary compounds in which the I_3^-, form linear chains.

Of note is that such nonlinear chains of I_3^- occur in the new series of organic conductors, (BEDT-TTF)$_2I_3$ which exists in two phases. There is an α form which undergoes a metal-semiconductor phase at 135 K. Another form, β, remains metallic on lowering the temperature even though an incommensurate structure is known to appear below 200 K[10,11]. The persistence of the metallic character in the β form is attributed to an increase of the 2-D character of the system as a result of increasing lateral sulphur-sulphur contacts between parallel chains. A superstructure is attributed to an ordering of end ethylene groups in the donor chain with a consequent reordering of the I_3^- ions[11]. The aim of the present work was to see whether the iodide ions can provide a measure of the superstructure in this latter compound.

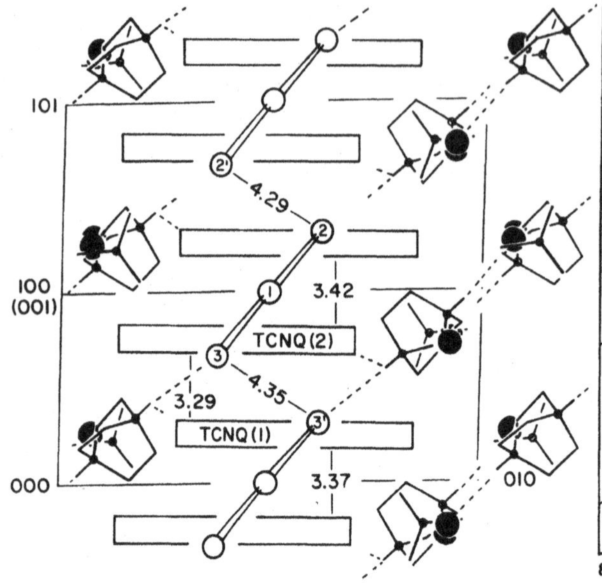

Figure 1. Crystal structure of $TEDAI_{2.5}TCNQ$ with I^- (●), I_3^- (o-o-o) TCNQ (▭) and TEDA (cage).

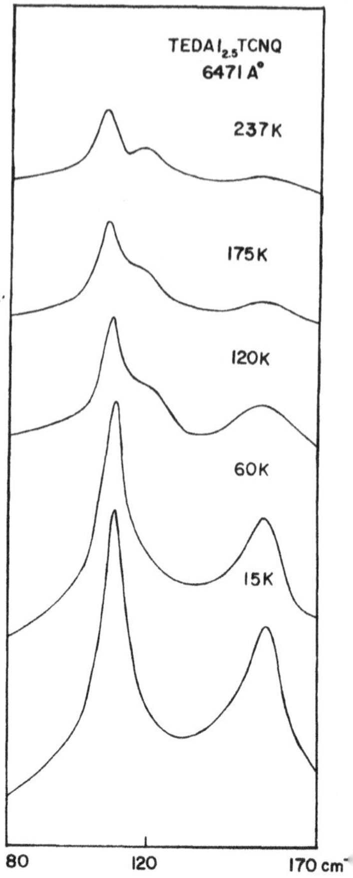

Figure 2. Raman low frequency modes of I_3^- in $TEDAI_{2.5}TCNQ$ as a function of temperature for excitation at 647.1 nm at 3 mw intensity.

RESULTS

In view of the similarity of the structure of the I_3^- anions in both compounds reported in the title, we have carried out resonance Raman scattering (RRS) experiments on single crystals of both compounds as a function of temperature. These measurements, following our previous experimental work[4-6], is done at very low laser powers, below 5 milliwatts in order to avoid damage to the surface of the crystal. Typical measurements involve recycling the measurements between low and high temperatures in order to ensure reproducibility. The relevant results are shown in figure 2 for the new ternary compound $TEDAI_{2.5}TCNQ$, and in figure 3 for the β-(BEDT-TTF)$_2I_3$ crystals.

We show results for excitation at the resonant wavelength 647 nm which is resonant with the $\sigma_g \rightarrow \sigma_u$ electronic transition of the I_3^-[12].

It is precisely for this resonant excitation energy that one expects to pick out selectively any deformation of the I_3^- such as asymmetry or bending of the usually linear anion. The symmetric ν_1 vibration is resonantly excited only by higher energy towards the UV[13], producing generally high overtones of this mode[4], with no excitation of nonsymmetric modes.

Perusal of figure 2 allows one to recognize three low frequency modes at 47, 115 and 152 cm^{-1} at low temperatures. The first, lowest mode (47 cm^{-1}) is usually assigned to the IR active ν_3 bending mode of I_3^-. The second peak is due to the symmetric ν_1 stretching mode whereas the high frequency mode is the ν_2 asymmetric stretch of I_3^-, implying a shortening-lenghtening of the I-I bonds[15-16]. Clearly at low temperatures the I_3^- anion is decidedly nonlinear, which explains the enhancement of ν_2 and ν_3 at 647 nm excitation. We reiterate that with other excitations, eg. 514.5 nm, ν_2 and ν_3 are not present, since these vibrations couple most strongly with the lowest electronic absorption band of I_3^- at ∿ 700 nm[12]. As one increases the temperature, both ν_2 and ν_3 disappear, implying a more linear structure of I_3^-. Our interpretation is therefore, that as one lowers the temperature, contraction of the lattice as a whole induces a slight bending of the I_3^- in order to accomodate itself more readily in the smaller lattice. Interestingly enough, we observed also a change in conduction activation energy at these lower temperatures. Correlation of these different physical effects will be reported elsewhere. We emphasize that in this new ternary compound where we have nonlinear chains of I_3^- and I^- ions, bending of the I_3^- anions occurs at lower temperatures. This behaviour has not been observed in linear chain compounds of I_3^- such as TMAITCNQ[4].

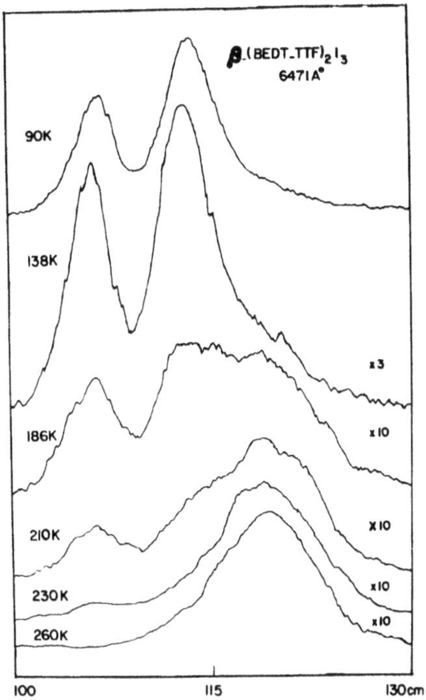

Figure 3. Raman low frequency modes of I_3^- in β-(BEDT-TTF)$_2I_3$ as a function of temperature.

We turn next to figure 3 where we now illustrate the low frequency mode of β-(BEDT-TTF)$_2I_3$ compound as a function of temperature. Variation of the laser excitation energy showed that the 647 nm line gives a different spectrum than from spectra excited at higher energies. In particular excitation of 488 and 514.5 nm essentially reproduced the spectrum of

Sugai and Saito on the same compound[16]. Excitation at 647 nm, figure 3, revealed new features, a splitting to lower frequencies of the ν_1 symmetric and a shifting from 120 cm^{-1} to 104 and 116 cm^{-1}, with the high frequency peak having a higher intensity. We observed no mode at 152 cm^{-1} as in the previous ternary compound. This is at least confirmation that in the β-compound, the I_3 anion remains linear. The splitting of the ν_1 symmetric mode is reminiscent of solid C_sI_3[17], with the difference that in the latter compound, a split 150 cm^{-1} line is also observed. This is assigned as a factor group effect in C_sI_3. We infer from the result in figure 3 that at high temperature, where the doublet disappears, we are dealing with equivalent I_3^-'s. It is known that at lower temperatures a freezing out of the ethylene puckering motion occurs with a concomitant ordering of these groups in the lattice[10-11], which also results in a slight dimerization of the I_3^-. The ν_1 mode downshifts in frequency and then splits by 12 cm^{-1}. However it is unclear at this moment whether we are dealing with a factor group splitting or different environments. We reiterate that we did not observe in the β compound the 150 cm^{-1} mode usually attributed to asymmetries. We would suggest on the basis of the present measurements that we are dealing with inequivalent I_3^- ions, i.e., the I_3^- ions are in different ethylene environments as one lowers the temperature due to the ordering of the end ethylene groups of the organic chain[10-11].

Acknowledgements - We thank Dr. D. Jerome, Laboratoire de Physique Solide, Orsay, for furnishing the BEDT-TTF samples. Finally we thank the Department of Defence for supporting this work under contract # 8SD5-00313.

REFERENCES

1. P. Coppens, in Extended Linear Chain Compounds, ed. J.S. Miller (Plenum Press, N.Y. 1982), p. 333.
2. P.C.W. Leung, T.J. Emge, A.J. Schultz, M.A. Beno, K.D. Carlson, H.H. Wang, M.A. Firestone, J.M. Williams, Solid State Commun. 57, 93 (1986).
3. A.D. Bandrauk, K.D. Truong, C. Carlone, A.W. Hanson, J. Phys. Chem. 91, 2063 (1987).
4. A.D. Bandrauk, K.D. Truong, C. Carlone, S. Jandl, Chem. Phys. Lett. 95, 78 (1983); J. de Phys. C3, 44, 1473 (1983).
5. A.D. Bandrauk, K.D. Truong, C. Carlone, S. Jandl, K. Ishii, J. Phys. Chem. 89, 434 (1985).
6. A.D. Bandrauk, K. Ishii, K.D. Truong, M. Aubin, A.W. Hanson, J. Phys. Chem. 89, 1478 (1985).
7. M.A. Abkowitz, A.J. Epstein, C.H. Griffiths, J.S. Miller, M.L. Slade, J. Am. Chem. Soc. 99, 5304 (1977).
8. C. Coulon, S. Flandrois, P. Delhaes, C. Hauw, P. Dupuis, Phys. Rev. B23, 2850 (1981).
9. A.D. Bandrauk, K.D. Truong, C. Carlone, A.W. Hanson, submitted to J. Phys. Chem.
10. P.W. Leung, T.J. Emge, M.A. Beno, H. Wang, J. Williams, V. Petricek, P. Coppens, J. Am. Chem. Soc. 107, 6184 (1985).
11. Y. Nogami, S. Kagoshima, T. Sugamo, G. Saito, Synthetic Metals 16, 367 (1986).
12. F. Mulazzi, I. Pollini, L. Piseri, R. Tubino, Phys. Rev. B24, 3555 (1981).
13. W. Kiefer, H.J. Bernstein, Chem. Phys. Lett. 16, 5 (1972).
14. F. Inagaki, I. Harada, T. Shimanouchi, M. Tasumi, Bull. Chem. Soc. (Japan) 45, 3384 (1972).
15. E.M. Mour, L.H. Chen, J. Laane, J. Phys. Chem. 90, 2481 (1986).
16. S. Sugai, G. Saito, Solid State Commun. 58, 759 (1986).
17. V. Gabes, H. Gerding, J. Mol. Structure 14, 267 (1972).

IMPORTANCE OF INTERMOLECULAR HYDROGEN - HYDROGEN AND HYDROGEN - ANION CONTACTS FOR THE LATTICE SOFTNESS AND THE SUPERCONDUCTIVITY OF β-$(ET)_2X$ ($X^- = I_3^-$, AuI_2^-, IBr_2^-)

Myung-Hwan Whangbo

Department of Chemistry
North Carolina State University
Raleigh, North Carolina 27695-8204, USA

Jack M. Williams, Arthur J. Schultz, and Mark A. Beno

Chemistry and Materials Science Divisions
Argonne National Laboratory
Argonne, Illinois 60439, USA

Organic donor molecule ET, 1, yields ambient-pressure superconductors β-$(ET)_2X$ ($X^- = I_3^-$, AuI_2^-, IBr_2^-). These organic salts contain corrugated

1

layers (2) of ET stacks, which are separated by layers of anion X^- (3) such that layers of partially oxidized donors $ET^{0.5+}$ and those of anions X^- alternate along the c-axis (4). These superconducting β-$(ET)_2X$ salts are two dimensional metals, as illustrated by the Fermi surface 5. Under ambient

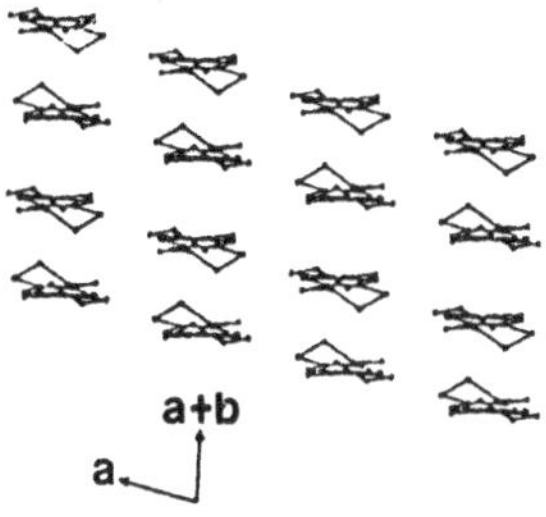

2

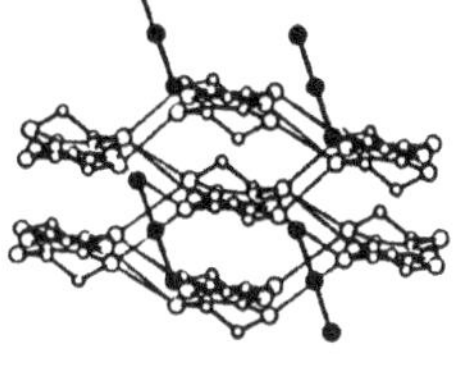

3

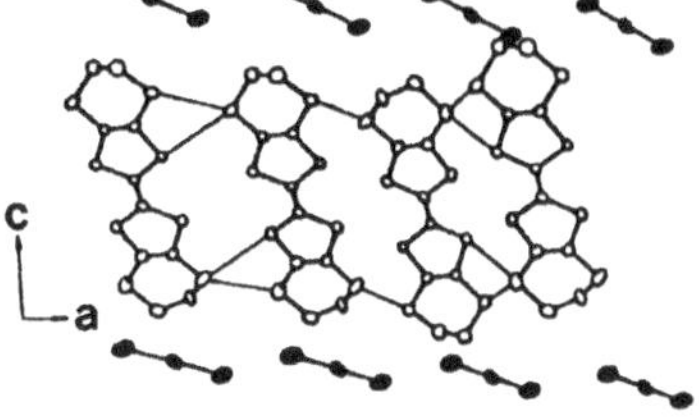

4

pressure, the superconducting transition temperatures T_c of β-$(ET)_2X$ are 1.4, 5.0 and 2.8K for $X^- = I_3^-$, AuI_2^- and IBr_2^-, respectively. As shown in **6**, the T_c of β-$(ET)_2X$ decreases gradually with increasing applied pressure P for $X^- = AuI_2^-$ and IBr_2^-. When P ≈ 0.5 kbar the T_c of β-$(ET)_2I_3$ jumps to 8K and decreases as P increases further.[1,2] This dependence of T_c upon X^- and P in β-$(ET)_2X$ is related to the change in the lattice softness as a function of X^- and P.

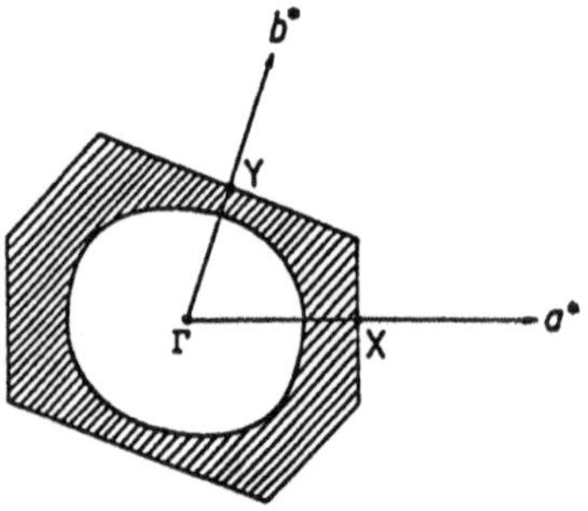

5

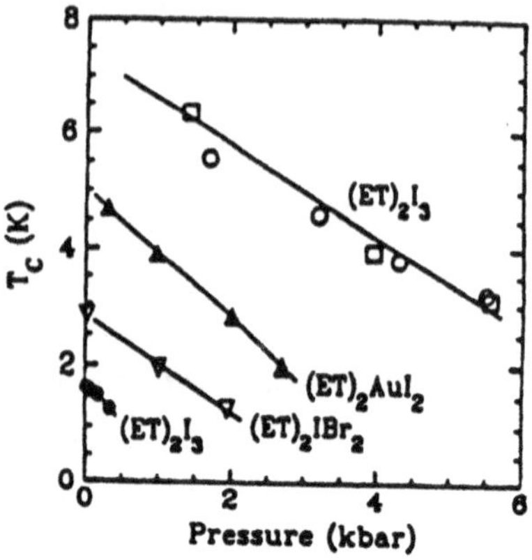

6

Within the BCS mechanism,[4] what is essential for superconductivity is electron-phonon coupling. The extent of electron-phonon coupling is measured by the electron-phonon coupling constant λ, which is related to T_c as[5]

$$T_c = (\theta_D/1.45)\ \exp[-1.04\ (1 + \lambda)/(\lambda - \mu^*(1 + \lambda \langle W\rangle/W_o))].$$

By employing the observed T_c values of β-$(ET)_2X$ in this equation along with the estimates $\mu^* = 0.1$, $\langle W\rangle/W_o = 0.3$ and $\theta_D = 200K$, one obtains[3] the values listed in Table 1, where β*-$(ET)_2I_3$ refers to the structure of β-$(ET)_2I_3$ under P > 0.5 kbar. It is evident that λ increases with T_c. Since λ is proportional to $1/\langle W^2\rangle$, a large λ (and hence a high T_c) results when the crystal lattice has a low-frequency phonon spectrum, i.e., when the lattice is soft toward certain vibrational modes (most likely to be translational and/or vibrational modes) effective for superconductivity. Thus the increase in the calculated values from 0.37 to 0.62 reflects a corresponding increase in the lattice softness.[3]

Table 1

Salt	T_c (K)		ET arrangement	H...X^- (A)	H...H (A)
β-$(ET)_2I_3$	1.4	0.37	eclipsed	2.842	2.477
			staggered	2.988	2.152
β-$(ET)_2IBr_2$	2.8	0.43	eclipsed	2.887	2.169
β-$(ET)_2AuI_2$	5.0	0.52	eclipsed	2.996	2.224
β*-$(ET)_2I_3$	8	0.62	staggered	3.014	2.261

Within each layer of $ET^{0.5+}$ donors, there exist a number of short intermolecular S...S contacts less than the van der Waals radii sum. However, all such S...S interactions are essentially nonbonding in nature since the overlap populations calculated for the contacts are essentially zero.[3] Consequently, the principal forces providing overall lattice cohesiveness are attractive Coulomb interactions bewteen layers of $ET^{0.5+}$ donors and those of X^- anions,[3] which leads to a number of short hydrogen... anion (H...X^-) contacts. As shown in **7**, each X^- anion is enclosed in a hydrogen pocket made up of 12 ET molecules. In such hydrogen pockets there occur short intermolecular hydrogen...hydrogen (H...H) contacts as well. Since ET molecules are anchored around X^- anions via the H atoms of their ethylene groups, how soft the motions of ET molecules are can be correlated with the lengths of the shortest H...X^- and H...H contacts: The longer these contact distances, the softer the crystal lattice.[3]

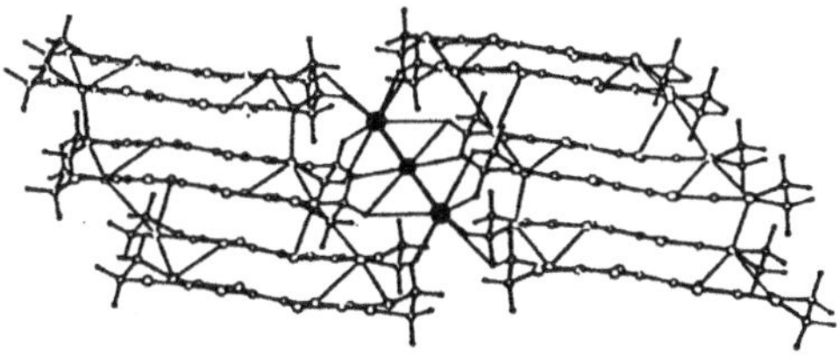

7

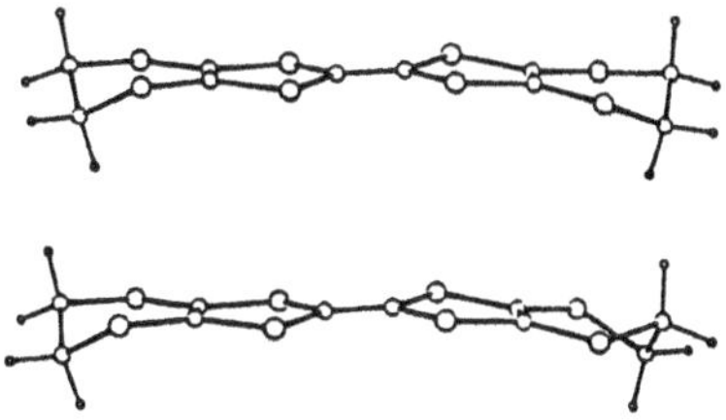

8

The occurrence of short H...X^- and H...H contacts in β-$(ET)_2X$ sensitively depends upon the relative arrangements of the two ethylene groups in each ET[3]. These groups twist out of the molecular π-framework so that, when viewed along the long molecular axis, they can be either eclipsed or staggered with respect to each other (**8**). All ET molecules have eclipsed arrangements in β-$(ET)_2AuI_2$ and β-$(ET)_2IBr_2$, both eclipsed and staggered ET molecules are found in nearly equal proportions in β-$(ET)_2I_3$, and all ET molecules adopt staggered arrangements in β^*-$(ET)_2I_3$. Thus, the most significant structural change that β-$(ET)_2I_3$ undergoes under $P \simeq 0.5$ is this ordering of the ethylene group arrangements.

Table 1 lists the shortest H...X^- and H...H contact distances present in β-$(ET)_2X$.[3] The T_c and λ values increase in the order β-$(ET)_2IBr_2$ < β-$(ET)_2AuI_2$ < β^*-$(ET)_2I_3$, which is correlated with the corresponding increase in their H...X^- and H...H distances and hence with the increase in their lattice softness. Shown in 9 are perspective views of the ethylene groups

9a

9b

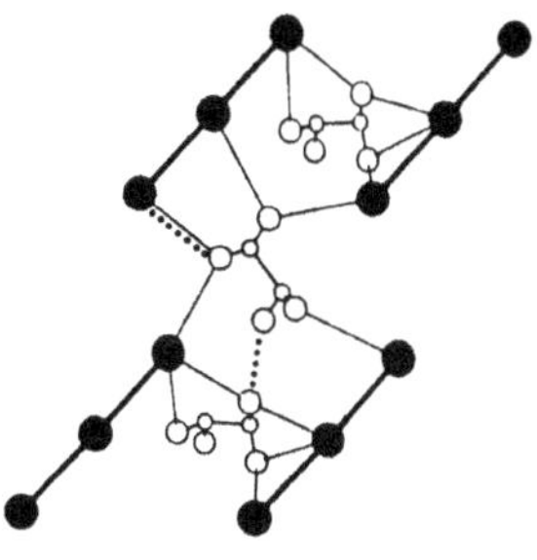

9c

of ET molecules positioned below a layer of I_3^- anions in β-$(ET)_2I_3$ and β^*-$(ET)_2I_3$ where short H...X$^-$ contacts are shown by their lines while the shortest H...X$^-$ and H...H are indicated by the dotted lines. Two kinds of ethylene sites are present in **9**, one between two I_3^- anions (Site A) and the other surrounded by four I_3^- anions (Site B). **9a** and **9b** refer to the cases when the ET molecules of β-$(ET)_2I_3$ have eclipsed and staggered arrangements, respectively. Note that only the ethylene group at Site B undergoes a conformational flipping on proceeding from **9a** to **9b**. The unfavorably short H...X$^-$ contact of **9a** is removed by the conformational flipping, but the shortest H...H contact of the resulting structure **9b** becomes unfavorably short. An effective way of reducing such short contacts is to introduce a structural modulation,[6] which is observed below 200K. The presence of the extremely short H...X$^-$ and H...H contacts in β-$(ET)_2I_3$ suggests that its lattice is stiff, and hence its T_c is low. The ethylene group arrangement around I_3^- ions in β^*-$(ET)_2I_3$, shown in **9c**, is similar to that in **9b** except that the ethylene group at Site B slips under applied pressure P ≃ 0.5 kbar to increase the shortest H...H and H...X$^-$ contact distances beyond the corresponding values of β-$(ET)_2AuI_2$. A further increase in applied presure P makes the β^*-$(ET)_2I_3$ lattice stiffer thereby lowering its T_c, just as in the case of β-$(ET)_2AuI_2$ and β-$(ET)_2IBr_2$.

In summary, the dependence of the T_c of β-$(ET)_2X$ upon the anion X$^-$ and applied pressure P is explained in terms of how the softness, and hence the electron-phonon coupling constant λ, of the β-$(ET)_2X$ lattice varies as a function of X$^-$ and P. What really governs the lattice softness is the short hydrogen...anion and hydrogen...hydrogen contacts, which occur as a consequence of the attractive Coulombic interactions between layers of $Et^{0.5+}$ donors and those of X$^-$ anions. In agreement with our analysis,[3] the recent point-contact measurement[7] on β-$(ET)_2X$ indiates the presence of a very soft phonon strongly coupled to the electrons.

Acknowledgement

Work at North Carolina State University and Argonne National Laboratory were supported by the U.S. Department of Energy, Office of Basic Energy Sciences, Division of Materials Sciences under Grant DE-FG05-86-ER45259 and under Contract W31-109-ENG-38, respectively. We express our appreciation for computing time made available by DOE on the ER-Cray X-MP computer.

References

1. J.E. Schirber, L.J. Azevedo, J.F. Kwak, E.L. Venturini, P.C.W. Leung, M.A. Beno, H.H. Wang, and J.M. Williams, Phys. Rev. B 33:1987 (1986).
2. J.E. Schirber, L.J. Azevedo, J.F. Kwak, E.L. Venturini, M.A. Beno, H.H. Wang, and J.M. Williams, Solid State Commun. 59:525 (1986).
3. M.-H. Whangbo, J.M. Williams, A.J. Schultz, T.J. Emge, and M.A. Beno, J. Am. Chem. Soc. 109:90 (1987).
4. J. Bardeen, L.N. Cooper, and J.R. Schriefer, Phys. Rev. 106:162 (1957).
5. W.L. McMillan, Phys. Rev. 167:331 (1968).
6. P.C.W. Leung, T.J. Emge, M.A. Beno, H.H. Wang, J.M. Williams, V. Petricek, and P. Coppens, J. Am. Chem. Soc., 106:7644 (1984); 107:6184 (1985).
7. A. Nowack, U. Poppe. M. Weger, D. Schweizer and H. Schwenk, Z. Phys., in press.

SYNTHESIS AND CHARACTERIZATION OF TTF SALTS OF PLANAR PLATINUM, PALLADIUM NICKEL AND COPPER (1,2 DITHIOOXALATO S,S') ANIONS AND THE CRYSTAL AND MOLECULAR STRUCTURE OF BIS(TETRATHIAFULVALENIUM)BIS(1,2 DITHIOOXALATO S,S')PALLADATE(II)

C. Bellitto, M. Bonamico, V. Fares, P. Imperatori, and S. Patrizio

Istituto di Teoria, Struttura Elettronica e Comportamento Spettrochimico dei Composti di Coordinazione del C.N.R., Area della Ricerca di Roma, C.P. 10,00016 Monterotondo Staz., Italy

INTRODUCTION

We have studied the metathesis reaction between Wudl's salt (1), i.e. $(TTF)_3(BF_4)_2$ and several inorganic metal(II) 1,2 dithiooxalate anions in the attempt to synthesize new molecular conductors.

In this communication we wish to report the chemical synthesis, the characterization of the donor-acceptor compounds formed by the $M(S_2C_2O_2)_2^{2-}$ complexes, II, (M=Ni,Pd,Pt and Cu) with the donor molecule tetrathiafulvalene, TTF, I, and the X-ray diffraction structure determination of $(TTF)_2Pd(S_2C_2O_2)_2$.

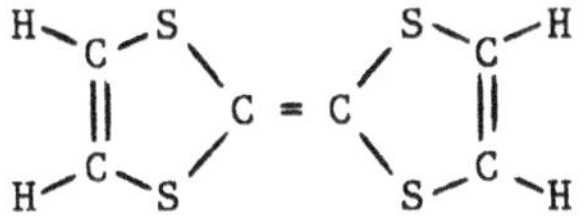

I

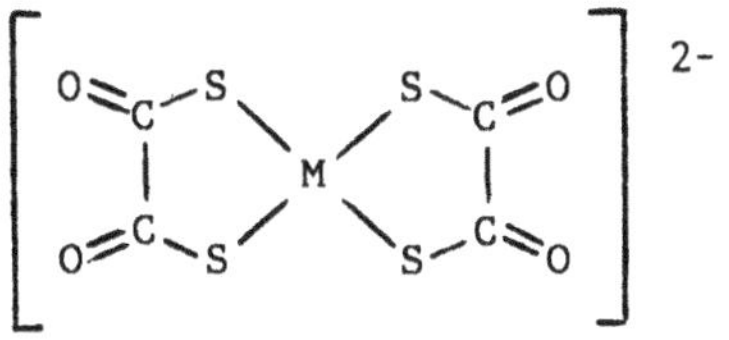

II

RESULTS AND DISCUSSION

The compounds presented here were all synthesized by a direct mixing of an acetonitrile solution of $(TTF)_3(BF_4)_2$ and of a solution of the square planar $<(C_2H_5)_4N>_2M(S_2C_2O_2)_2$ complexes, where M = Ni,Pd,Pt and Cu, in the same solvent.

In the case of the platinum derivatives, two different species were isolated :

$$<(C_2H_5)_4N>_2Pt(S_2C_2O_2)_2 + (TTF)_3(BF_4)_2 \rightarrow (TTF)_3Pt(S_2C_2O_2)_2 \ \ \mathbf{1}$$

$$\rightarrow (TTF)_2Pt(S_2C_2O_2)_2 \ \ \mathbf{2}$$

Both compounds were characterized by X-ray diffraction studies.The compound 2 is always present in higher yield (∿ 95%). In the case of Pd,Ni and Cu, only one type of compound could be isolated,i.e.$(TTF)_2Pd(S_2C_2O_2)_2$ $(TTF)_{1.5}Cu(S_2C_2O_2)_2$ and $(TTF)Ni(S_2C_2O_2)_2$. $(TTF)_2Pd(S_2C_2O_2)_2$ and $(TTF)_2Pt(S_2C_2O_2)_2$ have been found to be isomorphous. All the compounds are insoluble in the common solvents.

Single-crystals of these compounds were obtained by a solution diffusion technique. The nickel derivative was obtained only as a microcrystalline product.

Description of the crystal structure of $(TTF)_2Pd(S_2C_2O_2)_2$

The crystal structure is reported in figure 1 . It consists of mixed stacks of $(TTF^+)_2$, D ,dimers and $Pd(S_2C_2O_2)_2^{2-}$ anions, A, stacking along the b axis of the unit-cell according to the scheme :

...... A - D - A - D - A - ......

The A and D molecular best planes are almost parallel to each other with a dihedral angle of 4°, but their normals are tilted by 18° with respect to the b axis .

$Pd(S_2C_2O_2)_2^{2-}$ anion

The metal ion lies on a centre of symmetry in a $[PdS_4]$ square planar arrangment. The anion is not planar, the sulphur atoms S(1) and S(2) moving away from the best plane by 0.122 and 0.098 Å ,respectively. An overall "chair" conformation is therefore present with the dihedral angle between the $[PdS_4]$ central plane and the ligand $S_2C_2O_2$ plane of 8°.Bond distances and angles within the dithiooxalate group are comparable with the literature data (2) .

TTF molecule

The organic molecules form fully oxidized $(TTF^+)_2$dimers,which act as large divalent cations. The TTF^+ cation is distorted from planarity with the sulphur atoms below, and the carbon atoms above, the least-squares best plane (boat conformation). The C=C intramolecular bond length is expected to increase as the electron density is removed from the HOMO (3).

The observed value is 1.405(3) Å, which is characteristic of ionic TTF^+ (4). The intradimeric separation is 3.48 Å, comparable to that found in $(TTF)\cdot I_3$ (5) and shorter than that found in the segregated mixed-valence $TTF\cdot I_{0.71}$ (6).

Crystal Packing

The molecular packing in $(TTF)_2Pd(S_2C_2O_2)_2$ is reminiscent of that for $TTF\cdot I_3$ (5) and $TTF\cdot Br$ (4). In these compounds integrated stacks of $(TTF^+)_2$ dimers and halide occur, although they are more inclined with respect to the unit-cell axis than in the present case. Thus, the crystal structure of $(TTF)_2Pd(S_2C_2O_2)_2$ shows that $(TTF^+)_2$ dimers are virtually isolated.

Physical Measurements

Two different types of IR spectra are observed in the compounds studied. In the first type, a broad, intense band dominates from 4000 cm^{-1} to 2000 cm^{-1}, and is observed in both nickel and copper derivatives. In the second type this band is absent : this is the case of $(TTF)_2M(S_2C_2O_2)_2$ M = Pd, Pt. In the latter, the infrared spectra are superposable and have been assigned on the basis of the IR assignments of the constituent species (2,7). The presence of characteristic vibrations of $(TTF^+)_2$ are consistent with the formation of strong dimers (7). We have also obtained preliminary, r.t., d.c., electrical conductivity data (Silver paint contacts, van derPauw 4-probe method), by using pressed pellets on these polycrystalline materials. Results are reported in Table I.

TABLE I

COMPOUND	Colour	r.t., electrical conductivity ($S\ cm^{-1}$)
$(TTF)_2Pd(S_2C_2O_2)_2$	dark-brown	insulator
$(TTF)_2Pt(S_2C_2O_2)_2$	"	"
$(TTF)_{1.5}Cu(S_2C_2O_2)_2$	"	1.1×10^{-6}
$(TTF)Ni(S_2C_2O_2)_2$	dark-blue	1.2×10^{-4}

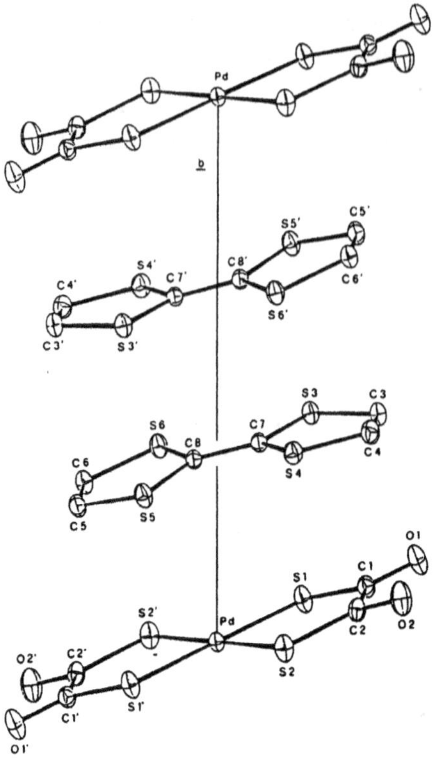

Figure 1. Side view of the stacking of $(TTF)_2Pd(S_2C_2O_2)_2$ along the b axis. Crystal Data. Monoclinic ,space group $P2_1/c$, a = 9.447(2), b = 11.044(4), c = 12.299(1) Å, β =113.52(2)°, V =1176.6 $Å^3$, Z = 2, D_c=2.133 gcm^{-3}, D_m=2.13 gcm^{-3}.

CONCLUSIONS

Several new organic/inorganic charge-transfer salts have been synthesized. In $(TTF)_2M(S_2C_2O_2)_2$,M = Pd,Pt , the experimental data confirm that the ground state is essentially ionic . In $(TTF)_3Pt(S_2C_2O_2)_2$, IR spectra and preliminary X-ray data suggest the presence of both TTF^0 and $(TTF^+)_2$ units. Further experiments and studies on the copper and nickel derivatives are currently under way.

REFERENCES

1. F.Wudl,J.Amer.Chem.Soc.,97,1962,(1975).
2. N.C.Baenzinger,S.M.Johnson,D.Coucouvanis,J.Amer.Chem.Soc.,95,3875,(1973).
3. S.S.Shaik,M.H.Whangbo,Inorg.Chem.,25,1201,(1986).
4. B.A.Scott,S.J.La Placa,J.B.Torrance,B.D.Silvermann,B.Welber,J.Amer.Chem.Soc., 99,6631,(1977).
5. R.C.Teitelbaum,T.J.Marks,C.K. Johnson,J.Amer.Chem.Soc.,102,2986,(1980).
6. J.J.Daly,F.Sauz,Acta Crystallogr.,31 B,620,(1975).
7. R.Bozio,I.Zanon,A.Girlando,C.Pecile,J.Chem.Phys.,71,2282,(1979).

FINE STRUCTURE TO THE ν_1 AND $2\nu_1$ BANDS ($\nu_1 = \nu(Cl\text{-}Pt^{IV}\text{-}Cl)$, THE SYMMETRIC CHAIN STRETCHING MODE) OF $[Pt(en)_2][Pt(en)_2Cl_2][ClO_4]_4$

Robin J.H. Clark and Vincent B. Croud

Christopher Ingold Laboratories
University College London
London WC1H 0AJ, U.K.

INTRODUCTION

Mixed-valence linear chain complexes of platinum and palladium of the Wolffram's red type, $[Pt(EtNH_2)_4][Pt(EtNH_2)_4Cl_2]Cl_4.4H_2O$, are of interest owing to their potential as one-dimensional semiconductors.[1,2] X-ray and XPS studies have identified two types of metal (M) site in the chains.[1,2] One, assigned to the II oxidation state, has pseudo square-planar geometry, and the other, assigned to the IV oxidation state, has tetragonal bipyramidal geometry with *trans* halogen atoms along the chain axis. Thus the M^{II} and M^{IV} moieties, which alternate along the chain, are bridged by halogen atoms which are displaced from the mid-point between the metal sites towards the M^{IV} site. The ratio of the M^{IV}-X distance to the M^{II}-X distance has been shown to correlate with the magnitude of the intervalence ($M^{IV} \leftarrow M^{II}$) transition energy, with the thermal activation energy for conductivity, and with the wavenumber of the symmetric (ν_1, Raman active) and asymmetric (ν_2, infrared active) X-M^{IV}-X chain-stretching modes.

The Raman spectra of halogen-bridged mixed-valence complexes obtained with an exciting line (ν_0) resonant with the intervalence band (resonance Raman spectra) are characterised by enormous intensification of the ν_1 band together with the development of an intense overtone progression, $v_1\nu_1$, where v_1 is the vibrational quantum number of ν_1. As many as 17 harmonics have been detected in the resonance Raman spectra of some such complexes (Fig. 1a). The excitation profile (EP) of the ν_1 band of powdered samples has been shown to maximise at or near the absorption edge (E_g). The wavenumber of ν_1 has been shown to be dependent on the wavenumber of the exciting line, with the former apparently increasing as ν_0 moves to the blue.[3,4] No dispersion is observed for $\nu_0 < E_g$.[4]

It has recently been demonstrated[5] that, for single crystals of $[Pt(en)_2][Pt(en)_2Br_2][ClO_4]_4$ (for which $\nu_0 > E_g$) the apparent dispersion of ν_1 with change in ν_0 can be ascribed to the fact that the ν_1 band, under high resolution, can be shown to consist of at least five components whose relative (and overall) intensities depend on ν_0. The lowest wavenumber components have greatest intensity for ν_0 to the red and the highest wavenumber components have greatest intensity for ν_0 to the blue. Additionally the overall intensification of the lower wavenumber components is greater than that of the higher ones. The maximum intensification of the lowest wavenumber component occurs close to or at the absorption edge. Hence the

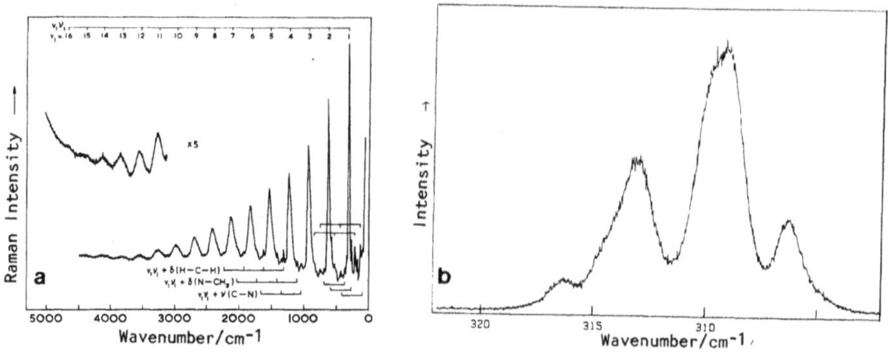

Fig. 1. a) Resonance Raman spectrum of Wolffram's red.
b) The ν_1 band profile for $[Pt(en)_2][Pt(en)_2Cl_2][ClO_4]_4$.

EP of a ν_1 band for which the components are unresolved maximises at or near the absorption edge. However only partial resolution of the profile was achieved for $[Pt(en)_2][Pt(en)_2Br_2][ClO_4]_4$ and so it was not possible to ascertain the total number, the half-band width (FWHM), or accurate wavenumbers for the components, nor was it possible to determine whether the wavenumbers of the components had any ν_0 dependence (although the results indicate that, if this is the case, it occurs to the extent of $< 2\ cm^{-1}$ over a range of ν_0 from 752.5 to 457.9 nm). Additionally the behaviour of the component bands in the overtone region was not able to be determined because these profiles are unresolved (though non-symmetric) due to component broadening.

We have demonstrated, as have Tanaka and Kurita,[6] that, for a single crystal of $[Pt(en)_2][Pt(en)_2Cl_2][ClO_4]_4$, better resolution of the ν_1 profile could be achieved than for the analogous bromide. Initially we resolved (at *ca.* 20 K) 3 components which we assigned to chlorine isotopic splitting based on a model of isolated $X\text{-}M^{IV}\text{-}X$ units for which a good fit was obtained.[7] Later, however, Tanaka and Kurita managed to identify five components (at *ca.* 2 K) and these, as well as our own results on the bromide-bridged complex,[5] encouraged us to reinvestigate the spectroscopy of the chloride-bridged complex.

EXPERIMENTAL DETAILS

Raman spectra were recorded on a Spex 14018 R6 spectrometer, in conjunction with Coherent Radiation CR 3000 K and CR 3 Kr^+ and Ar^+ lasers. Samples were in the form of single crystals, at orientations (Porto nomenclature) of y(zz)x + y(zy)x, at *ca.* 80 K using a liquid-nitrogen-cooled Dewar assembly. Very low laser powers were used (< 10 mW). Band wavenumbers were calibrated by reference to the emission lines of neon.

The complexes were synthesised by co-crystallizing stoichiometric quantities of $[Pt(en)_2]^{2+}$ and $[Pt(en)_2Cl_2]^{2+}$ in dilute perchloric acid solution.

RESULTS

Similar to the results of Tanaka and Kurita (at *ca.* 2 K) our spectra (at *ca.* 80 K) of the ν_1 fundamental show four resolved components and one or two partially resolved components to the high wavenumber side of these. In fact the quality of the spectra are comparable, showing that it is not necessary to cool to 2 K to achieve good resolution. For the overtones the high wavenumber components increase in intensity relative to the low wavenumber components as the vibrational quantum number increases. The relative intensities of the components of our sample compared with the Japanese sample are virtually identical confirming that the vibrational structure is intrinsic. However by looking more closely at the shape and FWHM of the resolved peaks, especially that of the most intense peak (for $v_1 = 1$) we believe that the profile is in fact even more complex than previously thought. By inspection eight peaks can be identified from the profile (Fig. 1b). In order to fit this profile more systematically, we have taken the resolved peak of lowest wavenumber (at *ca.* 306.5 cm^{-1}) as being a single component peak, due to its having the most symmetrical shape and the narrowest FWHM. We have used this value to define the FWHM and Gaussian/Lorentzian mix for the remaining components to the profile. We have thus fitted the complete ν_1 profile with a set of components having the same FWHM and Gaussian/Lorentzian mix, *i.e.* the components are assumed to have the same physical origin (whatever that may be).

The fit over the lower wavenumber part of the fundamental profile is excellent, especially for the region of the maximum of the most intense resolved peak. Because of the complexity of the higher wavenumber part of the profile an unambiguous fit is not possible, although given the starting parameters only two fits are attractive (given the usual constraint that the minimum number of peaks necessary to fit the profile are considered). The peak FWHM giving the best fit is 1.2 cm^{-1} which is very narrow and implies a long lifetime for the process giving rise to the components. The Gaussian/Lorentzian mix for the best fit is 20% Lorentzian. A 100% Lorentzian bandshape is ideally expected but the deviation is probably mainly due to the Gaussian perturbation caused by the spectrometer (decreasing the spectrometer slit width should, therefore, increase the Lorentzian character). A total of sixteen component bands have been fitted to the ν_1 band profile. This gives an average separation of 0.8 cm^{-1} between components (Fig. 2a, Table 1).

The first overtone is much more difficult to fit than the fundamental due to the greater complexity of the $2\nu_1$ band profile. A reasonable fit can be obtained with a FWHM of 1.8 cm^{-1} and the same Gaussian/Lorentzian mix as for the fundamental (Fig. 2b). A good wavenumber correspondence between the components of ν_1 and $2\nu_1/2$ is obtained, except at the high wavenumber end where the component bands are less intense. Table 1 also lists the $2\nu_1/2$ components and also the $2\nu_1/2$ components modified by way of a 1 cm^{-1} (assumed) "anharmonicity". Although both lists give good wavenumber correspondence with the observed ν_1 components, the latter tabulation leads to a better intensity correlation, with the lower wavenumber components losing intensity with respect to the higher ones in a

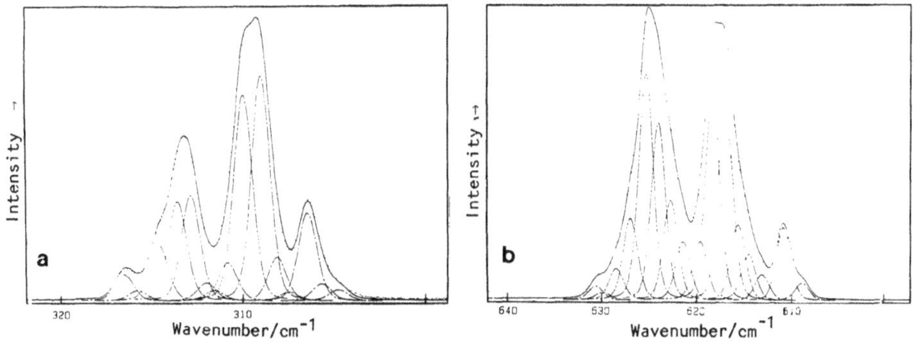

Fig.2 Fit of the sixteen component bands to the a) ν_1 and b) $2\nu_1$ band profile for $[Pt(en)_2][Pt(en)_2Cl_2][ClO_4]_4$.

Table 1. Wavenumbers of the Bands fitted to the ν_1 and $2\nu_1$ Profiles.[a]

ν_1		$2\nu_1/2$			
Band/cm^{-1}	height[b]	Band/cm^{-1}	height[b]	Band/cm^{-1} [c]	height[b]
304.7	3.7	304.5	6.0		
305.6	5.8	305.4	25.0	305.4	6.0
306.4	30.0	306.6	8.7	306.4	25.0
307.4	3.0	307.2	16.0	307.5	8.7
308.1	15.0	307.9	31.0	308.2	16.0
308.9	76.5	308.5	65.0	308.9	31.0
		309.2	70.0	309.5	65.0
309.9	69.5	309.8	20.0	310.2	70.0
310.7	13.0	310.7	20.0	310.8	20.0
311.5	3.5	311.5	34.0	311.7	20.0
311.9	6.0	312.0	60.0	312.4	34.0
312.7	36.0	312.7	76.5	313.0	60.0
313.5	34.0	313.4	28.0	313.7	76.5
314.5	19.2	314.2	11.1	314.5	28.0
315.7	3.4	314.9	1.6	315.2	11.1
316.4	9.0	315.3	4.7	315.9	1.6
317.6	1.0			316.3	4.7

[a]λ_0 = 514.5 nm. [b]Arbitrary scale. [c]Modified for 1 cm^{-1} "anharmonicity".

smoother fashion. One interesting point is that whichever $2\nu_1/2$ tabulation is used there is no corresponding ν_1 fundamental component between 308.9 and 309.9 cm^{-1}, suggesting that the ν_1 fit is incomplete in this region and that another band needs fitting here with a consequent adjustment to the intensities of adjacent bands being required. If this band can be satisfactorily fitted then we will be able to make a more positive statement on the component intensity changes on going from ν_1 to $2\nu_1$. However we are at least now in a position to state firmly that there is no evidence for combination bands present in the $2\nu_1$ profile formed from the ν_1 components.

Work is continuing to improve the fit and to extend the work to the higher overtones in order that the intensity change of the components with vibrational quantum number (v_1) be fully understood. At present we have no explanation for the number of components in the profile by way of correlation splitting, neither are we satisfied that chain length effects are the answer due to the virtual invariability of the profiles for many samples for a given ν_0 and due to the apparent complexity of the component intensity dependence on v_1.

REFERENCES

1. R.J.H. Clark, in "Advances in Infrared and Raman Spectroscopy", eds. R.J.H. Clark and R.E. Hester, Heyden, London, 1975, Vol.1, p.143, and 1983, Vol.11, p.95.
2. R.J.H. Clark, Chem. Soc. Rev., 1984, 13, 219.
3. R.J.H. Clark and M. Kurmoo, J. Chem. Soc., Faraday Trans. II, 1983, 79, 519.
4. R.J.H. Clark and V.B. Croud, J. Phys. C: Solid State Phys., 1986, 19, 3467.
5. S.D. Conradson, R.F. Dallinger, B.I. Swanson, R.J.H. Clark and V.B. Croud, Chem. Phys. Lett., in press.
6. M. Tanaka and S. Kurita, J. Phys. C: Solid State Phys., 1986, 19, 3019.
7. S.D. Allen, R.J.H. Clark, V.B. Croud and M. Kurmoo, Phil. Trans. Roy. Soc. London, 1985, A314, 131.

CORRELATION OF SPECTROSCOPIC AND MAGNETIC DATA OF TWO CHARGE TRANSFER COMPOUNDS OF TCNQ WITH CATIONIC COPPER CHELATES

Michael Schwartz and William E. Hatfield

Department of Chemistry, The University of North Carolina
Chapel Hill, North Carolina 27514

Introduction

In view of the initial work[1] on charge transfer compounds of TCNQ, one of the advantages of using copper(II) chelates as acceptors in charge transfer compounds with $TCNQ^-$ would appear to be the high probability of forming stacked $TCNQ^-$ segarated chains in preference to mixed CuL_2-TCNQ stacked chains. Such cationic copper chelate-TCNQ charge transfer compounds have unusual properties and have continued to attract attention.[2-6]

Experimental

Solid state electronic absorption spectra were obtained by using powdered samples mixed with KBr and pressed into pellets.[7] Magnetic susceptibility data were obtained by using a Faraday balance system. The balance was calibrated with $HgCo(NCS)_4$[8] and with Gd_2O_3.[9] A value of -1.21×10^{-4} emu mol^{-1} was used as the diamagnetic correction for TCNQ,[10] Pascal's constants were used for diamagnetic corrections,[8] and a value of 60×10^{-6} emu mol^{-1} was used for TIP of copper(II).

Results

Electronic spectral data for $[Cu(en)_2][TCNQ]_2$ and $[Cu(phen)_2][TCNQ]_2$ along with assignments are given in Table I. Features in the electronic spectra of the $[Cu(L)_2][TCNQ]_2$ compounds may be understood in terms of the electronic structures of $[TCNQ]_n^{q-}$ (q = 0, 1, 2; n = 1, 2). The spectral features probably arise from transitions at defects and chain ends, and from $TCNQ^- + TCNQ^- \rightarrow TCNQ^0 + TCNQ^{2-}$ and $TCNQ^- + TCNQ^0 \rightarrow TCNQ^0 + TCNQ^-$ charge transfer.

An SCF-LCAO-MO calculation of $TCNQ^-$ by Lowitz[11] provides the features necessary to understand the absorption spectrum of isolated $TCNQ^-$. More sophisticated calculations, such as those by Jonkman and Kommandeur[12] support the spectral assignments.

Where there is overlap, the magnetic data are in good agreement with the data reported by Inoue and coworkers.[4] The data for $[Cu(en)_2][TCNQ]_2$ may be fit by the Curie-Weiss law with S = 1/2, g = 2.10, and θ = 8.32. Since contributions from $TCNQ^-$ molecules are not necessary to explain the

Table I. Electronic Spectral Data[a]

$[Cu(en)_2][TCNQ]_2$ RT	$[Cu(en)_2][TCNQ]_2$ 77 K	$[Cu(phen)_2][TCNQ]_2$ RT	$[Cu(phen)_2][TCNQ]_2$ 77 K	Comments
365 i	355 i	370 i	370 i	$^2B_{2g} \rightarrow {}^2A_u$
			415 sh	
		500	500	
617 i	612 i	610 i	610 i	$^2B_{2g} \rightarrow {}^2A_u$
		685	685	
		773	765	
			780	
850 i	850 i	875 i	870 i	$^2B_{2g} \rightarrow {}^2B_{1u}$
		1000 sh	985 sh	
1220 vb	1220 vb	1170	1170	$TCNQ^- \rightarrow TCNQ^-$ CT
		~2500 vb	~2500 vb	$TCNQ^- \rightarrow TCNQ^0$ CT

[a]Abbreviations: RT - room temperature; i - intense; sh - shoulder; vb - very broad; CT - charge transfer; band positions given in nm.

observed data, it may be concluded that the $TCNQ^-$ molecules are strongly coupled. The magnitude of the effective magnetic moment indicates that a negligible amount of electron transfer has occurred from $TCNQ^-$ to the copper(II) cations.

$[Cu(phen)_2][TCNQ]_2$ was assumed to consist of copper(I) and copper(II) complex cations and $TCNQ^0$ and $TCNQ^-$ molecules within clusters or chains. The low values for the magnetic moments indicate that a substantial amount of electron transfer has occurred. If there is site preference for the electrons on the TCNQ chain, then the cluster models may be appropriate. The simplest cluster model consists of a pair, and the appropriate equation for the magnetic susceptibility is

$$\chi = 2(\rho - 0.5)\cdot(C/T) + \rho\cdot \frac{2N\mu_B g^2}{3kT}\left[1 + \frac{1}{3}\exp(-2J/kT)\right]$$

where the first term is the Curie contribution of the copper(II) complex cations, C is the Curie constant, and ρ is the degree of band filling. The quantity $2(\rho - 0.5)$ represents the fraction of copper(II) in the sample. The g value in the first term was held constant at the experimental value of 2.07. The second term is the contribution from the TCNQ clusters (pairs) with an energy gap of $2J = \Delta E_{ST}$. The g value was taken to be 2.0 for the organic radicals. The exchange coupling constant is given by the following expression:[13]

$$J = \left[\frac{t^2}{U + (4t/\pi)\sin\pi\rho}\right]\left[1 - \frac{\sin 2\pi\rho}{2\pi\rho}\right]$$

where U and t are Coulomb and transfer integrals. This model was applied yielding $\rho = 0.71$ and $2J = -0.077$ eV. Fitting calculations were made with non-linear least squares fitting program.[14]

Discussion

The experimental results suggest strong intermolecular interactions between the TCNQ molecules in both copper complexes. If these molecules stack to form chains, as is common in TCNQ charge transfer compounds, then strong intermolecular interactions may not be confined to pairs of TCNQ molecules. The Hubbard model[15] may be used to account for both delocalization and electron-electron repulsions (Coulomb correlations) associated with the formation of doubly occupied sites. In this model, delocalization effects are accounted for by the t_{ij} term, and localization effects through the U term.

Band Structure in $[Cu(en)_2][TCNQ]_2$. The Hubbard model does not take site energy differences into consideration, and it should apply to $[Cu(en)_2][TCNQ]_2$, since, by stoichiometry, all sites have one electron. Tanaka, et al.[16] have shown that the energy of the charge transfer transition is given by $(U - \sqrt{2}t)$. With this relationship, the energy of the charge transfer transition (1.02 eV), and results of the calculations of $\underline{t} = 0.2$ for a variety of TCNQ compounds,[17] a value of 1.3 eV may be estimated for U. The effect of U is to cause a band gap at the Fermi level. Hubbard[18] has determined that the gap develops for values of $4t/U < 1.15$. $[Cu(en)_2][TCNQ]_2$ meets this criterion since $4t/U$ equals 0.62.

Soos and Strebel[19] have determined that the singlet-triplet splitting is given by $4t^2/U$. Using the values of t and U, ΔE_{ST} is estimated to be 0.12 eV. A system with a ΔE_{ST} of 0.12 eV would exhibit a magnetic susceptibility on the order of 10^{-5} emu/mole at 300 K. A susceptibility this small could not be detected in the presence of the molar magnetic susceptibility of the $S = 1/2$ copper complex cation.

Band Structure of $[Cu(phen)_2][TCNQ]_2$. When there is less than one electron per site, conductivity can occur without the formation of doubly occupied sites. Because of electron transfer from $TCNQ^-$ to the copper(II) complex, $TCNQ^0$ sites are created, and $[Cu(phen)_2][TCNQ]_2$ should exhibit properties consistent with empty-site statistics. The key to understanding the electronic structure and conductivity is provided by the low energy absorption band which arises from $TCNQ^- + TCNQ^0 \rightarrow TCNQ^0 + TCNQ^-$ charge transfer. The presence of this transition reflects a site preference, and when double occupation can be neglected, the difference in site energies dominates the conductivity process.

Soos and Klein[20] have modified the Hubbard model to take into account site energy differences. The modified Hubbard model yields a band structure in terms of t and E. In the case of $t \gg E$, cosine dispersion curves are obtained. In the case of $E \gg t$, the bands are flat with a gap dependent on E. The percentage of copper(II) in $[Cu(phen)_2][TCNQ]_2$, as determined from the magnetic susceptibility measurements, shows that, on the average, there is about two-thirds of an electron per site. As an approximation, for this site occupancy, the energy of the charge transfer is 3E, and therefore the site energy difference in $[Cu(phen)_2][TCNQ]_2$ is 0.17 eV (0.5 eV/3). Because E and t are comparable in magnitude in $[Cu(phen)_2][TCNQ]_2$, the bandwidth is still appreciable, and high electrical conductivity and a low activation energy are expected, and observed.[3]

Thus, the energies of the charge transfer bands $TCNQ^- + TCNQ^- \rightarrow TCNQ^0 + TCNQ^0$ and $TCNQ^- + TCNQ^0 \rightarrow TCNQ^0 + TCNQ^-$ yield data which provide an explanation of the magnetic properties of $[Cu(en)_2][TCNQ]_2$ and $[Cu(phen)]_2[TCNQ]_2$ in terms of the Hubbard model and its modified variation.

Acknowledgements

This work was supported in part by the Office of Naval Research.

References

(1) Melby, L. R.; Harder, R. I.; Hertler, W. R.; Mahler, W.; Benson, R. E.; Mochel, W. E. J. Am. Chem. Soc. 1962, 84, 3374.
(2) Hoffmann, S. K.; Corvan, P. J.; Singh, P.; Sethulekshmi, C. N.; Metzger, R. M.; Hatfield, W. E. J. Am. Chem. Soc. 1983, 105, 4608.
(3) Inoue, M.; Inoue, M. B. Inorg. Chim. Acta 1980, 45, L129.
(4) Inoue, M. B.; Inoue, M. Chem. Phys. Lett. 1981, 80, 585.
(5) Inoue, M.; Inoue, M. B.; Seto, T.; Nakamura, D. Mol. Cryst. Liq. Cryst. 1982, 86, 139.
(6) Inoue, M.; Inoue, M. B. J. Chem. Soc., Chem. Commun. 1984, 790.
(7) (a) Oohashi, Y.; Sakata, T. Bull. Chem. Soc. Jpn. 1973, 46, 3330.
(b) Michaud, M.; Carlone, C.; Hota, N. K.; Zauhar, J. Chem. Phys. 1979, 36, 79.
(8) (a) Brown, D. B.; Crawford, V. H.; Hall, J. W.; Hatfield, W. E. J. Phys. Chem. 1977, 81, 1303.
(b) Weller, R. R.; Hatfield, W. E. J. Chem. Ed. 1979, 56, 652.
(9) Hines, W. A.; Moeller, C. W. Rev. Sci. Inst. 1973, 44, 1544.
(10) Scott, J. C.; Garito, A. F.; Heeger, A. J. Phys. Rev. B 1974, 10, 3131.
(11) Lowitz, D. A. J. Chem. Phys. 1967, 46, 4698.
(12) Jonkman, H. T.; Kommandeur, J. Chem. Phys. Lett. 1972, 15, 496.
(13) Klein, D. J.; Seitz, W. A. Phys. Rev. B 1974, 10, 3217.
(14) (a) Spendley, W.; Hext, G. R.; Himsworth, F. R. Technometrics 1962, 4, 441.
(b) Nedler, J. P.; Mead, R. Computer J. 1965, 7, 308.
(c) O'Neill, R. Appl. Stat. 1971, 20, 338.
(15) (a) Hubbard, J. Proc. Roy. Soc. Ser. A 1963, 276, 238.
(b) Hubbard, J. Proc. Roy. Soc. Ser. A 1963, 277, 237.
(16) Tanaka, J.; Tanaka, M.; Kawai, T.; Takube, T.; Maki, O. Bull. Chem. Soc. Jpn. 1976, 49, 2358.
(17) van Smaallen, S.; Kommandeur, J. Phys. Rev. B 1985, 31, 8056.
(18) Hubbard, J. Proc. Roy. Soc. Ser. A 1964, 281, 401.
(19) Soos, Z. G.; Strebel, P. J. J. Am. Chem. Soc. 1971, 55, 3284.
(20) Soos, Z. G.; Klein, D. J. J. Chem. Phys. 1971, 55, 3284.

COUNTER-ION INFLUENCES ON THE ELECTRICAL CONDUCTION PROPERTIES OF 1,2-DITHIOLENE COORDINATION COMPOUNDS

L.R. Groeneveld*, G.J. Kramer**, T.B.L.W. von Marinelli*, H.B. Brom**, J.G. Haasnoot* and J. Reedijk*

* Department of Chemistry, Gorlaeus Laboratories, Leiden University, P.O. Box 9502, 2300 RA Leiden, The Netherlands
** Department of Physics, Kamerlingh Onnes Laboratory, Leiden University, P.O. Box 9506, 2300 RA Leiden, The Netherlands

INTRODUCTION

The intention of the present study is to investigate the coordination chemistry and the magnetic and conducting properties of polynuclear coordination compounds containing 1,2-dithiolenes. DMIT-based (H_2DMIT = 4,5-dimercapto-1,3-dithiol-2-thione) complexes have become of particular interest since the discovery of superconductivity in (TTF)[Ni(DMIT)$_2$]$_2$[1]. With DMIT and related ligands square planar compounds are obtained for M = Ni, Cu, Pd and Pt, see Figure 1.
An important and general feature of the above-described compounds is the variety of apparent oxidation states for the metal ions, as well as the fact that these can easily be converted into each other.

RESULTS, DISCUSSION AND CONCLUSIONS

The compound $(Et_4N)_1$[Ni(DMIT)$_2$], prepared by oxidation by iodine of $(Et_4N)_2$[Ni(DMIT)$_2$][2,3], crystallizes in the space group $P2_1/n$, its structure has been solved by X-ray diffraction methods. See Figure 1. for crystallographic parameters and distances. It is a zig-zag chain with the shortest intermolecular distances 3.8 Å.
Oxidation of $(Et_4N)_2$ Pd(DMIT)$_2$ with air afforded crystals of $(Et_4N)_{0.5}$ [Pd(DMIT)$_2$], of which the crystal structure was solved as well. Its structural parameters are tabulated in Figure 1. This compound is a two-dimensional S....S network with nearest S....S contacts in the range of 3.26 to 3.72 Å.
Both compounds are semiconductors with a linear behaviour of ln σ vs. $1/\sqrt{T}$, see Figure 2. From the single crystal conductivity measurements of several compounds with DMIT can be concluded that the conductivity is strongly dependent on the cation, as well as on the metal atom and its oxidation state (Figure 3).

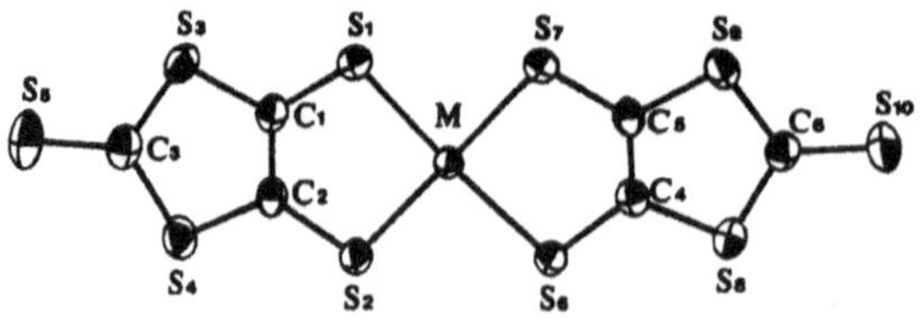

	$[(C_2H_5)_4N]_1Ni(DMIT)_2$	$[(C_2H_5)_4N]_{0.5}Pd(DMIT)_2$
Space group	P $2_1/n$	P $\bar{1}$
a, Å	7.333(1)	18.873(6)
b, Å	25.734(3)	8.515(3)
c, Å	12.798(4)	6.295(2)
α, (°)	90.0	117.77(3)
β, (°)	104.95(2)	96.66(3)
γ, (°)	90.0	90.78(3)
Z	4	2
R/R_w	4.49/5.36	2.89/2.97
M-S1, Å	2.151(4)	2.294(1)
M-S2	2.158(4)	2.308(1)
M-S6	2.159(4)	2.292(1)
M-S7	2.163(4)	2.302(1)
C1-C2	1.34(1)	1.375(5)
C4-C5	1.36(1)	1.401(6)
S1-M-S2, (°)	93.3(1)	90.41(5)
S1-M-S7	87.1(1)	87.84(5)
S2-M-S6	86.0(1)	91.44(5)
S6-M-S7	93.8(2)	90.07(5)

Fig. 1. Molecular structure and structural parameters of $(Et_4N)_1$ $Ni(DMIT)_2$ and $(Et_4N)_{0.5}$ $Pd(DMIT)_2$.

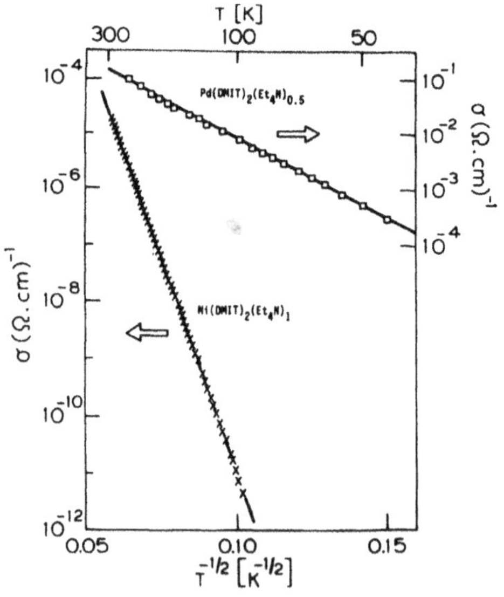

Fig. 2. Conductivity of $(Et_4N)_1[Ni(DMIT)_2]$ and $(Et_4N)_{0.5}[Pd(DMIT)_2]$ vs. $1/\sqrt{T}$.

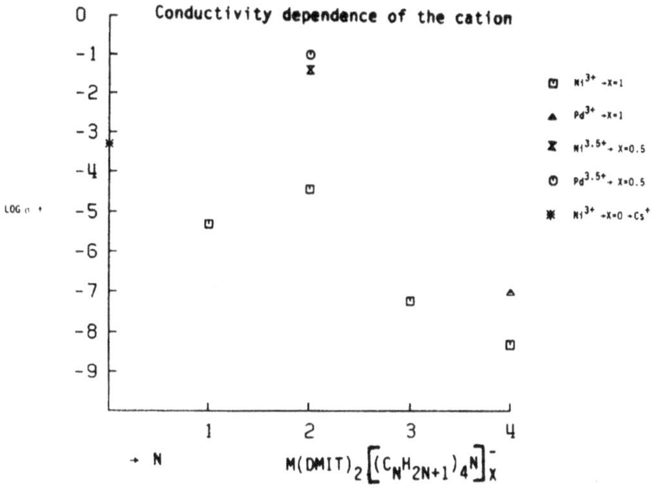

Fig. 3. Conductivity dependence on the cation of $(cat)_x[M(DMIT)_2]$. For $Ni^{3.5+}$ see ref. 4.

The magnetic measurements of the Ni and Pd compounds indicate 1-D behaviour and possibly a small trace of impurities. ^{1}H-NMR reveals that at low temperature molecular motion of the tetraethylammonium group is still present (Figure 4). This is of particular importance, since the anomalous conductivity behaviour is in general an indication of disorder. In the present compounds disorder may be caused by cationic motion. Because linewidth studies reveal that two cation conformations occur, static disorder might be expected at the lowest temperature.

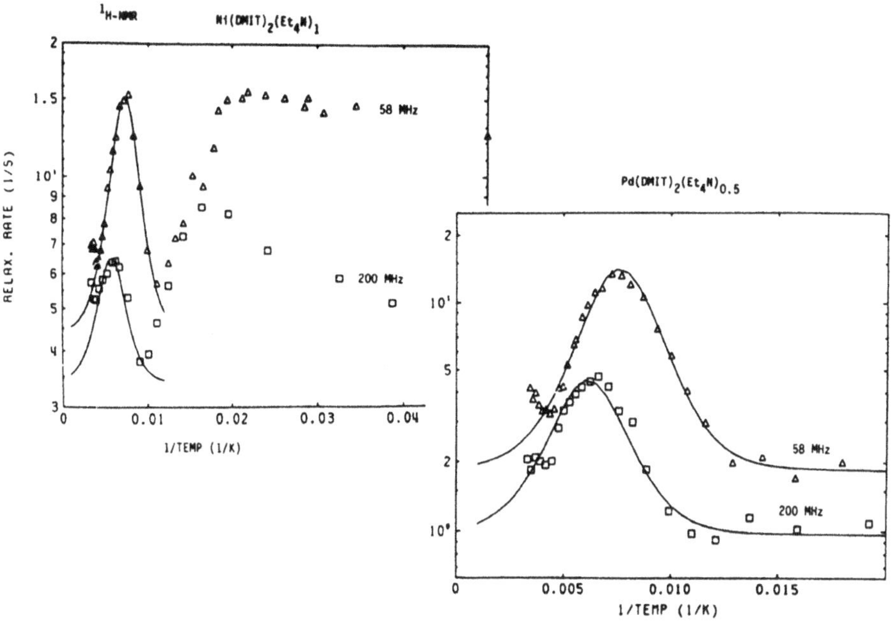

Fig. 4. Proton NMR relaxation rates vs. 1/T of $(Et_4N)_1[Ni(DMIT)_2]$ and $(Et_4N)_{0.5}$ $Pd(DMIT)_2$

ACKNOWLEDGEMENT

The autors acknowledge the support of the Leiden Materials Science Centre (Werkgroep Fundamenteel Materialenonderzoek).

REFERENCES

1. M. Bousseau, L. Valade, J-P. Legros, P. Cassoux, H. Garbauskas and L.V. Interrante, J. Am. Chem. Soc. 108: 1908 (1986)
2. L.R. Groeneveld, B. Schuller, G.J. Kramer, J.G. Haasnoot and J. Reedijk, Recl. Trav. Chim. Pays Bas 105: 507 (1986)
3. G.J. Kramer, L.R. Groeneveld, J.L. Joppe, H.B. Brom, L.J. de Jongh and J. Reedijk, Synthetic Metals 19: 745 (1987)
4. R. Kato, T. Mori, A. Kobayashi, Y. Sasaki and H. Kobayashi, Chem. Letters, 1 (1984)

SINGLE CRYSTAL ESR STUDIES OF n-TETRABUTYLAMMONIUM$^+$NICKEL(o-BENZENEDISELENOLATE)$_2^-$ AND RELATED TRANSITION METAL COMPLEXES

M. Thomas Jones, James H. Roble, Megh Singh,
and Toshio Maruo

University of Missouri-St. Louis
Department of Chemistry
St. Louis, MO 63121

INTRODUCTION

Previously reported temperature dependent studies of the ESR spectral envelopes of polycrystalline samples of n-tetrabutylammonium salts of nickel(o-benzenediselenolate(bds))$_2$ and nickel(o-benzenedithiolate(bdt)$_2$ reveal that the g-tensors for the bds and bdt salts are axial and pseudo-axial, respectively[1-5]. However, there was concern that the g-tensor for the bds might actually be pseudo-axial because of the poorer resolution as a result of the increased ESR spectral linewidth in the selenium salt due to the enhanced spin lattice rate caused by the heavier selenium atom. In addition, an abrupt change in $g_{||}$ was observed between 150 and 160 K for the Ni(bds)$_2^-$ salt, which, based on preliminary low temperature X-ray diffraction studies, has been confirmed as a crystalline phase change[6]. ESR studies of single crystals of the n-tetrabutylammonium salts of nickel(bds)$_2^-$ and nickel(bdt)$_2^-$ which have been synthesized in our laboratory and which are large enough to allow single crystal ESR studies to test the above points are reported. An additional reason for this study is that these compounds may serve as the building blocks for the synthesis of a new series of conducting salts based on small cations (e.g., alkali metals) in analogy with the recent work of Underhill et al[7] on the malonitrile complexes.

EXPERIMENTAL SECTION

The two n-butylammonium salts were prepared according to literature procedures[5,8]. Large, well formed needles, up to two to three centimeters in length and suitable for ESR studies, were prepared by slow solvent diffusion using a methylene chloride/ethanol solvent system. Typical crystal dimensions are 12 mm x 0.8 mm x 0.4 mm. Single crystals, which were to be rotated perpendicular to their long axes, were mounted with epoxy on the end of a small diameter pyrex tube and enclosed within a large diameter pyrex tube to protect the very fragile crystals. Single crystals, which were to be rotated parallel to their long axes, were mounted in narrow, thin wall tubing for protection against mechanical damage. ESR spectra were recorded at X-band with a Varian E-12 spectrometer equipped with a dual cavity using techniques that have been previously described in detail[9].

ESR STUDIES

Single crystal ESR spectra as a function of temperature and crystal orientation relative to the magnetic field direction have been studied for both $Ni(bds)_2^-$ and $Ni(bdt)_2^-$ salts. The results are summarized below.

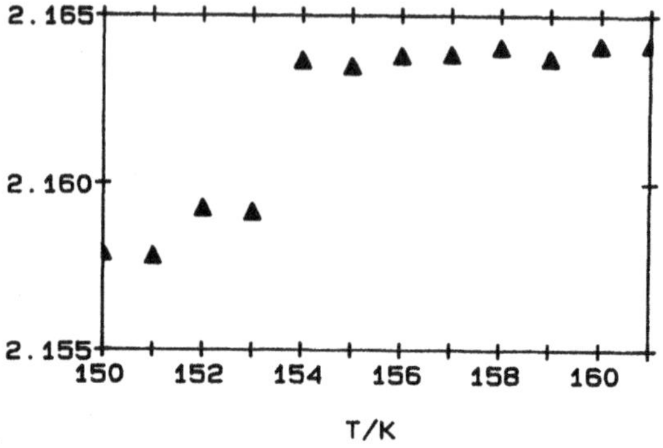

Figure 1

The Sharpness of the Phase Transition

Previously reported ESR studies of polycrystalline $Ni(bds)_2^-$ revealed a discontinuity in $g_{||}$ between 150 and 160 K. Preliminary low temperature single crystal x-ray measurements confirm that this discontinuity is the result of a crystalline phase change. Single crystal ESR studies on this salt allow the resolution of the two slightly different g-tensor components (g_1 and g_2) which make up the value for $g_{||}$ which is observed in the polycrystalline spectrum. The change in the values of both components is very sharp as shown in Figure 1 upon going through the phase transition. It takes place at 156 K over a temperature range of only 1°! The principal components of the g-tensor were found to be relatively independent of temperature above and below the transition temperature. There is also a change in the single crystal ESR spectral linewidth which, for example, changes from 23.8 G below the transition to 18.9 G above it over a 10° change in temperature when the magnetic field is oriented along the g_1 tensor component. No discontinuities in the principal components of the g-tensor and only a slight temperature dependence of the g-tensor components and spectral linewidth are observed for single crystals of the $Ni(bdt)_2^-$ salt over the temperature range of 95 to 300 K.

The Effect of a Temperature Gradient Across the Sample

An interesting phenomenon was observed while studying the single crystal ESR spectra of $Ni(bds)_2^-$ when the crystal is aligned such that the long axis is perpendicular to the magnetic field. The single crystals studied were rather long (i.e., 1.2 cm in length). It is well known by ESR spectroscopists that a temperature gradient of a degree or two exists along the length of an ESR sample in the standard experimental setup for variable temperature. Ordinarily, that does not create any serious problems. However, it can create an interesting phenomenon, if the material being studied undergoes a very sharp phase transition, the temperature range over which this transition occurs is very narrow, the spectral properties (e.g., the linewidth, g-tensor, etc.) of the material are different at the two temperatures, and the two ends of the crystal are at temperatures above and below the phase transition. Figure 2 shows ESR spectra taken at three different sample temperatures: A) The entire sample is above the temperature of the phase transition, B) Part of the sample is at a temperature above the phase transition and part is at a temperature below the phase transition, and C) The entire sample is at a temperature below the phase transition. With care, it is possible to arrange conditions consistent with B in the Figure 2. These phenomena are reproducible. However, the relative amounts of the two different species that are superimposed upon each other depend upon the relative temperatures of the crystal ends. We have checked carefully to assure that we are not looking at broken crystals which are misaligned and have repeated the experiment on the same crystal from below to above the transition temperature and back.

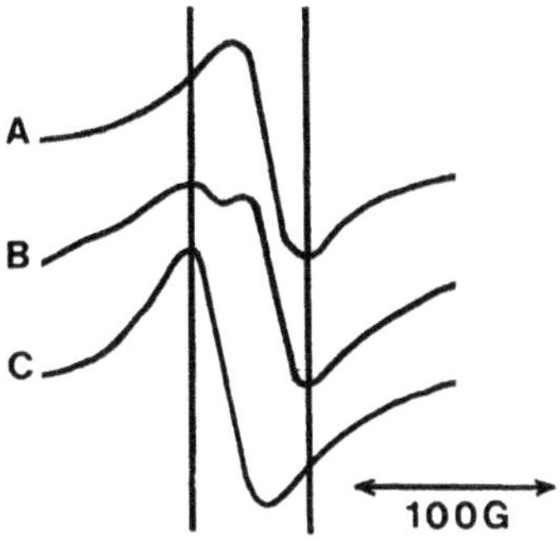

Figure 2

g-Tensor Anisotropy

The single crystal ESR measurements on both salts are in good agreement with the results obtained from the polycrystalline ESR spectral envelopes as previously reported[1-5]. The single crystal $Ni(bds)_2^-$ is observed to be psuedo-axial which brings it into agreement with both the single crystal and polycrystalline ESR data observed for $Ni(bdt)_2^-$. The results are summarized in Table 1. Based on the single crystal ESR measurements, we have learned that g_1 and g_2 essentially lie in the plane of the transition complex whereas g_3 is perpendicular to it. This is consistent with our preliminary low temperature single crystal x-ray measurements which suggest that the changes in the atomic coordinates occur in that plane also. Such changes would be expected to led to changes in g_1 and g_2 but not in g_3.

Table 1. Principal components of the g-tensors at T=300 K and T=100 K for $Ni(bds)_2^-$ and $Ni(bdt)_2^-$.

Ligand	bdt		bds	
T/K	100	300	100	300
g_1	2.125111	2.128725	2.155942	2.175880
g_2	2.106870	2.113524	2.155883	2.171676
g_3	2.028804	2.032326	2.079699	2.079521

REFERENCES

1. D. J. Sandman, J. C. Stark, L. A. Acampora, L. A. Samuelson, G. W. Allen, S. Jansen, M. T. Jones and B. M. Foxman, Mol. Cryst. Liq. Cryst. 107, 1(1984).

2. D. J. Sandman, J. C. Stark, G. Allen, L. S. Acampora, M. T. Jones, S. Jansen and G. J. Ashwell, Mol. Cryst. Liq. Cryst. 120, 405(1985).

3. M. T. Jones, S. Jansen, D. J. Sandman, B. M. Foxman and G. J. Ashwell, Mol. Cryst. Liq. Cryst. 125, 429(1985).

4. M. T. Jones and S. Jansen, Physica 143B, 515(1986)

5. D. J. Sandman, G. W. Allen, A. Acampora, J. C. Stark, S. Jansen, M. T. Jones, G. J. Ashwell, and B. M. Foxman, Inorg. Chem. 26, 1664(1987).

6. M. T. Jones, E. Schlemper, M. Singh, J. H. Roble and T. Maruo, manuscript in preparation.

7. P. I. Clemenson, A. E. Underhill, M. B. Hursthouse, R. L. Short, G. J. Ashwell, I. M. Sandy, M. T. Jones, C. Jacobsen, K. Carneiro, Synthetic Metals 19, 579(1987).

8. M. J. Baker-Hawkes, E. Billig, and H. B. Gray, J. Amer. Chem. Soc. 88, 4870(1966).

9. M. T. Jones, R. Ahmed, R. Kastrup, and V. Rapini, J. Phys. Chem. 83, 1327(1973).

ESR OF SOME $M[Pt(mnt)_2]$ and $M[Pd(mnt)_2]$ POLYCRYSTALLINE COMPLEXES

M. Thomas Jones and Joseph M. Patane[a]

University of Missouri-St. Louis
Department of Chemistry
St. Louis, MO 63121
USA

Peter I. Clemenson and Allan E. Underhill

University College of North Wales
Bangor, Gwynedd
Great Britain

INTRODUCTION

It has been shown that substitution of small cations(e.g., alkali metal ions) for large bulky cations such as the tetra-n-butylammonium ion in salts prepared from the square planar complex, $[X(mnt)_2]^{-n}$, where (X = Ni, Pt, Pd; mnt = 1,2-dicyano-1,2-ethylenedithiolato), leads to a series of highly conducting materials[1-6]. Briefly, large bulky cations favor the alternate stacking arrangement(i.e., cation/anion/cation/anion/...) in the crystal structure which is an arrangement which is not highly conductive. On-the-other-hand, in some instances, the small cations appear to support segregated stacking patterns in which the square planar transition metal complex anions are stacked pancakewise and which is an arrangement which is much more conducive to high electrical conductivity. We report here an ESR study of a series of the platinum and palladium complexes based on small counterions. A goal of these studies is to better understand those physical properties which lead to better electrical conductivity. A preliminary report of our work on the $Cs[Pd(mnt)_2].0.5H_2O$ is given reference 6.

EXPERIMENTAL SECTION

The salts were prepared as small black plate like crystals by electrocystallization of aqueous/acetone solutions as outline in references 1-6. ESR spectra of polycrystalline samples were recorded at X-band with a Varian E-12 spectrometer equipped with a dual cavity using techniques that have been previously described in detail[7]. Owing to limitations of space, discussion shall be limited mostly to room temperature ESR measurements. A more detailed report will be forthcomming.

---

[a]Present Address: U. S. Army Aviational Systems, St. Louis, MO, USA

$M[Pt(mnt)_2]$

ESR studies of polycrystalline salts of $M[Pt(mnt)_2]$, where M = Li^+, Mg^{++}, Na^+, H_3O^+, K^+, N_4H^+, and Et_4N^+, have been conducted as a function of temperature from 100 to 300 K. All of these salts display rather weak polycrystalline anisotropic g-tensor ESR spectra from which the principal components of the anisotropic g-tensors have been measured. The ESR signal intensities for all of the Pt salts are so weak that all the salts are either diamagnetic or very weakly paramagnetic. The ESR signal intensity for all samples follows Curie Law from 100 to 300 K. The principal components of the g-tensors and the overall spectral widths are summarized in Table 1.

Since the concentration of the paramagnetic species observed in the ESR studies of all of the Pt complexes is low, the magnetic ground state of these salts is a singlet and not doublet or triplet. Thus, it is concluded that there is a strong pairwise interaction of the unpaired electrons on adjacent Pt complex ions.

We believe the paramagnetic species observed in the ESR spectra of these salts are due to dislocations, fractures, odd numbers of paramagnetic species in a chain, etc. within the crystal structure. Thus, the structure of the paramagnetic species observed as deduced by the ESR studies is representative of the bulk sample even though we are not observing the ESR of the bulk of the individual complex ions.

In addition to the spectra described above, the K^+ salt exhibits an additional low field ESR spectrum at room temperature and the Na^+ salt exhibits two additional ESR spectral lines at low temperature. In the case of the Na^+ salt, one resonance line is at low field and the other is at high field. The species giving rise to these latter resonances are present at very minor concentrations relative to the predominate species.

Table 1. Summary of g-Tensor Data for $M[Pt(mnt)_2]$

Salt	g_1	g_2	g_3	$\langle g \rangle$	Temp./K	H/G
$Li^+_x Pt(mnt)_2^-$	2.114	2.055	1.860	2.010	293	418
$Mg^{++}_{0.5} Pt(mnt)_2^-$	2.111	2.051	1.877	2.013	293	387
$Na^+ Pt(mnt)_2^-$	2.111	2.059	1.874	2.015	293	391
$H_3O^+ Pt(mnt)_2^-$	2.227	2.057	1.800	2.028	297.5	705
$K^+ Pt(mnt)_2^-$	2.233	2.057	1.800	2.030	297.5	705
$NH_4^+ Pt(mnt)_2^-$	2.239	2.063	1.803	2.035	297.5	720
$Et_4N^+ Pt(mnt)_2^-$	2.223	2.064	1.831	2.039	225	650

Although we do not understand the physical significance of the observation, at present, we note that the data in Table 1 fall neatly into two categories based on the size of the Pauling ionic radius. Those salts whose ionic radii are equal to or less than that of sodium fall into one class and those whose radii are equal to or larger than that of H_3O^+ into a second class.

$M[Pd(mnt)_2]$

ESR studies of polycrystalline salts of $M[Pd(mnt)_2]$, where M = Li^+, Ca^{++}, K^+, Rb^+, Cs^+, N_4H^+, and Et_4N^+, have been conducted as a function of temperature from 100 to 300 K. The ESR spectra observed for these salts is varied. For example, at room temperature single line polycrystalline ESR spectra are observed for Li^+, Ca^{++}, K^+, and Cs^+ salts. Whereas, the other salts exhibit characteristic, polycrystalline, anisotropic g-tensor spectra. The room temperature g-tensor data obtained from these spectra are summarized in Table 2. Two of the samples(Li^+ and Cs^+) were lossy to microwaves upon being introduced into the ESR cavity at room temperature. This is a phenomenon which is observed for rather highly electrically conducting materials.

ESR spectral intensities vary rather strongly as a function of temperature for the Pd salts. The ESR spectrum of $Li[Pd(mnt)_2]$ is very weak and it was not possible to obtain reliable data as a function of temperature for the Li^+ salt. However, in the case of all of the other salts with the exception of the K^+ and the Cs^+ salts the ESR spectra were observed to follow Curie Law. The ESR signal intensity in the K^+ salt is activated. At low temperatures($T < 163$ K), the K^+ spectrum is a resolved anisotropic g-tensor spectrum. As the temperature is increased the

Table 2. Summary of g-Tensor Data for $M[Pd(mnt)_2]$

Salt	g_1	g_2	g_3	<g>	Temp./K	H/G
$Li^+{}_xPd(mnt)_2{}^-$	Single Weak Line				298	~50
$Ca^{++}{}_{0.5}Pd(mnt)_2]^{-a}$	-	2.044	-	2.044	293	178
$Rb^+Pd(mnt)_2{}^-$	2.067	2.040	1.955	2.021	293	185
$NH_4{}^+Pd(mnt)_2{}^-$	2.069	2.043	1.956	2.023	293	187
$Et_4N^+Pd(mnt)_2{}^-$	2.064	2.045	1.960	2.023	296	173
$Et_4N^+Pd(mnt)_2{}^-$	Solution measurement			2.024	296	---
$K^+Pd(mnt)_2{}^{-a}$	-	2.022	-	2.022	293	380
$Cs^+Pd(mnt)_2{}^{-a}$	-	2.028	-	2.028	298	636

[a]Single ESR line observed.

concentration of the paramagnetic species increases until the anisotropic spectrum collapses into a single spectra line. Whether this phenomenon is simply due to spin-spin exchange processes between spatially fixed electron spins as their concentration increases with temperature or due to spin-spin exchange as a result of spin-spin collisions owing to motional effects is being studied. Finally, the temperature dependence of the ESR spectral intensity of the Cs^+ salt is complicated but appears to be consistent with that expected from the Bonner-Fisher model[8].

REFERENCES

1. A. Kobayashi, Y. Sasaki, H. Kobayashi, A. E. Underhill, and M. M. Ahmad, J. Chem Soc., Chem. Commun. 1982, 390(1982).

2. A. Kobayashi, T. Mori, Y. Sasaki, H. Kobayashi, M. M. Ahmad and A. E. Underhill, Bull. Chem. Soc. (Jpn) 57, 3262(1984).

3. M. M. Ahmad, D. J. Turner, A. E. Underhill, A. Kobayashi, Y. Sasaki, H. Kobayashi, J. Chem. Soc., Dalton Trans. 1984, 1759 (1984).

4. A. Kobayashi, Y. Sasaki, H. Kobayashi, A. E. Underhill, and M. M. Ahmad, Chem. Letters 1984, 305(1984).

5. A. E. Underhill and M. M. Ahmad, J. Chem Soc., Chem. Comm. 1981, 67(1981).

6. P. I. Clemenson, A. E. Underhill, M. B. Hursthouse, R. L. Short, G. J. Ashwell, I. M. Sandy, M. T. Jones, C. Jacobsen, K. Carneiro, Synthetic Metals 19, 579(1987).

7. M. T. Jones, R. Ahmed, R. Kastrup, and V. Rapini, J. Phys. Chem. 83, 1327(1973).

8. J. C. Bonner and M. E. Fisher, Phys. Rev. 135, A640(1964).

A NEW FAMILY OF ONE-DIMENSIONAL CONDUCTORS DERIVED FROM MAGNUS GREEN-TYPE SALTS $[Pt(NH_3)_4][PtX_4]$ (X=Cl,Br) : CLASS II AND III MIXED-VALENCE CHAINS IN A DOUBLE CHAIN STRUCTURE

Jean-Pierre Catinat, Toussaint Robert and Georges Offergeld

General and Inorganic Chemistry
Faculty of Medicine, University of Mons
B - 7000 Mons

A family of coloured compounds having metallic lustre and crystallizing as needles has been synthetized by controlled oxidation of Magnus green salt $[Pt(NH_3)_4][PtCl_4]$ (MGS) and Magnus green-type salt $[Pt(NH_3)_4][PtBr_4]$ in aqueous $HClO_4$ or HBF_4. Experimental evidences show that all these compounds are single-phase materials having tetragonal structure and containing two types of mixed-valence platinum chains (of class IIIb and class II respectively[1]) with chain axis as the c axis : (a) Class IIIb, valency delocalized Pt···Pt chains where Pt is in a non-integral (presumably 2.33) oxidation state. These chains (α chains) are partially oxidized MGS-like chains with a Pt-Pt intrachain distance (= c/2) ranging from 2.85 to 2.92 Å depending on the kind of complex ; (b) Class II, with partially delocalized valencies, halogen-bridged Pt(II)···Y-Pt(IV) chains (Wolffram's type β chains with Y=Cl,Br,I). The Pt-Pt distance is given by the value of the c parameter (we assume statistical disorder for the Pt(II)-Pt(IV) sites in the planes perpendicular to the c axis) :

$$Pt(NH_3)_4 \cdots PtCl_{4-x}Br_x \cdots Pt(NH_3)_4 \cdots PtCl_{4-x}Br_x \quad (x=0,2,4) \quad (\alpha)$$

$$\cdots Y \text{———} Pt(NH_3)_4 \text{—} Y \cdots\cdots\cdots\cdots\cdots Pt(NH_3)_4 \cdots \quad (\beta)$$

$$\longleftarrow\ c\ \longrightarrow$$

The materials were characterized by several techniques : chemical analysis, oxidimetric titrations, X-ray powder diffraction, density measurements, electrical conductivity and spectroscopic measurements. All the results suggest a$[\alpha_3\beta]$structure. So we propose the following unit cell ((00h) projection) and the general formula giving the content of this cell:

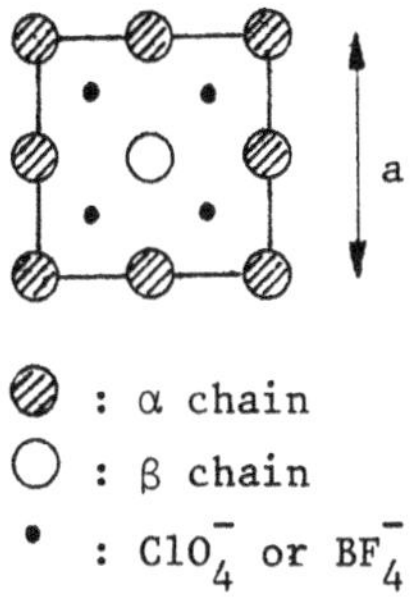

$$[\overset{2.33}{Pt}(NH_3)_4\overset{2.33}{Pt}Cl_{4-x}Br_x]_3[\overset{II/IV}{Pt}(NH_3)_4Y]A_4 \cdot zH_2O$$

($x = 0,1,2,4$; Y = Cl,Br,I ; $A^- = ClO_4^-$ or BF_4^- ; $0 \leqslant z \leqslant 4$ for samples dried under 10^{-3} torr at 30° C)

This formula will be abbreviated as $Cl_{4-x}Br_x$-Y-A. As formely pointed out in the case of the chloro-compound[2], density measurements indicate the presence of seven Pt atoms in the unit cell, in disagreement with a model of lattice containing exclusively chains with direct Pt-Pt contacts[3].

Preparation and elemental analysis

Eight different compounds have been prepared as follows :

$$\text{MGS} \xrightarrow[\text{in } HClO_4 \text{ 1M}]{PtCl_4,K_2[PtCl_6],[Pt(NH_3)_4Cl_2](NO_3)_2 \cdots H_2O_2 \text{ or } KMnO_4} Cl_4\text{-}Cl\text{-}ClO_4$$

$$\text{MGS} \xrightarrow[\text{in } HClO_4 \text{ 1M or } HBF_4 \text{ 3M}]{[Pt(NH_3)_4YOH](NO_3)_2} Cl_4\text{-}Y\text{-}\begin{cases} ClO_4 & (Y{=}Cl,Br,I) \\ BF_4 & (Y{=}Cl,Br) \end{cases}$$

$$\text{MGS} \xrightarrow[\text{in } HClO_4 \text{ 1M}]{[Pt(NH_3)_4Br_2](NO_3)_2} \text{"}Cl_3Br\text{-}Br\text{-}ClO_4\text{"} \; [(Cl_4)_{0.5}(Cl_2Br_2)_{0.5}\text{-}Br\text{-}ClO_4 ?]$$

$$\text{MGS} \xrightarrow{K_2[PtBr_6] \text{ in } HClO_4 \text{ 3.5M}} Cl_2Br_2\text{-}Br\text{-}ClO_4$$

$$[Pt(NH_3)_4][PtBr_4] \xrightarrow[\text{in } HClO_4 \text{ 3.5M}]{KMnO_4 \text{ or } [Pt(NH_3)_4BrOH](NO_3)_2} Br_4\text{-}Br\text{-}ClO_4$$

Table 1. Experimental and calculated formula units (H values by difference ; calculated O values omitted in view of the variable water content)

Compound	Elemental analysis	Proposed structure (calculated)
Cl_4-Cl-BF_4	Pt $N_{2.26}Cl_{1.85}B_{0.54}F_{2.28}$ (H_2)	Pt $N_{2.29}Cl_{1.86}B_{0.57}F_{2.28}$
Cl_4-Br-BF_4	Pt $N_{2.29}Cl_{1.69}Br_{0.14}B_{0.55}F_{2.37}$ (H_3)	Pt $N_{2.29}Cl_{1.71}Br_{0.14}B_{0.57}F_{2.28}$
Cl_4-Cl-ClO_4	Pt $N_{2.33}Cl_{2.44}O_{2.84}$ (H_8)	Pt $N_{2.29}Cl_{2.43}$
Cl_4-Br-ClO_4	Pt $N_{2.31}Cl_{2.30}Br_{0.19}O_{2.66}$ (H_9)	Pt $N_{2.29}Cl_{2.29}Br_{0.14}$
Cl_4-I-ClO_4	Pt $N_{2.25}Cl_{2.28}I_{0.16}O_{2.74}$ (H_8)	Pt $N_{2.29}Cl_{2.29}I_{0.14}$
Cl_3Br-Br-ClO_4	Pt $N_{2.23}Cl_{1.88}Br_{0.61}O_{2.24}$ (H_6)	Pt $N_{2.29}Cl_{1.86}Br_{0.57}$
Cl_2Br_2-Br-ClO_4	Pt $N_{2.26}Cl_{1.47}Br_{1.00}O_{2.45}$ (H_6)	Pt $N_{2.29}Cl_{1.43}Br_{1.00}$
Br_4-Br-ClO_4	Pt $N_{2.28}Cl_{0.57}Br_{1.84}O_{2.67}$ (H_4)	Pt $N_{2.29}Cl_{0.57}Br_{1.86}$

Experimental evidences for the presence of α(Pt···Pt) chains

1) The electrical conductivity in the starting MGS increases through partial oxidation by at least a factor of 10^4. Measurements on pressed pellets of Cl_4-Cl-ClO_4 give a σ value of about $10^{-2}(\Omega\ cm)^{-1}$ at room temperature and indicate semiconducting behaviour between 273 and 70 K with activation parameter $E_a \simeq 0.02$ eV[2]. However preliminary measurements on Cl_2Br_2-Br-ClO_4 needles show that σ values higher than $1(\Omega\ cm)^{-1}$ are to be expected for monocristals ;

2) The XPS (ESCA) spectrum[2] indicates a shift of + 0.9 eV of the Pt $4f_{7/2}$ and $4f_{5/2}$ lines of Cl_4-Cl-ClO_4 and Cl_4-Br-ClO_4 with respect to MGS (mean Pt(II) → Pt(IV) shift ≃ + 2.3 eV) ;

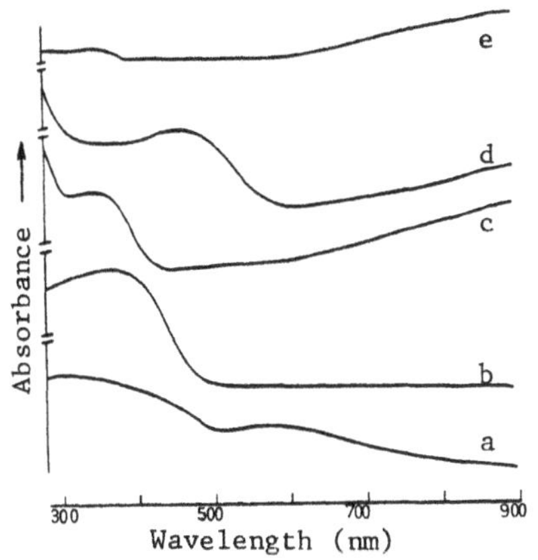

Fig. 1. UV-visible absorption spectra (powders) : a) MGS (for comparison) ; b) $[Pt^{II}(NH_3)_4]_{0.65}[Pt^{IV}(NH_3)_4Cl_2]_{0.35}(ClO_4)_2$ (for comparison) ; c) Cl_4-Cl-ClO_4 ; d) Cl_4-Br-ClO_4 ; e) $K_2[Pt(CN)_4]Br_{0.3}\cdot 3H_2O$(KCP) (for comparison).

3) The UV-visible absorption spectra of powder samples of our compounds show a regular increase of the absorbance from 600 nm up to the instrumental limit of 900 nm (Fig 1-c,d). This effect (actually a decrease of the transmission) appears also in the spectrum of a $K_2[Pt(CN)_4]Br_{0.3}\cdot 3H_2O$(KCP) powder sample (Fig 1-e). In this case it is related to the plasma edge at 625 nm observed in the reflectance spectrum of a monocrystalline sample (light polarized ∥ to the chain axis), which is considered as a test for metallic character[4].

Experimental evidences for the presence of β(Pt(II)-Pt(IV)) chains

1) The UV-visible absorption spectra of our complexes display below 600 nm a broad and intense band exhibiting a red shift in the Y sequence Cl → Br (Fig 1-c,d). We attribute this band to the Pt(II) → Pt(IV) intervalence electronic transition (mixed-valence band, cfr Fig 1-b) ;

2) Resonance enhancement is found in the vibrational Raman spectra for the band corresponding to the ν_1(Y-Pt(IV)-Y) fundamental symmetric stretching mode and for its overtones when these compounds are irradiated in the

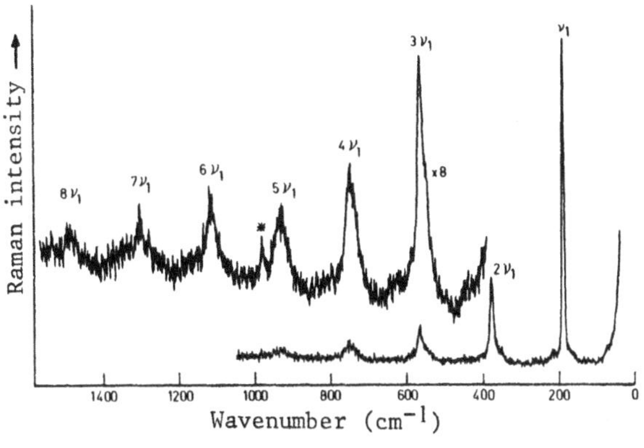

Fig. 2. Resonance Raman spectrum of Br_4-Br-ClO_4 as K_2SO_4 disk (λ_o = 457.9 nm, spectral slit width ≃ 2.2 cm^{-1}, the asterisk indicates the 983 cm^{-1} band of the $SO_4^=$ ion).

Table II : Cristallographic parameters (a,c) and ν_1 band in the Raman spectrum (estimated at $\pm$ 0.4 cm^{-1} ; λ_o = 457.9 nm)

Compounds	a(Å)	c(Å)	ν_1(Y-Pt(IV)-Y)(cm^{-1})
(1) Cl_4-Cl-BF_4	14.92	5.69	326.2
(2) Cl_4-Br-BF_4	14.95	5.70	186.9
(3) Cl_4-Cl-ClO_4	15.08	5.71	324.9
(4) Cl_4-Br-ClO_4	15.09	5.70	184.7
(5) "Cl_3Br-Br-ClO_4"	15.11	5.72	186.4
(6) Cl_2Br_2-Br-ClO_4	15.22	5.78	187.3
(7) Br_4-Br-ClO_4	15.29	5.83	189.5
(8) Cl_4-I-ClO_4	15.12	5.70	134.7

range 457.9 nm to 600 nm (Fig. 2). Irradiation within the contour of the mixed-valence band[5] leads to characteristic excitation profiles ;

3) Concerning these two effects there is a remarkable similitude with the optical and Raman spectra of the corresponding salts containing only β chains (Wolffram's type salts), which we have also prepared. In view of this similitude, we have checked that Raman resonance is an intrinsic property of our conductive compounds (not an impurity effect).

Some trends in structural and spectroscopic properties

1) The increase of the lattice parameters, and in particular of the c parameter, results primarily from the size of the ligands (Cl,Br) in the anionic units of the α chains (see compounds (4) to (7) in table II) ;

2) On the other hand the value of the c parameter depends very little on the nature of the bridging halogen Y along the β chains (see compounds (3), (4) and (8) in table II) ;

3) In accordance with the evolution of the c parameter, we observe an increase of the frequency of the ν_1(Br-Pt(IV)-Br) stretching mode (compounds (4) to (7) in table II), indicating that the ratio of the Pt-Br distances d(Pt(IV)-Br)/d(Pt(II)-Br) decreases in the series. This is also consistent with the blue shift of the mixed-valence Pt(II) → Pt(IV) band (475 nm → 445 nm in the series Cl_4-Br-ClO_4 → Br_4-Br-ClO_4) and with a concordant effect for the Raman resonance excitation profile.

4) As expected we observe a red shift of the Raman resonance excitation profile related to the decrease of the mixed-valence transition energy for the Y series Cl → Br → I ;

5) As in the starting MGS salts a factor limiting the Pt···Pt interaction is obviously the size of the halogens in the α chains. A circumstance in favour of a mixed-chain structure lies in the fact that the Pt(II)-Pt(IV) spacing in the salts containing only β chains is comparable to the c parameter in this family of compounds. In fact the looseness of the Pt(II)···Y bond allows the β chain some flexibility in adjusting to the spatial constraint imposed by the α chains.

References

1. M.B. Robin and P. Day, Adv. Inorg. Chem. and Radiochem. 10:247 (1967).
2. J.-P. Catinat, T. Robert and G. Offergeld, J. Chem. Soc., Chem. Commun. 1310 (1983).
3. M. Tanaka and I. Tsujikawa, Bull. Chem. Soc. Jpn 59:2773 (1986).
4. J.S. Miller and A.J. Epstein, Prog. Inorg. Chem. 20:1 (1976).
5. R.J.H. Clark, M. Kurmoo, A.M.R. Galas and M.B. Hursthouse, J. Chem. Soc., Dalton Trans. 2505 (1982).

CONDUCTIVE MATERIALS BASED ON RHODIUM AND IRIDIUM COMPLEXES OF CARBOXYAZOLES

J.C. Bayón and G. Net

Departament de Química, Unitat Química Inorgánica
Universitat Autònoma de Barceloma
Bellaterra, 08193 Barcelona, Spain

INTRODUCTION

Square planar d^8 metal complexes, such as Krogmann salt, have been widely investigated because they offer the opportunity of studying the influence of structural modifications on the conductivity of these materials. However there is a limited number of compounds in which the metallic properties are due to a spine of interacting metal atoms with short intrachain metal-metal distances (<3 Å). Reviews on this field[1] cite compounds based on tetracyanoplatinates, platinum oxalates and some iridium carbonyl species. Considerable efforts have been made in order to synthesize new conductive systems of this type, because their structural simplicity makes them specially appropiate to study structure-properties correlations. Some of the attempts have been based on the design of new sophisticated ligands which could match the properties of those used in known systems[2]. Another possibility is the use of new combinations of metals and ligands which form part of well known conducting compounds[3].

Since the observation of superconductivity in several seleno- and thiofulvalenes most of the research in the field of the molecular conductors has dealt with molecules containing a large number of polarizable atoms (i.e. sulfur or selenium)at their periphery. This strategie produces short interchain contacts which prevent in some cases the metal-insulator transition observed in one dimensional systems.

Rh(I) AND Ir(I) COMPLEXES

We are engaged in the synthesis of a new type of d^8 metal complexes which combine both the simplicity of one-dimensional chains of atoms and the possibility of interstack electronic interactions. These requirements can be satisfied by dinuclear, planar complexes which have strong interactions within the stacking unit. Electrochemical partial oxidation has been extensively used to provide the system with the charge carriers required to achive the metallic state. For that reason many synthetic precusors of highly conducting metal complexes are anionic species. The uncommon conditions of dinuclearity, planarity and anionic character are met by some Rh(I) and Ir(I) complexes of the ligands 4,5-dicarboxyimidazole (H_3Dcbi), 4,5-dicarboxytriazole (H_3Dcbt) and 3,5-dicarboxypyrazole (H_3Dcbp). We report here the preliminary results on the synthesis of these compounds and their electrochemical oxidation.

Reaction of $[Ir_2(COD)_2Cl_2]$ with the triprotic ligand and three mols of NR_4OH affords $NR_4[Ir_2(COD)_2L]$, (L= Dcbi, Dcbt, Dcbp) in good yields. X-ray structure of NBu_4 $Rh_2(COD)_2$Dcbi corroborates the coplanarity of the ligand and the two square--planar coordination spheres of the metal atoms[4]. By bubbling CO through solutions of these complexes the salts of the anions I, II and III are formed as fine dichroic needles.

The planarity of the dinuclear carbonyl anions favours the presence of extended intermolecular metal-metal interactions in the solid state. The colour of the salts of these anions strongly depends on the cation. As a general trend, smaller cations produce absorptions at lower frequencies in the visible spectrum. Pressed pellet conductivity for the NMe_4^+ salts of the iridium complexes are about 10^{-7} S/cm.

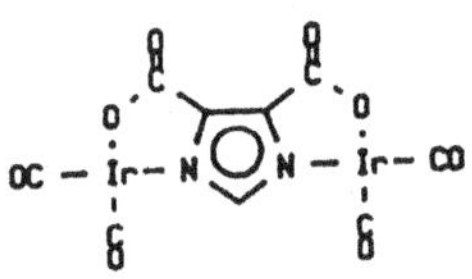

I

$[Ir_2(CO)_4Dcbi]^-$

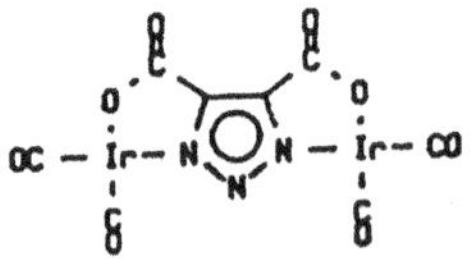

II

$[Ir_2(CO)_4Dcbt]^-$

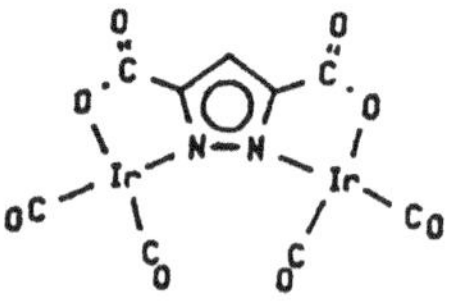

III

$[Ir_2(CO)_4Dcbp]^-$

Electrochemical oxidations were carried out in a modified H-cell under inert atmosphere using acetonitrile as solvent. Because of their higher solubilities NBu_4^+ salts of the complexes were used.

The oxidation of the anion $[Rh_2(CO)_4Dcbp]^-$ produces a dark purple insoluble solid on the anode which analyzes as $[Rh_2(CO)_4Dcbp]$. The IR spectrum reveals the lack of the characteristic absorptions of the cation and a shift to higher energies of the carbon monoxide stretches (2109 and 2017 cm^{-1} vs. 2087 and 2005 $cm^{-1}$of the unoxidazed compound). The compound can be also obtained by chemical oxidation (eg. $(C_7H_7)BF_4$, $(Ph\text{-}N_2)BF_4$). This formally Rh(I)-Rh(II) mixed valence complex behaves as an insulator as revealed by the pressed pellet conductivity ($< 10^{-8}$ S/cm).

Electrochemical oxidation of the iridium salts NBu_4 $Ir_2(CO)_4L$, yields black insoluble materials on the surface of the anode. These solids show a band edge type absorption in the IR spectrum and preliminary room temperature conductivity measurements (pressed pellets) are in the range 10^{-2} - 10^{-3} S/cm. Analytical data for C, H and N indicate the composition: $(NBu_4)_x[Ir_2(CO)_4L]$ (x = 0.4-0.5), the value of x depending on the carboxyazolic ligand, but being consistent for different synthesis within the experimental error. Unfortunately elemental analysis results are not very sensitive to the DPO since the same elements are present both in the anion and the cation. The previously reported DPO values for monomeric iridium carbonyl halides[5] are in the same range than the ones found in these dinuclear compounds, which would suggest that only one metal atom per anion is involved in the stack. However, a lower DPO value (0.33) has been found in an iridium carbonyl complex with a N heterocyclic ligand[2] and therefore further spectroscopic and crystallografic evidence is required in this line.

Finally, the electrochemical oxidation of the salts in the presence of TTF does not cause incorporation of the radical cation, but the same partially oxidized material with NBu_4^+ is formed, indicating lower oxidation potential for the dinuclear anion than for TTF. Consistenly, the complex anion is oxidized by $(TTF)_3(BF_4)_2$ to form the $(NBu_4)_x[Ir_2(CO)_4L]$

species. Conversely, the rhodium compound (TTF)$[Rh_2(CO)_4L]$ can be prepared either by electrochemical oxidation or by metathesis with $(TTF)_3(BF_4)_2$. We are currently exploring the use of TTF derivatives in order to match the redox potential of the iridium complex anion.

ACKNOWLEDGEMENTS

This work was supported by C.A.I.C.Y.T. (PB85-0008-C02-02) and by the C.I.R.I.T. G.Net thanks a scholarship from the M.E.C.

REFERENCES

1. "Extended Linear Chain Compounds" ed. J.S. Miller, Plenum, New York, 1982, vol.I.

2. P.G. Rasmussen, O.H. Bailey, J.C. Bayón; Inorg. Chem. (1984), 23, 338.

   P.G. Rasmussen, J.E. Anderson, O.H. Bailey, J.B. Kolowich, J.C. Bayón; Mol. Cryst. Liq. Cryst. (1985), 126, 87.

3. For instance, we are exploring the chemistry of the anion $Ir(CO)_2(ox)^-$.

4. J.C. Bayón, G. Net, P.G. Rasmussen, J.B. Kolowich; J. Chem. Soc. (Dalton) (in press).

5. A.P. Gingsberg, J.W. Koepke, J.J. Hauser, K.W. West, F.J. Di Salvo, C.R. Sprinkle, R.L. Cohen; Inorg. Chem. (1976), 15, 514.

ONE-DIMENSIONAL ELECTRONIC SYSTEMS: METALLIC AND SYNTHETIC WIRES

H.B. Brom and G.J. Kramer

Kamerlingh Onnes Laboratory
Nieuwsteeg 18, 2311 SB Leiden
The Netherlands

INTRODUCTION

Although a one-dimensional world (1D) seems confined enough there are two research fields, in which one-dimensional conducting systems are studied while the meaning of one-dimensionality is not the same.

In one-dimensional electronic systems (wires) surprising theoretical and experimental results[1,2,3] are recently obtained on the transport properties. For the description of the deviation from normal metallic behaviour, i.e. quantum interference effects and universal conductance fluctuations, a meaningful division can be made between disordered electronic systems on one hand and "metallic" samples on the other[1]. The classification bears on two main aspects in this field: Anderson localization - which deals with the nature of the wave function of a electron in the presence of a random potential - and the effects of interactions between electrons - the fact that electrons are diffusive instead of propagating leads to a profound modification of the traditional view based on Fermi liquid theory of metals[2].

The limit, within which quantum interference effects, like the Aharanov-Bohm effect, can be expected to be visible, is set by the phase breaking length $L_\phi = (D\tau_\phi)^{\frac{1}{2}}$, in which D is the diffusion constant and τ_ϕ the time between phase breaking collisions. Because elastic collisions change deterministically the phase (but relaxes the momentum) L_ϕ can be made much longer than the mean free path (L). Only if inelastic impurity or magnetic scattering becomes dominant the coherence length is destroyed.

Conducting stacks of organic molecules or polymers (one-dimensional synthetic metals) form the other 1 D-research field. Here it is well-known that the one-dimensionality is the origin of a conductor to semi-conductor transition (the Peierls transition).

In this contribution we indicate what the requirements are for making "wire"-experiments possible in 1 D-synthetic metals. We will show that the Peierls instability in one-dimensional conducting chains of stacked organic molecules or polymers can be surpressed, by increasing the number of bonds crossing the Fermi surface. The so obtained metallic chains are suited for a study of quantum coherence and localization effects, seen in metallic wires. The possibility of the opening of a Peierls (pseudo)-gap in submicron structures is also addressed: the opening of the gap is expected to occur if the number of bands at the Fermi level reduces to unity.

ONE-DIMENSIONAL ELECTRONIC SYSTEMS (SUBMICRON WIRES)

In most one dimensional (1D) electronic systems the sample size exceeds the localization length, which is the limit of disordered systems. At low temperatures there will be deviations from the extrapolated T=0 resistance value (R_0) due to 1D-localization.

An understanding of what happens is based on the evaluation of the 1 D-Landauer formula for the dimensionless conductance g (conductance in units of e^2/h) of a quantum resistor $g = 2|t|^2/(1 - |t|^2)$, in which $|t|^2$ is the transmission coefficient. The relative variance of the resistance actually diverges exponentially as the length of the system goes to infinity. A simple reason for the presence of these large fluctuations in the conductance is the possibility of resonant tunnelling. If the incoming electron is closely matched to the energy of the localized state, the transmission probability, t, will approach unity, while in the other circumstances t will be very small.

Following Lee, Stone and Fukuyama (LSF)[1] the regime, where the sample dimensions are much greater than the electron mean free path but less than the localization length is called metallic ($g > 1$). To characterize both cases in another equivalent way: in the metallic regime the resonance width is larger than the level spacing while in the localized regime the resonances are narrow and well separated. Using a generalization of the Landauer formula LSF show that in systems of "macroscopic" size quantum tunnel fluctuations are easily observable at low temperatures: the fluctuation effects of any given sample are typically as large or larger than the average effects. These fluctuations are not noise in the usual sense (i.e. random time dependent changes in the resistance) but are time-independent, stochastic (magneto)resistance patterns, which vary between samples, but are reproducible within a given sample. Apart from L_ϕ another important length scale for the description of these fluctuations at finite temperatures appears to be the thermal diffusion length $L_T = (hD/kT)^{\frac{1}{2}}$. LSF have tabulated their finite temperature results for the conductance fluctuations for $L_T < L_\phi < L_Z$ (= the length of the sample) and $L_T < L_Z < L_\phi$. Experimental verification is mostly done in MOSFETS, where one has the advantage of a variable band filling

CHAIN AND PLANAR SYNTHETIC COMPOUNDS

Most uniform chain compounds with a partially filled band are known to be unstable for a Peierls distortion: the electronic system can lower its energy at the expense of an increase in energy of the 3D lattice. The fact that not all chain compounds end up as semiconductors or insulators is mostly due to the non-negligible 3D or 2D coupling which makes the Peierls mechanism less efficient (the requirement of an appreciable nesting in k-space at the Fermi surface can no longer be fulfilled).

Because in real one-dimensional systems no phase transition can occur at all (at any finite temperature the entropy gain in the distribution of the ordered segments over the chain will destroy such a coherent state) it is not surprising that a mean field calculation of the Peierls temperature, T_{PM}, is far above the experimental value, T_{PE}. Above T_{PM} the amplitude of the order parameter (or energy gap) Δ, fluctuates around $|\Delta| = 0$. Below T_{PM} the phase of the order parameter rotates with $|\Delta|$ = constant. The absence of long range order smears the gap in the electronic system, but, due to the well defined short range order, the density of states at the Fermi-level is strongly reduced: the so-called "pseudogap"[4,5]. The treatment is normally restricted to one energy band systems: the width of the sample, W, which determines the quantization in k-space in the perpendicular direction has to be so small that no degeneracy is left within the conduction band, associated with k-parallel.

SUBMICRON WIRES COMPARED TO SYNTHETIC CHAINS

Two questions merge from the above: under what conditions submicron structures will suffer a Peierls distortion and is it possible to stabilize synthetic chains and layers to study the effects predicted for metallic or strongly localized electronic systems. We will focus on chain structures and submicron wires. Critical for the Peierls dimerization to occur is the requirement that the electronic free energy gain outweighs the elastic energy loss of the lattice. To make the entropy term negligible, let us consider the energy gain at $T = 0$.

In a system of one-dimensional chains the energy gap will be proportional to $\exp(-1/N(0)\ V)$, in which the electron phonon interaction term V is inversely proportional to the force constant C of the chain and N(0) denotes the density of states at the Fermi level[6].

If more chains are combined into one instead of keeping them separated, in practice by using larger molecules in the stack, the force constant will grow linearly with the number of chains combined into one (n). So using n elementary units to built the larger molecule and for not too high a symmetry, the resulting band structure will have approximately n different bands crossing E_F at $k_1 \ldots k_n$. The average energy gap will now be proportional to $\exp(-n/\overline{N(0)}V)$ in which $\overline{N(0)}$ is an average density for the various bands crossing the Fermi level. Obviously the energy gain by a Peierls distortion will be diminished by this configuration. Phtalocyanine (Inabe et al.] might be a possible example of the stabilizing influence of the combination of "more chains" into one: the system remains metallic at least down to 1 K. The slight increase in resistivity at the lowest temperatures might be even a manifestation of one-dimensional localization. A similar increase in resistivity is found in TSeT-iodine[8], possibly due to the same mechanism. According to very recently published band structure calculations[9] TTF-$[Ni(dmit)_2]_2$, which remains metallic at least down to 3 K (and becomes superconductive under pressure of 7 kbar at 1.62 K) seems to be another example of a stabilized one-dimensional chain compound. The explanation given by Kobayashi et al.[9] for the absence of the Peierls instability in this compound is also based on the "multi-Fermi surface", but the authors allow only a single modulation wave, instead of the here considered multi-modulation.

The inverse argument for submicron wires leads to the expectation that if one approaches the situation in which only one electron band crosses the Fermi level, the Peierls pseudo gap must become visible. Suited systems to look for these effects must have a low band filling to benefit from a small k_F-value ($\lambda_F/a \gg 1$) e.g. Bi-metal[10] or GaAs-AlGaAs-inversion layers.

ACKNOWLEDGEMENTS

This research is supported by the werkgroep voor Fundamenteel Materialen Onderzoek of the University of Leiden. It is a pleasure to acknowledge Prof. L.J. de Jongh for fruitful discussions.

REFERENCES

1. P.A. Lee, A. Douglas Stone and H. Fukuyama, Phys. Rev. B35.1039 (1987).
2. P.A. Lee and T.V. Ramakrishnan, Rev. Mod. Phys. 57:287 (1985).
3. S. Washburn and R.A. Webb, Adv. Phys. 35:375 (1986).
4. D. Jerome and H.J. Schulz, Adv. Phys. 31:299 (1982).
5. T.D. Schultz in The Physics and Chemistry of Low Dimensional Solids ed. Luis Alcacer, D. Reidel Publ. Co., Dordrecht (1980), 1-30.

6. C. Kittel, Introduction to solid state physics, John Wiley & Sons Inc., New York (1976), 315.
7. T. Inabe, T.J. Marks, R.L. Burton, J.W. Lyding, W.J. McCarthy, C.R. Kannewurf, G.M. Reisner and F.H. Herbstein, Solid State Comm. 54:501 (1985).
8. P. Delhaes C. Coulon, S. Flandrois, B. Hilti, C.W. Mayer, G. Rihs and J. Rivory, J. Chem. Phys. 73:1452 (1980).
9. A. Kobayashi, H. Kim, Y. Sasaki, R. Kato and H. Kobayashi, Solid State Comm. 62:57 (1987).
10. P.H. Woerlee, G.C. Verkade and A.G.M. Jansen, J. Phys. C16:3011 (1983).

SUPERCONDUCTIVITY IN RAPIDLY COOLED $Gd_3Ba_3Cu_4O_z$

D. A. Robinson, A. Morrobel-Sosa, C. Alexander, Jr.[#], J. S. Thrasher, C. Asavaroengchai, and R. M. Metzger*

Departments of Chemistry and [#]Physics
University of Alabama
Tuscaloosa AL 35487-9671 USA

INTRODUCTION

Recently Bednorz and Mueller [1] initiated an almost explosive growth in the field of superconductivity by detecting indications that in the quaternary mixed-valence system $La_xBa_{2-x}CuO_{4-y}$ superconductivity with critical temperatures T_c in excess of 30 K could be achieved. Since the previous record-holder was Nb_3Ge at 23.3 K, the report generated immense interest, and within a few months it was found in laboratories around the world that there were several compounds with critical temperatures in the 90-100 K range, i.e. well above the boiling temperature of liquid nitrogen [2-7].

The superconductivity is obtained in compounds $R_xA_yCuO_z$ in which a certain amount (x) of trivalent rare earth ion (R) (or its analog) coexists with a divalent alkaline earth ion (A) and also a certain fraction of the oxide sites in the lattice are vacant; all this makes possible a mixed-valence Cu(II)-Cu(III) state, in which the copper (a Jahn-Teller ion) can be either tetrahedrally coordinated or octahedrally coordinated in a two-dimensional array. The crystal structures of single crystals of some of these compounds have been determined [8]: they are either of the layered perovskite type (K_2NiF_4 structure, space group I4/mmm) or an orthorhombic distortion of this phase. Typically, **c** = 13.2 Å is the longest axis, with Cu and O ions in the (001) plane. In the tetragonal phase **a** = **b** = 3.787 Å, whereas the orthorhombic phase has a doubling of the tetragonal **a** x **b**, with **a** = 5.378 Å, **b** = 5.371 Å [8].

The Meissner effect has been measured, and there seems to be no isotope effect. There is no theoretical consensus about the exact physical model that is applicable to these new compounds. However, the superconductivity seems not to be simply of the BCS (Bardeen - Cooper - Schrieffer) type (where phonons mediate the Cooper pairing of electron states), but may be excitonic or polaronic.

RESULTS AND DISCUSSION

Powdered samples of Gd_2O_3, $BaCO_3$, and CuO in the mole ratio 3:3:4 were ground together in an agate mortar and pestle, then heated at 950°C in a Lindberg tube furnace in a ceramic boat in air for 15 hours; the samples were quickly cooled by dropping the boat into liquid nitrogen. The powder was then ground again, formed into a pellet at 1.2 kbar in a Carver press, heated at 1000°C in the furnace for 26 hours in air, and then cooled again in liquid nitrogen. The resulting pellet was mostly black on the outside but green on the inside. A small specimen was studied by X-ray diffraction (Phillips Model 3100 generator, with Cu K_α radiation); the powder pattern data are given in Table 1.

Table 1. X-ray powder diffractogram for $Gd_3Ba_3Cu_4O_z$ (nominal composition).

d/Å	I/I_0	d/Å	I/I_0	d/Å	I/I_0	d/Å	I/I_0
11.883	5	2.719	13	2.001	37	1.586	38
4.464	4	2.692	9	1.957	17	1.571	18
3.940	9	2.524	8	1.939	33	1.549	6
3.878	9	2.435	12	1.883	14	1.515	20
3.438	6	2.350	19	1.854	3	1.505	10
3.242	4	2.316	8	1.791	13	1.493	17
3.011	80	2.247	17	1.774	7	1.457	1
2.946	56	2.227	17	1.754	15	1.436	3
2.849	30	2.192	10	1.734	4	1.423	8
2.813	20	2.167	6	1.673	20	1.381	10
2.761	100	2.144	3	1.662	17	1.370	9
2.741	56	2.042	9	1.598	15		

Since there are both a green (normal) phase and a black (superconducting) phase, the data have not yet been analyzed.

Four-probe DC resistance measurements were made by reading the voltage across the sample (a pellet with four strips of conductive silver paint, Dupont No. 7713) baked at 540°C) using a Keithley Model 140 nanovolt amplifier, while maintaining a sample current of 2 mA. The sample was attached to the cold finger of a CTI-Cryogenics Model 22C closed-cycle refrigerator using GE 7031 varnish. Temperature measurements were made with a Scientific Instruments digital temperature indicator and a chromel vs. Au + 0.07% Fe thermocouple attached to the sample with the varnish. Data were collected with a Hewlett-Packard Model 9816 computer and Model 3497A data system.

The results are shown in Figs. 1 and 2. The onset temperature for superconductivity was 71 K; the midpoint temperature was 67.5 K; the temperature where minimum resistance was reached was 62 K. It should be noted that a residual current is observed, i.e. the fraction of the sample that superconducts is small, and resistive pathways persist. It should also be noted that Fisk et al. obtained superconducting onset temperatures of 93 ± 1 K for a sample of the same nominal composition [7]. However, Fisk et al. did not rapidly cool their samples, as was done in the present study.

No electron paramagnetic resonance spectrum was observed, either at 9.5 GHz or at 25 GHz.

Of course, other work is in progress and will be reported on in due course.

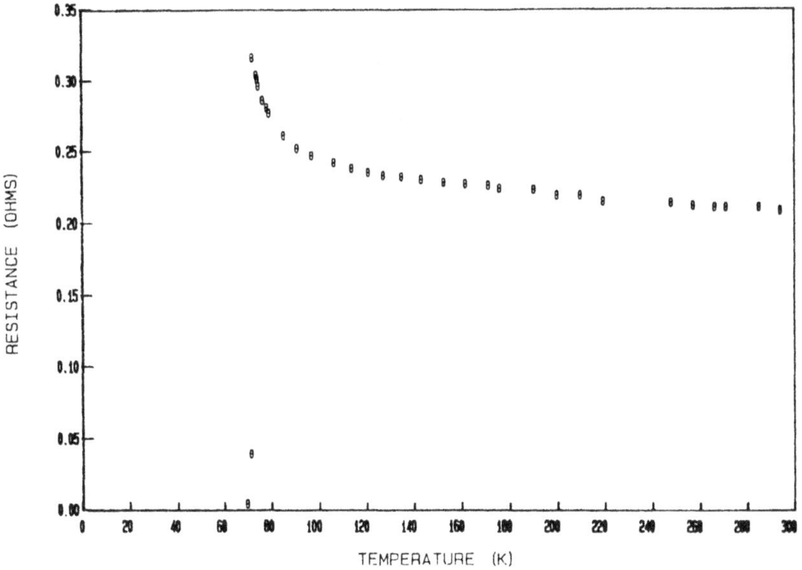

Fig. 1. Resistance-temperature data for $Gd_3Ba_3Cu_4O_z$ obtained during cooling (one datum every 9 seconds).

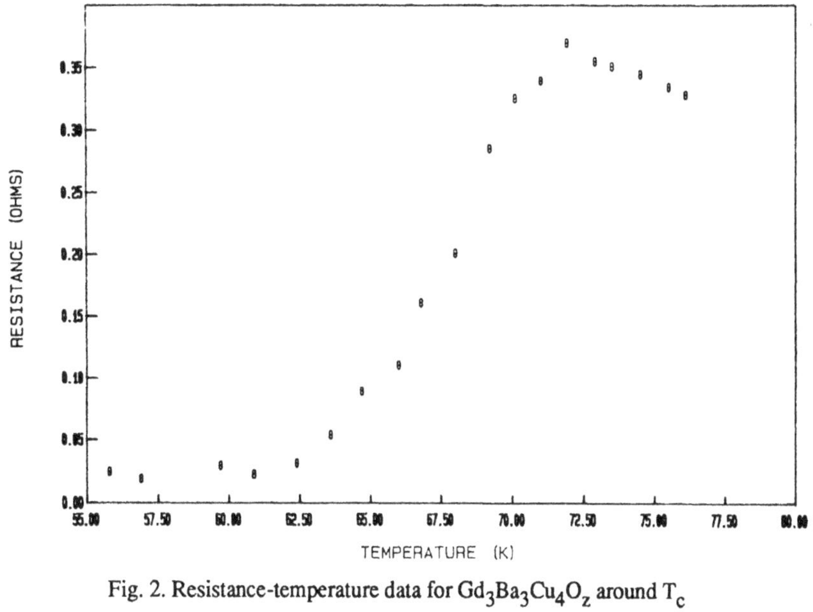

Fig. 2. Resistance-temperature data for $Gd_3Ba_3Cu_4O_z$ around T_c

ACKNOWLEDGMENT

We are grateful to Prof. J. H. Fang of the Department of Geology for providing the X-ray data.

REFERENCES

[1] J. G. Bednorz and K. A. Mueller, Z. Phys. B **64**, 189 (1986).
[2] M. K. Wu, J. R. Ashburn, C. J. Torng, P. H. Hor, R. L. Meng, L. Gao, Z. J. Huang, Y. Q. Wang, and C. W. Chu, Phys. Rev. Lett. **58**, 908 (1987).
[3] R. J. Cava, B. Batlogg, R. B. van Dover, D. W. Murphy, S. Sunshine, T. Siegrist, J. P. Remika, E. A. Rietman, S. Zahurak, and G. P. Espinosa, Phys. Rev. Lett. **58**, 1676 (1987).
[4] A. M. Stacey, J. V. Badding, M. J. Geselbracht, W. K. Ham, G. F. Holland, R. L. Hoskins, S. W. Keller, C. F. Millikan, and H.-C. zur Loye, J. Am. Chem. Soc. **108**, 2528 (1987).
[5] H. Takagi, S.-I. Uchida, K. Kishio, K. Kitazawa, K. Fueki, and S. Tanaka, Japan. J. Appl. Phys. **26**, L320 (1987).
[6] Z. Fisk, J. D. Thompson, E. Zirngiebl, J. L. Smith, and S.-W. Cheung, Phys. Rev. Lett., in press.
[7] Z. Zhao, L. Chen, Q. Yang, Y. Huang, C. Chen, R. Tang, G. Liu, C. Cui, L. Chen, L. Wang, S. Guo, S. Li, and J. Bi, Kexue Tongbao, in press.
[8] M. Onoda, S. Shamoto, M. Sato, and S. Hosoya, Japan. J. Appl. Phys. **26**, L363 (1987).

DIMENSIONAL CROSSOVER IN THE MOLECULAR FERROMAGNET DECAMETHYLFERROCENIUM TETRACYANOETHENIDE

Sailesh Chittipeddi, Arthur J. Epstein*, and Joel S. Miller**

Department of Physics, The Ohio State University
Columbus, Ohio 43210-1106, USA
*Department of Physics and Department of Chemistry
The Ohio State University
Columbus, Ohio 43210-1106, USA
**Central Research and Development Station
E.I. Du Pont de Nemours and Company, Inc.
Wilmington, DE 19898, USA

INTRODUCTION

Much attention has been focussed on the study of quasi-one-dimensional conducting systems which form segregated stacks (1). The report (2) that (DMeFc)(TCNQ), which consists of parallel stacks made up of alternating donor and acceptor ions, forms a metamagnetic system has led to increased interest in the magnetic properties of these metallocene complexes. In addition, there has been extensive experimental and theoretical investigation of low-dimensional magnetic systems, such as $CsNiF_3$ and the 2-D ferromagnet K_2CuF_4. Such studies have revealed detailed information on the magnetic excitations in these systems (3). In this paper we discuss the paramagnetic to ferromagnetic phase transition for (DMeFc)(TCNE) and examine the issue of a dimensional crossover and evaluate the critical exponents for this phase transition.

STRUCTURE

The $[Fe(C_5Me_5)_2]^+[TCNE]^-$ charge transfer complex, i.e., (DMeFc)(TCNE), posseses a structural motif where the cation metallocene donors, D, and anion acceptors, A, form linear chains of alternating D's and A's, i.e., ···DADADA···, Fig. 1. There are no intermolecular attractions less than the sum of the van der Waal radii (4). (DMeFc)(TCNE) is composed of $S = \frac{1}{2}$ $(DMeFc)^+$ ions with an anisotropic g value of $g_{\parallel}^{DMeFc} \sim 4$ and a $g_{\perp}^{DMeFc} \sim 1.3$, and $TCNE^-$ ions $(S = \frac{1}{2})$ with a $g^{TCNE} \simeq 2$.

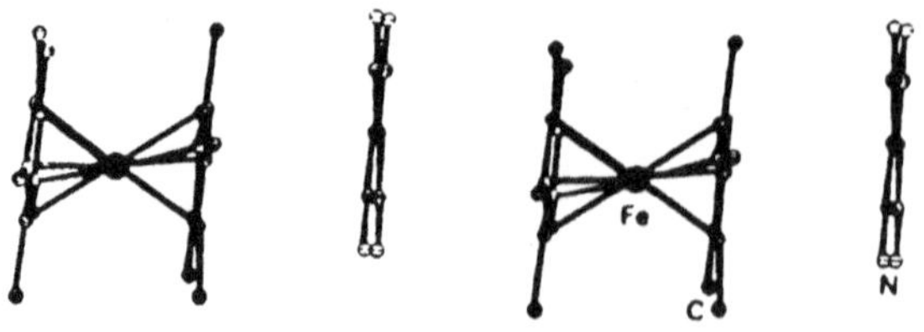

Figure 1. Alternating donor/acceptor structure of 1-D (DMeFc)(TCNE)

MAGNETIC SUSCEPTIBILITY

The magnetic susceptibility of (DMeFc)(TCNE) above 1.9K has been measured using a previously reported (5) Faraday technique. Data reported here are for a single crystal of the material with the needle axis being parallel to the applied field. In Fig. 2, we plot the reciprocal of the experimentally measured spin susceptibility as a function of the temperature. Above 100K, the data fits a Curie-Weiss expression (dashed line) showing that the system has dominant ferromagnetic interactions. Assuming independent Curie spins with H parallel to the stacking axis

$$\chi^{spin} = \chi^{spin}_{DMeFc} + \chi^{spin}_{TCNE} = N[(g_{\parallel}^{DMeFc})^2 + (g^{TCNE})^2]\,\frac{S(S+1)\mu_B^2}{3k_B T}$$

where N is Avagadro's number, μ_B is the Bohr magneton and k_B Boltzmann's constant, we obtain using $g_{\parallel}^{DMeFc} = 4$ and $g^{TCNE} = 2$ a spin susceptibility of 6.45×10^{-3} emu/mol which is in good agreement with the experimental value of 6.67×10^{-3} emu/mol. For $T > 16K$, the data fit a model for a 1-D Heisenberg ($S = \frac{1}{2}$) ferromagnetic chain (solid line) obtained using the Padé series approximation (6). A fit to the series approximation yields $J = 27.4K$ and a prefactor consistent with one spin of $g = 3.9$ and one spin of $g = 2$ per repeat unit. The curve derived from the theoretical model well approximates the data despite the model assumption that the donor and acceptor g values are identical. A plot of M vs. H for $T < T_c$ is shown in Fig. 3. The hysteresis loops become increasingly rectangular with decreasing temperature. This is in accord with domain formation in this material. At $T = 2.77K$, the value of the saturation moment, M, is in agreement with the value calculated for aligned spins of $g_{\parallel}^{DMeFc} = 4$ and $g^{TCNE} = 2$. The experimentally obtained value for magnetization is 1.63×10^4 emuG/mol.

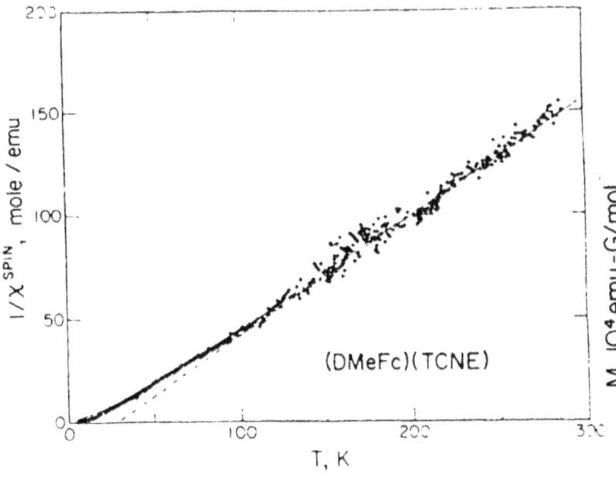

Figure 2. $(\chi^{spin})^{-1}$ vs. T for (DMeFc)(TCNE). The solid line is a fit to a 1-D Heisenberg ferromagnetic chain. The dashed line is a fit to the Curie-Weiss expression with $\theta = 30$K.

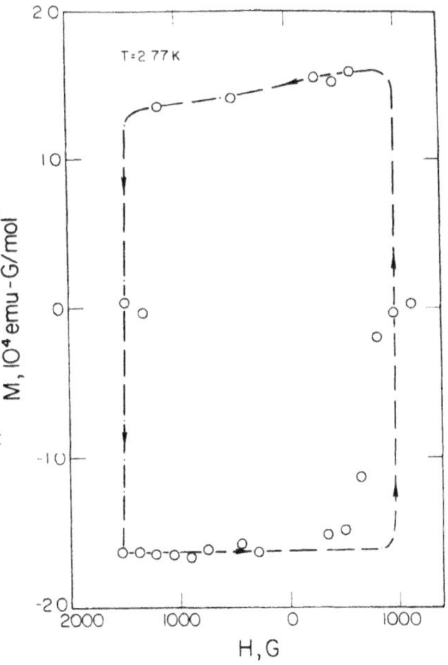

Figure 3. Magnetization M vs. applied magnetic field H at 2.77K (note H is parallel to the stacking axis).

The deviation from the one-dimensional behavior can be seen in Fig. 4, which shows the susceptibility measured at several low fields together with an extrapolation of the high temperature one-dimensional behavior plotted as log χ^{spin} vs. log $((T/T_c)-1)$. Below 16K χ^{spin} increases far more rapidly than that predicted by the 1-D Heisenberg ferromagnetic model, increasing as $((T/T_c)-1)^{-\gamma}$, with $\gamma = 1.2$. Attempts to model the behavior below 16K in terms of a 2-D system (7) were unsuccessful.

The spontaneous magnetization (H=0) increases for $T<T_c$ as $M=M_0(T_c-T)^\beta$, Fig. 5. The solid line in Fig. 5 is a fit to the above expression with a $\beta=0.5$ and $M_0=8000$ emuG/mol. A $\beta=0.5$ is a mean field result and is larger than that obtained for 3-D magnetic systems. Near T_c, at low fields (96G<H<1100G) we observe $M \propto H^{1/\delta}$ with a $\delta=4.4$. The experimental δ is nearly identical to that obtained for the anisotropic ferromagnet $CrBr_3$ (6).

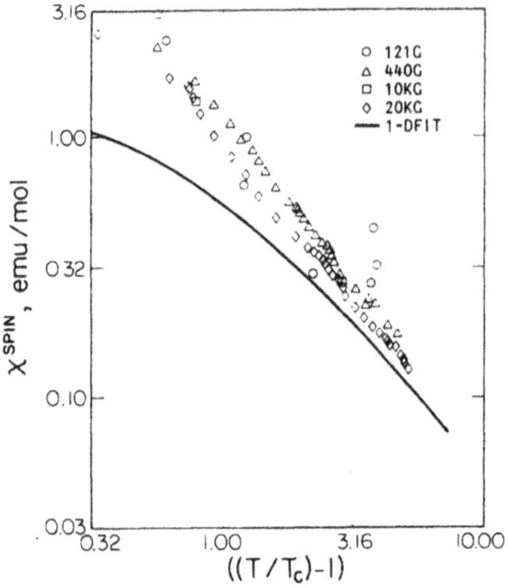

Figure 4. Log χ^{spin} vs. Log $((T/T_c)-1)$ with $T_c = 4.8$K.

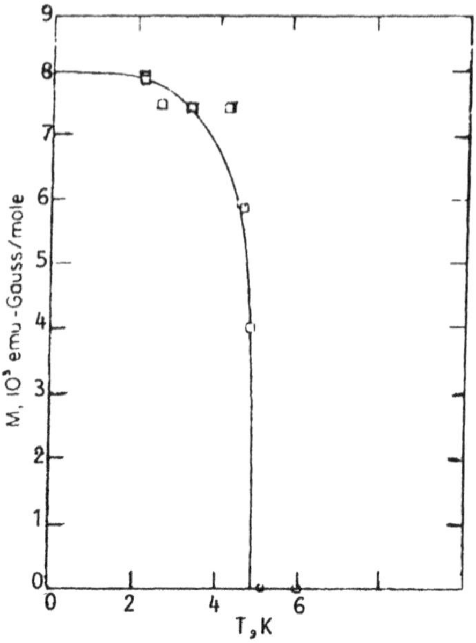

Figure 5. Spontaneous magnetization vs. T; the solid curve is a fit to $M_0(T_c-T)^{0.5}$.

CONCLUSION

The magnetic susceptibility and magnetization experiments results have been summarized for the molecular ferromagnet (DMeFc)(TCNE). The phase transition to 3-D ferromagnetic behavior can be described by the critical exponents, β,γ,δ. A crossover from 1-D to 3-D behavior is noted in this system.

REFERENCES

1. See for example, Proc. Int. Conf. Synth. Met., Kyoto, Japan, June 1986, Synth Met. 17, 18, 19 (1987).
2. G.A. Candela, L. Swartzendruber, J.S. Miller and M.J. Rice, J. Am. Chem. Soc., 101, 2755 (1979).
3. M. Steiner and A.R. Bishop, in "Solitons", ed. by V. L. Pokrovskii, et al. (North-Holland, 1986).
4. J.S. Miller, J.C. Calabrese, H. Rommelmann, S. Chittipeddi, J.H. Zhang, W.M. Reiff and A.J. Epstein J. Am. Chem. Soc. 101, 769 (1987).
5. E. Gebert, A.H. Reis, Jr., J.S. Miller, H. Rommelmann, and A.J. Epstein, J. Am. Chem. Soc. 104, 4403 (1982).
6. G.A. Baker, Jr., G.S. Rushbrooke, H.E. Gilbert, Phys. Rev. 135A, 1272 (1965) and D.D. Swank, C.P. Landee, and R.D. Willet, Phys. Rev. B 20, 2154 (1979).
7. J.M. Kosterlitz, J. Phys. C: Solid State Phys. 7, 1046 (1974); J.M. Kosterlitz and D.J. Thouless, J. Phys. C: Solid State Phys. 6, 1181 (1973).

SYNTHETIC PATHWAYS TO LOW-DIMENSIONAL COMPOUNDS CONTAINING TRANSITION METAL IONS AND NITROXIDE RADICALS

A. Caneschi[3], D. Gatteschi[3], J. Laugier[2], P. Rey[1], R. Sessoli[3], and C. Zanchini[3]

Laboratoires de Chimie[1] et Service de Physique[2]
Centre d'Etudes Nucléaires de Grenoble
85 X, F-38041 Grenoble Cedex, France
and
Department of Chemistry[3], University of Florence
Florence, Italy

Nitroxyl free radicals are weak Lewis bases which can function as ligands for a variety of transition metals. The nitroxyl group has been observed to bind to transition-metal ions only under rather special circumstances. In particular, enhancement of the Lewis acidity of the metal center by electron-withdrawing ligands seems to be required. Among the ligands employed for this purpose has been the hexafluoroacetylacetonate ion (hfac).

Much of the work devoted to the coordination chemistry of the nitroxyl free radicals involved copper(II) ions since the complexes afforded good models for two S = 1/2 interacting spins. Magneto-structural correlations were obtained, in good agreement with the previously published results on the dependence of the magnetic interactions on the geometry of the magnetic orbitals (1). Thus equatorially short- bonded complexes are antiferromagnetically coupled (2) while axially long- bonded ones exhibit ferromagnetic behaviours (3,4). Complexes of nitroxyls with other first-raw transition-metal ions are scarce and the only structurally and magnetically characterized ones are two bis-adducts of $Mn(hfac)_2$ (5). Although in these cases the nitroxyl is axially coordinated, the Mn-nitroxyl interaction is antiferromagnetic ; this behaviour is consistent with the presence of unpaired electrons in the Mn orbitals which can give non-zero overlap with the π^* orbital of the nitroxide.

Wishing to obtain extended interactions using the properties of these bulding blocks we thought of studying the coordination chemistry of the nitronyl nitroxides.

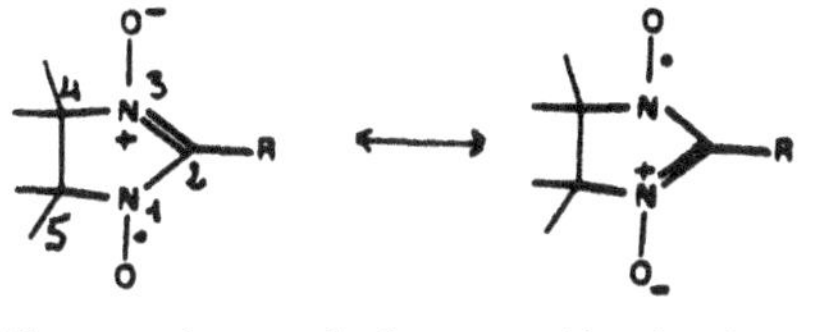

1 R = Ph, NITPh
2 R = Et, NITEt
3 R = Me, NITMe

These nitroxyl free radicals have two conjugated equivalent sites of coordination. Therefore they are expected to be bridging ligands, and owing to their unsaturated chemical structure, good transmitters of the magnetic information. Moreover changes in the R group allow to modifiy the base strength of the NO groups as well as the steric hindrance of the nitroxides. Thus, compared to the piperidinyl nitroxides, one expects new modes of coordination and new properties. In particular, owing to the single substitution of the carbon atom in position 2, release of the steric hindrance could lead to μ-1,1 coordination as shown in scheme 1.

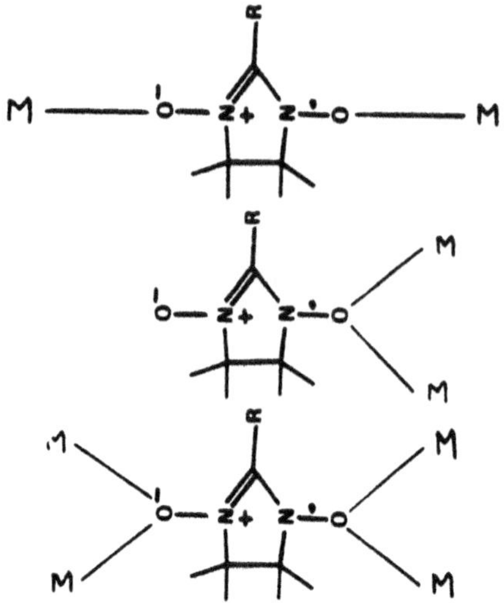

Adducts of the phenyl substituted nitroxide

In this ligand, the two NO groups are hindered by the phenyl ring and discrete adducts are obtained with all the metal hexafluoroacetyl acetonates. However in non-polar solvents we have characterized some other compounds with Mn(II).

We present here some exemples of adducts of the nitronyl nitroxides with the hexafluoroacetylacetonates of the first row transition-metal ions. Special emphasis is made on the relation between the C_2-substitution in the nitroxide, the solid state structure of the adduct and the observed magnetic properties.

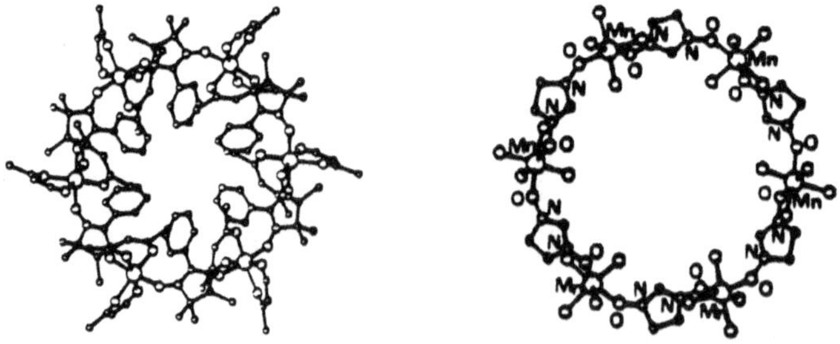

Figure 1 - X-Ray structure of $(Mn(hfac)_2NITPh)_6$

One, which has a 1/1 stoechiometry, has been characterized by X-ray crystallography. Figure 1 shows the unexpected cyclic hexameric structure in which the Mn(II) ions are coordinated by two cis-nitroxyl ligands. This is the first example of an equatorially bound nitroxyl to manganese.

Although cyclic, this compound shows that NITPh is indeed a bridging ligand for Mn(II) ions and affords one of the few known examples of a large, but limited number of interacting spins. In particular, since the rings are well shielded the magnetic properties are well interpreted using a simplified model in which, at low temperature, strong antiferromagnetic Mn-NO coupling leads to 6 interacting parallel spins S = 2. The value of XT = 12.7 cm^3 $mole^{-1}$ T observed at 6 K is indeed close to the theoretical value of 13 predicted by the model. Therefore the antiferromagnetic nature of the Mn(II)-nitroxyl coupling is confirmed whether the coordination is axial or equatorial.

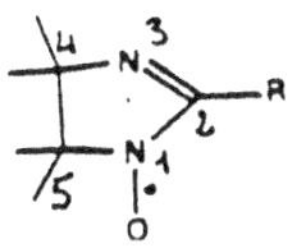

Reaction of $Mn(hfac)_2$ with this radical at higher temperature yields a compound which analyzed for $Mn(hfac)_2(IMPh)_2$ NITPh. The presence of IMPh is confirmed by chemical procedures. The redox reaction between Mn(II) and NITPh which leads to IMPh is not yet understood. Spectroscopic data strongly suggest that in this adduct the nitroxyl group is u-1,1 coordinated. The overall magnetic behaviour of the compound is antiferromagnetic, no simple model for the interpretation of the magnetic data has yet been built up.

Adducts of the Ethyl substituted nitroxyde

This ligand is less hindered and 1/1 steochiometry are obtained not only with large ions like Mn(II), but also with Cu(II) and Co(II).

The structures of these adducts can be tentatively understood from the magnetic data. The copper adduct essentially similar to that of the NITMe derivative (vide supra) is believed to have a linear chain structure. On the contrary, the Co(II) adduct, which shows an antiferromaghetic behaviour might be a chain of u-1,1 bridged dimers interacting through terminal NO groups.

Adducts of the Methyl derivatives

With the methyl derivative the steric hindrance of the two coordination sites is low and polynuclear adducts have been obtained with all the first raw transition metal ions.

The copper derivative has been structurally characterized and shows extended linear chains of copper ions bridged by axially coordinated NITMe molecules.

The structure is in agreement with the magnetic data since it is interpreted using a linear model of ferromagnetically interacting s = 1/2 spins.

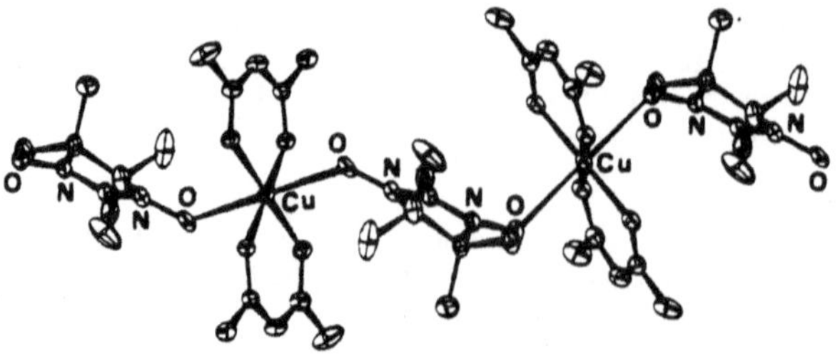

Figure 2 - X-Ray structure of $(Cu(hfac)_2NITMe)_n$

Although no suitable crystals have been obtained for the derivatives of the other metal ions, the 1/1 stoechiometry, and the magnetic properties, agree well with extended linear structures similar to that found for the Cu(II) adduct.

CONCLUSION

These few examples show that the nitronyl nitroxides are versatile bridging ligands for transition metal ions. Depending on the steric hindrance of the coordination sites of the free radical-size of the C-2 substituant- and the size of the metal ion, very different structural arrangements have been observed. The phenyl derivative seems to be the bulkiest group that can be accomodated by the largest Mn(II) ion. Eventhough in that case, redox reaction or intermolecular interactions lead to unexpected adducts. Noteworthy is the μ-1,1 coordination of the nitronyl nitroxides to two metal ions. It affords models in which metal ions with large spin like Mn(II) are close and polarized by the $S = 1/2$ spin of the free radical.

The last expected mode of coordination -μ-1,1-μ-3,3- has not been observed with the hexafluoroacetylacetonates. However a few adducts of NITMe with metal Lewis acids show the requiered stoechiometry of one radical to two metal ions.

References

1. Kahn, O ; Charlot, M.F. Nouv. J. Chim. 1980, 4, 567.

2. Lim, Y.Y. ; Drago, R.S. Inorg. Chem. 1972, 11, 1334.

3. Grand, A. ; Rey, P. ; Subra, R. Inorg. Chem. 1983, 22, 391.

4. Bencini, A. ; Benelli, C. ; Gatteschi, D. ; Zanchini, C. J. Am. Chem. Soc. 1984, 106, 5813.

5. Dickman, M.H. ; Porter, L.C. ; Doedens, R.J. Inorg. Chem. 1986, 25, 2595.

MAGNETIC PROPERTIES OF COUPLED LANTHANIDE-RADICAL SPECIES

Cristiano Benelli, Andrea Caneschi, and Dante Gatteschi,

Department of Chemistry
University of Florence
Florence, Italy

Jean Laugier, and Paul Rey

Centre d'Etude Nucleaires
Grenoble, France

Systems containing rare-earth ions interacting with paramagnetic centers different from the lanthanide themselves are not common. Beyond garnets, orthochromites, and orthoferrites, only recently reports of exchanging trimers constituted by Cu_2-Gd units were published[1,2,3]. In these compounds a weak ferromagnetic interaction whithin the Cu-Gd pairs was observed while an antiferromagnetic one was reported for the Cu-Cu pair. The nature of the observed exchange interactions is not fully understood especially for the spin coupling involving f and d electrons. As lanthanides can play an important role in detremining the properties of new magnetic materials, it is in our opinion a relevant task to attempt a detailed study of the interactions through f orbitals. An useful tool for this work could be to use stable organic radicals as paramagnetic ligands toward rare-earth ions. As, for instance, in stable nitroxide radicals the unpaired electron is largely delocalized on the oxygen atoms which is the donor one, in case of magnetic interaction, this should be a direct one, giving first hand informations on the bonding situation and the coupling pathway.

In the recent past we observed that the reactions of stable nitroxides with transition metal ions give a variety of compounds with magnetic interactions different either in sign and in intensity. So we decided to use this type of radicals to attempt the synthesis of similar complexes with rare earth ions.

By reacting $RE(hfac)_3$ (hfac = 1,1,1,5,5,5-hexafluoropentanedione, RE = rare-earth ions) with the phenyl (NIT-Ph) and ethyl (NIT-Et) derivatives of 4,4,4,5,5,5-tetramethyl-imidazoline-3-oxide-1-oxy, we obtained crystalline products whose elemental analysis fit quite well the formula $RE(hfac)_3(NIT\text{-}R)_2$.

The X-ray structure[4] of $Gd(hfac)_3(NIT\text{-}Ph)_2$ showed that the two radicals are directly bound through an oxygen atom to the Gd(III) ion in a dodecahedral arrangement with the six oxygen atoms of three molecules of hfac and two oxygen atoms of two water molecules.

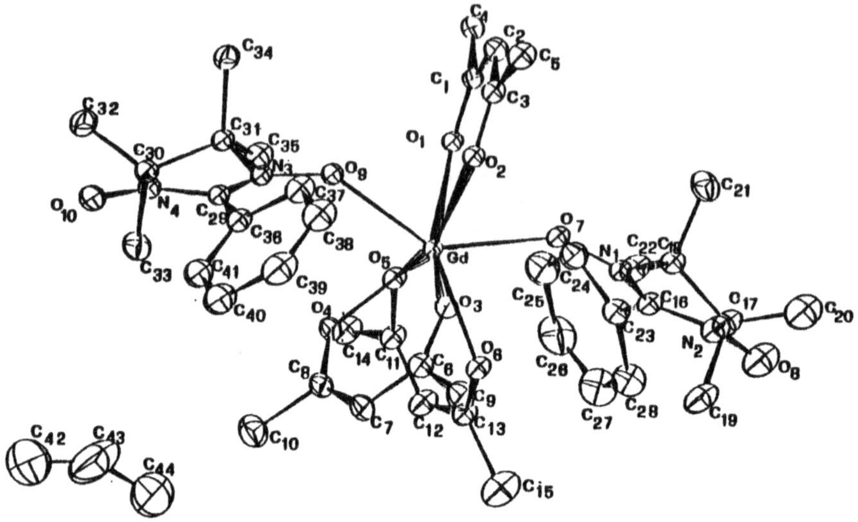

Figure 1. ORTEP drawing of $Gd(hfac)_3(NIT\text{-}Ph)_2$.

The magnetic susceptibility of the gadolinium derivatives, measured in the 6 - 300 K range, are reported in Figure 2 in the χT versus T fashion.

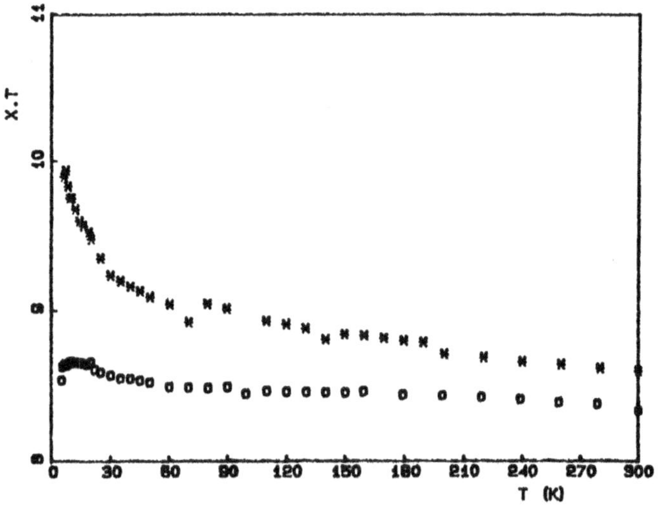

Figure 2.The experimental X.T values of $Gd(hfac)_3(NIT\text{-}Ph)_2$ (*) and of $Gd(hfac)_3(NIT\text{-}Et)_2$ (o) in the 6-300 K range.

The data relative to the gadolinium(III) compounds show an increase in the T values on decreasing the temperature, reaching at 6 K a value higher than this expected for a gadolinium(III) ion and the two unpaired electrons of two radicals without any exchange coupling. As the ground state of gadolinium(III) is ${}^{8}S_{7/2}$, namely a pure S state, the interpretation of the magnetic data may be attempted in a straightforward way assuming that the isotropic term is the leading one in the interaction hamiltonian. Putting the two Gd-radicals interaction as equivalent, it is possible to write the hamiltonian:

$$\mathbf{H} = J(\mathbf{S}_{Gd}\mathbf{S}_{rad1} + \mathbf{S}_{Gd}\mathbf{S}_{rad2}) + J'\mathbf{S}_{rad1}\mathbf{S}_{rad2}$$

and the relative formula for magnetic susceptibility is:

$$\chi = N\mu_{\beta}^{2}g^{2}\,[105 + 252\exp(-9x/2 + x') + 252\exp(-7x/2) + 445\exp(-8x)]/(6kTZ)$$

with $Z = 6 + 8\exp(-9x/2 + x') + 8\exp(-7x/2) + 10\exp(-8x)$. The quantities x and x' are J/kT and J'/kT, respectively.

The best fit parameters for the experimental data are g = 2.010(1), J = -1.022(1) cm^{-1}, J' = 1.65(2) cm^{-1} for the NIT-Ph derivative and g = 2.037, J = -0.42 cm^{-1}, J' = 3.06 cm^{-1} for the NIT-Et one. These results confirm the previous finding for the $Gd-Cu_2$ trimers of a ferromagnetic interaction between the S = 7/2 and S = 1/2 spins and an antiferromagnetic one between the S = 1/2 spins. At the moment it is not clear why for the NIT-Et complex the interaction is so small. As no X-ray analisys for this compound is available, it is not possible to correlate the different magnetic behavior to any structural difference.

The data for the europium(III) derivatives show a decrease of the T values on decreasing the temperature, starting, at room temperature, from a value close to that expected for two uncoupled radical and one europium(III) ion. At 6 K the χT values are lower with respect to those corresponding to the above mentioned situation. This analisys is based on the fact that the energy level diagram for Eu(III) ion consist of a ground state with J = 0 and the first excited state at about 400 cm^{-1} with J = 1. So, if there were no magnetic interactions in the system, the susceptibility should derive from the sum of two terms: a Curie-like contribution due to the two electrons of radicals and one coming from the europium, whose suscebtibility decreases as the population of the J = 1 magnetic state decreases on lowering the temperature. Below 50 K this contribution consist simply of a TIP term of about $6x10^{-3}$ emu. Attempt to fit the observed values of the magnetic susceptibility are in progress and the first results are indicative of an antiferromagnetic coupling of the two radicals via europium ion. Another model which we are developping is based on the possibility of interactions between radicals and the J = 1 state of the europium.[5]

REFERENCES

1. A. Bencini, C. Benelli, A. Caneschi, R. L. Carlin, A. Dei, D. Gatteschi, "Crystal and Molecular Structure of and Magnetic Coupling in Two Complexes Containing Gadolinium(III) and Copper(II) Ions.", J. Am. Chem. Soc., 107: 8128 (1985).
2. A. Bencini, C. Benelli, A. Caneschi, A. Dei, D. Gatteschi, "Magnetic Properties of a Trinuclear Complex Containing Exchange Coupled $GdCu_2$ Species.", J. Magn. Magn. Mat., 54-57: 3457 (1986).
3. A. Bencini, C. Benelli, A. Caneschi, A. Dei, D. Gatteschi, "Crystal and Molecular Structure and Magnetic Properties of a Trinuclear Complex Containing Exchange Coupled $GdCu_2$ Species.", Inorg. Chem., 25: 572 (1986).
4. C. Benelli, A. Caneschi, D. Gatteschi, J. Laugier, P. Rey, "X-ray Crystal Structure and Magnetic properties of a Gadolinium Hexafluoroacetylacetonate Adduct with the Nitrone-nitroxide Radical 2-Phenyl-4,4,5,5-Tetramethylimidazoline-3-oxide-1-oxy", Angew. Chem., in press.

5 M. E. Foglio, J. H. Van Vleck, Proc. Roy. Soc., 336A: 115 (1970).

MAGNETIC BEHAVIOR OF 1D HELICOIDAL CHAINS IN $Ba_7CuFe_6F_{34}$

A NEW KIND OF FERRIMAGNETIC 1d SYSTEM

J. Renaudin[a], G. Ferey[a], M. Zemirli[b],
F. Varret[b], A. de Kozak[c], and M. Drillon[d]

[a] Laboratoire des Fluorures,
Université du Mans 72017 Le Mans(France)
[b] Laboratoire de Spectométrie Mössbauer,
Université du Mans, 72017 Le Mans(France)
[c] Laboratoire de Chimie Minérale, 4 Place Jussieu,
Université Paris VI, 75700 Paris(France)
[d] EHICS, DSM, 1 rue Blaise Pascal, 67008 Strasbourg(France)

INTRODUCTION

One way of obtaining new low-dimensional magnets is to study pseudo-binary diagrams, in which one limit corresponds to a 1d magnet and the other one to a 2d compound. We previously developed this method and chose the system $BaFe_5$ (1d) - $BaCuF_4$ (2d) which gathers S = 5/2 and S = 1/2 spins, of special interest for magnetic investigations. In this system appears a new phase, $Ba_7CuFe_6F_{34}$ (1), whose structure was recently solved (2); its description in terms of isolated complex chains corresponds to the first part of this paper; the second part will deal with magnetic properties and, finally, the corresponding Mössbauer data will be given in the last section.

DESCRIPTION OF THE STRUCTURE

$Ba_7CuFe_6F_{34}$ is monoclinic (SG: C 2/m) with a = 16.982(2) Å, b = 11.372(2) Å, c = 7.663(1) Å, β= 101.47(1) . The structure is built up from complex entities of copper(II) and iron(III) octahedra, isolated one from each other by barium atoms. These units (Fig.1) have the formulation $[CuFe_6F_{34}]$. They form chains, running along the b axis, which are built up from two cis-blocks of corner sharing iron(III) octahedra. The central and terminal Fe^{3+} ions belong to two distinct crystallographic sites, respectively 4i and 8j of the space group C 2/m. The two cis blocks which deduce one from the other by a center of symmetry, are linked together by a copper(II) octahedron which shares its four equatorial fluorine ions with four cis blocks. The building unit can then be described as a double helix, whose branches intersect on the copper(II) ions. The

Fe^{3+}-Fe^{3+} and Cu^{2+}-Fe^{3+} intrachain distances are 3.682 and 3.941 A, respectively, while the shortest distance between Fe^{3+} ions of adjacent chains is 5.134 A. This leads to a completely new type of bimetallic chain the magnetic characteristics of which are described in the next part.

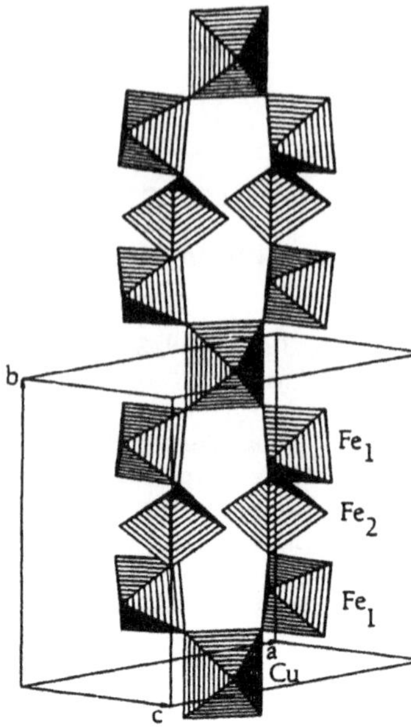

Fig. 1 - Perspective view of the double helix

MAGNETIC BEHAVIOR

The magnetic susceptibility of $Ba_7CuFe_6F_{34}$ was measured in the temperature range 4 K - 300 K; it is plotted in figure 2 as X.T = f(T) which is more revealing than X to describe such a coupled system. The main features to be emphasized are:

(i) the regular decrease of X.T, upon cooling down, from 22 emu (at 300 K) to 11.5 emu (at 40 K); the asymptotic behavior of X.T at higher temperature agrees with the expected Curie constant for uncoupled ions (X.T = 26.6).

(ii) the rounded minimum around T = 40 K, and the significant divergence of X.T at lower temperature

(iii) the sharp maximum at 8 K suggesting a phase transition to a magnetically ordered state.

Clearly, the high temperature variation (T > 8 K) of the susceptibility shows the typical features of 1d-ferrimagnets. So far, 1d-ferrimagnetism has been reported in bemetallic complexes characterized by alternating spins or/and Landé factors in a linear network[3,4]. Recently, it has been shown that intertwinning double chain may equally entail the same magnetic features[5].

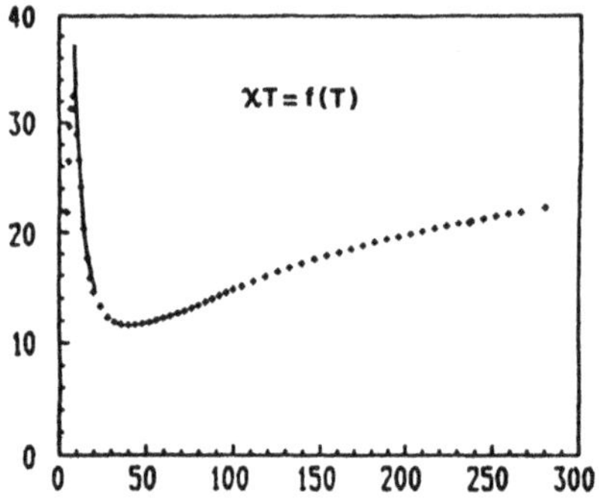

Fig. 2 - Magnetic behavior of $Ba_7CuFe_6F_{34}$

The above system is a new example of bimetallic chain characterized by a peculiar stacking of the metal ions. It may be described as a chain of metal rings, each one built up from six iron(III), brought together in [Fe_3] units, and two copper(II) ions. The [Fe_3] units are connected through copper(II) ions in order to form an infinite 1d system.

In view of the topology of this chain and the spin multiplicity of the interacting ions, an analysis of the data cannot be done from finite link computations, as developed for regular or alternating Heisenberg chains[6,7]. In turn, a relevant approch would be to consider a (classical-quantum) spin system in which classical spins stand for iron(III) ions and quantum ones relate to copper(II) ions, but the solutions of this problem are not as yet available.

On closer examination of the magnetic sites and the metal-metal distances, it appears in fact that, at low enough temperature, each $[Fe_3]$ unit behaves as a S = 5/2 spin coupled on each side to 1/2-spin of copper(II) ions. When only the low-lying states are thermally populated, it can be assumed that 5/2-spins are roughly colinear, so that a fictitious chain model with alternating spins (s=1/2, S=5) should be closely approximated. Owing to the large difference of spin values on both sites, the classical spin limit seems well-adapted to describe the above chain. Then, the magnetic susceptibility is shown to reduce to the closed formula[3]:

$$X = Nu_B^2/3kT\,[g_+^2(1+U)/(1-U) + g_-^2(1-U)/(1+U)]$$

where $U = \coth(J/kT) - kT/J$, $g_\pm = (g \pm G)/2$.

Obviously, the parameters J, g and G of this expression must be renormalyzed by the appropriate scaling factors $[s(s+1)]^{1/2}$ and $[S(S+1)]^{1/2}$ to account for the quantum nature of the spins. Assuming that Landé factors take the usual values g = 2.20 (Cu^{2+}) and G = 2.0 (Fe^{3+}), the best agreement between theory and experiment is obtained, in the range 8-25 K, for J = -6.1 K.
It thus appears that this simplified model gives some insight into the exchange coupling between copper(II) and iron(III) units. Obviously, the quality of the fit is limited in the temperature scale, as anticipated above. This results from the rough approximation made by neglecting the upper spin-states related to the decoupling within $[Fe_3]$ units. A more complete analysis is in progress.

MOSSBAUER STUDY

In the paramagnetic state, the spectra can be described by the superposition of two quadrupolar doublets (Fig. 3) with rather narrow lines which correspond to the two crystallographic sites occupied by Fe^{3+} ions. The fitted Mössbauer parameters are listed in table 1:

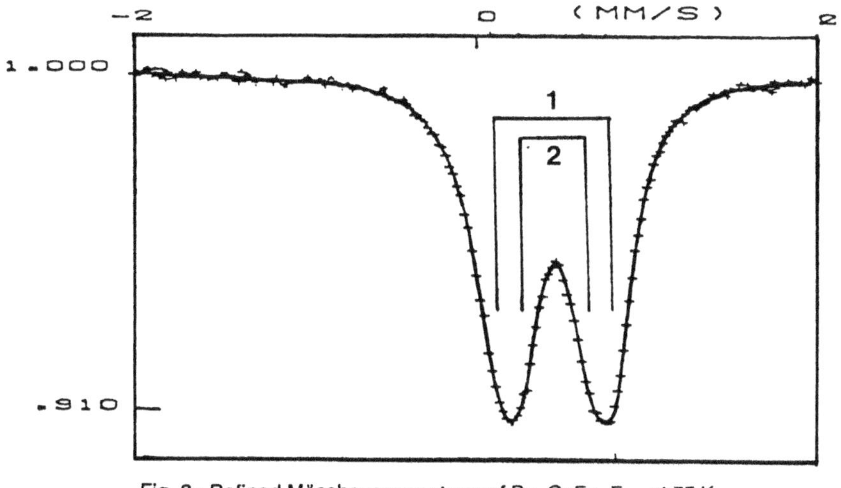

Fig. 3 - Refined Mössbauer spectrum of $Ba_7CuFe_6F_{34}$ at 77 K

TABLE I

Mössbauer parameters in the paramagnetic state

T/K	Site 1 (8j)				Site 2 (4i)			
	IS	T	QS	P%	IS	T	QS	P%
300	0.437(2)	0.302(4)	0.66(1)	63	0.437(2)	0.302(4)	0.37(1)	37
77	0.532(2)	0.330(4)	0.66(2)	67	0.534(2)	0.272(4)	0.37(2)	33

IS: isomer shift relative to metallic iron ($mm.s^{-1}$)
T : linewidth($mm.s^{-1}$)
QS: quadrupole splitting ($mm.s^{-1}$)
P : percentage of the total iron in each site

It is clear that site 1 corresponds to Fe^{3+} (8j) while site 2 relates to Fe^{3+} (4i). In the vicinity of T_N, the thermal evolution of the paramagnetic fraction R observed by the thermal scanning technique leads to a Néel temperature of 8.7 K, in good agreement with the value deduced from susceptibility measurements.
Below T_N, hyperfine fields H_{hf} appear at the Fe nucleus on both sites, with rather different values, as indicated in table II.

TABLE II

Refined Mössbauer data in the antiferromagnetic state, at 4.2 K.

Site	IS ($mm.s^{-1}$)	2ε($mm.s^{-1}$)	H_{hf}(kOe)
1 (8j)	0.53(1)	-0.49(2)	450(1)
2 (4i)	0.55(2)	0.03(5)	405(2)

$2\varepsilon=eQV_{HH}/2$, V_{HH}=EFG along hyperfine field

REFERENCES

1. M. SAMOUEL, A. DE KOZAK, J. RENAUDIN, M. LEBLANC, G. FEREY
Rev. Chim. Min. 22, 64(1985)
2. J. RENAUDIN, G. FEREY, A. DE KOZAK, M. SAMUEL
Rev. Chim. Min. (in press)
3. M. DRILLON, E. CORONADO, D. BELTRAN and R. GEORGES
Chem. Phys. 79, 449(1983)
4. A. GLEIZES and M. VERDAGUER, J. Am. Chem. Soc., 106, 3727(1984)
5. M. DRILLON, M. BELAICHE, J.M. HEINTZ, G. VILLENEUVE, A. BOUKHARI and J. ARIDE, same issue
6. J.C. BONNER and M.E. FISHER, Phys. Rev., 135, A640(1964)
7. M. DRILLON, J.C. GIANDUZZO and R. GEORGES, Phys. Lett.,96A, 413(1983)

MAGNETIC ANISOTROPY STUDIES OF 1d BIMETALLIC FERRIMAGNETIC COMPOUNDS

Yu Pei[1], Michel Verdaguer[1≠], Olivier Kahn[1], Jean Pierre Renard[2] and Jorunn Sletten[3]

[1] Laboratoire SET, U.A. 420, Université de Paris Sud, 91405 Orsay, France
[2] Laboratoire I.E.F., U.A. 022, Université de Paris Sud, 91405 Orsay, France
[3] Department of Chemistry, University of Bergen, 5007 Bergen Norway
[≠] Permanent adress: E.N.S. Le Parc 92211 Saint-Cloud, France

INTRODUCTION

We report magnetic anisotropy studies performed on two bimetallic ferrimagnetic chains of formula $MnCu(pba)(H_2O)_3.3H_2O$ (1) and $MnCu(pbaOH)(H_2O)_3$ (2), where pba is 1,3-propylene bisoxamato and pbaOH the 2-hydroxy derivative.

Perspective views of the two structures are shown in Figure 1. Those

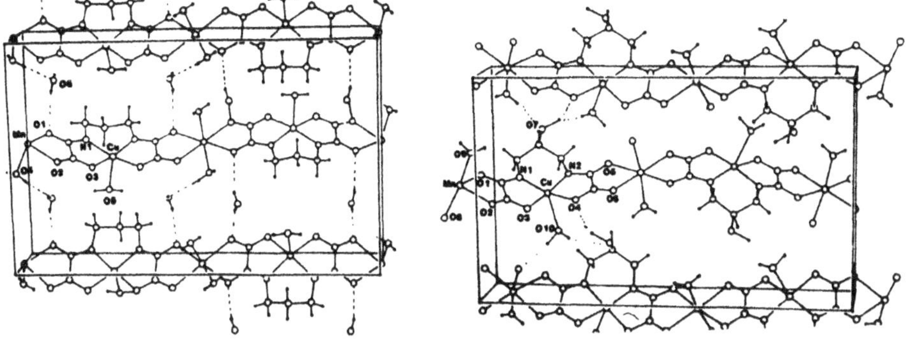

Figure 1 - Perspective views of $MnCu(pba)(H_2O)_3.2H_2O$, 1 (left) and $CuMn(pbaOH)(H_2O)_2$, 2 (right).

structures and the bulk magnetic properties have already been discussed[1-3]. Both compounds present the characteristic features of 1D ferrimagnetism, but the relative positions of the chains are slightly different, which leads to totally different magnetic properties in the low temperature range. 1 orders antiferromagnetically (AF) and 2 exhibits a spontaneous magnetization which can be described as resulting either from a ferrimagnetic order of the metal ions of from a ferromagnetic order of the ferrimagnetic chains (F). We show hereunder how anisotropy experiments provide new insights on the mechanism of the 3d ordering.

EXPERIMENTAL SECTION

Magnetic anisotropy measurements for 1 and magnetization measurements for 2 were carried out with a Laboratory made low field SQUID magnetometer[4]. The crystals used were a well shaped distorted hexagonal-based prism, weighting about 300μg for 1, and a well shaped parallelepiped weighting only 3.7 μg for 2. This latter crystal was the only available. The crystals were stuck on pure silicon that could be orientated in the SQUID. EPR measurements, not discussed here[5], show that the axes of the orthorhombic lattice are actually the principal magnetic axes.

SUSCEPTIBILITY DATA FOR 1

The principal magnetic susceptibilities χ_a, χ_b and χ_c along the crystal axes in the 1.8-5 K temperature range are shown in Figure 2. Down to 2.6 K,

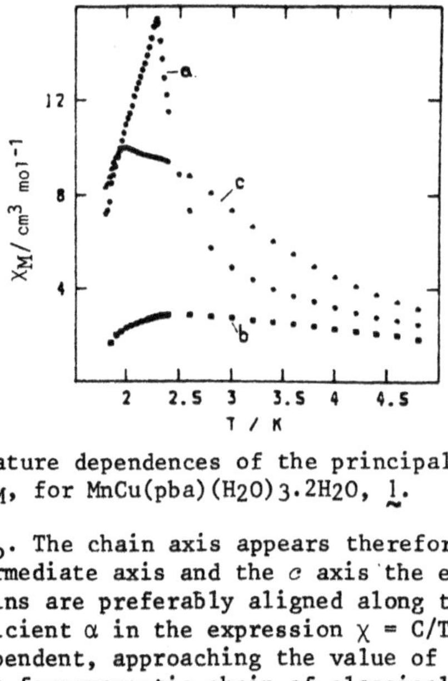

Figure 2 - S Temperature dependences of the principal magnetic susceptibilities χ_M, for $MnCu(pba)(H_2O)_3.2H_2O$, 1.

we have $\chi_c > \chi_a > \chi_b$. The chain axis appears therefore to be the hard axis, the *a* axis the intermediate axis and the *c* axis the easy axis. In other words, the local spins are preferably aligned along the *c* axis. Our data show that the coefficient α in the expression $\chi = C/T^{\alpha}$ when $T \to T_{Neel}$ is very orientation dependent, approaching the value of 2 along the easy axis *c*, as expected for a ferromangetic chain of classical spins[6]. This dependence upon orientation points out the impossibility to extract a physically significant value of α from powder measurements. In the 2.3-1.9 K range, the approach of the 3d ordering appears a complicated process. χ_b presents a maximum at 2.4 K and a break at 1.95 K whereas χ_c shows a break at 2.4 K and a maximum at 1.95 K. Moreover, below 2.5 K, χ_a becomes larger than χ_c and presents a sharp maximum at 2.3 K, which reveals a canting of the spins in the a direction, with a non-zero resulting moment along this direction. Such a weak ferromagnetism has already been reported for $CuMn(dto)(H_2O)_4$. $CuMn(dto)(H_2O)_4.3.5\ H_2O$ [7,8]. The observed phenomena are consistent with the spin structure in the AF phase shown in Figure 3, where the canting of the spins with the weak resulting moment along the *a* axis has not been represented.

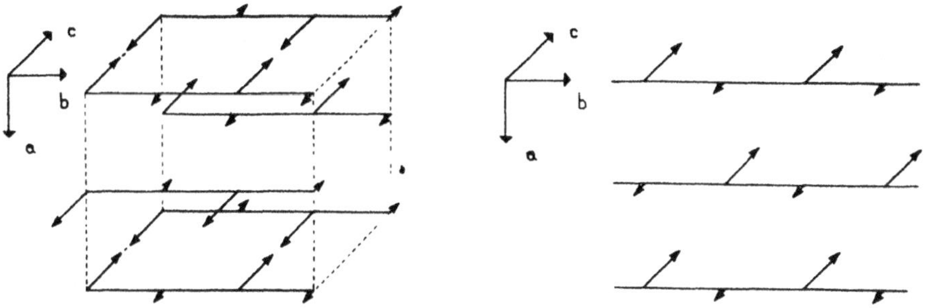

Figure 3 - Schematic spin structures in the 3d ordered phase for 1 (left) and for 2 (right) in the *a*,*b* planes.

MAGNETIZATION DATA FOR 2

Powder magnetization data in weak field and hysteresis loop are reported elsewhere in this book[3]. The thermal variation of the molar magnetizations M_a, M_b and M_c along the crystal axes, in the 2-5 K temperature range is shown in Figure 4. We have $M_c >> M_a \geqslant M_b$. As for 1, the *c* axis is the

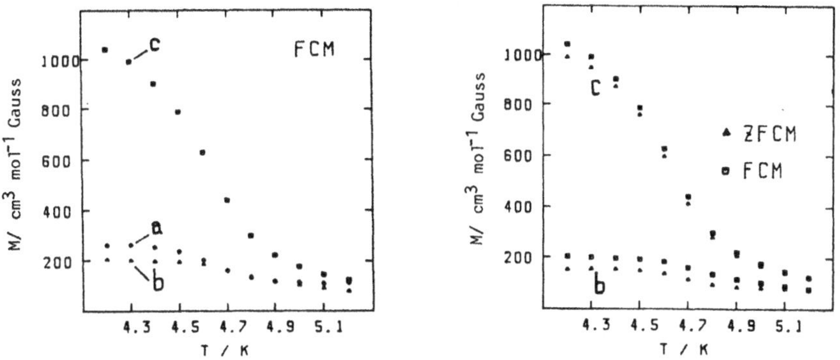

Figure 4 - Anisotropy magnetization curves for $MnCu(pbaOH)(H_2O)_3$, 2 : field-cooled thermal variation along the principal axes M_a, M_b M_c (left) ; field cooled (FCM, ▫) and zero-field cooled magnetization (ZFCM, ▵) along the *b* and *c* axes ; data for the *a* axis are not shown for clarity (right). The applied field is 3 Oe.

easy magnetization axis. These measurements give a clear confirmation of the 3d F transition. The spin structure in the *ab* plane is given in Figure 3. Field cooled magnetization (FCM) and zero field cooled magnetization (ZFCM) curves along *b* and *c* are reported in Figure 4 (right). In the ZFCM measurements, the sample was first cooled in zero field down to 3.5 K, then warmed up within the field. The striking feature, compared to the powder data is the almost similar values of the FCM and the ZFCM at a given temperature. This is due to the fact that the crystal has obviously much less defects than the powder sample, so that the domain walls move more freely.

The variation of the average magnetization M at 3.0 K and 4.2 K is given in Figure 5. There is practically no field range where M varies linearly versus H. The average zero field susceptibility $(dM/dH)_{H=0}$ is large and equal to 75(3) cm^3mol^{-1} at 4.2 K and even larger at 3.2 K, which confirms that we are in the F phase. Finally, the saturation magnetization well corresponds to S = 5/2 - 1/2 = 2 per MnCu unit. At 3.0 K, half of the saturation is reached at 600 G and 2/3 at 900 G.

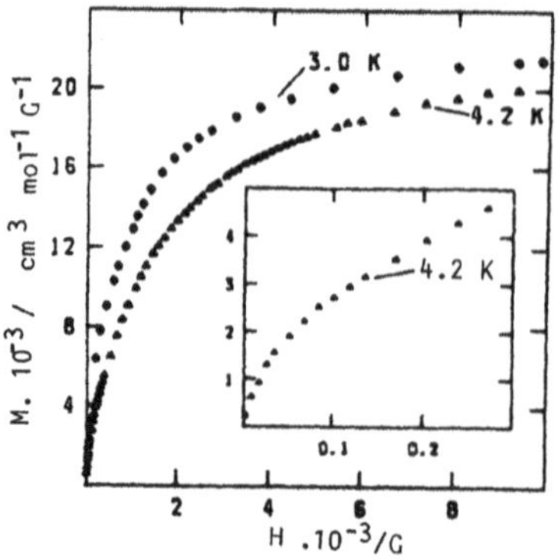

Figure 5 - Polycrystalline magnetization data vs field for 2 ; low-field data are detailed in insertion.

DISCUSSION AND CONCLUSION

Our data clearly show that small variations in the crystal packing of otherwise almost identical 1d ferrimagnetic chains induce drastic differences in the ordered 3d magnetic phases. The antiferromagnetic interaction between nearest neighbor ions belonging to the same chain or to two adjacent chains is fairly well understood in the frame of a model based on the overlap (or the orthogonality) of localized magnetic orbitals[3,7]. This model explains both the AF character of the *ab* planes in 1 and the F character of these *ab* planes in 2. The F interaction between the *ab* planes along *c* in 2 is a more difficult problem and deserves further comments. Indeed, for both 1 and 2, the shortest separations along *c* are Mn...Mn and Cu...Cu, which should favor an antiferromagnetic coupling of the *ab* planes. This is observed in 1 but not in 2. We suggest that in 2, the dominant interaction between *ab* planes is of dipolar nature. To confirm this assumption, we have performed point dipole calculations including all the Mn and Cu local spins with antiferromagnetic interactions between nearest neighbors within spheres up to 100 Å of radius. Those calculations show that the dipolar field created by a Mn site favors a F ordering, *c* being the easy magnetization axis.

Owing to magnetic anisotropy measurements, we have been able to specify the mechanism of the 3d ordering of our ferrimagnetic chains. The AF ordering in 1 may be qualitatively interpreted by considering only antiferromagnetic interactions between nearest neighbor ions, which leads to a cancellation of the spin at the scale of the lattice. In contrast, for 2 we need to take into account the dipolar interaction mainly due to the Mn local spins aligned along the *c* direction.

The main goal of this work is to design molecular systems with a 3d F ordering. In 2, this goal has been achieved owing to the dipolar interactions driven by the magnetic anisotropy. In conclusion, we must confess that we are still far from controling all the subtleties of these effects.

REFERENCES

1. Y. Pei, M. Verdaguer, O. Kahn, J. Sletten and J.P. Renard, J. Am. Chem. Soc. 108:7428 (1986) ; Inorg. Chem. 26:138 (1987).
2. O. Kahn, Y. Pei, M. Verdaguer, J. Sletten and J.P. Renard, to be submitted to J. Am. Chem. Soc.
3. O. Kahn, these proceedings.
4. P. Beauvillain, C. Chappert and J.P. Renard, J. Phys. E. 18:839 (1985).
5. Work in progress in collaboration with D. Gatteschi and C. Zanchini.
6. M. Fisher, Am. J. Phys. 32:343 (1964).
7. A. Gleizes and M. Verdaguer, J. Am. Chem. Soc. 106:3727 (1984).
8. M. Verdaguer, A. Gleizes, J.P. Renard and J. Seiden, Phys. Rev. B : Condens. Matter 29:5144 (1984).
9. Work in progress in collaboration with P. Beauvillain.

CRYSTAL CHEMISTRY AND MAGNETIC PROPERTIES

OF NEW FLUORIDES WITH USOVITE-TYPE STRUCTURES

J. Darriet, Xu Qiang, A. Tressaud, R. Georges and
J.L. Soubeyroux*

Laboratoire de Chimie du Solide du CNRS, 351, cours de la
Libération, 33405 Talence Cedex, France
*I.L.L. B.P. 156X, 38042 Grenoble Cedex, France

The natural usovite $Ba_2CaMgAl_2F_{14}$ crystallizes in the monoclinic system with space group C2/c. A characteristic feature of the structure is the presence of $|CaMgAl_2F_{14}|$ layers parallel to the (100) plane and separated by Ba atoms which are surrounded by twelve fluorine atoms (Fig. 1) |1,2|.

In each layer the cations occupy three different crystallographic sites : the Ca atoms (i.e. the M^{2+} cations in a $Ba_2M^{2+}M'^{2+}M''^{3+}F_{14}$ formulation) are linked to eight F atoms in a distorted square antiprism, while Mg (M'^{2+} ions) and Al (M''^{3+} ions) occupy two different types of octahedra (Fig. 1).

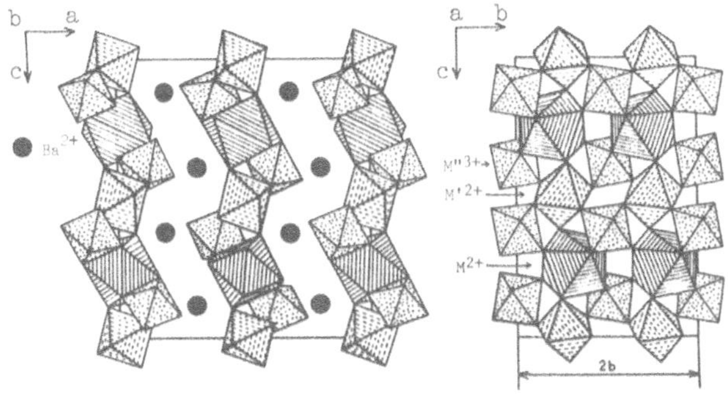

Fig. 1. Schematic representation of the usovite-type structure.

If M^{2+} and M'^{2+} are paramagnetic species and M''^{3+} is a diamagnetic one, the magnetic cations constitute isolated chains parallel to the [101] direction (Fig. 1). this is the case of the $BaMnM''^{3+}F_7$ (M" = Al or Ga) compounds, in which Mn^{2+} ions are located in chains constituted by corner-linked polyhedra, which are alternately octahedra and distorted square antiprisms. The model used to fit the data is based on classical spins and takes into account of the differenciation between the two types of sites |3|. The expression used to fit the susceptibility data is :

$$\chi = \frac{N\mu_B^2}{3kT}\left(G^2\ \frac{1+F(x)}{1-F(x)} + \delta G^2\ \frac{1-F(x)}{1+F(x)}\right) \text{ where } x = \frac{J}{kT}\,(S(S+1)),$$

$G = 1/2(g_A+g_B)(S(S+1))^{1/2}$, $\delta G = 1/2(g_A-g_B)(S(S+1))^{1/2}$ and $F(x)=\coth(x)-1/x$.

The exchange interaction is antiferromagnetic J/k=-5K (Fig.2). The 3D magnetic ordering has been observed by neutron diffraction (T_N = 1.65 K).

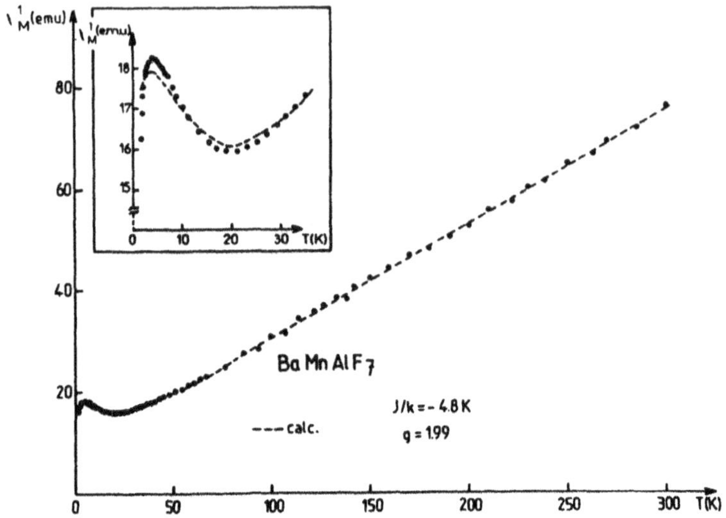

Fig.2. Thermal variation of reciprocal susceptibility of $BaMnAlF_7$.

When M^{2+} is a diamagnetic cation such as Ca^{2+} or Cd^{2+}, the M'^{2+} and M''^{3+} paramagnetic ions form a new arrangement of complex chains parallel to the [010] direction and separated by M^{2+} ions occupying square antiprims (Fig. 1). In these chains the paramagnetic species M'^{2+} and M''^{3+} have respectively two and four nearest magnetic neighbors. This type of cationic distribution has been achieved in $Ba_2CaM'^{2+}Fe_2F_{14}$ (M'^{2+} = Mn^{2+}, Co^{2+} and Cu^{2+}) |4,5|.

In the case of $Ba_2CaCoFe_2F_{14}$ and $Ba_2CaMnFe_2F_{14}$ the magnetic behavior is characteristic of ferrimagnetic chains with a minimum of the product $\chi.T = f(T)$ (Fig. 3). The thermodynamic behavior of the exchange

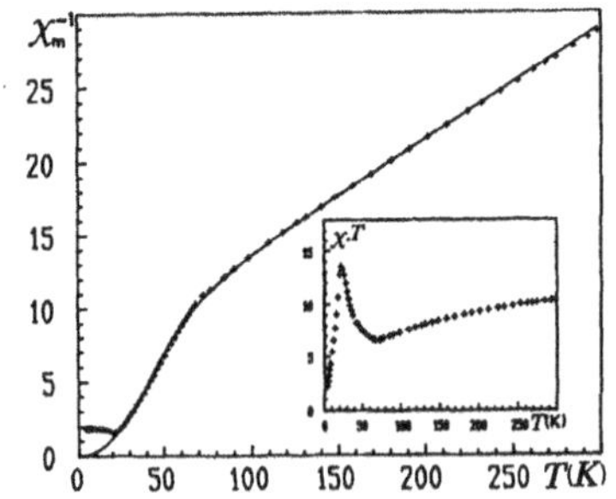

Fig.3. Thermal variation of reciprocal susceptibility of $Ba_2CaMnFe_2F_{14}$.

couplings in this type of chain has been investigated from a classical point of view. For $Ba_2CaMnFe_2F_{14}$ the best fitting between experimental and theoretical data leads to $J_1/k = -6.7$ K and $J_2/k = -2.7$ K with $g_1 = g_2 = 1.99$ (Fig. 3).

Neutron diffraction measurements on $Ba_2CaCuFe_2F_{14}$ which is isostructural with $Ba_2CaCuGa_2F_{14}$, confirm that the Cu^{2+} atoms occupy axially elongated octahedron(M'^{2+} sites) with a longer distance corresponding to the F_4 atom (Cu-F_4 = 2.33 Å), the two other distances being shorter (Cu-F_7 = 1.87 Å and Cu-F_2 = 2.00 Å) (Fig. 4). For Cu^{2+} ions (i.e. a d^9 configuration) the case of an axial elongation corresponds to the occupancy of the $d_{x^2-y^2}$ orbital by the single electron in the crystal field approach. One may assume that the main exchange interaction will be Cu-F_2-Fe since the $d_{x^2-y^2}$ Cu^{2+} orbital is involved (Fig. 4). Cu-F_4-Fe coupling will be very weak, as they correspond to completely filled d_{z^2} orbitals and therefore in a first approximation the magnetic behavior can be interpreted in a larger temperature range in terms of $|Fe^{3+}-Cu^{2+}-Fe^{3+}|$ linear trimeric units.

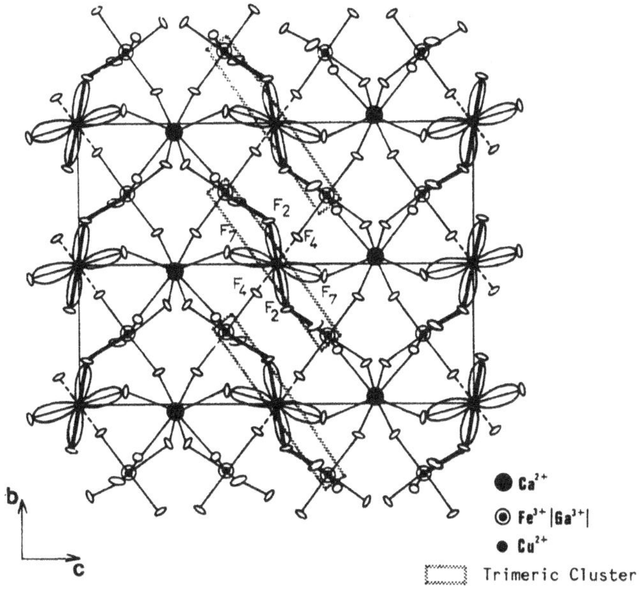

Fig. 4. Schematic representation of the $d_{x^2-y^2}$ orbitals of Cu^{2+} ion in $Ba_2CaCuM''F_{14}$ (M'' = Fe^{3+}, Ga^{3+}).

The model is based on a Heisenberg Hamiltonian with different g-values for Fe^{3+} and Cu^{2+}. The intratrimeric exchange constant is antiferromagnetic $J/k \simeq -18$K. The ground state corresponds to S=9/2 and the first four excited states correspond to a decrease of the spin multiplicity, the magnetic behavior therefore becomes similar to that of a trimer with a ferromagnetic interaction. At higher temperature the remaining excited states become populated and the susceptiblity follows now a Curie-Weiss law with a negative θ_p-value. Magnetization measurements at T = 1.8K with applied magnetic fields up to 60 kOe confirm that the exchange

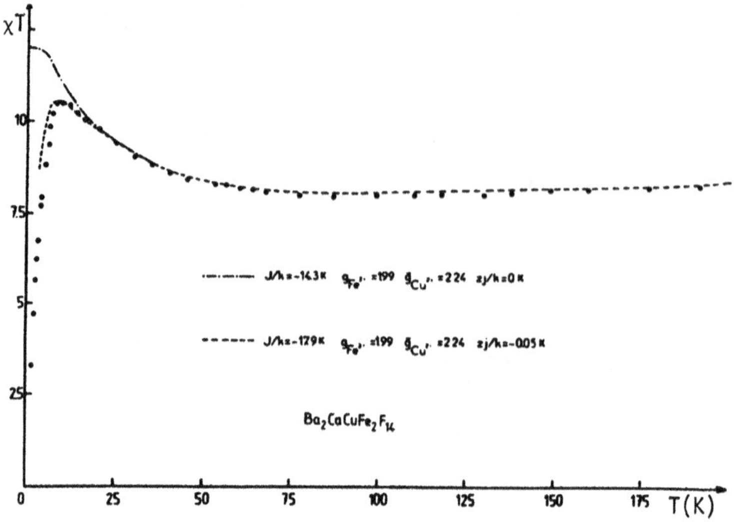

Fig.5. Thermal variation of reciprocal susceptibility of $Ba_2CaCuFe_2F_{14}$.

interaction is antiferromagnetic. However at low temperature intercouplings between trimers arise and have to be taken into account to fit the susceptibility data (Fig.5).

REFERENCES

|1| A.L. Litvin, A.A. Petrunina, S.S. Ostapenko and A.S. Povarennykh, Dopov. Akad. Nauk. Ukr, RSR Ser.B., n03, 47 (1980).

|2| H. Holler, J. Pebler and D. Babel, Z. Anorg. Allg. chem. 522:189 (1985).

|3| M. Drillon, E. Coronado, D. Beltran and R. Georges, Chem. Phys. 79:449 (1983).

|4| J. Darriet, Xu Qiang and A. Tressaud, C. R. Acad. Sc. Paris, 303:1545 (1986).

|5| J. Darriet, Xu Qiang, A. Tressaud and P. Hagenmuller, Mat. Res. Bull., 21:1351 (1986).

DESIGN OF LOW-DIMENSIONAL FERRIMAGNETIC COMPOUNDS: NEW MAGNETIC LATTICES IN THE "EDTA-FAMILY"

Eugenio Coronado[a], Antonio Barba[a], Daniel Beltrán[a], Ramon Burriel[b], and Richard Carlin[c]

[a]Departament de Química Inorgánica, Universitat de Valencia Burjassot, Valencia Spain
[b]Instituto de Ciencia de Materiales de Aragón, Zaragoza, Spain
[c]Department of Chemistry, University of Illinois, Chicago

INTRODUCTION

The ability of EDTA to wrap around one metal atom (M') while still providing bridging carboxylate groups to a second atom (M) is the basis for the obtainment of a wide variety of low-dimensional bimetallic ordered systems.

In this way, the series MM'EDTA.$6H_2O$ [where M, M'= Mn, Co, Zn, Mg(II)] provides an ideal structural support to isolate 1-d ferrimagnets. Here, two of the four carboxylate groups of the $M'(EDTA)^{2-}$ anion are bridging to two $M(H_2O)_4^{2+}$ cations, remaining the other two as terminal ligands. The result is a bimetallic ordered chain arrangement[1-2].

If we find a way to incite the terminal carboxylate groups to link to a third atom an increase of the dimensionality of the ferrimagnetic sysytem would be achieved. A method to do that is the extrusion of some water from the MM'EDTA structure. Thus by a hydrothermal treatment of the parent hexahydrate, a new dihydrate phase can be obtained.

STRUCTURE

The polymeric structural arrangement shown by these dihydrates has been determined elsewhere on the CoCoEDTA.$2H_2O$ compound[3].

Such a structure offers three different sites to the metal ions: one, denoted as "chelated", is an octahedral site in which the metal M' is five chelated by the EDTA and co-ordinated to one water molecule; the second, "hydrated" site, is also an octahedral one in which the metal M is co-ordinated to two water molecules and linked by four bridging carboxylates to four M' atoms. Finally, the third is a tetrahedral site in which M^t is

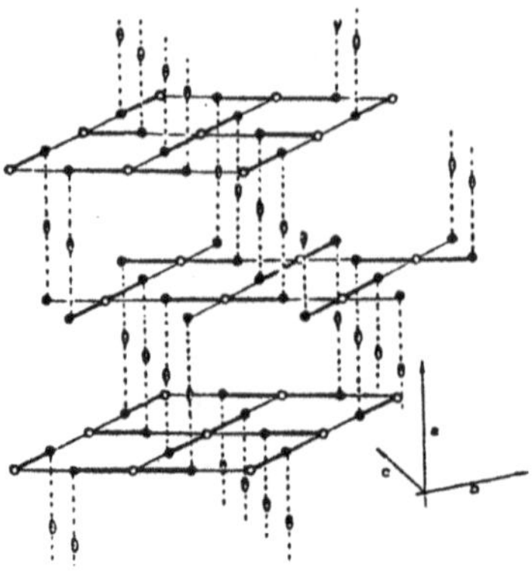

Fig.1. Scheme of the layers of regularly alternating octahedra (o,M ; •,M') connected by tetrahedral (θ,M^t) sites.

coordinated to two bridging carboxilate and two imidoacetate groups from neighboring EDTA moieties.

The polymer can be viewed as a succession of puckered layers built up by an alternating network of "chelated" and "hydrated" sites. These layers are intercalated between layers of tetrahedral sites in such a way that each tetrahedral site connects two "chelated" sites of neighboring bimetallic layers as shows figure 1.

According to the structural findings this new class of compounds may be conveniently formulated as $M^tM\,M'_2(EDTA)_2.4H_2O$; in short (M^t M M').

MAGNETIC LATTICE ENGINEERING

This kind of structural support offers us a great flexibility in the design of new magnetic lattices. The occupational order of the "hydrated" and "chelated" sites by various metal ions has shown to be controlled by the stability constants of the discrete metal-EDTA complexes in aqueous solution[4]. This fact may equally guarantee here the obtention of bimetallic ordered systems of variable magnetic dimensionality.

Through a selective accommodation of diamagnetic and paramagnetic ions in the appropiate sites,magnetic lattices ranging from isolated trimers to 3-d bimetallic systems would appear. Thus,if a diamagnetic ion occupies the M site magnetic bimetallic trimers of the type M'-M^t-M' could be created. A more interesting situation would appear when M^t is diamagnetic; the exchange pathways are,in this case,restricted to the bimetallic layer (M,M'),so that a two-dimensional ferrimagnetic system could be isolated. Notice that this layer is formed by twice more M' than M sites; then,even when M and M' were equal, the two magnetic sublattices would be uncompensated leading to a ferrimagnetic layer. Finally, if the diamagnetic M^t ion is substituted in a continuous way by increasing amounts of a magnetic ion, a controlled passage from two to three dimensions would be espected in the magnetic lattice.

Bear in mind the above ideas we have introduced the metals Ni, Co, Mg and Zn as divalent ions in the structure (M^t M M'). The introduction by hydrothermal treatment of other interesting ions as Cu(II) and Mn(II) was unsuccessful due to redox unstability problems of the EDTA-M aqueous system at the working temperatures.

The following ordered compounds have been prepared:

i) (Co Mg Co) and (Co Mg Ni), that consist of isolated trimers

ii) (Zn Ni Ni), that consists of bimetallic layers

iii) (Co Ni Ni), (Co Co Ni) and (Co Co Co); here, the bimetallic layers (Ni Ni), (Co Ni) and (Co Co) are connected by tetrahedral Co(II).

MAGNETIC RESULTS

We report in figure 2 the magnetic susceptibility data of (Zn Ni Ni) and (Co Co Co). In the former we observe a continuous decrease of the χT product when cooling down. A quite different behavior is observed in (Co Co Co): χT exhibits a minimum around 0.3 K and a divergence at lower temperatures down to 0.08 K where a three dimensional ordering transition seems to appear (a peak is observed in the χ *vs*. T plot).

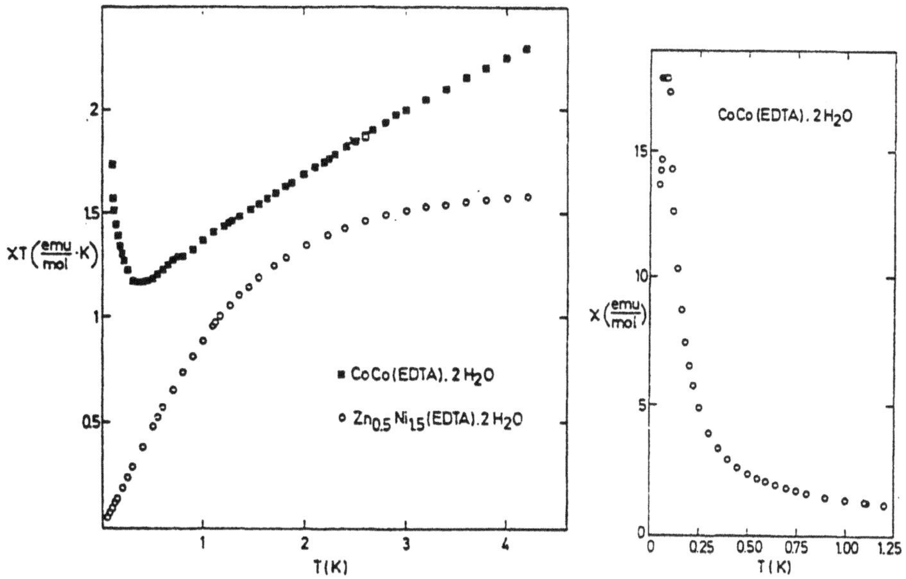

Fig.2. Magnetic behavior of (Zn Ni Ni) o and (Co Co Co) ■ .

Owing to the nature of the magnetic lattice associated to (Zn Ni Ni) the "antiferromagnetic" behavior exhibited by this system seems to be surprising. This could be related to large zero field splittings (D) of the Ni(II) ions, which in this kind of systems may be comparable to or larger than the nearest neighbor exchange interaction (J). In fact, a value of D around 9 K has been found for the "chelated" site in the family MNi(EDTA).$6H_2O$, while the J values are in the range 1 to 10 K.[1]

In turn, the (Co Co Co) compound exhibits the expected magnetic behavior for low-dimensional ferrimagnetic systems. Owing to the very anisotropic nature of the Co(II) ion, this system may be viewed as a succession of ferrimagnetic Ising layers connected by tetrahedral Co(II). Then, the 3-d ordering is expected to be ferromagnetic if an antiferromagnetic exchange coupling between the Co^t and the Co' sites is assumed.

REFERENCES

1. E.Coronado, M.Drillon, A.Fuertes, D.Beltrán, A.Mosset and J.Galy, Structural and Magnetic Study of $Ni_2(EDTA)(H_2O)_4$ $2H_2O$. Alternating Landé Factors in a Two-Sublattice 1D System, J.Am.Chem.Soc.108:900 (1986)
2. C.P.Landee, Insulating Magnetic Chains: Recent Models and Materials, in this book.
3. P.Gómez-Romero, G.B.Jameson, N.Casañ-Pastor, E.Coronado and D.Beltrán. Low Dimensional Bimetallic Ordered Systems: Synthesis and Characterization of the Isomorphous Series Co_xNi_{2-x}EDTA $2H_2O$. Crystal Structure of Co_2EDTA 2 H_2O and Preferential Site Occupation in CoNiEDTA $2H_2O$, Inorg.Chem.25:3171 (1986).
4. E.Escrivá, A.Fuertes and D.Beltrán, Order and Disorder in Mixed Metal Linear Chains: The Homo and Heterobimetallic EDTA($(M(H_2O)_4O_IO_{II})(M'(EDTA))$ $2H_2O$) Complexes (M=Mg, Mn, Co, Zn and Ni; M'=Co, Ni, Cu and Zn), Transition Met. Chem. 9:184 (1984).

AKNOWLEDGEMENTS

This work was supported by the Comisión Asesora de Investigaciones Científicas y Técnicas .

LOW TEMPERATURE INVESTIGATION OF THE THERMAL AND MAGNETIC PROPERTIES OF 1-d FERRIMAGNETIC SYSTEMS

Eugenio Coronado[a], Paul Robert Nugteren[b], Marc Drillon[c], Daniel Beltrán[a], L.J. de Jongh[b] and Roland Georges[d]

[a] Departament de Química Inorgánica, Universitat de Valencia. Burjassot, Valencia. Spain.- [b] Kamerlingh Onnes Laboratorium, Leiden. The Netherlands.- [c] Departement Science des Materiaux, Strasbourg France.- [d] Laboratoire de Chimie du Solide, Bordeaux. France.

INTRODUCTION

The series of structurally ordered bimetallic chains formulated as MM'(EDTA).$6H_2O$, in short (MM'), provides the most versatile known example of 1-dimensional ferrimagnets, since on the same structural support many different magnetic ions (Mn, Co, Ni and Cu(II)) can be selectively accommodated.[1] Consequently, many choices of alternating magnetic moments with Heisenberg or Ising exchange couplings can be investigated. Moreover, taking into account that the M-M' distances along the chain are alternating, this series could show at the same time an alternation of the exchange parameter[2].

So far, all the analyses of the magnetic and thermal data have been satisfyingly explained ignoring the J-alternation. Here we show that this effect is quite noticeable in the measurements at very low temperatures.

MAGNETIC CHARACTERIZATION

Properties of several members of this family have been reported and interpreted taking into account the spin-symmetry of the interacting ions. Some significant results are displayed in the table. These emphasize that some drastic differences among these systems exist. The main features to be noted are:

i) The spin-anisotropy of the Co(II) systems

Owing to the very anisotropic groundstate of the Co(II) ions, we have assumed anisotropic exchange models to explain the magnetic and thermal data of the materials containing this ion ((CoNi), (CoCo), (CoCu) and (MnCo)), and isotropic ones (Heisenberg model) in the other cases. A somewhat surprising Ising behavior has been observed in the susceptibility of (NiNi).[2] That has been attributed to the large zero field splitting of the "chelated" Ni(II) ion.

Table. Magnetic characterization of the MM'(EDTA).$6H_2O$ family. The letters inside the parentheses refer to the symmetry of the interaction hamiltonian : (H)=Heisenberg; (I)=Ising; (A)=Anisotropic exchange model. Note that in the hamiltonian the exchange constant is written as -J. Scaling factors have been considered in J to account for the effective spin doublet of Co(II). T_m and T_c are defined in the text.

COMPOUND	EXCHANGE J/k (K)	COMMENTS	REF
(NiNi)	-8.35 (H) -8.24 (I)	Susceptibility measurements (χ) to 4K no T_c to 0.07K (in C_p) uniform chains (J=j)	2
(CoNi)	-4.5 (H) -7.5 (I)	χ data to 2K T_c=0.5K uniform chains	3
(CoCu)	-2.7 (A)	χ data to 0.15K. T_m=0.2K no T_c to 0.07K "quasi" isolated dimers ($j/J < 0.01$)	4
(CoCo)	-6.3 (A)	χ data to 0.15K. T_m=0.2K no T_c to 0.07K "quasi" isolated dimers	4
(MnCo)	-1.14 (I)	χ data to 1.20K. T_m=1.6K T_c=1.06K uniform chains	5
(MnNi)	-1.5 (H)	χ data to 0.15K. T_m=2.8K T_c=0.66K uniform chains	5
(MnCu)	-0.5 (H)	χ data to 0.15K. T_m=0.28K T_c=0.20K J-alternating chains	This work

In fact, the separation between the singlet $|0\rangle$ groundstate and the doublet $|\pm1\rangle$ (estimated to be D/k=8.7K) is very close in energy to the exchange constant.

ii) The 1-d ferrimagnetic behavior and the 3-d magnetic ordering of the Mn(II) systems

In Figure 1 we plot the magnetic behavior of (MnCu). The product χT of this compound is characterized by i) a minimum around T_m=0.28K, ii) a divergence down to T_c=0.20K, and iii) a sharp decrease at lower temperatures ($T<T_c$). Notice that i and ii are the characteristic features of 1-d ferrimagnets.[5] On the other hand, the observation at T_c of a λ-type anomaly in the specific heat measurements (figure 2) agrees with a transition to a magnetically ordered state. Then, iii is to be attributed to an antiferromagnetic 3-d ordering.

This is the general behavior of the (MnM') family (M' = Co, Ni and Cu). One expects a similar behavior in the other compounds (even in the homometallic case) by the virtue of the fact that the two magnetic moments involved are non-compensated. Nevertheless, due to the similar magnitude of these moments (case of (NiNi) and (CoNi)), or to the dimerization of chains (case of (CoCu) and (CoCo)), the ferrimagnetic behavior is located at too low temperatures to be observed (in the later situation a little minimum has been observed at 0.2K).

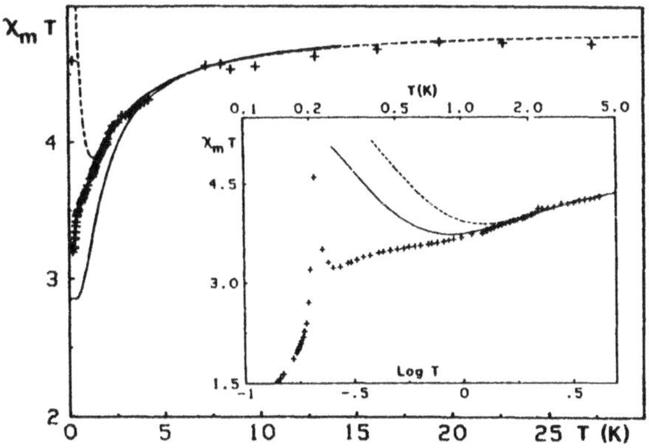

Figure 1. Magnetic behavior of (MnCu). Dashed line gives the result of the fit to a (S,S')-ferrimagnetic Heisenberg chain with S = 5/2, S' = 1/2, J = -0.5K and g_{Cu}/g_{Mn} = 1.15. The behavior of an isolated dimer is reported in solid line. In the inset the solid line corresponds to the fit at low temperatures (T < 5K); it results J = -0.35K.

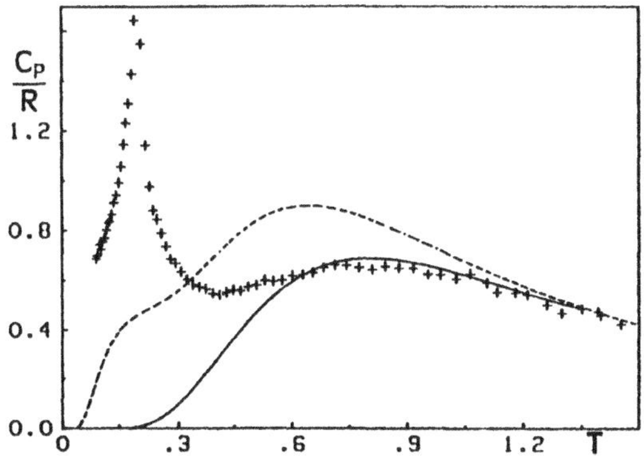

Figure 2. Specific heat data for (MnCu). Dashed line gives the behavior of a closed chain of 3 pairs (5/2,1/2) for J = -0.47K. Solid line gives the behavior of an isolated dimer (J = -0.34K).

With regard to the interchain interactions, the appearance of a 3-d ordering in (MnM') and (CoNi) only, may be closely related to the ground spin configuration of the chain. Thus, while (MnM') and (CoNi) systems are characterized by containing two distinct spins (high-spin groundstate), the others are described by a singlet-spin groundstate.

iii) The J-alternation

A wide variation in the J-alternation has been noted (see table). Thus, while the (MnNi) system behaves down to T_c=0.66K as a uniform chain (j/J=1), (CoCo) and (CoCu) exhibit a "quasi" total dimerization (j/J<0.01). On the other hand, we note that while systems involving Mn or Ni(II)ions behave as uniform chains, those involving Co or Cu(II) ions show significant dimerizations. The effect of the J-alternation is clearly noticeable in the magnetic behavior of the (MnCu) system (see inset of fig.1). Thus, while the theoretical behavior of an uniform chain predicts a divergence of the product χT below 1K, experimental data show a continuous decrease down to 0.3K.

The observed variability in the J-alternation could be related with the variability found in the alternating intermetallic distances (structural factor), but also with the electronic nature of the interacting ions (electronic factor). Let us discuss both factors.

With regard to the former, structural data indicate that the inequivalence in the M-M' distances is larger in the bimetallic chains ((MnCo) and (MnCu)) than in the homometallic ones ((NiNi) and (CoCo)).

With regard to the latter, one expects that the (electronically) isotropic Mn(II) ion will be lesser sensible to small variations in the geometry of the chain than the (electronically) anisotropic Cu(II) ion . This is due to the fact that while in Mn(II) the five d-orbitals are magnetically active, in Cu(II) only the $d_{x^2-y^2}$ is.

From the above remarks, we can conclude that the degree of J-alternation is mainly dependent on the electronic nature of the interacting ions.

AKNOWLEDGEMENTS

This work was supported in part by the European Economic Community (grant ST2/164), and the Comisión Asesora de Investigaciones Científicas y Técnicas (2930/83). E.C. thanks a grant from the Generalitat Valenciana.

REFERENCES

1. C.P.Landee, Insulating Magnetic Chains:Recent Models and Materials, in this book, and references therein.

2. E.Coronado, M.Drillon, A.Fuertes, D.Beltrán, A.Mosset and J.Galy, Structural and Magnetic Study of $Ni_2(EDTA)(H_2O)_4.2H_2O$. Alternating Landé Factors in a Two-Sublattice 1D System, J.Am.Chem.Soc. 108:900 (1986)

3. M.Drillon, E.Coronado, D.Beltrán and R.Georges, Ferrimagnetic Heisenberg Chain. Influence of a Random Exchange Interaction, J.Appl.Phys. 57:3353 (1985).

4. E.Coronado, M.Drillon, P.R.Nugteren, L.J.de Jongh and D.Beltrán, Anisotropic Exchange in The $Co_2(EDTA)6H_2O$ and $CoCu(EDTA)6H_2O$ Bimetallic Ordered Chains; Low Temperature Investigation of The Thermal and Magnetic Properties, submitted to J.Am.Chem.Soc.

5. M.Drillon, E.Coronado, D.Beltrán, J.Curely, R.Georges, P.R.Nugteren, L.J. de Jongh and J.L.Genicon, J.Mag.Mag.Mater. 54-57:1507 (1986).

FERRIMAGNETISM IN HOMOMETALLIC LINEAR CHAINS: $CuX_2 \cdot 4/3TMSO$, X = Cl, Br

C.P. Landee, A. Djili, M. Newhall, D.F. Mudget
R.D. Willet*, H.Place* and B. Scott*

Department of Physics
Clark University
Worcester, Massachusetts,
U. S. A.

*Department of Chemistry
Washington State University
Pullman, Washington
U. S. A.

1. INTRODUCTION

The development of ordered bimetallic chains is an exciting advance in coordination chemistry. The possibility of alternating transition metal ions of different magnetic moments leads to uncompensated moments and ferrimagnetic behavior. A review of these chains is presented elsewhere in this volume [1]. We present here the first report of ferrimagnetism in homometallic compounds, where the ferrimagnetism arises from an odd number of metal ions within the unit cell.

2. EXPERIMENTAL

The title compounds, full name tetrakis(tetramethylenesulfoxo)copper(II) hexahalodicuprate(II), are prepared by evaporation of saturated solutions, yet care must be taken to avoid the presence of alternate phases [2]. The chloride crystallizes readily from methanolic solutions of a stoichiometric ratio of anhydrous copper chloride and TMSO. The bromide analog, $CuBr_2 \cdot 4/3TMSO$, which grows as black, triclinic crystals, is best grown from ethanol; it cannot be obtained from methanol.

The crystal structures were studied at room temperature using a Nicolet R3m/E system for the data collection. The structures were solved by a combination of direct methods and difference Fourier maps. Magnetic measurements were made on powders between 1.8 and 300 K in fields of 10 kilo-oersted using a vibrating sample magnetometer. Temperatures were measured using a calibrated carbon-glass resistance thermometer[3]. Full details will be published elsewhere.

3. DESCRIPTION OF STRUCTURE

The crystal structures of the two compounds are identical and consist of alternating chains of tetrakis(tetramethylenesulfoxo) copper(II) cations and dimeric, bibridged $Cu_2X_6^{=}$ anions which extend in chains along the crystallographic c axis, Fig. 1. The unit cell therefore contains an odd number of copper ions, one from the monomeric cation and two from the

TABLE 1. Structural Properties of $CuX_2 \cdot 4/3TMSO$

Unit Cell Dimensions

COMPOUND	a (Å)	b (Å)	c (Å)	α(°)	β (°)	Γ (°)
Chloride	8.250(4)	9.330(5)	11.131(6)	67.59(4)	73.29(6)	74.67(6)
Bromide	8.448(2)	9.630(2)	11.655(2)	65.42(1)	71.59(2)	73.52(1)

Bond Distances (Å)

	Chloride	Bromide		Chloride	Bromide
Cu(1)-X(1)	2.213	2.340	Cu(1)-X(2)	2.183	2.315
Cu(1)-X(3)	2.27	2.410	Cu(1)-X(3a)	2.325	2.447
Cu(2)-X(1)	2.894	3.063	Cu(2)-O(1)	1.947	1.939
Cu(2)-O(2)	1.957	1.942	Cu(1)-Cu(1a)	3.351	

Bond Angles (°)

Cu(1) - X(3) - Cu(1a)	93.4 (Cl)	91.1 (Br)
Cu(1) - X(1) - Cu(2)	112.3 (Cl)	112.4 (Br)

dimeric anion; the presence of odd number of magnetic atoms is the structural origin of the ferrimagnetic behavior in these compounds. The coordination geometry of the copper atom in the cation, Cu(2) is 4+2, with the four oxygens from the sulfoxide groups arranged around the copper in a square plane with all angles within one degree of 90°. The axial sites of this momomeric copper are occupied by the terminal halides X(1) of the dimeric anions via long bonds of 3.062 Å for the bromide and 2.895 Å for the chloride. The Cu(2)-X(1) bond is normal to the $Cu(2)O_4$ plane. Since the Cu(2) atom sits on an inversion center, it is also connected to the X(1a) terminal halide of a second anion, forming the chains.

The geometry about the Cu(1) atom within the dimer is intermediate between square planar and tetrahedral. The Cu(1) atoms are bibridged through the X(3) halides with bridge angles of 91.1° (Br) and 93.4° (Cl).

The chains are well separated by the bulk of the sulfoxide ligands, which prevent any interchain metal-metal contacts closer than the unit cell translation along a, which is 8.448 Å for the bromide and 8.250 Å for the chloride.

4. MAGNETIC RESULTS

The magnetic results for the powder susceptibilities of the two compounds are presented as plots of χT versus temperature in Fig. 2. For both analogs the high temperature data show a gradual rise in the effective moment which is attributed to the presence of ferromagnetic interactions within the dimers, characterized by the parameter J_d. Below 20 K, the two data sets differ greatly, χT of the bromide undergoing a steep decline, Fig. 2a, while χT of the chloride ascends sharply, Fig. 2b. This change of behavior reflects the the onset of the weaker coupling J_m between the ferromagnetic dimers and the monomers. The systems are best described as alternating chains in which the alternation pattern involves three exchange strengths, two of which are identical by symmetry••• $-J_d-J_m-J_m-$••• where the temperature dependence of the data indicates that $J_d \gg J_m$. If both the exchange constants are positive, at low temperatures the chain becomes an alternating S=1, S=1/2 ferromagnetic chain, with a consequent divergence in susceptibility. Even if J_m is antiferromagnetic, the material acts as an alternating S=1, S=1/2 antiferromagnet with the subsequent ferrimagnetic behavior at temperatures low compared to J_m.

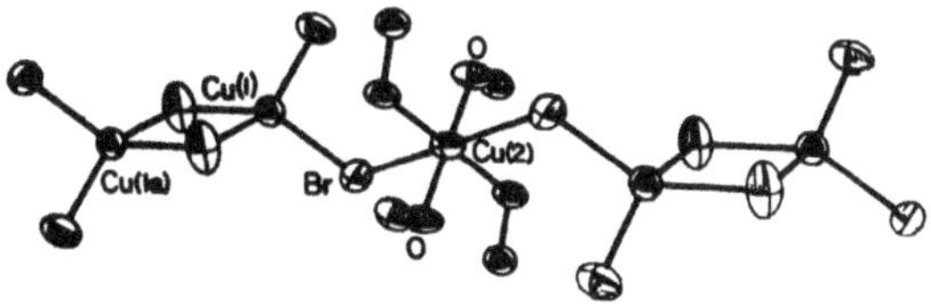

Fig. 1. Crystal structure of $CuBr_2 \cdot 4/3$ TMSO, showing the chains of alternating copper dimers and monomers. Only the oxygens of the TMSO groups are shown, for clarity.

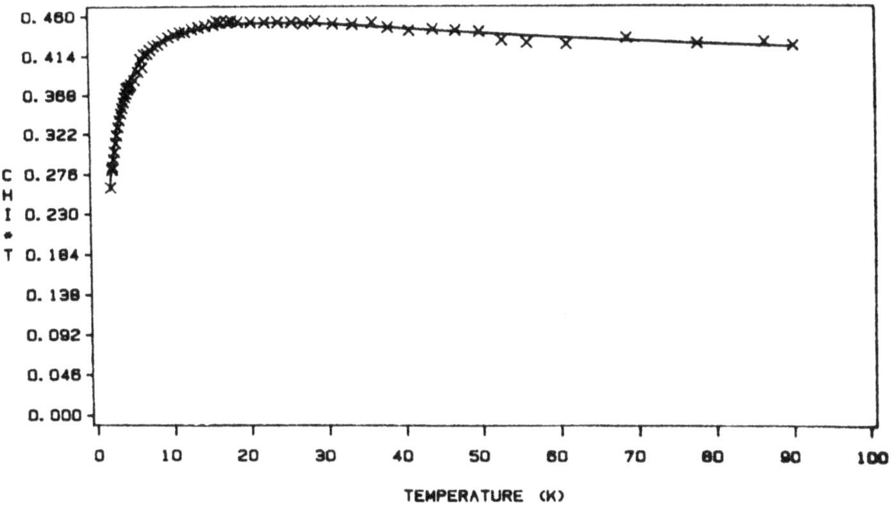

Fig. 2a

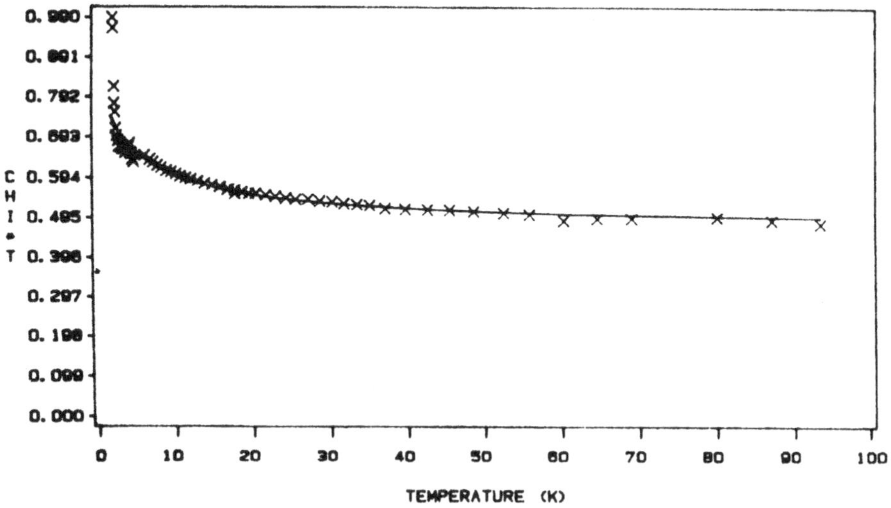

Fig. 2b

Fig. 2. Susceptibility data (plotted as Chi*T versus T) for $CuBr_2 \cdot 4/3$TMSO (Fig. 2a) and $CuCl_2 \cdot 4/3$TMSO (Fig. 2b). Solid curves correspond to the theoretical fits described in the text.

In order to explain the data and extract exchange strengths, approximations are necessary. Although closed form expressions exist for antiferromagnetic alternating chains with an exchange strength pattern of $\cdots J, \alpha J, J, \alpha J, \cdots$, no such expressions exist when there are three exchange parameters in the sequence, even two of them are identical. We have therefore initially compared the data to the predictions for the susceptibility of isolated trimers with two exchange constants. This approximation will of course break down at temperatures comparable to the weaker of the two exchange strengths since the product χT of a trimer approaches a constant value at T goes to zero. In contrast, the product χT of a ferrimagnetic chain with an odd number of $S=\frac{1}{2}$ moments per unit cell diverges at low temperature, no matter what combination of ferro and antiferromagnetic exchange constants is used.

The results of the fitting procedure are shown as the solid lines in Fig. 2. The solid line for the bromide, Fig. 2a corresponds to $J_d(Br)/k = 18(4)$ K and $J_m(Br)/k = -1.5$ K. The value obtained from the fit for J_m have been divided by two to adjust for the two exchange paths between the monomers and the adjacent dimers. The rapid dropoff of χT below 20 K corresponds to the antiferromagnetic alignment of the monomeric $S=\frac{1}{2}$ ions with regard to the effective S=1 copper dimer. The characteristic divergence of χT in the $S=\frac{1}{2}$, S=1 antiferromagnetic chain is not seen for temperatures above 1.8 K but the minimum in χT is only predicted to occur at a temperature of $1.36\ |J_m|/k$ [4], i. e. near 2 K for the bromide. That the minimum is not yet observed may reflect the uncertainty of the value obtained for J_m from the approximate model.

The corresponding values for the exchange strengths of the chloride are $J_d(Cl)/k = 16(3)$ K and $J_m(Cl) = 1.3$ K, both with positive sign. Under these conditions, there will be no minimum in χT, only a rapid divergence at low temperatures. Experiments are planned at lower temperatures for both compounds, to confirm the presence of the minimum in χT for the bromide and to follow the rate of divergence of the susceptibility in the chloride.

There are two previous reports of isotropic $S=\frac{1}{2}$, S=1 ferrimagnetic chains, both bimetallics of nickel and copper ions. In both $NiCu(ox)\cdot 4H_2O$ [4] and $NiCu(pba)(H_2O)\cdot 2H_2O$ [5], the antiferromagnetic exchange strengths are strong, due to the efficiency of the oxalato and oxamato ligands at transmitting the antiferromagnetic interaction. More details of comparison are found in reference 1.

ACKNOWLEDGEMENTS

The support of the National Science Foundation under grant DMR-8306432 is gratefully acknowledged. We thank R. E. Greeney for his assistance in the data collection.

REFERENCES

1. Landee, C. P., this volume.
2. Landee, C. P. and Greeney, R. E. 1986 Inorg. Chem. 25, 3771.
3. Landee, C. P., Greeney, R. E. and Lamas, A. C. 1987, Rev. Sci. Inst., accepted for publication.
4. Verdaguer, M., Julve, M., Michalowicz, A. and Kahn, O. 1983 Inorg. Chem. 22, 2624.
5. Pei, Y., Verdaguer, M., Kahn, O., Sletten, J. and Renard, J-P. 1987 Inorg. Chem. 26, 138.

MAGNETISM OF QUANTUM-CLASSICAL FERRIMAGNETIC CHAINS SHOWING ALTERNATE HEISENBERG COUPLING: APPLICATION TO MnCu (Obp)$(H_2O)_3$, H_2O

J. Curély, R. Georges, J.C. Gianduzzo and Q. Xu

Laboratoire de Chimie du Solide du C.N.R.S., Université de Bordeaux I, 33405, Talence Cédex, France

O. Kahn and Y. Pei

Laboratoire de Spectrochimie des éléments de transition, Université de Paris-Sud, Orsay, France

J. Sletten

Department of Chemistry, University of Bergen, 5000 Bergen, Norway

INTRODUCTION

For a long time considerable litterature has been devoted to one-dimensional spin systems. The possibility of exactly solving various non trivial cases has been a strong motivation in this field. More recently the increasing number of new compounds showing quasi one-dimensional magnetic behavior gave a further impulsion to this subject. In the past, mainly so-called ferromagnetic and antiferromagnetic chains were considered from the point of view of their static properties[1]. Then, dynamic properties were considered[2], and later on much attention was paid to ferrimagnetic chains,[3] showing two types of alternating cations. A large variety of situations are involved in this last category, giving rise to new specific problems[4]. In the present article we are concerned with the static magnetic properties of a large class of ferrimagnetic chains defined by the following two conditions : -i) Two types of cations alternate along the chain. One of them, at least, must be characterized by a spin number large enough to be treated classically ; ii) The hamiltonian is fully isotropic (Heisenberg coupling, no spin-lattice anisotropy) within the spin space.

The first condition is generally considered to be well fitted for spin numbers as large as 5/2. The isotropy condition is often fulfilled due to first order quenching of the orbital momentum. Seiden[5] has solved this problem in the special case of a regular chain involving 1/2 - spins as the non-classical ones. The present work extends his calculations to any quantum spin number. Moreover, it includes the very general case where this number as well as the Landé factors on both type of cations, and the exchange interactions are randomly distributed. Correlations may be introduced between any two of these distributions. Of course this includes the simple case where part only, or none, of these parameters shows a random distribution. Also, the finite length effects may be treated by mean of the present model.

Let us start with the hamiltonian

$$H = \sum_i J_i \vec{s}_i((1+\alpha_i)\vec{S}_i+(1-\alpha_i)\vec{S}_{i+1}).$$

Here, $\vec{s}_i$ refers to the quantum spin of the ith unit cell, isotropically coupled with its classical neighbors $\vec{S_i}$ and $\vec{S}_{i+1}$. The corresponding magnetic moments are $g_i\vec{s}_i$ and $G_i\vec{S}_i$. Without loss of generality, all the S_i's will be given a unit amplitude. In the fully isotropic case, the magnetic susceptibility χ is known to be simply related to spin-spin correlations through the Landé factors :

$$\chi = \frac{\beta}{3}\sum_{ij}(g_i g_j<\vec{s}_i.\vec{s}_j>+g_i G_j<\vec{s}_i.\vec{S}_j>+G_i g_j<\vec{S}_i.\vec{s}_j>+G_i G_j<\vec{S}_i.\vec{S}_j>),$$

where β is the Boltzmann factor. Let dS_i denote the integration over the spherical coordinates (say θ_i and ϕ_i) of the spin S_i, and by Tr_i the sum over the states of a complete basis for s_i. The partition function Z_{n+1} of a chain beginning at S_o and ending at S_n is given by :

$$Z_{n+1} = \int dS_o Tr_o \int dS_1 Tr_1 \ldots \int dS_{n-1} Tr_{n-1} \int dS_n \exp(-\beta H).$$

Let us describe the sum Tr_i in terms of the spherical harmonics of the angles θ_i, ϕ_i, θ_{i+1} and ϕ_{i+1}.

$$Tr_i \exp(-\beta H_i) = \sum_l A_l(\lambda_i,\eta_i,s_i)\sum_{m=-1}^{+1}(Y_l^m(\theta_i,\phi_i)Y_l^{-m}(\theta_{i+1},\phi_{i+1}),$$

where :

$$\lambda_i = -\beta J_i\sqrt{2(1+\alpha_i^2)}\ ,\quad \eta_i=\sqrt{(1-\alpha_i^2)/(1+\alpha_i^2)},$$

$$A_l(\lambda,\eta,s) = 2\pi\sum_{\sigma=-s}^{+s}\int_{-1}^{+1}\exp(\sigma\lambda\sqrt{1+\eta u})\ P_l(u)du.$$

Here $P_l(u)$ refers to the l-index Legendre's polynomial. Due to orthogonality of the spherical functions, integrating over the directions of the S_i's leads to the very simple result :

$$Z_{n+1} = 4\pi\sum_{i=0}^{n} A_o(\lambda_i,\eta_i,s_i).$$

In a similar way we get the correlations

$$<\vec{S}_o.\vec{S}_n> = \sum_{i=0}^{n-1}(A_1(\lambda_i,\eta_i,s_i)/A_0(\lambda_i,\eta_i,s_i)).$$

Introducing now the development

$$Tr_i(S_i^z\exp(-\beta H_i)) = \sum_l B_l(\lambda_i,\eta_i,s_i)\quad Y_l^m(\theta_i,\phi_i)Y_l^{-m}(\theta_{i+1},\phi_{i+1}),$$

with

$$B_l(\lambda,\eta,s) = 2\pi\sum_{\sigma=-s}^{+s}\int_{-1}^{1}(1+\eta u)^{-1/2}\exp(\sigma\lambda\sqrt{1+\eta u})P_l(u)du.$$

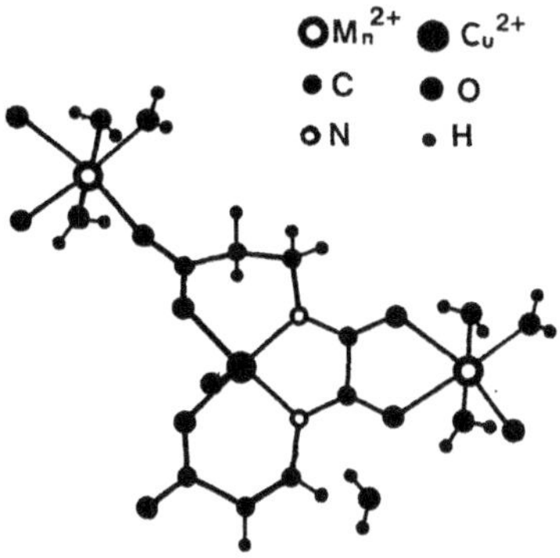

Fig. 1. Structure of the ferrimagnetic chains in $MnCu(obp)(H_2O)_3$, H_2O.

We readily show that the remaining correlations reduce to

$$\langle \vec{s}_o . \vec{S}_n \rangle = -\beta J_o (A_o(0))^{-1}((1+\alpha_o)B_1(0)+(1-\alpha_o)B_o(0)) \sum_{i=1}^{n-1} (A_1(i)/(A_o(i)),$$

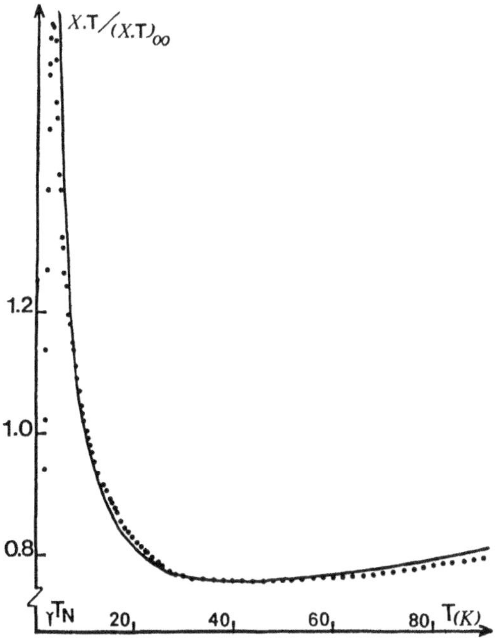

Fig. 2. Thermal variation of the powder magnetic susceptibility of $MnCu(obp)(H_2O)_3$, H_2O. The full line shows the theoretical fitting.

and analogous expressions for $<\vec{s}_o.\vec{s}_n>$ and $<\vec{S}_o.\vec{s}_n>$. Here, the unique argument (i) for A_1 and B_1 is a condensation for (λ_i,η_i,s_i). We notice that only A_o, A_1, B_o and B_1 have to be evaluated. The corresponding integrals may be performed analytically. For instance, we get

$$A_o(\lambda,\eta,s) = 4\pi/(\lambda^2\eta) \sum_{\sigma=-s}^{s} \sum_{\varepsilon=\pm 1} \varepsilon\sigma^{-2}(\sigma\lambda\sqrt{1+\varepsilon\eta}-1)\exp(\sigma\lambda\sqrt{1+\varepsilon\eta}),$$

and similar expressions for A_1, B_o and B_1. Let us now introduce the new notations ($\varepsilon=\pm 1$) :

$$P_i = A_1(i)/A_o(i) \ , \ Q_{i,\varepsilon} = -\beta J_i((1+\varepsilon\alpha_i)B_o(i)+(1-\varepsilon\alpha_i)B_1(i))/A_o(i) \ ,$$

and denote $\overline{Y}$ the average value of any randomly distributed i-dependant quantity Y_i, we obtain the general expression for the infinite random chain susceptibility (referred to the "unit cell") :

$$\chi_{cell} = (\beta/3)\ (\overline{g^2 s(s+1)+G^2}+2\overline{gGP}+(2/(1-\overline{P}))(\overline{gQ_+}.\overline{gQ_-}+\overline{gQ_-}.\overline{G}+\overline{GQ_+}.\overline{gP}+\overline{G}.\overline{GP})).$$

Of course, whenever necessary any suitable average value has to be computed with reference to the specific random distribution

APPLICATION TO THE COMPOUND MnCu(obp)$(H_2O)_3$,H_2O

In the present part we merely consider the regular chain with alternating exchange constants. Namely, α_i will take a non vanishing constant value, α. Such a situation conveniently describes the molecular compound MnCu(obp) $(H_2O)_3$, H_2O we have recently synthetized. The crystal structure shows an assembly of very well separated molecular chains containing Mn^{2+} and Cu^{2+} cations (fig. 1). Each manganese interacts with its two neighboring copper cations through quite different organic links, so that two different exchange interactions are expected ($J(1\pm\alpha)$). Actually, we got the best fitting of the experimental susceptibility curves with :

$$J(1+\alpha) = 46\pm 1K,\ J(1-\alpha) = 9{,}8\pm 1K,\ g = 2{,}32\pm 0{,}05$$

The strongest coupling must be affected to the shortest bridging, which is oxamido-type and involves two Oxygen and tow Nitrogen atoms. The exchange constants between Mn^{2+} and Cu^{2+} have also been evaluated when similar oxamato- (one N, three O) and oxalato- (four O) bridgings are involved[6]. They are respectively 26K and 36K. This reveals a strict linear increase of the coupling with the number of Nitrogen present in the bridge - Such a behavior should, no doubt, be observed for various pairs of coupled cations. It also indicates that the basic features of the present model should be convenient for a large number of one dimensional ferrimagnets.

REFERENCES

1. L. J. de Jongh, A. R. Miedema, Adv. Phys. 23, 1 (1974).
2. M. Steiner, J. Villain, C.G. Windsor, Adv. Phys. 25, 87 (1976).
3. M. Drillon, J. C. Gianduzzo, R. Georges, Phys. Letters 96A, 413 (1983).
4. J. Curély, R. Georges, M. Drillon, Phys. Rev. B33, 6243 (1986).
5. J. Seiden, J. Physique Lettres 44, L947 (1983).
6. M. Verdaguer, M. Julve, A. Michalowicz, O. Kahn, Inorg. Chem 22, 2624 (1983). Y. Pei, M. Verdaguer, O. Kahn, J. Sletten, J. P. Renard, Inorg. Chem. 26, 138 (1987).

DIMERS, REGULAR AND DOUBLE CHAINS IN VANADYL (IV) PHOSPHATES

Gérard Villeneuve, Pedro Amoros*, Daniel Beltran* and Marc Drillon **

*Laboratoire de Chimie du Solide du CNRS, Universite de Bordeaux I, Talence, France and **Department of Science des Marieux, EHICS, Strasbourg, France

INTRODUCTION

Vanadyl phosphates offer a large variety of crystal structures favorable to low dimensional magnetic interactions. According to the connection of VO_6 octahedra and PO_4 tetrahedra, one can observe isolated vanadyl dimers in $VO(HPO_4)$-$0.5H_2O$[1], regular vanadyl chains in $VO(PO_3)_2$[2] and $VO(H_2PO_4)_2$[3], double chains in $(VO)_2P_2O_7$[4] and $VO(HPO_4)$-$4H_2O$[5]. Furthermore, we have isolated new vanadyl hydrogenphosphates hydrates $VO(HPO_4)$-nH_2O (n = 1.5, 2 and 3), the structure of which is not fully known.

All of these materials contain vanadyl, VO^{2+} cations, characterized by a short V-O distance (typically 1.6 A) due to the existence of double bond V=O. The single electron occupies the bonding vanadium d_{xy} orbital ($^2B_2$ state in C_{4v} symmetry) lying in the plane perpendicular to the V=O bond. MO calculation[6] and UV-visible spectroscopic study show that the first excited state is 12000 cm^{-1} higher than the fundamental one.

In this article we present the analysis of the magnetic properties of these low dimensional vanadyl phosphates and we correlate the values of the exchange interactions involved with the stacking of vanadyl entities.

EXPERIMENTAL

We have perfected an unified method which allows to obtain, besides the known mono and dihydrogenphosphates $VO(HPO_4)$-$0.5H_2O$ $VO(HPO_4)$-$4H_2O$, $VO(H_2PO_4)_2$, the three new monohydrogenphosphates hydrates cited above. The control of the process is carried out by the proper choice of temperature, pH and solvent. The starting materials are V_2O_5 and H_3PO_4 ; iodhydric acid acts as reductor of V(V) and donor of protons, the ratio acetone/water (solvent) is fixed between 0 and 50, depending on the desired degree of hydration. The anhydrous materials $(VO)_2P_2O_7$ and $VO(PO_3)_2$ are obtained by dehydration of $VO(HPO_4)$-$0.5H_2O$ and $VO(H_2PO_4)_2$ respectively.

The products are characterized by chemical analysis, thermogravimetry and X-ray diffraction. Their magnetic properties are measured on a SQUID magnetometer operating between 0 and 6T at temperature down to 1.8 K. The susceptibility is determined from the slope of the magnetization vs H between 0 and 0.1 T. Diamagnetism and Van Vleck temperature independent paramagnetism are substracted before analysis of the data.

DIMERS IN $VO(HPO_4)$-0.5H_2O

The crystal structure of $VO(HPO_4)$-0.5H_2O consists of layers containing pairs of face shared VO_6 octahedra linked by PO_4 tetrahedra[1]. The layers are held together by hydrogen bonding. The dimeric entity is shown in Fig. 1. As pointed out by Johnson et al[7], the behavior of mgnetic susceptibility is well described in terms of isolated dimers with an intradimer isotropic interaction $= -2JS_1S_2$. At low temperature (T < 14K), the main contribution is due to the Curie behavior of a small fraction f of paramagnetic defects. The magnetic data of Fig. 2 are fitted with J = -43 K and f = 2% (assuming S=1/2 defects). The mean g value used for the fit ($\langle g \rangle = 1.97$) is determined by ESR.

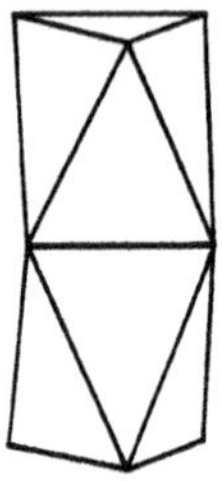

Fig. 1. Binuclear vanadyl entity (schematic)

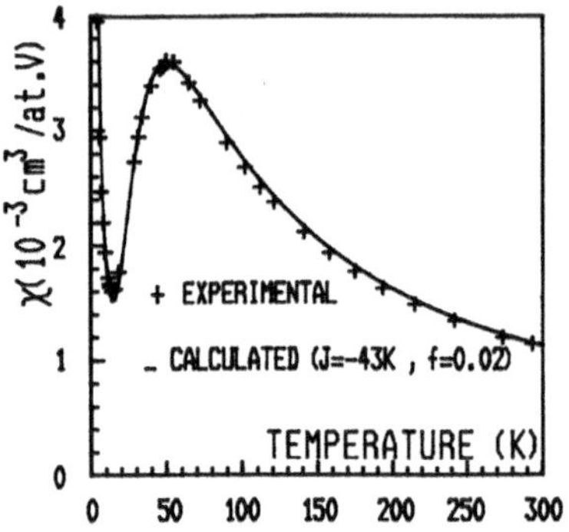

Fig. 2. Magnetic susceptibility of $VO(HPO_4)$-0.5H_2O

A similar mode of condensation of vanadyl ions exists in $Cs_3V_2O_2F_7$, in which the common face only comprises fluorine atoms. In spite of its smaller V-V distance (2.99 A compared to 3.09 Å in $VO(HPO_4)$-0.5H_2O), the J value determined by Darriet et al. is only -13.5 K,[4] suggesting that the anions take part in the exchange mechanism.

REGULAR CHAINS IN $VO(PO_3)_2$ AND $VO(H_2PO_4)_2$

The structure of $VO(PO_3)_2$ is formed by infinite chains of corner sharing VO_6 octahedra[2]; two adjacent chains are connected by two edge sharing PO_4 tetrahedra[2] (Fig. 3). The magnetic properties were investigated by Tofield et al. who found a quasi perfect paramagnetic behavior[9]. Our data disagree with these results : (i) the high temperature χ^{-1} vs T (T < 10 K) follows a Curie-Weiss law with $\theta \simeq -5$ K, (ii) the susceptibility exhibits a broad maximum arount 4 K and (iii) a sharp peak is observed at 3.30K (Fig. 4), as a consequence of the onset of an antiferromagnetic ordering. Above T_N, the susceptibility is analysed considering Heisenberg linear chains with an intrachain coupling J and an interchain coupling j. If we treat the latter in the molecular field approximation, the susceptibility is given by[10] :

$$\chi = \chi_c/(1 - zj\,\chi_c/Ng^2\beta^2)$$

where χ_c is the susceptibility of isolated chains as calculated numerically by Bonner and Fisher[11], z the number of magnetic neighbors in the adjacent chains (here z = 4). The best fit of the data gives J = -3.0 K and j =-0.3 K with $<g>$= 1.96 obtained by ESR.

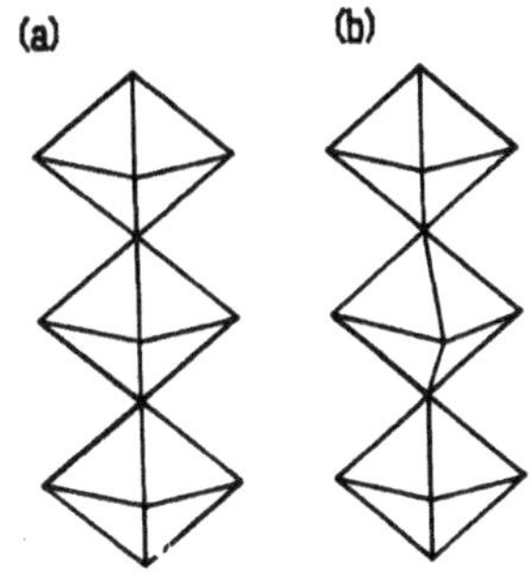

Fig. 3. Vanadyl chains in $VO(PO_3)_2$(a) and $VO(HPO_4)_2$ (b)

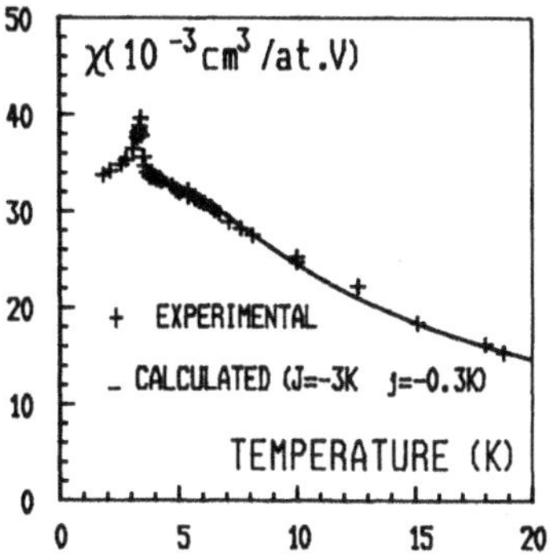

Fig. 4. Magnetic susceptibility of $VO(PO_3)_2$

The stacking of VO_6 octahedra in $VO(H_2PO_4)_2$ differs only by the fact that their equatorial planes are alternatively twisted by + and -18° with respect to the axis of the chain (Fig.3). No long range magnetic ordering is observed down to 1.8 K, but a broad maximum occurs around 2.5 K. The susceptibility is well fitted assuming isolated chains with J = -2.0 K. This result is consistent with a less effective (direct and indirect) overlapping between the d_{xy} orbital due to the twisting of the octahedra.

LADDER LIKE CHAINS IN $(VO)_2P_2O_7$

$(VO)_2P_2O_7$ can be regarded as a S = 1/2 ladder structure : infinite chains of corner sharing VO_6 octahedra along the V=O direction form the uprights while edge sharing VO_6[4] octahedra in a direction perpendicular to the uprights act as the rungs (Fig. 5). At low temperature (T < K), the susceptibility exhbits a Curie-Weiss behavior, as does its precursor $VO(HPO_4)$-$0.5H_2O$, due to paramagnetic defects in the structure. Then, a broad maximum occurs around 70K, characteristic of low dimensional antiferromagnetic interactions. No calculation are available at present for infinite ladder of spin 1/2 with isotropic Heisenberg interactions. We have tried to fit the experimental susceptibility assuming isolated dimers (J_c=0) and isolated chains (J_d =0). The best agreement is obtained in the latter case with J_c =-52K as shown in Fig. 6. However, Johnston et al.[12] by fitting their data as alternating antiferromagnetic chains found J_1 = -65.7K and J_2 = -46K. Their procedure is questionable but it clearly confirms that theoretical calculation of thermodynamic properties of ladder spin configuration are needed. We have solved this problem for the Ising case using matrix transfert technique, but the results cannot apply to vanadyl ion.

Fig. 5. Ladder chain in $(VO)_2P_2O_7$

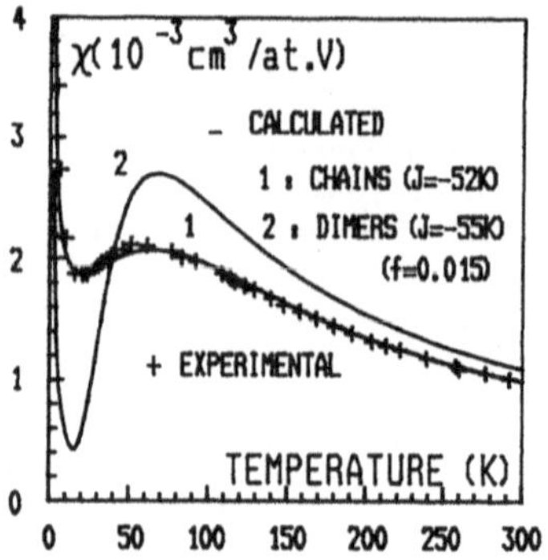

Fig. 6 – Magnetic susceptibility of $(VO)_2P_2O_7$

MISCELLANEOUS

The structure of $VO(HPO_4)$-$4H_2O$ can be viewed[5] as double chains of alternating VO_6 octahedra and PO_4 tetrahedra isolated by water molecules. In spite of the complexity of the structure the magnetic behavior seems to correspond to a linear antiferromagnetism. The structures of the hydrates $VO(HPO_4)$-$1.5H_2O$ and $VO(HPO_4)$-$2H_2O$ are not yet known. The magnetic susceptibility reveals the presence of vanadyl chains in the former (J = - 5K) and vanadyl dimers in the latter (J = -23K).

* Departamento de Quimica Inorganica, Universidad de Valencia, Burjasot, Spain

** Departement Sciences des Matériaux, IHECS, Strasbourg, France.

REFERENCES

1. C.C. Torardi and J.C. Calabrese, Inorg. Chem. 23:1308 (1984).
2. A. Erragh, J. Darriet, G. Villeneuve and J.P. Chaminade, Rev. Chim. Min. (in press).
3. S.A. Linde, E. Yu Gorbunova, A.L. Lavrov and V.G. Kunsetsov, Dokl. Akad. Nauk. SSSR 224:1411 (1979)
4. E. Yu Gorbunova and S.A. Linde, Sov. Phys. Dokl, 24:138 (1979).
5. M.E. Leonowicz, J.W. Johnson, J.F. Brody, H.F. Shannon Jr and J.M. Newsam, J. Solid State Chem., 56:370 (1985).
6. C.J. Ballhausen and H.B. Gray, Inorg. Chem., 1:111 (1962).
7. J.W. Johnson, D.C. Johnston, A.J. Jacobson and J.B. Brody, J. Am. Chem. Soc. 106:8123 (1984).
8. J. Darriet, E. Bonjour, D. Beltran and M. Drillon, J. Mag. Mag. Mat., 44:287 (1984).
9. B.C. Tofield, G.R. Crane, G.A. Pasteur and R.C. Sherwood, J. Chem. Soc. Dalton, 4:1806 (1975).
10. See for example.
11. J.C. Bonner and M.E. Fisher, Phys. Rev. 135:A640 (1964).
12. D.C. Johnston, J.W. Johnson, D.P. Goshorn and A.J. Jacobson, Phys. Rev. B25:219 (1987).

NEW 1d-SYSTEMS MADE OF INTERACTING COPPER(II) TRIMERS;

EXPERIMENTAL AND THEORETICAL INVESTIGATION

M. Drillon[a], M. Belaiche[a], J.M. Heintz[a],
G. Villeneuve[1b], A. Boukhari[1c] and J. Aride[1d]

[a] EHICS. Département Science des Matériaux
1, rue Blaise Pascal
67008 Strasbourg (France)

INTRODUCTION

Much effort has been devoted, in the last few years, to the magnetic properties of 1d-systems with ferro- or antiferromagnetic exchange coupling. Several of them have been solved rigorously, as for instance Ising or XY quantum chains[2], classical spin ones with arbitrary spin dimensionality[3], and more recently so-called ferrimagnetic chains for several configurations of alternating spins or/and Landé factors[4,5]. Some examples of ladder-like chains have equally been reported[6]. We discuss here the behavior of a new series of copper(II) compounds, the structures of which consist of discrete trimers [Cu_3] stacked above each other to form infinite and parallel ribbons. The aim is to show that for peculiar topologies of the exchange pathways, the spin multiplicity in the ground-state may be strongly enhanced, thus stabilyzing a 1d ferrimagnetic system, although only copper(II) ions are involved.

STRUCTURAL FEATURES

The three basic salts of copper, $Cu_3(PO_4)_2$, $A_3Cu_3(PO_4)_4$ where A=Ca,Sr, and $Cu_2O(SO_4)$ may be described through the same model.
SHOEMAKER et al first reported that $Cu_3(PO_4)_2$ is isostructural with the Stranskiite $Zn_2Cu(AsO_4)_2$[7]. The crystal structure corresponds to infinite chains of $(Cu_3O_4)^{2-}$ trimers linked by bridging oxygen atoms along the b triclinic axis. The middle copper(II) ion (on an inversion center) is in slightly distorted square planar coordination while the other two lie in irregular polyhedra of five oxygen atoms. Two trimers of adjacent chains are fastened by bridging oxygen atoms. Thus, copper(II) units are observed in the crystallographic plane ($10\bar{1}$) joined together by phosphate tetrahedra $(PO_4)^{3-}$. In these layers, the 1d character results, in a large extent, from the Jahn-Teller distortion of copper(II) polyhedra, which induces the localization of the unpaired electron in the basal plane of the square-pyramids. On closer examination of the Cu-O bonds, we can assume that the system reduces to a chain of alternatingly spaced metal ions with exchange interactions obeying the sequence J-J-j-J-J-j.

In $A_3Cu_3(PO_4)_4$ where A=Ca,Sr, the copper atoms show coordination polyhedra that are very similar to those of the above system[8]; one (Cu(1)) exhibits a square-planar surrounding, while the other two (Cu(2)) are in square-pyramidal surrounding (Fig. 1). The average bond lengths are 1.943 Å (square-planar polyhedron) and 2.029 Å (square-pyramid) as compared to 1.953 Å and 2.025 Å, respectively, in the orthophosphate. These polyhedra are connected through oxygen atoms in order to form infinite ribbons spreading along the b axis. As a result, the metal network may be described as intertwinning double chains of copper(II) ions connected through Cu(1) kinks. In view of the metal-oxygen distances and bond angles, two different exchange pathways should be considered in the discussion, one within trimer units (J_1), and a weaker between adjacent trimers (J_2), as sketched in figure 1.

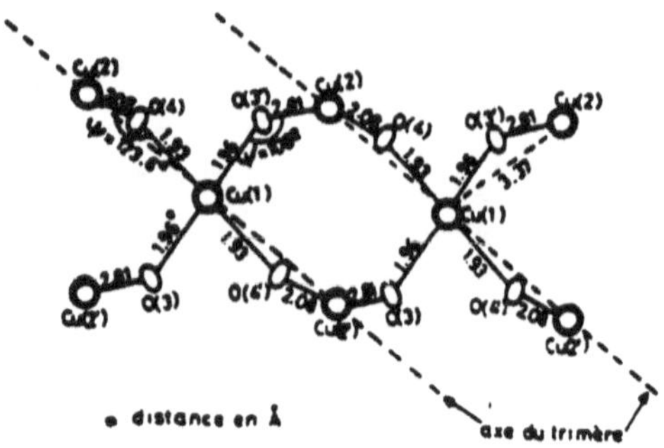

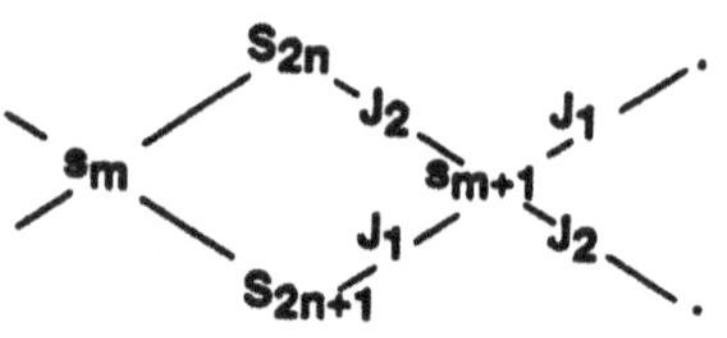

Fig.1 - Relevant exchange pathways in $A_3Cu_3(PO_4)_4$; A = Ca, Sr

As a common structural feature, $Cu_2O(SO_4)$ contains two oxygen polyhedra coordinating copper(II) ions[9]. One corresponds to a distorted octahedron (4+2 coordination), the other to a trigonal bipyramid surrounding. These polyhedra form a two-dimensional network parallel to the (001) plane, and separated from each other by (SO_4) tetrahedra. By focusing on the metal ions and bridging oxygen network, $Cu_2O(SO_4)$ may be viewed as zig-zag chains of triangular units spreading along the a axis. It may also be described as linear chains of copper(II) octahedra coupled to perpendicular dimeric units built from edge-sharing trigonal bipyramids. Accordingly, it can be anticipated that the above model, with an additional coupling between Cu(2) ions, should approximate the magnetic behavior of $Cu_2O(SO_4)$

THEORETICAL OF THE 1d ISING SYSTEM

We consider in the following a 1/2-spin system made of interacting trimers as sketched in figure 1. Further, in order to describe the above copper compounds with a unique model, we assume that spins S_{2n} and S_{2n+1} are coupled by the interaction J_3. Then the system can be viewed as an intertwinning double chain (IDC model).
If local moments on both sites are identical, the full Hamiltonian is written as:

$$H = - \sum J_1 s_m(S_{2n-2}+S_{2n+1})+J_2 s_m(S_{2n-1}+S_{2n})+J_3 S_{2n}S_{2n+1}+g\mu_B H(s_m+S_{2n+1}+S_{2n+2})$$

Consider, at first, that only Z-components of the spins are coupled. It may be argued that this model is questionable for copper(II) ions, but correcting factors may be introduced in a second step to account for the isotropic character of the interaction.
This problem can be solved exactly by the transfer matrix technique if the external magnetic field is assumed to be along the quantization axis. It is then straightforward to show that the transfer operator (T) for a trimer inserted into the infinite chain results from the appropriate product of a 4x2 matrix by a 2x4 one. The elements of the 2x2 matrix are given by:

$$a_{11} = 2h\,(c.\cosh(K^+ + L) + c^{-1}) \qquad a_{12} = 2h\,(c.\cosh(2L) + c^{-1}.\cosh(K^-))$$
$$a_{22} = 2h^{-1}(c.\cosh(K^+ - L) + c^{-1}) \qquad a_{21} = 2h^{-1}\,(c.\cosh(2L) + c^{-1}.\cosh(K^-))$$

where $L = g\mu_B H/2kT$, $h = \exp(L)$, $K^{\pm}=(J_1 \pm J_2)/2kT$ and $c = \exp(J_3/4kT)$

In the limit of very long chains, we show that the effective partition function is related to the largest eigenvalue $U_{H,max}$. The zero-field parallel susceptibility is then easily derived from the second derivative of $\log(U_{H,max})$ with respect to H, giving:

$$X = (Ng^2\mu_B{}^2/2kT).[U_{0,max}.c.(4\exp(K^+)+\cosh(K^+)+c^{-2})-8(\cosh(K^+)-\cosh(K^-))]$$
$$/[U_{0,max}.(U_{0,max}-2c.(\cosh(K^+)+c^{-2}))]$$

where $U_{0,max} = 2(c.\cosh(K^+)+c^{-1}.\cosh(K^-)+c+c^{-1})$

Of course, this expression agrees fully with the well-known formulae derived in the limiting cases. For $J_2=J_3=0$, it describes the behavior of isolated trimers, while, in turn, for $J_1=J_3$ and $J_2=0$ it reduces to the susceptibility of the regular Ising chain as derived by FISHER[2]. Finally, for $J_2=0$ or $J_3=0$ (the largest interaction being J_1) we obtain a ribbon of coupled trimers stacked above each other in two different ways. For the former, that is to say the (J_1,J_3) coupled chain, it is to be noted that the exchange alternation does not alter basically the antiferromagnetic character of the chain; the ground-state is a spin singlet. Conversely, the (J_1,J_2) coupled chain is characterized by two non-compensated spin sublattices, resulting in a high-spin ground-state and a ferrimagnetic-like behavior.

DISCUSSION OF THE RESULTS

- $Cu_3(PO_4)_2$

This compound exhibits a broad maximum of susceptibility around 30 K characteristic of low-dimensional magnetic systems. Such a behavior differs drastically from that expected for discrete copper(II) trimers, since X should diverge as T tends to 0 K. In fact, we show here that 1/2-spins are strongly coupled within trimers (Table 1), but the non-compensated spins order antiparallely at absolute zero, so that the behavior looks like a regular antiferromagnetic chain. The significant coupling between adjacent trimers (J_3), obtained from the fit (Table 1), shows that the 1d character is noticeable in a large temperature range. As expected, the Ising model becomes a poor approximation at low temperature. The result of the fit by the regular Heisenberg chain model is given for comparison (J = - 68.8 K)

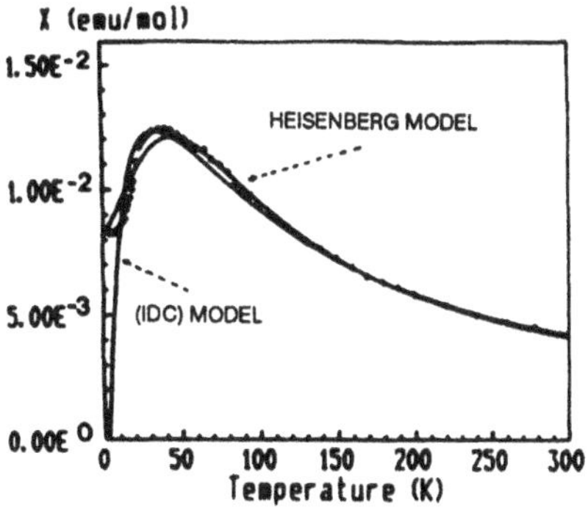

Fig.2 - Magnetic behavior of $Cu_3(PO_4)_2$

- $A_3Cu_3(PO_4)_4$ with A = Ca, Sr

The magnetic susceptibility of these systems was, in a first step, described as arising from independent antiferro-magnetically coupled $[Cu_3]$ trimers (J = -100 to -110 K). In fact, the magnetic behavior shows for both systems the typical features of 1d ferrimagnets, namely a minimum of X.T around 25 K and a strong divergence at lower temperature, as displayed for $Ca_3Cu_3(PO_4)_4$ in figure 3. Recent measurements, performed in the very low temperature range, showed that $Ca_3Cu_3(PO_4)_4$ orders ferromagnetically at 0.8 K, while in the case of $Sr_3Cu_3(PO_4)_4$ an antiferromagnetic ordering occurs below 0.9 K. In the paramagnetic region, the above (IDC) model gives a very good description of the data (see figure 3). The best fit parameters, listed in Table 1, clearly indicate that the leading interaction (antiferromagnetic) occurs within Cu(II) trimers (J_1). Accordingly, it is convenient to use, in the expression of X, the scaling factor $X_{Heisen.}/X_{Ising}$ corresponding to isolated trimers to determine the right exchange parameters.

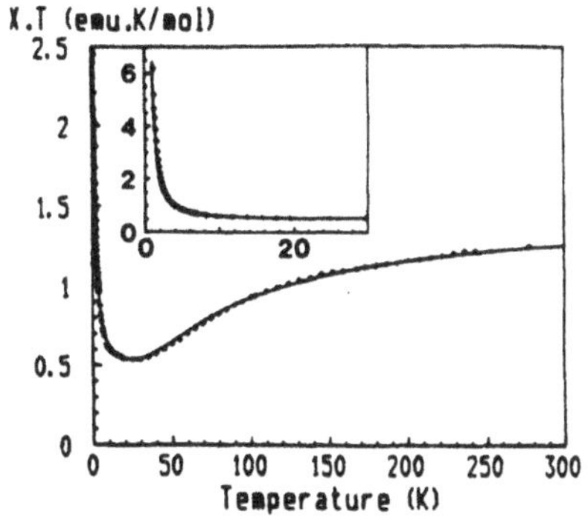

Fig.3 - Magnetic behavior of $Ca_3Cu_3(PO_4)_4$

Due to the relative stacking of the trimers, the non-compensated spins are parallel in the ground-state, giving a ferromagnetic-like behavior below 10 K. Thus, although only one metal is involved and all the couplings are antiferromagnetic, the chain is characterized by a high-spin ground-state.

- $Cu_2O(SO_4)$

The magnetic susceptibility of $Cu_2O(SO_4)$ shows evidence for 1d ferrimagnetism (Fig. 4). Below 25 K, the susceptibility increases rapidly (and also the X.T product) and finally becomes dependent of the applied field below 16 K. Then, we have a transition to a long-range ordering with spontaneous magnetization. At 2 K, the net magnetization corresponds to 0.17 u_B.

In the paramagnetic region, the experimental data were fitted by using the IDC model. Very close exchange constants (J_1,J_2,J_3) were found as shown in Table 1. Then, a frustrated situation occurs, due to competing exchange interactions in the triangle units of the chain. Owing to the ratio between J1 (or J2) and J3, the 1d system is characterized by a non-compensation of the two spin sublattices. Finally, below 16 K, the interchain couplings give rise to a canted ferrimagnetic structure.

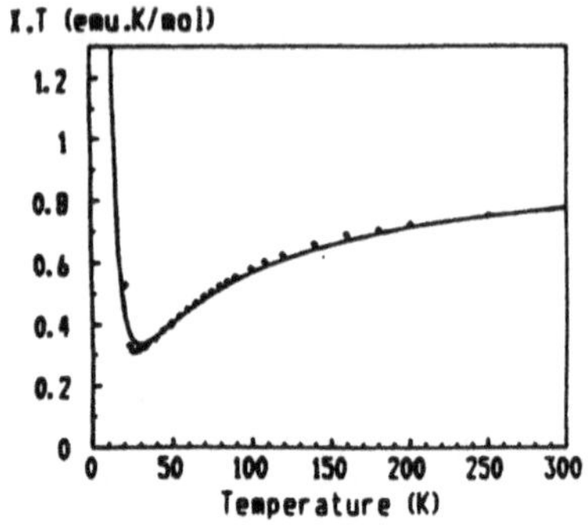

Fig.4 - Magnetic behavior of $Cu_2O(SO_4)$

From these three examples, it is clear that the topology of the chain plays a major role in the spin configuration of the ground-state. According to the exchange pathways, it may induce for the same ion the stabilization of a ferri- or antiferromagnetic 3d ordering, as displayed in the series $A_3Cu_3(PO_4)_4$.

Table 1: Best fit parameters from the IDC model

samples	J_1/K	J_2/K	J_3/K	g
$Cu_3(PO_4)_2$	-124.7		-38.3	2.30
$Ca_3Cu_3(PO_4)_4$	-137.6	-2.6		2.28
$Sr_3Cu_3(PO_4)_4$	-150.3	-2.5		2.25
$Cu_2O(SO_4)$	-66.3	-71.6	-68.4	2.23

REFERENCES

1.b) Laboratoire de Chimie du Solide, 33405 Talence (France); c) Applied Solid State Chemistry Laboratory, Rabat (Morocco); d) Ecole Normale Supérieure de Takaddoum, Rabat (Morocco)
2. M.E. FISHER, J. Math. Phys., 4, 124 (1963)
3. M.E. FISHER, Am. J. Phys., 32, 343 (1964)
4. M. DRILLON, J.C. GIANDUZZO and R. GEORGES, Phys. Lett., 96A, 413 (1983)
5. J. CURELY, R. GEORGES and M. DRILLON, Phys. Rev., B33, 6243 (1986)
6. D.B. BROWN, J.A. DONNER, J.W. HALL, S.R. WILSON, R.B. WILSON, D.J. HODSON and W.E. HATFIELD, Inorg. Chem., 18, 2635 (1979)
7. G.L. SHOEMAKER, J.B. ANDERSON and E. KOSTINER, Acta Cryst., B33, 2969 (1977)
8. A. BOUKHARI, A. MOQINE and S. FLANDROIS, Mat. Res. Bull., 21, 395 (1986)
9. E. FLUGEL-KAHLER, Acta Crystallogr., 16, 1009 (1963)

MAGNETIC BEHAVIOR OF HYDRATED TETRAFLUOROMANGANATES(III)

Fernando Palacio and Mercedes Andrés

Instituto de Ciencia de Materiales de Aragón
University of Zaragoza - C.S.I.C.
50009 Zaragoza, Spain

INTRODUCTION

Fluoromanganate (III) derivatives present a rich variety of structural arrangements most of them showing low-dimensional magnetic behavior. The influence of a strong Jahn-Teller effect, present in these high-spin compounds[1], favors a tendency to form chains or layers of corner sharing octahedra. That is always the case in $A^{I}{}_{2}MnF_5\cdot H_2O$ (x=0 for A^I = Li, Na and NH_4; x=1 for A^I = K, Rb and Cs) and $A^{II}MnF_5\cdot H_2O$ (A^{II} = Sr and Ba) compounds, which are formed by zig-zag chains of *trans*-$[MnF_4F_{2/2}]^{2-}$ octahedra[2].

On the other hand, tetrafluoromanganate (III) derivatives can be formed by either layers, chains or isolated octahedra units, depending on the number of water molecules in the formula unit. Thus, $CsMnF_4$ is known to form layers of compressed *trans*-$[MnF_2F_{4/2}]^-$ where the four fluorine atoms placed at the layer plane are shared by four neighbor octahedra, its structure being of the $TlAlF_4$ type[3]; $AMnF_4\cdot H_2O$ (A = Rb and K) are formed by chains where alternated *trans*-$[MnF_4F_{2/2}]^{2-}$ and *trans*-$[MnF_2F_{2/2}(H_2O)_2]$ octahedra are conected with each other by sharing apical fluorine atoms[4,5] as depicted in Fig. 1 for the potassium derivative and, finally, compounds with two molecules of water in the formula, such as $CsMnF_4\cdot 2H_2O$, have a structure formed by isolated $[MnF_4(H_2O)_2]^-$ octahedra[6].

It is, therefore, reasonable to expect a very interesting and divers magnetic behavior in these compounds. So far, $CsMnF_4$ has been described as a planar ferromagnet[3] while $RbMnF_4\cdot H_2O$ posses 1-d antiferromagnetic properties[7]. Moreover, the coordination octahedra, which form the lattice skeleton in all these compounds, are tilted with respect to each other, what favours the appearence of magnetic canting phenomena. In this paper, the magnetic properties of $AMnF_4\cdot H_2O$, (A = Rb and K) and $CsMnF_4\cdot 2H_2O$ in the paramagnetic regime and the weak ferromagnetic behavior of the these compounds at the onset of 3-d ordering are reviewed.

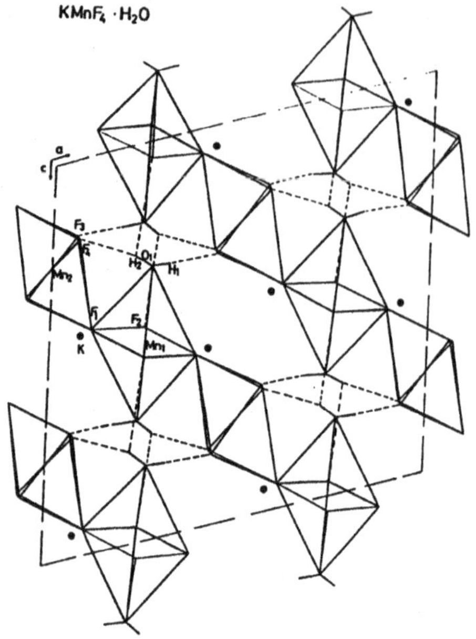

Fig. 1. Crystal structure of $KMnF_4 \cdot\cdot H_2O$ projected on the ac plane. Dotted lines indicate hydrogen bonds.

MAGNETIC PROPERTIES

<u>$AMnF_4 \cdot H_2O$ (A = Rb and K)</u>

Single crystal ac magnetic susceptibility measurements at zero dc field, from 4.2 K to 100 K are presented in Figs. 2 and 3 for the three principal axis of $RbMnF_4 \cdot H_2O$ and $KMnF_4 \cdot H_2O$, respectively. The axis labeled as *d* and *e* correspond to directions parallel and perpendicular to the chains within the *ac* plane, respectively. Two distinct regions can be clearly differentiated. At temperatures above 10 K, there appears a broad maximum corresponding to the one-dimensional character of the compounds. In the Rb derivative the maximum, at 42 K, can only be observed along the *e* axis, while in the K one a maximum is present along each of the three axis; the temperature at which the maximum appears is much lower for the *b* axis, 38 K, than for the *e* and *d* axis, 52 K. Measurements along the *b* axis give a constant value in the temperature region where χ_d peaks, whereas measurements perpendicular to the chain direction, χ_e, tend to drop to zero as $T \rightarrow 0$. Below 10 K, χ_d increases very rapidly presenting a peak at 7.30 K and at 8.45 K respectively in the Rb and K derivatives.

The behavior of the susceptibility in both compounds is characteristic of a 1-d antiferromagnet with the easy axis of alignement perpendicular to the chain direction within the *ac* crystallographic plane. A weak ferromagnetic moment appears along the *d* axis below 10 K. The moment is due to the canting of antiferromagnetic sublattices towards the direction parallel to the chains.

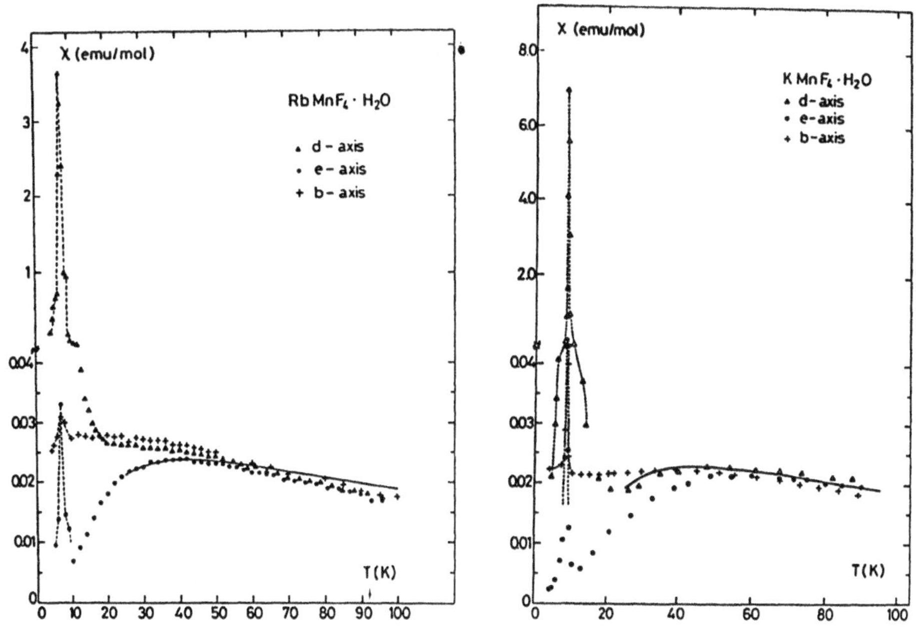

Figs. 2 and 3. Temperature dependence of zero-field magnetic susceptibility of $RbMnF_4 \cdot H_2O$ and $KMnF_4 \cdot H_2O$, respectively, along the three principal axes. For T < 10 a dotted line is used as eye-guide. Solid line is the theoretical calculation.

The fit from high temperature series (hts) expansions and Padé approximants for a Heisenberg S = 2 linear chain[8,9] gives for $RbMnF_4 \cdot H_2O$[7] the value of $|J|/k$ = 6.2 K with g = 2.03 and for the K analog $|J|/k$ = 6.5 K with g = 2.05.

<u>$CsMnF_4 \cdot 2H_2O$</u>

Magnetic ac susceptibility measurements at zero dc field are represented in Fig. 4. Susceptibility data in the temperature range from 1.2 K to 4.2 K correspond to a single crystal and those from 4.2 K to 100 K correspond to a powdered sample. Measurements show a smooth increment in the susceptibility down to 4 K where an abrupt increase appears presenting a very sharp peak of 0.4 K half width at 1.55 K. The behavior is characteristic of a 3-d antiferromagnet with a weak ferromagnetic moment due to a canting of the antiferromagnetic sublattices with respect to each other.

The structure contains separate $[MnF_4(H_2O)_2]^-$ units, interlinked by a net of hydrogen bonds extending throughout the unit cell. Magnetic interactions between Mn neighbors are

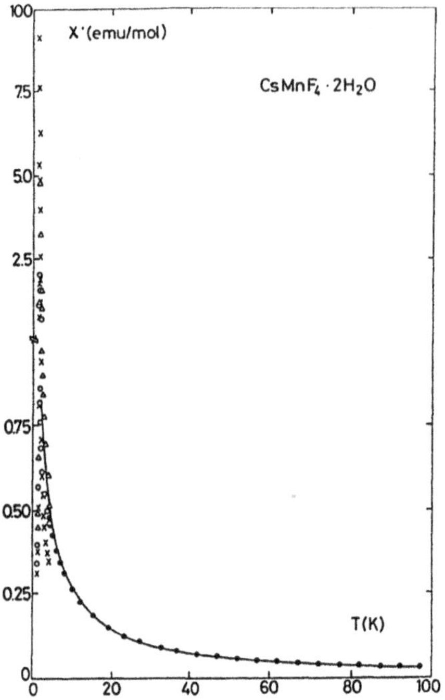

Fig. 4. Temperature dependence of zero-field magnetic susceptibility of $CsMnF_4{\cdot}2H_2O$. The solid line is the theoretical calculation.

transmited by a superexchange mechanism through double bridges Mn-O-H···F-Mn. Pathways of this type have been already described in related systems[10]. The fit from hts expansions for a Heisenberg S = 2 simple cubic lattice gives $|J|/k = 0.06$ K with g = 2.00.

Acknowledgements. The work was supported by "Comisión Asesora de Investigación Científica y Técnica (C.A.I.C.Y T.)", Ministerio de Educación y Ciencia, Spain, grant nº 3380/83.

REFERENCES

1. P.Köhler, W.Massa, D.Reinen, B.Hoffmann and R.Hoppe, Z. anorg. allg. Chem., 446:131 (1978).
2. W.Massa and J.Pebler, IIIrd European Conference on Solid State Chemistry, Regensburg (1986).
3. W.Massa and M.Steiner, J. Solid State Chem., 32:137 (1980).
4. V.Kaucic and P.Bukovec, J.C.S. Dalton, 1512 (1979).
5. F.Palacio, M.Andrés, C.Esteban-Calderón, M.Martínez-Ripoll and S.García-Blanco, to be published. A preliminar communication with the structure was presented at the XIX Biannual Meetting of the Real Sociedad Española de Física y Química, Santander(1982).
6. P.Bukovec and V.Kaucic, J.C.S. Dalton, 945 (1977).
7. F.Palacio, M.Andrés, R.Horne and A.J. van Duyneveldt, J. Magn. Magn. Mat., 54-57:1487 (1986).
8. G.S.Rushbrooke and P.J.Wood, Mol. Phys., 1:257 (1958).
9. R.Navarro, Thesis, University of Zaragoza (1976).
10. R.L.Carlin and F.Palacio, Coord. Chem. Rev., 65:141 (1985).

STRUCTURE AND MAGNETISM IN LINEAR CHAIN FLUOROMANGANATES(III)

W. Massa, J. Pebler, F. Hahn and D. Babel
Fachbereich Chemie der Universitat Marburg
Hans-Meerwein -Str., D-3550 Marburg, Germany

Due to the strong Jahn-Teller effect of the d^4 high-spin configuration of Mn(III) in fluoromanganates(III), the formation of linear chain anions of strongly elongated MnF_6-octahedra sharing trans-vertices is favorized. Dependent on the cations, the bridging angles ß (Mn-F-Mn) varie between 121.5 and 180°, the geometry of the MnF_6-octahedra remaining approximately constant, anyhow.

The linear trans-chain fluoromanganates(III) may be classified by the type of chain packing:

I. Tetragonal chain packing: $A_2MnF_5(\cdot H_2O)$ (A = K,Rb,Cs)
II. Pseudohexagonal chain packing: A_2MnF_5 (A = Li,Na,NH_4)
$AMnF_5 \cdot H_2O$ (A = Sr,Ba)

The compounds are suitable one-dimensional models for studies of the antiferromagnetic superexchange interaction as a function of bridge angle. In Table 1 the structural data and the intrachain exchange energies are compiled. The latter have been derived from magnetic powder susceptibility measurements by the High Temperature Series Expansion Method (HTE [1]) or the Scaled Fisher Model (SFM[2])). The results of Nuñez, Darriet, Bukovec, Tressaud and Hagenmuller[3]) which are also comprised in Table 1 are consistent with our data. Magnetic investigations of group II A_2MnF_5 compounds were reported as well by Emori et al.[14]) and by Kida[15]). The intrachain exchange energies varie with $\cos^2 ß$, approximately (Fig. 1), as is

Table 1. Chain geometry and intrachain exchange energies

Compound	Mn-F$_{bridge}$ mean values	Mn-F$_{term}$ in Å	ß [°] bridge angle	J/k [K] HTE	J/k [K] SFM	g-factor	references struct.	references magn.
Li_2MnF_5	2.123	1.847	121.5	-6.0	-6.3	2.02	(4)	(4)
Na_2MnF_5	2.109	1.849	132.5	-8.1	-8.4	1.84	(5)	(4)
$SrMnF_5{*}H_2O$	2.108	1.845	139.8	-9.9	-10.3	1.88	(6)	(4)
$(NH_4)_2MnF_5$	2.091	1.853	143.4	-10.0	-11.2	1.93	(7)	(4)
$BaMnF_5{*}H_2O$	2.127	1.854	147.7	-12.2	-13.5	1.95	(6)	(4)
$K_2MnF_5{*}H_2O$	2.07	1.83	163.3	-18.2		2.06	(8)	(3)
$Rb_2MnF_5{*}H_2O$	2.089	1.848	175.4	-18.8	-20.0	2.07	(9)	(4)
				-21.8		2.09		(3)
Rb_2MnF_5			180 ?	-22.6		2.10	(10)	(3)
$Cs_2MnF_5{*}H_2O$	2.130	1.836	180	-16.5	-19.0	2.01	(11)	(4)
				-17.9		2.04		(3)
Cs_2MnF_5	2.114	1.856	180	-19.4		2.05	(12)	(3)

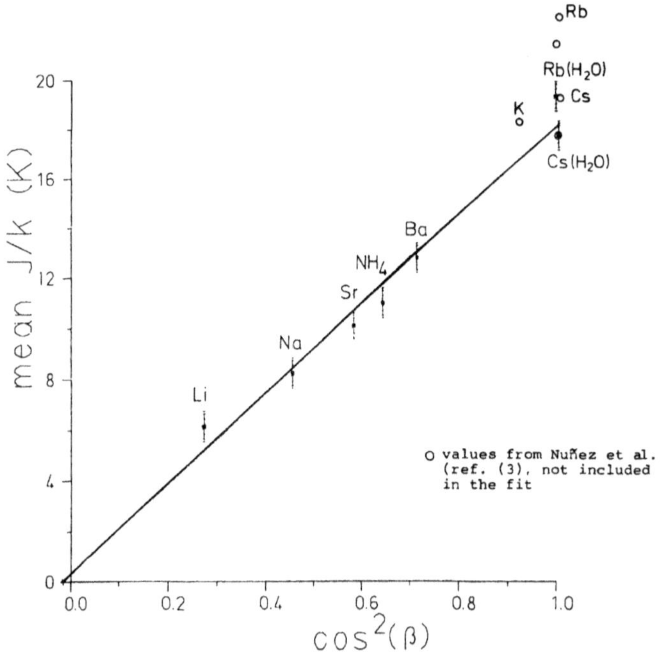

Fig. 1

expected for the decrease in σ-overlap of the d_{z^2}-orbitals of Mn(III) and the p_z-orbitals of the bridging fluoride ions with decreasing bridge angle β.

Recent work on mixed crystals $(NH_4)_2Mn_{1-x}Fe_xF_5$ by means of X-ray crystallography on powders and, in part, on single crystals, by Mössbauer and ligand field spectroscopy, suggests the paradoxon, that the tetragonal distortion of the MnF_6 - octahedra increases with increasing contents of the non-Jahn-Teller - ion Fe(III), in spite of a decreasing metal-metal distance (along the b axis) within the chain (Tab. 2). In contrast to a reverse behaviour in some comparable mixed crystals of layer structures (e.g. $A_2Cu_{1-x}Zn_xF_4$[16] or $Rb_2Cr_{1-x}Mn_xCl_4$[17]), the bands in the ligand field spectra (Fig. 2) are shifted towards higher energies when the Fe contents increase. This may be understood by the assumption of an asymmetry in the Mn....F..Fe bridges caused by the very different bond strengths of axial Fe-F and Mn-F bonds.

Table 2. Crystallographic data for mixed crystals $(NH_4)_2Mn_{1-x}Fe_xF_5$. Mean axial/equatorial bond lengths (Å)

x	space group	lattice constants (Å)			$M\text{-}F_{exp}$	$M\text{-}F_{calc}$[a]
0[b]	Pnma[7]	6.20(3)	7.94(1)	10.72(1)	2.091	
					1.840	
0.2[b]	Pnma	6.264(3)	7.862(5)	10.808(6)	2.085	2.08
					1.862	1.85
0.5[b]	$C222_1$?	12.562(5)	7.785(2)	21.752(9)		2.05
		(2x6.281)		(2x10.876)		1.87
0.59	?	6.290(3)	7.751(9)	10.898(7)		2.05
			superstructure ?			1.88
1			unknown			2.015
						1.901

a) calculated via bond valences[13]

b) single crystal investigations

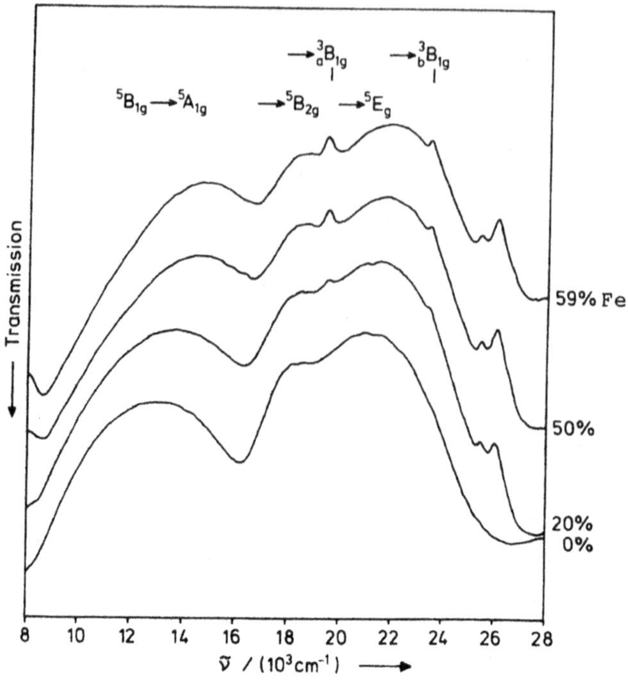

Fig. 2 Ligand field spectra of mixed crystals $(NH_4)Mn_{1-x}Fe_xF_5$

References

(1) G.S.Rushbrooke and P.J.Wood, Mol.Phys. 1, 257 (1958); 6, 409 (1963)
(2) M.E.Fisher, Am.J.Phys. 32, 343 (1964)
(3) P.Nuñez, J.Darriet, P.Bukovec, A.Tressaud and P.Hagenmuller, Mat.Res.Bull., in press
(4) J.Pebler, W.Massa, H.Lass and B.Ziegler, J.Solid State Chem., in press
(5) W.Massa, Acta Crystallogr. C42, 644 (1986)
(6) W.Massa and V.Burk, Z.Anorg.Allg.Chem. 516, 119 (1984)
(7) D.R.Sears and J.L.Hoard, J.Chem.Phys. 50, 1066 (1969)
(8) A.J.Edwards, J.Chem.Soc. (A), 2654 (1971)
(9) P.Bukovec and V.Kaucic, Acta Crystallogr. B34, 3339 (1978)
(10) J.R.Günter, J.P.Matthieu and H.R.Oswald, Helv.Chim.Acta 61, 328 (1978)
(11) V.Kaucic and P.Bukovec, Acta Crystallogr. B34, 3337 (1978)
(12) F.Hahn, Diplomarbeit, Marburg 1986
(13) W.Massa and D.Babel, Chem.Rev., in press
(14) S.Emori, M.Inoue, M.Kishita and M.Kubo, Inorg.Chem. 8, 1385 (1969)
(15) J.Kida, J.Phys.Soc.Japan 30, 290 (1973)
(16) D.Reinen and S.Krause, Inorg.Chem. 20, 2750 (1981)
(17) G.Münninghoff, W.Treutmann and E.Hellner, J.Solid State Chem. 34, 289 (1980)

MAGNETO-STRUCTURAL CORRELATIONS IN μ-HALO BRIDGED COPPER(II) CHAINS

D. Beltran[a], J.V. Folgado[a], R. Ibañez[a], E. Coronado[a],
J.L. Mesa[b], T. Rojo[b] and G. Villeneuve[c]

[a]Department de Química Inorgánica, Universitat de Valencia
Burjassot, Valencia, Spain
[b]Departamentos de Química y Cristalografía-Mineralogía
Universidad del Pais Vasco, Bilbao, Sapin
[c]Laboratoire de Chimie du Solide du C.N.R. S.
Université de Bordeaux, Talence, France

To design synthetic pathways to systems of desired properties is a growing challenge for inorganic chemist. Our current interest in this area is focused on copper(II) chemistry. Besides to advance in the understanding of the factors determining the conformation around copper(II) in the solid state, we intend to gain insight into the chemical and structural effects that govern exchange coupling interactions in condensed species. In this context, we have approached the synthesis and characterization of a wide set of pentacoordinated $Cu(L_{III})XY$ complexes (L_{III}=tridentate ligand, X=coordinating anion, Y=coordinating or non-coordinating anion) showing a great structural diversity (including mono, bi and polynuclear species)[1-4]. Tridentate rigid "quasi-planar" ligands such as terpy (2,2':6',2"-terpyridine), paphy (pyridine-2-aldehyde-2'-pyridylhydrazone) and TPT (2,4,6-tris(2-pyridil)-1,3,5-triazine) favour the stacking of $(Cu(L_{III})X)^+$ entities to give dimers or other condensed systems (simple or ladder-like chains) when X is a potentially bridging anion[1-3].

The monomeric complexes $Cu(L_{III})X_2$ may be considered as intermediates in the synthesis of condensed derivatives. Thus, the synthetic strategy from the monomers to the condensed species is based in the equilibria:

$$Cu(L_{III})X_2(s) \rightleftharpoons Cu(L_{III})X_2(aq) \rightleftharpoons (Cu(L_{III})X)^+(aq) + X^-(aq)$$

$$(Cu(L_{III})X)^+(aq) + Y^-(aq) \rightleftharpoons 1/n\ (Cu(L_{III})X)_n(Y)_n\ (aq\ or\ s)$$

involving a methatetic reaction is which: 1) a X ligand is extruded while its charge is balanced by the non-coordinating Y anion and 2) the remaining X might now bridge two (dimers or simple chains) or three (ladder-like

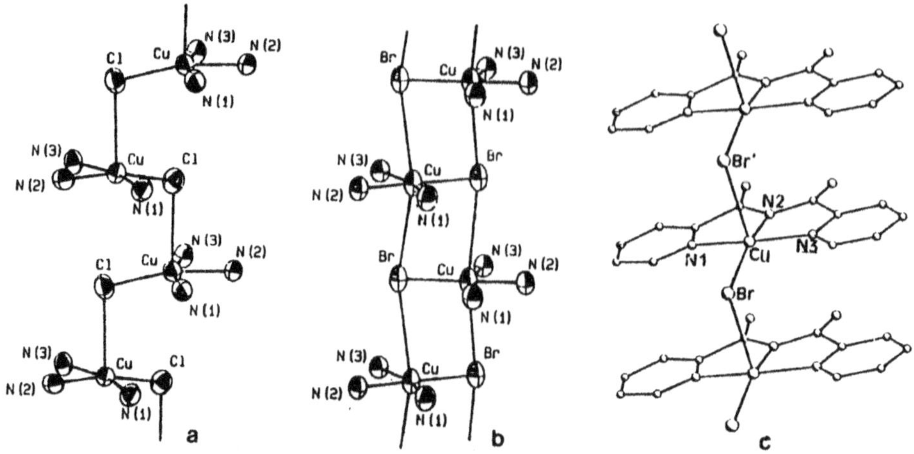

Figure. Molecular structure of (a) compound I, (b) compound II and (c) compound IV.

chains) metallic centers. Additionally, in contrast to the other tridentate rigid N-donors, BPCA (N-2-pyridinylcarbonil-2-pyridinecarboximidate) is a monoanionic ligand so that only one positive charge remains on the copper (II)-BPCA moiety. This should allow the formation of neutral Cu(BPCA)X entities. If these anions are also capable of bridging metallic centers, both the "quasi-planarity" of BPCA and the tendency of copper(II) to acquire a '4+1' coordination mode would favour the stacking of Cu(BPCA)X units to give dimers or chain structures [4].

In this work we report the crystal structure and magnetic properties of four new condensed systems: the μ-halo chains (Cu(paphy)X)(PF_6).H_2O and Cu(BPCA)X, with X=Cl or Br. The synthesis of these compounds has been carried out basis on the strategy described above and were reported[4,5].

The molecular structures of Compounds (Cu(paphy)Cl)(PF_6).H_2O (I), (Cu(paphy)Br)(PF_6).H_2O (II) and Cu(BPCA)Br (IV) are illustrated in the Figure. The molecular structure of Cu(BPCA)Cl (III) has not been resolved, but the complex is isotructural to IV and spectroscopic data (UV-V, e.s.r.) become similar in both cases. Compounds I and IV (and very likely III) exhibit mono-μ-halo bridges giving linear copper(II) chains, whereas II contains bromine atoms that bridge three different copper(II) centres to give a ladder-like arrangement. In the linear chains I and IV, the copper (II) is '4+1' coordinated (with a longer Cu-X' axial bond distance) and the surrounding geometry is close to distorted square pyramidal, lying the metallic centre above the least-squares basal plane. In II, the copper(II) environment is a tetragonal elongated octahedron with a Cu-Br equatorial bond distance of 2.331(1) Å and two Cu-Br' axial ones of 3.358(2) Å.

Magnetic behaviour The molar susceptibility of the four reported complexes has been measured in the temperature range 100 to 1 K. Whereas the effective magnetic moment stays practically constant for the ladder-like chain in the teperature range studied, in compound I increases at low temperatures (from 30 K) and in III and IV decreases from 5 K. The experimental data have been fitted to the Baker's equation for ferromagnetically coupled copper(II) chains in the case of compound I, and to the Hall's expression for antiferromagnetically coupled systems in III and IV [6]. Fixing the g values to 2.1 (from e.s.r. experiments) the best least-square fit to these expressions give the set of J-values shown in the Table, 2J being the singlet to triplet gap. In the Table, the magneto-structural parameters reported so far are also shown [6].

The experimental magnetic data accumulated for simple systems as parallel square-planar halo-bridged copper(II) dimers have shown to have only rough dependences on topological parameters [2]. From an inspection of data in Table it is evident that a simple magneto-structural correlation does not exist in the reported compounds. This is not surprising owing to the large number of possible (and independent) parameters to be considered in these polymeric compounds, when comparing with more simple systems.

It can be noted that J-values are weak in all cases. This is understood taking into account the actually geometry of the entities. In all the cases, the copper(II) orbital ground state is $d_{x^2-y^2}$ and the exchange path, involving the d_{z^2} orbital, is very unfavoured. Notwithstanding, some trends can

Table. Magneto-structural parameters in μ-halo bridged copper(II) chains[a].

Compound	$J(cm^{-1})$	R	θ/°	φ/°	θ/φ	θ/2R	θR/φ[b]	φ/R
$Cu(DMSO)_2Cl_2$	-6.1	2.702	146.1	144.6	1.01	27.0	36.6	53.5
$Cu(DMSO)_2Br_2$	-7.9	2.777	146.1	144.6	1.01	26.3	35.6	52.1
$Cu(IMH)_2Cl_2$	-2.1	2.751	166.9	117	1.43	30.3	25.5	42.5
$Cu(CAF)_2(OH_2)Cl_2$	+0.5	2.788	178.8	128.1	1.40	32.1	25.7	45.9
$Cu(MAEP)_2Cl_2$	+1.6	2.785	176.0	113.6	1.54	31.6	23.2	40.8
$(Cu(paphy)Cl)(PF_6)H_2O$	+0.5	2.805	167.6	102.3	1.64	29.8	21.8	36.4
Cu(BPCA)Br	-0.8	2.887	165.3	94.0	1.75	28.6	19.7	32.5
Cu(BPCA)Cl	-0.4							
$(Cu(paphy)Br)(PF_6)H_2O$	~0	3.358	178.4	81.7	2.18	26.6	13.6	24.3

[a] R= Cu-X' distance, θ= X-Cu-L(trans) angle, φ= Cu-X-Cu' angle.
[b] Values x 10^2.

be reported: the antiferromagnetic coupling is favoured as the θ angle is smaller than 180° (i.e. copper(II) surrounding geometry going towards trigonal bipyramidal involving a greater contribution of the z^2 orbital to the ground state) as well as the ϕ angle is greater than 90° (increasing of the effective overlapping between the magnetic orbitals). In fact, only the derivatives with small θ and large ϕ angles show significant antiferromagnetic J-values.

REFERENCES

1. J.V. Folgado, E. Coronado and D. Beltrán, Spectroscopic and Magnetic Properties of $Cu_2(terpy)Cl_4$ (terpy=2,2':6',2"-terpyridine). A Magnetic System constructed of Two Exchange-coupled Dimers, J. Chem. Soc. Dalton Trans. 1061, (1986).
2. T. Rojo, M.I. Arriortua, J. Ruiz, J. Darriet, G. Villeneuve and D. Beltrán, Magnetostructural Correlations in Parallel Square-planar Chloride Bridged Copper(II) Dimers: Structure, Dynamic Nuclear Magnetic Resonance Study and Magnetic Properties of $(Cu_2(terpy)_2Cl_2)(PF_6)_2$, J. Chem. Soc. Dalton Trans., 285 (1987).
3. J.V. Folgado, E. Escrivá, A. Beltrán and D. Beltrán, Mono and Dinuclear Copper(II) Complexes of 2,4,6-Tris(2-pyridyl)-1,3,5-triazine and Halide or Pseudohalide Ions: Synthesis and Spectral Studies, Transition Met. Chem., 11:485 (1986).
4. J.V. Folgado, E. Escrivá, A. Beltrán and D. Beltrán, Synthesis and Spectral studies of N-2-Pyridinylcarbonyl-2-pyridinecarboximidate copper(II) complexes, Transition Met. Chem., in the press.
5. J.L. Mesa, M.I. Arriortua, J.M. Savariault, T. Rojo, D. Beltrán and J. Galy, Hexafluorofostatos de μ-halo(piridin-2-aldehido-2'-piridilhidrazona) cobre(II), Proceed. XX Reunión Bienal Soc. Esp. Quim., Castellón, Spain, comm. 4-37 (1984).
6. W.E. Estes, W.E. Hatfield, J.A.C. van Ooijen and J. Reedijk, Magnetic Properties and Superexchange in Single Chloride-bridged Copper(II) Chain Compounds, J. Chem. Soc. Dalton Trans., 2121 (1980).

AKNOWLEDGEMENTS

We gratefully aknowledge support of this research by the Comisión Asesora de Investigación en Ciencia y Tecnología. J.L.M. and J.V.F. thank to the Spanish Ministerio de Educación y Ciencia for FPI fellowships.

CLUSTER CALCULATIONS ON THE CHAIN COMPOUND AQUA[N-(SALICYCLALDIMINATO)GLYCINATO]COPPER(II) HEMIHYDRATE

William E. Hatfield, Jeffrey H. Helms, and Brian R. Rohrs

Department of Chemistry, The University of North Carolina at Chapel Hill, Chapel Hill, North Carolina 27514

The spin analog of the regular Peierls distortion[1] of a quasi-one-dimensional system containing paramagnetic transition metal ions has recently been suggested for the linear chain compound aqua[N-(salicylaldiminato)glycinato]copper(II) hemihydrate, CuNSG.[2] Theoretically the spin-Peierls (SP) transition may arise in electrically insulating, antiferromagnetically coupled, uniformly spaced spin chains, but it had previously been restricted to those systems in which the unpaired spins are delocalized on large flat organic molecules.[3] The absence of SP phase transitions in the numerous antiferromagnetically coupled, uniform spin chains containing metal ion localized unpaired spin density can be attributed to long range 3-dimensional ordering at temperatures above the potential SP ordering temperature, T_{sp}. Cluster calculations on the temperature dependent magnetization at varying applied external fields verify the SP phenomenon in CuNSG.

Coordination of copper(II) in CuNSG is square pyramidal with the basal plane formed by the donor atoms of the tridentate Schiff base ligand and one water molecule. The apical position is filled by a second carboxylate oxygen from a neighboring molecule to form the chain link for superexchange (Fig. 1). Neighboring chains pack antiparallel with a resultant interweaving of molecular units. This interweaving allows for good interchain phonon interaction without the presence of interchain superexchange pathways. These structural features were considered important criteria for a SP candidate.

Magnetic susceptibility measurements were made in the range 1.8-60 K by using a vibrating sample magnetometer and an applied field of 0.1 T.[4] The low temperature magnetic susceptibility data (Fig. 2) exhibits a maximum near 4.4 K indicative of antiferromagnetic exchange interactions in a low-dimensional, magnetically condensed material. In view of the structure of CuNSG, these interactions are along a uniformly spaced chain of paramagnetic S = 1/2 copper(II) ions. At approximately 2.2 K, the susceptibility decreases rapidly with decreasing temperature. The value of the susceptibility at the lowest temperature obtained represents a reduction of more than fifty percent from the value at 2.2 K. Such a reduction excludes the possibility of 3-dimensional antiferromagnetic ordering since this state would have a low temperature limit of 66% of the maximum. The spin-Peierls transition leads to a 3-dimensional ordered state which is magnetoelastic in nature and is due to a progressive dimerization of the magnetic chains below T_{sp}.

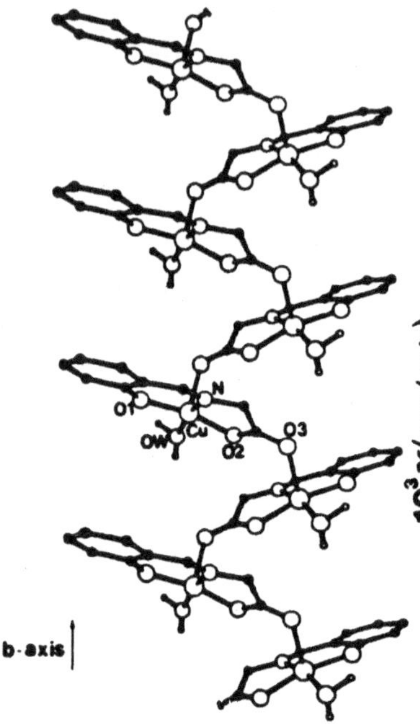

Figure 1. Chain Structure of CuNSG. Darkened bonds represent the superexchange pathways.

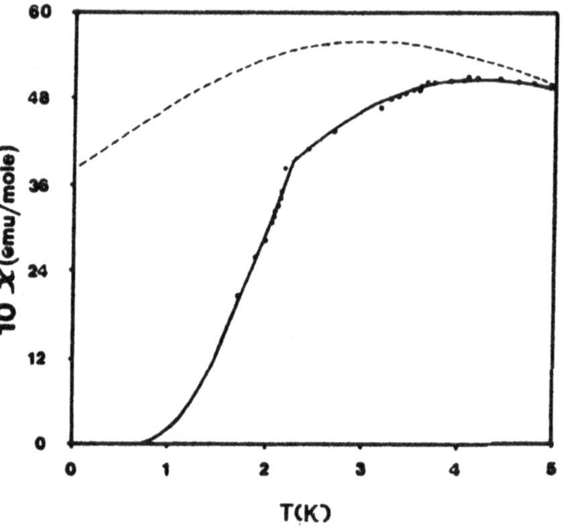

Figure 2. Theoretical fit line to CuNSG below 5 K illustrating considerable deviations from the uniform chain model below 5.5 K.

The magnetic susceptibility data may be fit by the S = 1/2 1-D Heisenberg theory[5] by using $J = -1.55$ cm^{-1}. The fit is improved if a slightly temperature dependent exchange coupling constant is used below T = 5.5 K. The low temperature susceptibility cannot be fit by either the static alternating chain model or the temperature dependent alternating chain model of Bulaevskii[6]. This is probably due to large spin-phonon coupling in CuNSG. The magnetic susceptibility below the "knee" can be fit by the expression

$$\chi = (Ng^2\beta^2/3kT)\cdot S(S+1)\cdot \exp[-2J(1 - T/T_c)/kT]$$

where the temperature parameter T_c in the Boltzmann term qualitatively accounts for the progressive dimerization in CuNSG. The best fit of T_c = 5.8 K is nearly the temperature at which the correction to the Bonner-Fisher chain result becomes important.

Magnetization data for CuNSG was taken in the range 1.3-4.2 K and in a maximum applied field strength of 15 T using a Foner type magnetometer[7]. The temperature was determined by the helium vapor pressure under vacuum. The data was analyzed using a ten spin S = 1/2 ring approximation to an infinite chain and an alternating chain Hamiltonian

$$H = -2J_1 \sum [S_{2i} \cdot S_{2i-1} + \alpha\, S_{2i} \cdot S_{2i+1}] + g\beta \sum H \cdot S_z$$

where $\alpha = J_2/J_1$ and the spin operators have their usual meanings. The constants J_1 and J_2 represent the coupling between alternating nearest neighbor S = 1/2 ions at a given temperature. The alternation parameter, α, describes the extent of dimerization present for the fitting temperature where a value of 1.0 represents a uniform chain and a value of 0.0 represents an isolated dimer. The program CAMSFIELD was run on the CRAY-X/MP at the Pittsburgh Supercomputing Center, grant number PSCA-57. The results of the fit calculations are given in Table 1. The calculations show the degree of alternation is virtually constant below 2.3 K with α = 0.28 and is still quite large at 4.2 with α = 0.5. At the current rate of increasing α, the uniform chain limit (α = 1.0) may not be realized until the temperature reaches 8 to 9 K. The susceptibility fits (Figure 2) show deviations from the uniform chain model and therefor clearly indicate the occurence of a transition above 5.5 K. Above 6 K, the theoretical magnetic susceptibility versus temperature plots using calculated fit parameters converge prohibiting the exact determination of the spin-Peierls onset. Additional magnetization data is needed to precisely identify this temperature.

Acknowledgement This work was supported in part by the National Science Foundation through Grant CHE 86 01438. High field magnetization measurements were made at the Francis Bitter National Magnet Laboratory, which is supported at M.I.T. by the National Science Foundation. We thank J. C. Bonner and L. J. de Jongh for helpful discussions, and L. G. Rubin and B. Brandt for their invaluable help.

Table I. Magnetization versus applied field experimental data with best fit parameters listed as a function of temperature. Energy levels for 10 spin ring calculated for applied fields up to 100000 Oe.

Temperature (K)	g	J (cm^{-1})	α
1.324	2.105	-2.204	0.257
1.530	2.10	-2.171	0.278
1.799	2.10	-2.193	0.277
2.101	2.12	-2.195	0.274
2.298	2.14	-2.16	0.279
2.489	2.17	-2.194	0.357
2.747	2.17	-2.186	0.400
3.000	2.17	-2.126	0.434
3.800	2.18	-2.106	0.490
4.200	2.16	-2.125	0.505

References

1. R. E. Peierls, "Quantum Theory of Solids", Oxford University Press, Oxford (1955).
2. W. E. Hatfield, J. H. Helms, B. R. Rohrs, and L. W. ter Haar, Observation of a Spin-Peierls Transition in the Linear Chain Compound Aqua[N-(salicylaldiminato)glycinato] copper(II) Hemihydrate, J. Am. Chem. Soc. 108:542 (1986).
3. J. W. Bray, L. V. Interrante, I. S. Jacobs, and J. C. Bonner, The Spin-Peierls Transition, in: "Extended Linear Chain Compounds Vol. 3," J. S. Miller, ed., Plenum, New York (1982).
4. P. J. Corvan, W. E. Estes, R. R. Weller, and W. E. Hatfield, Ferromagnetism in Copper(II) Oxydiacetate Hemihydrate, Inorg. Chem. 19:1297 (1980).
5. J. C. Bonner and M. E. Fisher, Linear Magnetic Chains with Anisotropic Coupling, Phys. Rev. A. 135:640 (1964).
6. A. I. Buzdin and L. N. Bulaevskii, Sov. Phys. Usp.(Engl. Transl.) 23:409 (1980).
7. S. Foner, Rev. Sci. Instrum. 30:548 (1959).

NON-LINEAR EXCITATIONS IN THE ANTIFERROMAGNETIC CHAIN OF $Rb_2Mn_{1-x}Fe_xF_5*H_2O$ ($X \leq 0.50$)

Jurgen Pebler and Dietrich Babel

Fachbereich Chemie der Universitat Marburg
Hans-Meerwein-Str., D-3550 Marburg, Germany

The soliton concept becomes of growing importance for the study of low-d magnetic systems [1]. With the aid of Mössbauer spectroscopy and magnetic susceptibility measurements the quasi 1-d magnetic chain of $Rb_2Mn_{(1-x)}Fe_xF_5*H_2O$ ($x \doteq 0.5$) was investigated.
The orthorhombic structure of $Rb_2MnF_5*H_2O$ (space group Cmcm, Z=4) is isotypic with that of $Rb_2FeF_5*H_2O$ [2,3]. Isolated infinite and but slightly linked chains of trans-linked (MnF_6) resp. (FeF_6) octahedra run along the c direction. Rubidium atoms and water molecules are located between the chains. The room temperature lattice parameters of the solid solutions continuously vary between x=0 and x=0.55 [4].
With the aid of a vibrating sample magnetometer (Foner) the magnetic susceptibility has been measured from 4.2 to 300 K in external magnetic fields up to 2 T. The observed broad maxima near 130 K are undoubtedly due to short-range antiferromagnetic interaction within the linear chain. With increasing iron concentration the minimum of the reciprocal magnetic susceptibility derived at lower temperatures becomes sharper, indicating the presence of a 3-d antiferromagnetic order at T_X (Fig.1).
From the results of Mössbauer spectroscopy we are able to distinguish between two Fe positions corresponding to the two different (Fe-F-Mn) and (Fe-F-Fe) interactions within the chain and two different crystallographically distorted Fe octahedra (see Fig. 2).
The saturation hyperfine fields show significant spin-reduction at low temperatures and are additionally reduced with increasing Fe concentration or decreasing easy-axis anisotropy energy (Table).
In the following we restrict our considerations to the case of x=0.01: Mössbauer spectra at 4.2 K under applied magnetic fields up to 5 T give evidence for an antiferromagnetic behaviour with a large anisotropy energy.
From crystallographical considerations we conclude that the electric field gradient is nearly axial symmetric along the chain direction. In this case it is the higher energy transition that broadens initially for internal magnetic field fluctuations collinear to V_{zz}, and therefore the sign of the

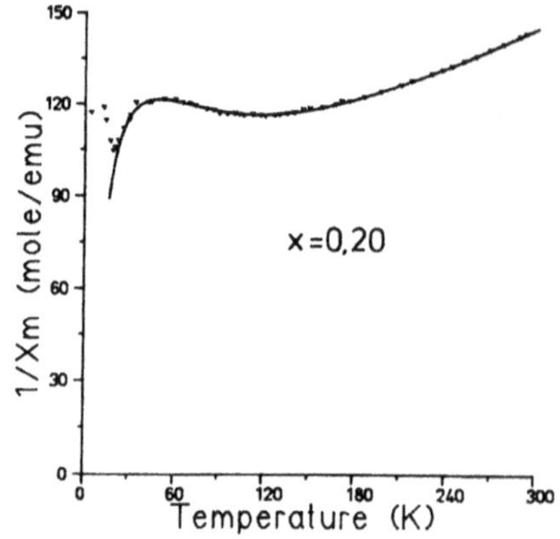

Fig. 1. Reciprocal magnetic susceptibility for x=0.20

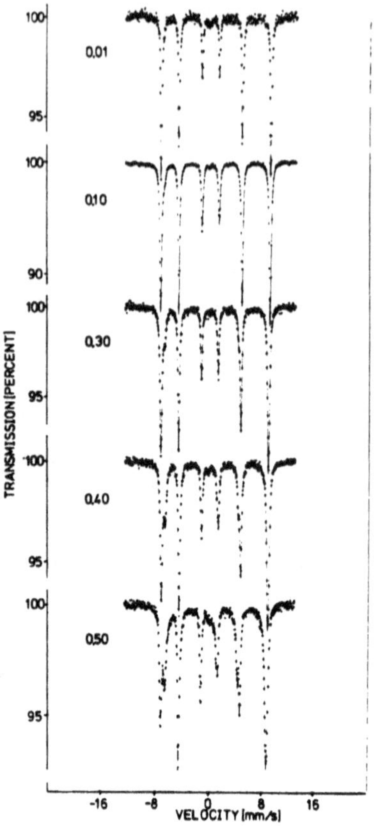

Fig. 2. Fe(57) Mössbauer spectra at 4.2 K

Table 1. Concentration x, Saturation Fields $H_0(1)$ and $H_0(2)$ at 4.2K, Quadruple Splittings EQ1 and EQ2 at 300K, Transition Temperatures T_M and T_X and mean Intra-Chain Exchange Energy J/k.

x	A1 (%)	$H_0(1)/H_0(2)$ (kOe)	EQ1/EQ2 (mm/s)	T_M/T_X (K)	J/k (K)
0.01	100	511(3)	.841		-21.0(8)
		-	-		
0.20	90	505(3)	.788(8)	23.5(8)	-17.5(8)
		483(3)	1.146(8)	19.0(1.0)	
0.30	84	502(3)	.777(8)	21.5(8)	-
		481(3)	1.127(6)	17.1(1.0)	
0.40	70	498(3)	.773(8)	19.5(8)	-16.8(8)
		480(3)	1.136(7)	15.5(1.0)	
0.50	59	493(3)	.786(8)	18.2(8)	-16.0(8)
		477(3)	1.146(8)	14.5(1.0)	
1.00	(ref. 2)	396(3)	1.360(10)	9.7	-13.1

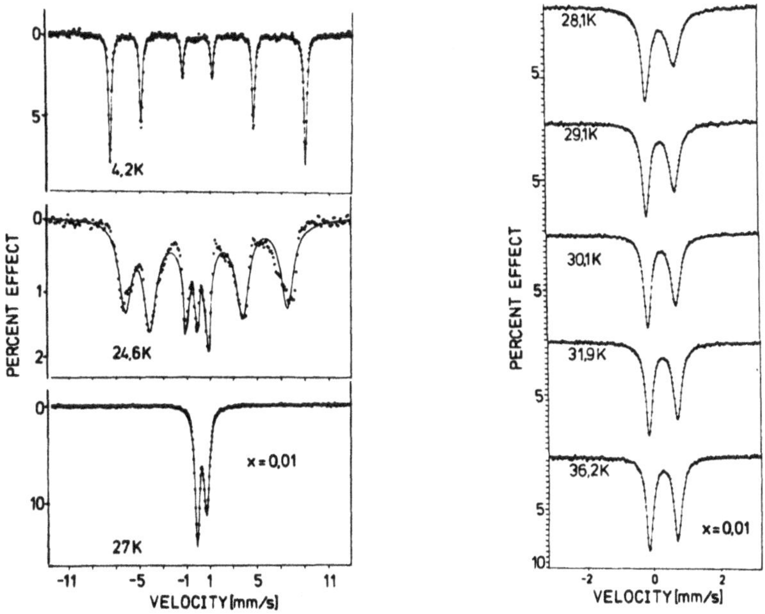

Fig. 3 and 4. Temperature dependent Mössbauer and relaxation spectra

quadrupole splitting is positive. To fit the 4.2 K spectra we assumed that the electric field gradient is not temperature dependent. By using for x=0.01 EQ=0.84 mm/s, $\eta < 0.3$ we found that the hyperfine field H_0 should be close to the V_{zz} gradient axis. In zero magnetic field the iron and presumably the manganese atoms are approximately oriented along the chain direction.

Although the easy-axis anisotropy may not be neglected, the temperature dependence of the magnetic susceptibilities was analyzed on the basis of the isotropic Heisenberg Hamiltonian [5]. For calculating the temperature dependence of the magnetic susceptibilities (see Fig. 1) we adopted the results of Mössbauer spectroscopy with respect to site occupation. The slightly higher transition temperatures found from the Mössbauer data in comparison with T_X, the transition temperatures derived from X measurements (see Table), can be related with the frequency dependence of the blocking of the magnetization in the magnetic chains.

The temperature dependence of the Mössbauer spectra for x=0.01 are shown in Fig. 3 and 4. The spectra above 42 K are essentially the same. There is some slight temperature -independent asymmetry over this range and we attribute this to sample texture. At lower temperatures between 42 and 27 K an additional temperature-dependent asymmetry of the absorption lines is observed. As the temperature is further decreased, the spectra exhibit magnetic hyperfine splitting corresponding to antiferromagnetic ordering. Somewhere between 27 and 26.5 K the quadrupole doublet of the paramagnetic phase disappears.

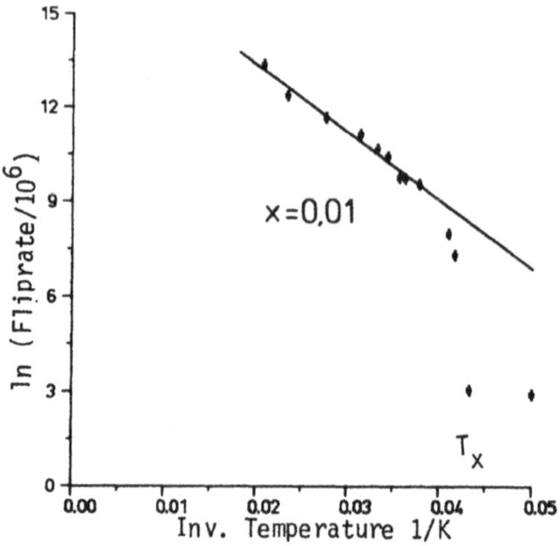

Fig. 5. Flip rate versus reciprocal temperature for x=0.01

The Mössbauer spectra below and above the 3-d ordering temperature could successfully be fitted adopting the soliton model developed for the description of non-linear excitations in quasi 1-d magnetic systems[6]. We used an relaxation model which is appropriate for our system[7]. It allows for flipping of the hf field into a reversed direction with a rate T. From the resolved spectra for x=0.01 it is apparent that in the temperature range near 24.6 K clusters with different relaxation frequencies must be present. At 23.1 K this shows up only in asymmetric lines. Between 24 and 26 K, however, there additionally appears a central asymmetric doublet (Blume effect) due to faster relaxation which is still observable up to 42 K.

In Fig. 5 we show the spin-flip rate for x=0.01. As indicated by the straight line it shows the exponential dependence on reciprocal temperature, characteristic for an activated soliton process[6]. The activation energy is 217(5) K. The temperature at which the drop occurs corresponds to the transition temperature T_x=23(2) K deduced by extrapolation T_x(X→ x=0.01) from the susceptibility measurements.

References

1. L.J. de Jongh, "Magneto-Structural Correlations in Exchange Coupled Systems", NATO ASI Series, D.Seidel Publishing Company, 1983.
2. J.L.Fourquet, R. de Pape, J.Teillet, F.Varret and G.C. Papaefthymiou, J. Magnetism and Magnetic Materials 27, 209 (1982).
3. J.Pebler, W.Massa, H.Lass and B.Ziegler, published in Solid State Chemistry, 1987.
4. J.Pebler, to be published.
5. M.Drillon, E.Coronado, D.Beltran and R.Georges, Chemical Physics 79, 449 (1983).
6. H.J.M. de Groot, R.C.Thiel and L.J. de Jongh, Proceedings of an Int. Workshop, San Miniato, Italy, p. 21-26, 1984.
7. J.Pebler, Phys. Stat. Solidi (a), 78, 589 (1983)

ONE DIMENSIONAL EXCITATION ENERGY TRANSFER IN TMMC AND RELATED COMPOUNDS

R. Knochenmuss and H.U. Güdel

Department of Chemistry
University of Bern
3000 Bern 9
Switzerland

INTRODUCTION

The "one-dimensional" nature of materials with linear chain structures is reflected not only in the ground state properties, but also in the characteristics of electronically excited states. Some insulating antiferromagnetic compounds of Mn^{2+} are "1-D", and the excitons formed by optically exciting from the $^6A_{1g}$ ground state to the $^4T_{1g}$ first excited state are confined to motion predominantly along the chains. The luminescence from the excitons can be used to determine the degree of anisotropy and other parameters of the motion. However, 1-D effects are apparent in luminescence only when small amounts of traps are introduced into the host lattice, in which case the transport of energy to the traps is limited by the topology of the crystal.

The hopping of the $^4T_{1g}$ excitons is a thermally activated process; as the temperature is increased they become increasingly mobile and are able to reach trap ions before they radiatively decay. This results in quenching of the emission and dramatic increases in trap luminescence (if the traps do not decay non-radiatively). Example host and trap emission time dependences are shown in Fig.1 for $CsMnBr_3$ doped with Er^{3+}. The average hopping time can be inferred from the time dependence of the Mn^{2+} decay and the rise in the trap emission. More sophisticated models which predict the detailed shape of the decay curves can be used to extract further information, such as on and off-chain hopping rates, trapping efficiency, coherence (or lack of it), etc.

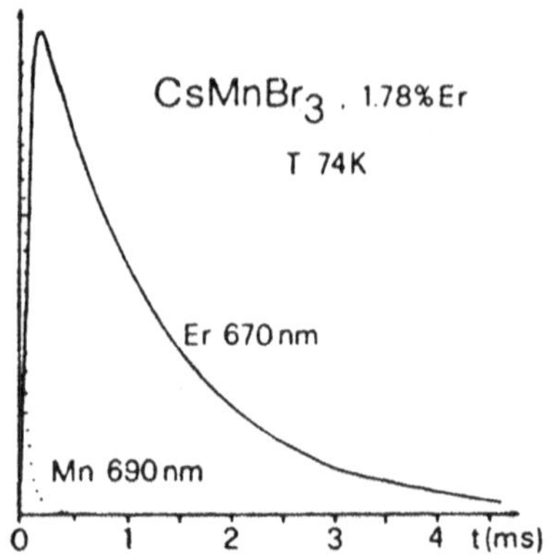

Fig.1 Time dependence of Mn^{2+} (dotted line) and Er^{3+} (full line) emission intensities after pulsed selective Mn^{2+} excitation.

RESULTS

As a detailed example, consider TMMC (tetramethyl ammonium manganese trichloride), a prototype 1-D material in its ground, and as we find, first excited state. We prepared crystals doped with Cu^{2+} and Co^{2+} traps, and analyzed the Mn^{2+} emission (these traps do not emit).

At low temperature (4K), these crystals all showed identical strong, exponentially decaying red emission. As the temperature was increased, the emission was quenched and the decay curves became non-exponential, as demonstrated in Fig. 2 (exponential decays would give straight lines). By about 150K trapping is significant but not overwhelming.

The (model independent) non-radiative contribution to the average exciton lifetime: $\langle t \rangle = \int_0^\infty t \, \delta N/\delta t \, dt$, N = number of excited Mn centers; was first used to determine the Mn-Mn hopping rates. It was found that the rates vs temperature and concentration were consistent with Zumofen's expression for partially reflecting traps in a 1-D system [2]. At 200K the Mn-Mn rates were about 10^{11} s^{-1} and the trapping efficiency was a surprisingly low 1 in 10^4, for both traps.

Three models were used to fit the detailed shapes of the decay curves. The simplest was that of Movaghar, Sauer and Würtz [3] (dot-dash line in Fig. 2b). It assumes strictly 1-D motion and perfect trapping and has only 1 adjustable parameter, the hop rate. Not too surprisingly, the fits are not good and get worse as higher temperature and trap concentration increase the number of (mostly reflecting) interactions with traps before radiative decay.

A more sophisticated model is that of Wieting, Fayer and Dlott [4] (dashed lines in Fig. 2), which includes off-chain transport and the effect of scattering as well as trapping centers. This model gave good fits except at higher temperatures and concentrations of traps. The reasons for the partial failure of the model are not clear. Auerbach and McPherson [5]achieved uniformly good results with the model on Cu^{2+} doped TMMC.

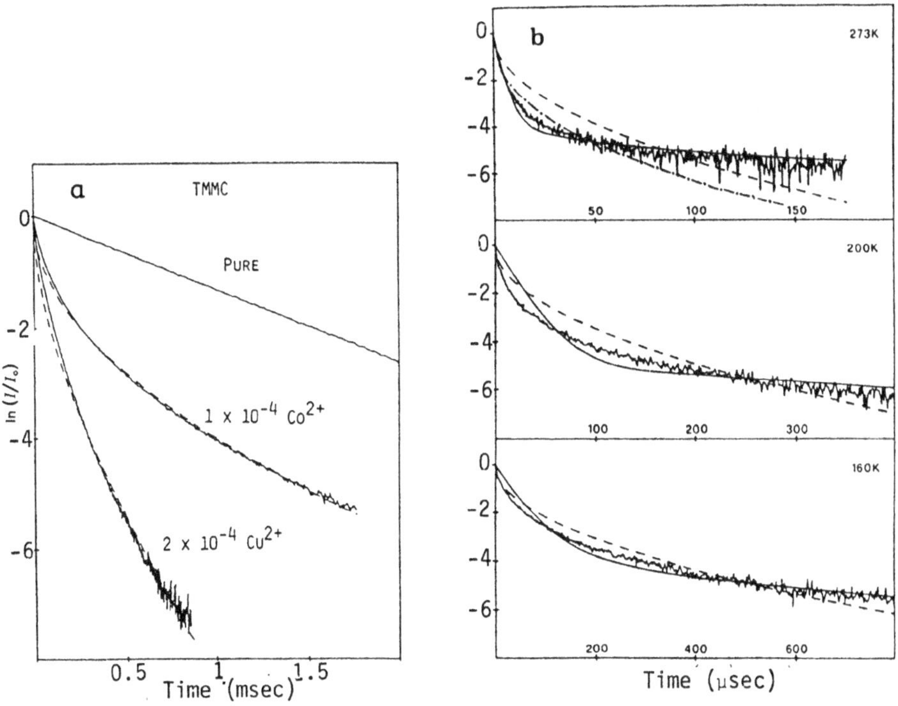

Fig. 2 a) Decay curves of pure and doped TMMC at 273K. The dashed lines are fits with the model by Wieting et al. (ref.4). b) Decay curves of TMMC doped with $2x10^{-3}$ mole fraction Cu^{2+}. Dot-dashed , dashed and solid lines are fits with the models of Movaghar et al. (ref.3), Wieting et al. (ref.4) and the incoherent trapping model of Kenkre (ref.6), respectively.

The on-chain hopping rates from this model were also about 10^{11} s^{-1} at 200K. In addition, off-chain hopping rates can be inferred. In ref.5 this was found to be 10^3 s^{-1}, but ignored the possibility of reflection at traps. Including this effect we estimated a rate of $<10^6$ s^{-1}. Unfortunately the model gives identical shapes for the decay curves regardless of trapping efficiency, resulting in some confusion in the interpretation of the fit parameters. In any case, it is clear that the excitons are moving in an extremely 1-D fashion.

The third model employed was the incoherent (hopping in contrast to wavelike motion) trapping model of Kenkre [6]. Arbitrary amounts of coherence may be included, but since the relaxed $^4T_{1g}$ is subject to geometric distortions and hence has an activation barrier of several hundred cm^{-1}, the excitons are highly localized and no coherence is expected (or inferred from the spectra). This model has as adjustable parameters the Mn-Mn hopping rate and the Mn-trap rate. Over the whole temperature and concentration range it gave the best performance (solid lines in Fig. 2 b), but remained not completely satisfactory due to somewhat excessive scatter in the best fit parameters. On-chain rates again averaged 10^{11} s^{-1} at 200K, and the Mn-trap rates were found to be about 10^7 s^{-1}, not varying strongly with temperature. This ratio of (Mn-Mn)/(Mn-trap) rates of 10^4 is in good agreement with the trapping efficiency determined by the model independent analysis.

CONCLUSIONS

Optical spectroscopy is a powerful tool for determining properties of 1-D systems in electronically excited states, although it is necessary to study deliberately impure materials to determine these properties. In the case of TMMC, the following has been learned:

- $^4T_{1g}$ excitons are mobile, and have a thermal activation barrier of 800-1000 cm^{-1}.

- The excitons move in an incoherent, hopping fashion, with a Mn-Mn rate of about 10^{11} s^{-1} between 150K and room temperature.

- The motion is extremely one-dimensional, the cross-chain Mn-Mn rate is at least 6 orders of magnitude less than the on-chain rate.

- Trapping by Cu^{2+} or Co^{2+} dopant ions is not efficient, about 1 in 10^4.

- Hopping over traps occurs at about 10^6 s^{-1}

- The Mn-trap rate is about 10^7 s^{-1} for these traps.

REFERENCES

1) R.Knochenmuss, H.U.Güdel, J.Chem.Phys. 86, 1104 (1987)

2) G.Zumofen, private communication

3) B.Movaghar, G.W.Sauer, D.Würtz, J.Stat.Phys. 27, 473 (1982)

4) R.D.Wieting, M.D.Fayer, D.D.Dlott, J.Chem.Phys. 69, 1996 (1978)

5) R.A.Auerbach, G.L.McPherson, Phys. Rev.B 33, 6815 (1986)

6) V.M.Kenkre, in "Exciton Dynamics in Molecular Crystals and Aggregates", ed. G.Höhler (Springer, Berlin, 1982)

QUANTUM ENERGY GAP IN THE S=1 HEISENBERG ANTIFERROMAGNET NENP: EXPERIMENTAL DATA AND HALDANE CONJECTURE.

J.P. Renard[1], M. Verdaguer[2*], L.P. Regnault[3], W.A.C. Erkelens[3**] J. Rossat-Mignod[3], and W.G. Stirling[4]

[1] Institut d'Electronique Fondamentale, CNRS UA 022, Bâtiment 220, Université Paris-Sud, 91405- Orsay (France)
[2] Laboratoire de Spectrochimie des éléments de transition, CNRS UA 420, Université Paris-Sud, 91405- Orsay (France)
[3] Centre d'Etudes Nucléaires de Grenoble, DRF/SPh-MDN, 85 X, 38041 - Grenoble - Cedex (France)
[4] Institut Laue Langevin, 156 X, 38042 - Grenoble - Cedex (France)
* Permanent address : ENS, Le Parc, 92211 - Saint-Cloud, (France)
** On leave from Kamerlingh Onnes Laboratory, University of Leiden, The Netherlands

INTRODUCTION

Since the early work of Ising [1], a lot of theoretical studies have been devoted to one-dimensional (1d) magnetism [2-6]. Recently, the interest for the 1d Heisenberg antiferromagnets (1d-HAF) with integer spin values has been renewed by Haldane [7] who predicted for these systems, an energy gap between the singlet ground state and the first excited states.

In a pioneering work, Botet, Jullien and Kolb (BJK) [8] investigated the properties of the S = 1, 1d Heisenberg-Ising antiferromagnet described by the Hamiltonian

$$(1)\quad \mathcal{H} = \sum_i [-J (S_i^x S_{i+1}^x + S_i^y S_{i+1}^y + \lambda S_i^z S_{i+1}^z) + D (S_i^z)^2]$$

where J represents the intrachain exchange interaction, 1 - λ the exchange anisotropy and D the single-ion anisotropy.

For a certain range of λ and D values, a novel phase with an unordered singlet ground state and an energy gap to the excited state continuum is found. Haldane predictions and the results of BJK were confirmed by the recent finite size scaling calculations up to N = 14 for S = 1 by Parkinson and Bonner [9] and the Monte Carlo calculation up to N = 32 by Nigthtingale and Blöte [10]. The gap is about equal to $|J|/4$ for the pure 1d-HAF (λ = 1, D = 0) and exhibits a non trivial variation with the anisotropy parameters λ and D (Fig.1).

We compare these predictions with experiments for the model system $Ni(C_2H_8N_2)_2\ NO_2ClO_4$, **NENP**.

EXPERIMENTAL STUDY OF NENP

MAGNETIC DATA

The existence of an energy gap E_G between the singlet ground state and the first excited states would produce drastic effects on the magnetic properties. At temperature below E_G/k, the magnetic susceptibility X(T) would exhibit a fast decrease with T and tends to a value X(0), much smaller than the maximum one, X_{max} at T=|J|/k, for all magnetic field orientations. In addition, since the correlation length does not diverge as T→0, the quasi 1d-HAF with a gap cannot develop 3d magnetic long range order if the interchain exchange interaction J' is sufficiently small. Indications for a quantum gap have been recently reported from inelastic neutron scattering experiments in $CsNiCl_3$ [12]. We discussed this point in Ref. [13].

A possible way to approach the ideal S=1, 1d-HAF is to increase J and the interchain distance. This is realized in the orthorhombic crystal of NENP where the nickel ions are covalently linked by nitrito groups along the **b** crystal axis (Fig. 2). This insures a large antiferromagnetic exchange interaction |J|/k = 50 K. The magnetic chains are well separated from each other by ClO_4 anions, which leads to a small value of the interchain interaction|J'/J|= 4×10^{-4}.

The magnetic susceptibility along the three orthogonal crystal axes a, b and c have been measured by means of a low field SQUID magnetometer [15] down to 1.7 K. For all field orientations, the susceptibility exhibits a rounded maximum at about 60 K, in agreement with the previous measurements [14], and falls down abruptly as the temperature is lowered below 15 K (Fig. 3). The limiting value X(0) of the susceptibility, (as T tends to zero), is only a few percent of its maximum value X_{max}. This temperature dependence of the susceptibility, is completely different from the expected behavior of a gapless 1d-HAF, for which $X(0)/X_{max}$ is slightly smaller than 1 ($X(0)/X_{max}$ = 0.69 for S = 1/2 and 0.83 for S = ∞).

No 3d long range magnetic ordering was detected by NMR [16] and specific heat measurements [17] down to 1.2 K. High-field magnetization reveals the vanishing of the gap at a critical field of 90 kOe [18]. All these features are clearly consistent with the existence of the Haldane gap.

1

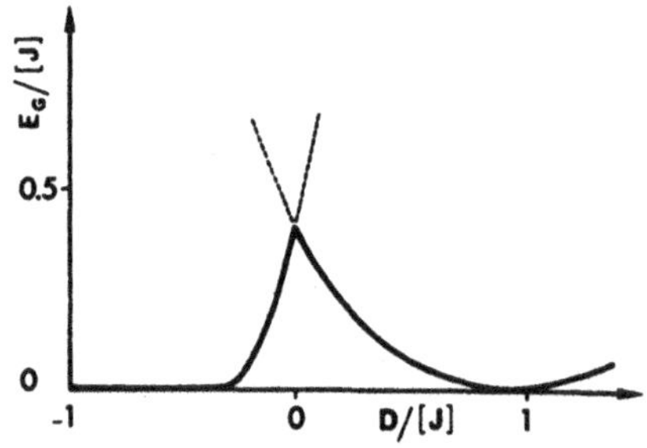

2

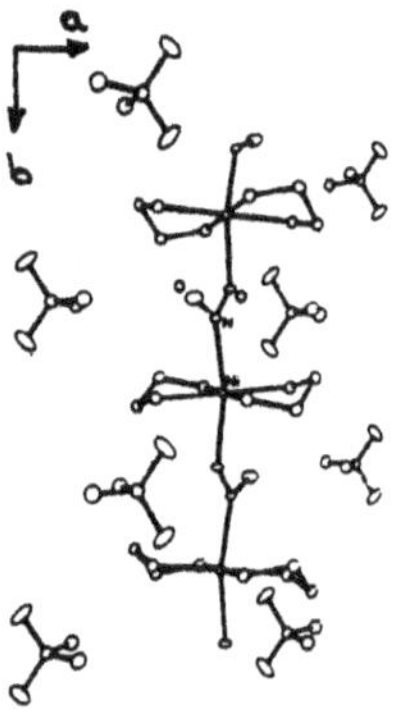

Figure 1 . Variation of the energy gap between the ground state 9and the first excited state of the S=1, 1d-HAF, with the single ion anisotropy D (λ=1). D=0 corresponds to the pure Heisenberg case, D<0 to XY type anisotropy and D>0 to Ising type anisotropy (Ref.[8]).

Figure 2 . Perspective view of NENP; chains run parallel to **b** axis [14].

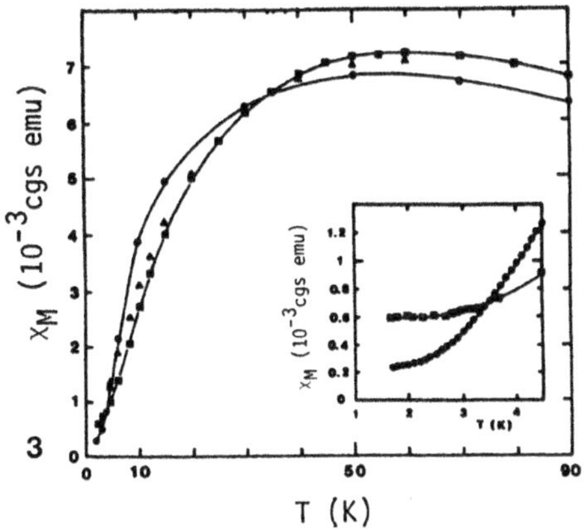

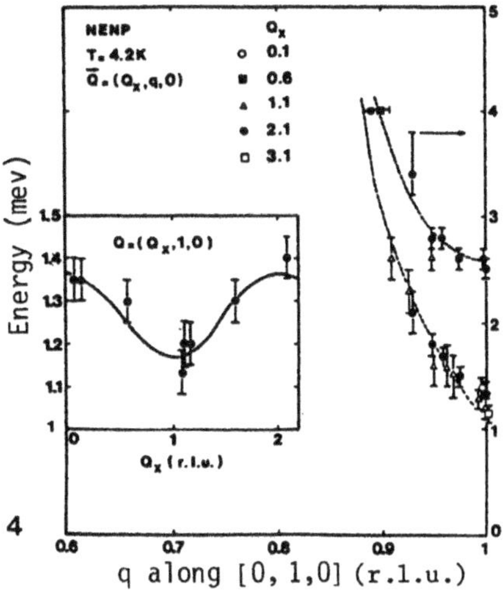

Figure 3 . Molar susceptibility of NENP along the three crystal axes a (■), b (●) and c (▲) vs temperature.

Figure 4 . Dispersion curves of the magnetic excitations along, and perpendicular to the chain axis b at T = 4.2 K. The full line is a fit with E_G = 1.17 meV and $|J'/J|$ = 4 x 10^{-4} [13].

INELASTIC NEUTRON SCATTERING.

From detailed inelastic neutron scattering (INS) experiments, the excitation spectrum of NENP was determined for different scattering vectors (Fig.4) [13].

These measurements show two energy gaps corresponding to spin fluctuations parallel and perpendicular to the chain axis, at respective values of $|E_{g\perp}/J|$ = 0.26 and $|E_{g\parallel}/J|$ = 0.58. The existence of a finite gap without quasi-elastic fluctuation, i.e. with a singlet ground state, corresponds to the prediction of Haldane [7]. However, the experimental situation in NENP, is more subtle since the gap appears to be anisotropic with values depending on the spin-component. This is likely due to the finite magnetic anisotropy of the Ni^{2+} ion. Effects of temperature on the gap were also studied which are also consistent with the existence of a quantum gap [18] .

DISCUSSION

Our magnetic susceptibility and neutron scattering experiments strongly support the Haldane conjecture, i.e. the existence of a well isolated singlet ground state in the pure S = 1, 1d-HAF, with a gap E_G = $0.4|J|$.
The absence of an observable 3d long-range magnetic order in NENP is a strong additional proof of the existence of the quantum gap. For a gapless Heisenberg model, 3d magnetic ordering is expected at $T_N = 2S^2 \sqrt{|J J'|}/k$, i.e. at T_N = 2.5 K in NENP, since $|J|/k$ = 50 K and an estimate of $|J'/J|$ is 4 x 10^{-4}. This T_N value of 2.5 K is a lower limit for NENP since the magnetic anisotropy certainly enhances the actual 3 D ordering temperature. There is no experimental evidence for such a transition neither in the magnetic susceptibility curves neither in the neutron diagrams,

nor in the NMR data. Indeed, powder neutron patterns at 2K and single crystal diffraction experiments down to 4.2 K show no magnetic reflection, and the proton magnetic resonance spectra are typical of a disordered magnetic phase down to 1.2 K.

CONCLUSION

All the experimental features presented by our model compound NENP are consistent with the presence of a quantum gap in this S=1 1d-HAF chain. We have no room here to discuss the other hypotheses able to present the same phenomena (spin-Peierls transition, alternating chain). We have discussed and ruled out them in Ref. 13. Nevertheless, other experimental characterization of the gap are clearly needed, phase diagram in an applied magnetic field, low temperature spin dynamics, direct measurement of the correlation length vs temperature, quasi-elastic neutron scattering on deuteriated compounds, etc.. Indeed, the study of S=1 1d-HAF is just at the very beginning and promised to interesting developments, if suitable model compounds can be designed (various J and D values, low ratio J'/J). We are working along these lines with nickel(II) chains slightly different from NENP, i.e. Ni(propanediamine)$_2$NO$_2$(ClO$_4$) (NINO) with nitrite and other molecular bridges [19].

REFERENCES

1. E. Ising, Zeit. Phys. 31: 253 (1925).
2. See e.g. J. Bonner, L.J. De Jongh, W. Hatfield, J.P. Renard, and R.D. Willett in "Magneto-structural correlations in exchange-coupled systems, NATO ASI Series, 140, D. Gatteschi, O. Kahn, R.D. Willett ed., Reidel (1985).
3. L.J. De Jongh, and A.R. Miedema, Adv. Phys. 23: 1 (1974).
4. M. Steiner, J. Villain, and C.G. Windsor, Adv. Phys. 25: 87 (1976).
5. L.J. De Jongh in "Recent developments in condensed matter physics", Vol. 1, J.T. Devreese ed., Plenum, 343 (1981).
6. See also C.P. Landee, J.P. Renard, M. Steiner in these proceedings.
7. F.D.M. Haldane, Phys. Lett. 93A: 464 (1983) and Phys. Rev. Lett. 50: 1153 (1983).
8. R. Botet, R. Jullien and M. Kolb, Phys. Rev. B 28: 3914 (1983); R. Botet and R. Jullien, Phys. Rev. B 27: 613 (1983).
9. J.B. Parkinson and J.C. Bonner, Phys. Rev. B 32: 4703 (1985).
10. M.P. Nightingale and H.W.J. Blöte, Phys. Rev. B 33: 659 (1986).
11. W.J.L. Buyers, R.M. Morra, R.L. Armstrong, M.J. Hogan, P. Gerlach and K. Hirakawa, Phys. Rev. Lett. 56: 371 (1986)
12. M. Steiner, K. Kakurai, J.K. Kjems, D. Petitgrand, and R. Pynn, private communication and these proceedings.
13. J.P. Renard, M. Verdaguer, L.P. Regnault, J. Rossat-Mignod, W.A.C. Erkelens, and W.G. Stirling, Europhys. Lett. 3: 945 (1987).
14. A. Meyer, A. Gleizes, J.J. Girerd, M. Verdaguer and O. Kahn, Inorg. Chem. 21: 1729 (1982).
15. P. Beauvillain, C. Chappert and J.P. Renard, J. Phys. E, Sci. Instr., 18, 839 (1985).
16. J.P. Renard, S. Clément and T.V. Hiep, to be published.
17. W.A.C. Erkelens, L.P. Regnault, J. Rossat-Mignod, P.R. Nugteren, L.J. de Jongh and J.P. Renard, to be published.
18. J.X. Boucherle, P. Zang, W.A.C. Erkelens, L.P. Regnault, J. Rossat-Mignod and J.P. Renard, to be published.
19. M. Monfort, J. Ribas, X. Solans, M. Verdaguer, and J.P. Renard to be published in Inorg. Chem.

NEUTRON DIFFRACTION STUDY OF MnRhAs NEAR T_N : THE CRITICAL REGIME

J. Bartolomé[1] , B. Chenevier[2] , M. Bacmann[3] , D. Fruchart[3]
C. Rillo[1] , R. Navarro[1] , J.P. Regnault[4] , P. Erkelens[4] ,
J. Rossat-Mignod[4] , and R. Fruchart[5]

[1] Instituto de Ciencia de Materiales de Aragon, CSIC-Univ. de Zaragoza, 50009 Zaragoza, Spain
[2] Institut Laue-Langevin, 156X, 38042 Grenoble Cedex, France
[3] Lab. de Cristallographie CNRS, Grenoble, France
[4] Centre d'Etudes Nucléaires de Grenoble, Grenoble, France
[5] ENSEP, INPG, St. Martin D'Heres, France

INTRODUCTION

The structural and magnetic properties of MnRhAs have been previously studied on powder samples. In particular the specific heat (C_p) experiments demonstrated the existence of an anomaly at T_p=(240.7±0.3)K with no corresponding singularity in the magnetic ac susceptibility(χ).[1] On the other hand a sharp increase of χ at T_c=(190.5±0.5) K indicated the onset of ferromagnetic order, without any accompanying anomaly in C_p. Moreover, at T_t=(156.9±0.1)K both χ and C_p show anomalies associated to the disappearance of the net moment in a ferro to antiferromagnetic transition.

Powder neutron diffraction data reported magnetic ordering below T_c=190.5 K and magnetization versus field experiments indicated that the behaviour between T_c and T_p was not that of a simple paramagnet.[2] Finally, at T<500 K χ showed a marked departure from the Curie Weiss behaviour as the temperature was reduced towards T_c. It was interpreted as a clue of some degree of two dimensional character, inspired as well by the magnetic structures encountered at low temperatures.[3]

This feature was very exciting since a separation in temperatures between T_c and T_p had been predicted for bidimensional XY systems. This is the well known vortex-antivortex formation model proposed by Kosterlitz and Thouless (KT).[4] Thus, given all the experimental evidence for MnRhAs and some calculations performed for its specific structure, the possibility of KT ordering was conjectured for this compound.[1]

The availability of a single crystal large enough to perform neutron experiments permitted a definitive check to these ideas. The present report along this line, unambiguously demonstrates the three dimensional (3d) character for this compound, thus contradicting the KT character previously proposed on basis of the existing data.

The neutron diffractogram verified the change in magnetic ordering at T_t reported previously. Indeed, peaks consistent with an antiferromagnetic structure with a propagation vector **k**=[0 0 1/4] below T_t disappeared at this temperature.[5] Above T_t a ferromagnetic ordering of the Mn μ_z component was seen as an extra contribution on the nuclear peaks. Moreover, the antiferromagnetic ordering between μ_{xy} components of adjacent layers gives rise to diffraction peaks indexed with the propagation vector **k**=[0 0 1/2].

The integrated intensity of the [1 1 1/2] reflection was explored as a function of temperature till 270 K (See fig. 1). It decreases continuously, disappearing at the Néel temperature T_N = 229 K. This feature shows unambiguously that a 3d magnetically ordered structure prevails in the intermediate region between T_c and T_p. The transition to paramagnetic regime derived from this experiment $T_p = T_N$ = (229±2) K, is in reasonable agreement with T_p = 240 K found in C_p for a powder [1], taking into account that the sample comes from a different batch.

In order to investigate the extent of magnetic correlations in the disordered region, its critical scattering was studied above T_N. The experiment was performed at the Siloe Reactor at the CENG on a double axis spectrometer DN3. The wavelength was λ=1.554 Å and a pyrolitic graphite filter was used to suppress second order contamination. To improve the instrumental resolution functions, horizontal and vertical collimation of 20′ was set. The best resolution achieved was $\Delta q_{//}$ = 0.042 rlu defined as full width at half maximum (FWHM) (1 rlu = $2\pi c^{-1}$ = 1.6913 $Å^{-1}$, and c = 3.715 Å), whereas in the perpendicular direction

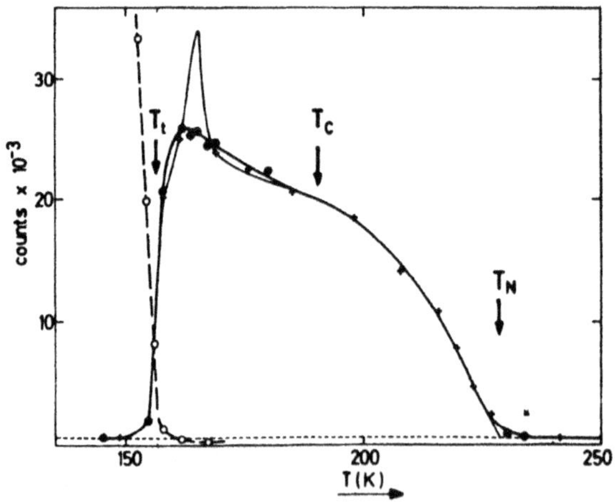

Fig. 1. Integrated intensity of the magnetic Bragg peaks (0 -([1 1 1/4) (•, + [1 1 1/2]). (- -) λ/2 contamination in the incident beam. [5]

$\Delta q_\perp$ = 0.054 rlu = 0.060 Å (1 rlu = 1.1195 Å). Such large resolution widths are inherent to a rather large mosaic spread within the sample.

The critical fluctuations near the antiferromagnetic Bragg peak **Q** = [1 1 1/2] were investigated. This point in reciprocal space corresponds to the propagation vector **k** = [0 0 1/2] from the Brillouin zone center $2\pi\tau$ = [1 1 0]. Scans across this point parallel to the c* direction $\mathbf{q}_{//}$ = [0 0 η], were performed to study the inverse correlation length $\kappa_{//}$ between planes. Perpendicular scans with an offset from the **k** position namely $\mathbf{q}_\perp$ = [ξ ξ -0.2], (see fig. 2), were done to study the inverse correlation length within the (0 0 1) planes, thus avoiding the actual Bragg peak position.

Several temperatures were studied above T_N (233 K < T < 269 K) and one below (202 K). While the $\mathbf{q}_{//}$ scans show a clear temperature dependence with a strong enhancement of the critical susceptibility, the $\mathbf{q}_\perp$ scans demonstrate less temperature dependence. Once that a small Bragg contribution due to λ/2 contamination of the incident neutron beam is subtracted, both scans could be fitted by a single Lorentzian function. For the $\mathbf{q}_{//}$ scans a q dependent background had to be subtracted which stems from a Lorentzian tail originating at the **Q** = [1 1 0] peak. The same contribution was subtracted at all temperatures. The HWHM values of the Lorentzians fitted are shown in figure 3.

The $\kappa_{//}$ data were fitted to the expression $\kappa_{//} \propto t^{\nu}$, with $t=(T/T_N)-1$ obtaining a value of $\nu=(0.5\pm0.1)$. This value is relatively low but the uncertainties involved do not allow to derive any strong conclusion. The scanning geometry chosen observes longitudinal and transverse fluctuations simultaneously thus it is not possible to derive one or the other without doing further experiments. On the other hand, the results can be compared to the reduced q model expressed in terms of the anisotropic inverse correlation lengths $\kappa_{//}$ and $\kappa_\perp$.

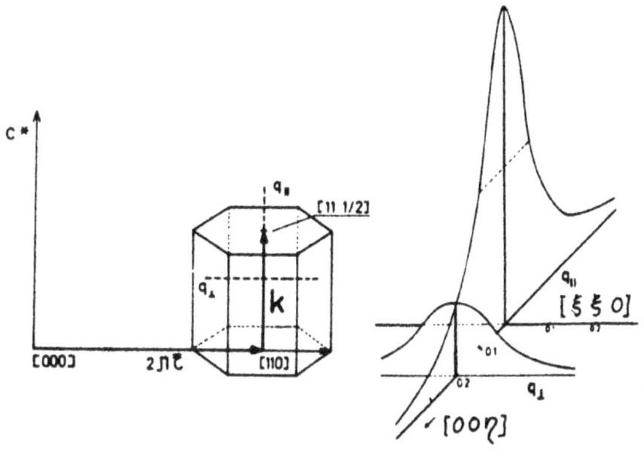

Fig. 2. Scan directions about the Brillouin zone boundary
Q = [1 1 1/2]
Q = 2πτ + **k** + **q**

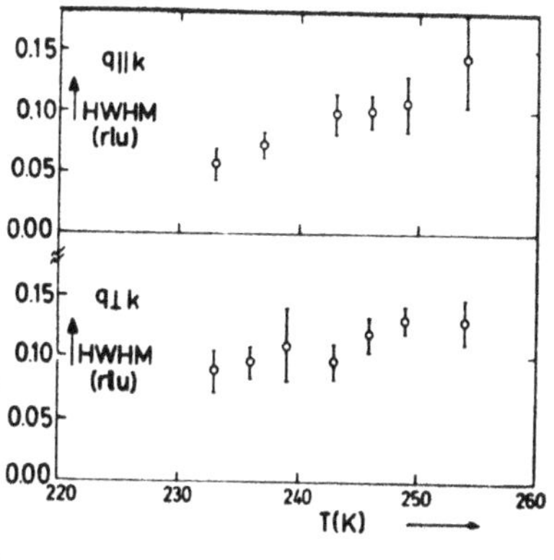

Fig. 3. HWHM values of the Lorentzians fitted to the critical scattering curves along and perpendicular to the q_c* axis.

Table 1. Temperature dependence of the inverse correlation lengths $\kappa_{//}$ and $\kappa_{\perp}$ in reciprocal lattice units (r.l.u.). The derived correlation lengths $\xi_{//}$ and $\xi_{\perp}$ are expressed in Å.

T(K)	$\kappa_{//}$ [a]	κ' [b]	$\kappa_{\perp}$ [c]	$\kappa_{//}/\kappa_{\perp}$	$\xi_{//}$(Å)	$\xi_{\perp}$(Å)
233	0.0570	0.0895	0.0245	2.33	10.4	36.5
236		0.0961	0.0325	2.22		27.5
237	0.0720				8.21	
239		0.1099				
243	0.0996	0.0963	0.0429	2.32	5.94	20.8
246	0.1018	0.1197	0.0543	1.92	5.81	16.4
249	0.1074	0.1306	0.0618	1.73	5.51	14.4
254	0.1437	0.1091	0.0636	2.26	4.11	14.0

[a] $\kappa_{//}$=HWHM of $q_{//}$ scan; [b] κ'=HWHM of $q_{\perp}$ scan; [c] $\kappa_{\perp}=\kappa'[1+(0.2/\kappa_{//})^2]^{-1}$

$$\chi(Q) = \chi(Q_0)\ [1+(\xi_{\perp}/\kappa_{\perp})^2 + (\eta/\kappa_{//})^2]^{-1} \text{ with } \xi_{\perp} = \sqrt{3}\,\xi$$

After due substitution of $|\eta|$=0.2 and the corresponding values of $\kappa_{//}$ one obtains $\kappa_{\perp}$. (See Table 1). The values of $\kappa_{//}/\kappa_{\perp} \approx 2$ for all temperatures, prove the existence of some anisotropy in the exchange interactions between and within the planes. The absolute values of the correlation lengths obtained are short, i.e. at 233 K it is about ten times the Mn-Mn distance.

Concluding, the neutron diffraction experiment has proved that the transition at T_N is due to the para to antiferromagnetic ordering in agreement with the entropy analysis of C_p.[1] Below T_c a ferromagnetic component in the c direction appears, while antiferromagnetic ordering remains in the perpendicular plane, thus explaining the χ behaviour. Moreover, the antiferromagnetic fluctuations among planes are shortly correlated. Anisotropic exchange interactions seem to be active which would be consistent with the anisotropic susceptibility observed. However, the clear 3-d character discards the presence of a KT phase transition in this compound.

The financial support of the Comision Asesora de Investigacion Cientifica y Tecnica (project 1423/82), and the Spanish-French cooperation (Action Intégrée 40/115) are acknowledged.

REFERENCES

1. - J. Bartolomé, J. Garcia, L.M. Floria, F. Falo, D. Fruchart, M. Bacmann, L.J. de Jongh. J. Magn. & Magn. Mat. 557, 1547-1548 (1986).
2. - B. Chenevier, D. Fruchart, M. Bacmann, J.P. Senateur, P. Chaudouet, L. Lundgren. Phys. St. Sol. 84, 199 (1984).
3. - B. Chenevier, M. Bacmann, D. Fruchart, J.P. Senateur, R. Fruchart. Phys. Stat. Sol. (a) 90, 331, (1985).
4. - J.M. Kosterlitz, D.J. Thouless. J. Phys. C6, 1181 (1973).
5. - B. Chenevier et al., in press.

NEW HIGHLY CONDUCTING SYSTEMS

H.J. Keller (Chairman/Heidelberg), A.D. Bandrauk (Quebec), J.C. Bayon (Barcelona), C. Bellito (Roma), R.J.M. Clark (London), P. Day (Oxford), J.G. Haasnoot (Leiden), W.E. Hatfield (Chapel Hill), M.T. Jones (Saint Louis), G. Papavassiliou (Athens), T. Robert (Mons), J.U. von Schültz (Stuttgart), A.E. Underhill (Bangor), and M.M. Whangbo (Rayleigh)

INTRODUCTION

Crystalline inorganic and organic solids with the potential for high electrical conductivity in one or two dimensions are discussed in this section which includes all crystalline materials which have metal-like behaviour especially if they indicate the possibility of superconductivity. Not included are the typically amorphous conducting organic polymers, such as polypyrrole, polyaniline etc. which have excited wide-spread interest over the last few years, but which raise quite different chemical and physical questions.

The materials under discussion can roughly be divided into three classes:

I. "Classical" Inorganics like metal oxides, sulfides, selenides or halides.

II. Coordination compounds of metal ions with mainly organic ligands such as metal dithiolates, phthalocyaninates.

III. Solids comprised of organic ions responsible for the conduction path and inorganic counterions.

I. "Classical" Inorganic Solids

(A) Oxides

The potentiality of classical inorganic solid-state chemistry for generating simple solids with only 3 or 4 different kinds of atoms has been demonstrated recently with the discovery of high temperature superconductivity in different classes of this type of compounds.

The first report of the onset of superconductivity at a high temperature (~30K) in the Ba-La-Cu-O system by Bednorz and Mueller in mid 1986 led to a number of studies aimed at determining the identity and the structure of the superconducting phase. These intense efforts showed in late 1986 that the superconducting phase has the composition $La_{2-x}M_xCuO_4$ (M = Sr or Ba; X $\sim$ 0.1–0.2; $T_c \sim$ 30-40K), the phase that has been known since 1979. In late 1986 Chu et al. found that the T_c of this Ba-La-Cu-O phase increases to a higher temperature, which suggested to them that La should be replaced by Y so as to induce an internal pressure. This effort led to the discovery of the new superconducting Y-Ba-Cu-O system in early 1987 that shows superconductivity at $T_c > 90K$ (higher than liquid nitrogen temperature) under ambient pressure. Intense studies on this Y-Ba-Cu-O system showed that it has the composition $Y Ba_2 Cu_3 O_{7-y}$ ($y \simeq 0.10-0.19$). The related phases of composition $L Ba_2 Cu_3 O_{7-y}$ (L = Sm, Eu, Gd, Dy, Ho or Yb) are also found to exhibit high temperature superconductivity ($T_c \simeq$ 90-100K).

Another recent and promising advance to higher T_c has been made in the $BaBiO_3$ system. By partially substituting K in the place of Ba transition temperatures around 20K have been reached so far.

A number of studies on low-dimensional molybdenum bronzes have shown that the blue bronze $K_{0.3}MoO_3$, the purple bronze $K_{0.9}Mo_6O_{17}$, and the Magnelli phase Mo_4O_{11} show charge density wave phenomena, and that the purple bronze $Li_{0.9}Mo_6O_{17}$ becomes a superconductor at 1.9 K.

These surprising new findings stimulated much new preparative chemical work. Many of the ternary and/or quaternary lanthanide/transition-metal/chalcogenide system should be explored in the future in order to find out whether the known superconducting phases are unique or whether a systematic path to higher transition temperatures can be found by chemical variations. At the moment it is very hard to predict in which direction these chemical efforts should be aimed. In this regard we should mention the frequent reports of materials with indications of superconducting behaviour at 240 K. Unfortunately the identity of these materials is unknown at the present time. It is clear that additional investigations of these materials will lead to exciting new results.

(B) Other Chalcogenides and Halides

There are many other simple classical inorganic solids with low-dimensional metal-like behaviour which have been studied in detail during the last decade because of their special physical properties. The metal di- and trichalcogenides need special mention in this respect. Highly conducting materials with layered structures have been found among metal disulfides while many triselenides show very interesting one-dimensional collective phenomena. Superconducting phases have even been found among the so-called Chevrel phases, which are transition metal sulfides. Some "linear" tetraselenides (Nb and Ta) with parallel spines of polyiodide chains have been added to these interesting solids recently. Though an overwhelming amount of work has been done in this area during the last two decades, this field still remains a fruitful area for further preparative efforts, e.g. by systematically changing the transition metals, their oxidation states, and their chalocogenide linkage in the lattice.

There have been various reports of high temperature (> 77 K) superconductivity in a number of other materials, but these have generally not been confirmed. An excellent example is provided by CuCl, for which giant diamagnetic anomalies have been observed in a number of laboratories. However, it has not been possible consistently to repeat the observations. Reports on the properties of CuCl samples are reminiscent of comments by Chu, for example, that the high T_c signals in YBa_2CuO_{6+x} were elusive, initially, and difficult to stabilize. Perhaps CuCl and other such materials should be reinvestigated in view of the early observations on YBa_2CuO_{6+x} and of the recent findings of high temperature superconductivity.

II. Coordination Compounds

We will divide this section into two groups of complexes.

(A) Those with classical "hard" bases as ligands such as CN^- or NH_3 and

(B) those with "soft" donor atoms in the ligands, especially with sulfur, selenium and phosphorus.

(A) "Classical" Complexes

The work in this area produced a number of 1d conductors and/or semiconductors which have been intensively investigated by physical methods during the last decades. The most famous examples are the mixed-valence compounds KCP, platinum-oxalates, $Hg_{3-x}AsF_6$ and metal porphyrins. The experiments in this field were motivated mainly by theoretical predictions. First of all there were the conjectures of Peierls, Kohn, Froehlich et al. predicting the behaviour of 1d metals and secondly there was the proposal of Little who suggested the detailed structure of an excitonic high temperature superconductor. Though the latter goal has not been achieved yet, the development in the field of 1d coordination compounds is a typical example of the fruitful collaboration between theoreticians and experimentalists in materials science.

(B) Sulfur, Selenium, Phosphorus, etc., as Ligands

Many transition metal compounds involving the above mentioned donor atoms, especially dithiolates, have been prepared recently with the aim of obtaining highly conducting materials. The efforts are twofold. It is planned,

(i) to obtain solids with networks of higher dimensionality and

(ii) to use the metal thiolate derivatives as anions in connection with other "stacking" organic cations such as the TTFs.

(i) Bis(dithiolato)metal complexes

Many conducting crystalline dithiolate metal complexes have been prepared recently and investigated with respect to their conductivity. The experiments in this field again are guided by theoretical predictions. For example Hofmann has provided guidelines for the selection of compounds - many of which are still unknown - which may be candidates for desirable highly conductive behaviour. Much further preparative work is needed before one is able to estimate the potential of these systems with respect to their conductive properties.

(ii) Dithiolate anions and donor cations

This latter aspect is connected with the "organic" materials. One can think of innummerable combinations between dithiolato-metal anions and planar organic radical cations crystallizing in very different lattice types: At this point we wish to call attention to superconductivity in $TTF[Ni(dmit)_2]_2$ (H_2dmit = 4,5-dimercapto-1,3-dithiol-2-thione), the first molecular superconductor containing a transition metal complex; the conductivity pathway is not known.

III. Organics

At least three classes of solids can be discerned under this heading:

(A) Crystals in which the conduction path is made up by stacks of negatively charged radical anions. Simple inorganic metal ions, classical complex cations or more "organic" ions such as ammonium or phosphonium cations act as counterions without contributing to the conduction directly. The TCNQ salts are a typical example.

(B) Solids made up of stacks of planar organic cations and non conducting counter anions. $(TMTSF)_2X^-$, the Bechgaard Salts, are mentioned as example here.

(C) Crystals containing segregated stacks of organic radical anions, counter-balanced by stacks of planar organic cation, one famous example being TTF-TCNQ.

(A) Radical Anion Chains

The age of organic conductors began with the discovery of TCNQ and the high electrical conductivity of some of its salts by the DuPont group in the early sixties. Since then many hundreds of solids containing TCNQ-anion stacks as the conducting unit have been investigated. No superconductors have been found under these materials so far.

Very recently a similar type of organic radical anions - obtained by chemical reduction of differently substituted N-dicyanobenzoquinonediimines - could be isolated as stacked materials. Some of them show much higher conductivities compared to those of most of the TCNQ salts especially at very low temperatures.

(B) Radical Cation Chains

The exploration of solids containing organic cation stacks counterbalanced by simple inorganic anions like perchlorate, triiodide, etc., resulted in the only organic superconductors so far known. Aromatic hydrocarbons, sulfur, oxygen, and selenium heterocycles have been used as donors in these substances. They are oxidized by different means - chemically or electrochemically - in the presence of different anions which results in the formation of highly conducting crystalline materials. Only a small number of anions have been used so far and it is suggested that the variation of these ions should be explored in a broad fashion. Some of these are layered compounds and the question of which kind of counterions fits best in this structure should be investigated. Very recently the range of anions built into these solids has been extended:Metal cluster anions and metaldithiolates have been introduced, the latter leading to another superconducting material.

Since so much work has been done on the chalcogenofulvalene donors chemists should be stimulated to provide other promising donor systems. To stimulate the development of new low-dimensional organic and inorganic based solids, to test concepts and observe anticipated phenomena theoretical insight described in such a way that enables the synthetic chemist to design and prepare new materials to test and extend the theories is necessary. Little's model was particularly good example of this, whereas, the Haldane conjecture was not.

(C) Segregated Stacks of Radical Cations and Radical Anions

The segregated "double" stack material TTF-TCNQ paved the way for the explosive development of organic conductors. It has been a fruitful area in the past. Though many combinations of different planar organic radical anions and radical cations have been isolated and investigated intensively none of them has turned out to be superconducting.

The only organic superconductors so far have been found in the group of hole conductors like $(TMTSF)_2X$ but none of them in the huge class of organic electronic conductors like the TCNQ salts. Certainly the question arises:Why is this so? So far this question has not been answered. Therefore, theoretical work in this direction is recommended.

The above discussion of highly conducting materials has only been concerned with the preparation and investigation of this wide variety of solids in bulk, which means as single crystals. The so-obtained solid state properties certainly can be further varied by using specially prepared samples, like LB films, sputtered films or by the inclusion of these compounds into other matrices to enhance the overall conductivity. Only minor efforts in these directions have been undertaken up to the present time.

Since future developments in science are difficult to predict this contribution more or less gives a summary of the "chemical" status quo in low-dimensional highly conducting materials hinting only at some clearly open questions.

FERROMAGNETIC COMPOUNDS

R.L. Carlin (Chairman/Chicago), D. Beltran (Valencia),
R. Burriel (Zaragoza), C. Benelli (Firenze),
A. Caneschi (Firenze), O. Kahn (Orsay),
C.P. Landee (Worcester) and J. Pebler (Marburg)

INTRODUCTION

Although antiferromagnets are more common, there is a number of ferromagnetic insulators known, and the rules for explaining ferromagnetic exchange behavior (Goodenough, Kanamori, etc) are available. Unfortunately it is not clear how to use these rules in the laboratory in order to construct new three-dimensional ferromagnets. There are also no good examples of organic ferromagnets known.

The preparation and characterization of new ferromagnets is important not only for the intellectual challenge presented but also for potential industrial applications. In this regard, the discovery of transparent ferromagnets is particularly desirable, as well as ferromagnets which have a transition temperature near room temperature.

A knowledge of both the thermodynamic and kinetic behavior of coordination compounds is essential in order to make reliable predictions of the kind of compounds that can be prepared. Furthermore, in characterizing a ferromagnet, it is important to bear in mind that the susceptibility of a ferromagnet depends on the method of measurement. The ac inductance method provides the best procedure for measuring zero-field susceptibilities.

A desirable feature of an ordered ferromagnet is the presence of anisotropy which will permit the formation of anisotropy is spin-orbit coupling which is absent in organic radicals. We thus expect that purely organic ferromagnets will not function adequately, and see the need for the inclusion of anisotropic transition metal ions such as iron(II) and cobalt(II). Nevertheless, a well-characterized organic ferromagnet, whose preparation is reproducible, is still required.

RECOMMENDATIONS

This workshop surveyed these phenomena and came to the following recommendations:

1. The exchange mechanism in oligomers is sufficiently well understood that one may plan in a rather rational way the synthesis of compounds in which ferromagnetic linear chain interactions will be

operative. The extension of the current work on ferrimagnetic bimetallic linear chains to the construction of both two- and three-dimensional ferromagnets is highly desirable, yet the synthetic strategy remains to be developed. Copper(II) is frequently a good source of ferromagnets because of the principles governing the orthogonality of orbitals; high spin manganese(II) and iron(III) are, on the other hand, unfavorable. One of the keys to success lies with the design of quite specific bridging ligands, able to accomodate metal ions according to a specific topology.

This seems the most likely way at the moment to prepare new ferromagnets, though the method is as likely to provide ferrimagnets instead.

2. Compounds containing metal ions interacting with stable organic free radicals may be good candidates for these studies, for radicals are paramagnetic ligands with a well-known ability to transmit exchange interactions. Furthermore it is possible to experiment with their organic chemistry so as to vary their Lewis base character and steric hindrance, in order to modify them for the planned synthesis of new compounds.

3. A possible synthetic strategy might be derived from the following suggestion. Compounds such as the A_2CuCl_4 and A_2CrX_4 series, where A is an alkylammonium ion, form extensive sets of well-known two-dimensional ferromagnets. Is it possible to learn how to retain the planar, ferromagnetic nature of these materials while converting them into three-dimensional ferromagnets?

4. Extending work such as that proposed above to compounds containing rare earth ions may provide an enhanced capacity of having a high net magnetic moment. Unfortunately, the superexchange interactions are usually small and so is the ordering temperature.

CONCLUSION

It is desirable to find new ferromagnets, especially those with a magnetic transition temperature at or near room temperature. Progress requires the preparation of new materials, but there is no guide beyond the empirical. It is interesting to note that future progress in understanding the solid state physics of ferromagnetism depends in large part on success in both synthetic inorganic chemistry and synthetic organic chemistry!

ROUND TABLE ON NEW LOW-DIMENSIONAL MAGNETIC MATERIALS

F. Palacio (Chairman/Zaragoza), E. Coronado (Valencia), J. Darriet (Bordeaux), M. Drillon (Strasbourg), G.Ferey (Le Mans), D.Gatteschi (Firenze), R. Georges (Bordeaux), K. Kopinga (Eindhoven), W. Massa (Marbourgh), P. Rey (Grenoble), T. Rojo (Bilbao), Z. Soos (Princeton), J. Torrance (San Jose), M. Verdaguer (Orsay) and G. Villeneuve (Bordeaux)

INTRODUCTION

The search for new low-dimensional magnetic materials, in short (l-dmm), has been an area of increasing interest in the past for which it is possible to foresee an intense activity in the next future. Both physical and chemical reasons, acting as driving forces, maintain the need of new materials.

From one side, there are new theoretical predictions which require new experimental magnetic model systems to be checked, new experiments are made on already known materials that demand a different synthetic approach in order to comply with experimental requirements (e.g., preparation of large deuterated crystals) or, else, further experimental confirmation of new physical properties. In all these cases parameters affecting both the quality of samples and the physics of the system have well defined boundaries and the problem usually is to find a suitable material that satisfies experimental demands. Although requirements on the sample conditions depend upon every experiment, there are several orientative exigencies that are worth to be listed:

- "Background" effects should preferably be small, which requires systems having as many magnetic moments per unit volume, or mass, as possible.

- An unambiguous interpretation of the data is often more feasible if only one type of chains (or layers) is present in the structure.

- A thorough analysis requires stable single crystals. For some experiments, large crystals are desired, i.e., approx. 100 mg. for heat capacity measurements and about 1 cm^3 for inelastic neutron scattering.

- For neutron scattering experiments no hydrogen should be present in the crystals, what implies the possibility that a deuterated version of the system can be synthesized.

On the other hand, chemical arguments demand more new 1-dmm to be synthesized and studied in order to acquire more information and to learn more about the rules that govern the physical properties of 1-dmm in a crystal. There is still a lot of work to do before the knowledge of these rules permit to control them and magnetic lattices can be built at will. Moreover, one can easily predict that, on the way, many new magnetic materials with very interesting physical properties will come out.

From the analysis of the *state of the art* in 1-dmm, which has been extensively reviewed along the plenary lectures on the subject, it seems possible to establish the demand for new 1-dmm and to foresee new developments as extrapolation of today's active synthetic lines. These two aspects of the search for new 1-dmm are commented on, respectively, in the first and second parts of this report. In a third, part a discussion on new synthetic approaches to new 1-dmm is also reported.In order to maintain the paper under recommended length either figures and references have not been included. Both can be easily found in the plenary lectures and poster papers contained in this volume.

DESIRED NEW MAGNETIC MATERIALS

Many theoretical and experimental studies have been devoted to the physical properties of 1-dmm. Despite this large activity, various interesting theoretical predictions have not yet been tested experimentally because of the lack of relevant model systems. A few cases will now be commented on.

The magnon dispersion of a uniform $S = 1/2$ antiferromagnetic chain with isotropic (Heisenberg) nearest neighbor interaction is predicted to show pronounced quantum effects in the presence of an applied field. To verify this prediction, systems with an as isotropic as possible intrachain interaction J/k of the order of 5K and a relative interchain coupling $J'/J < 0.01$ are desirable.

The so-called Haldane conjecture implies that the zero-field magnon dispersion relation of uniform antiferromagnetic Heisenberg chain shows a gap for integer S values and no gap for half-integer values. At present there is some evidence that this conjecture holds, but a more conclusive test requires magnetic chain systems with $S = 1$ and eventually $S = 2$, an interaction J that should be isotropic, or having a small and variable XY-like anisotropy, and a relative interchain coupling J'/J as small as possible. Apart from this, it would be desirable to have a comparable $S = 3/2$ system for reference, since this system should have no Haldane gap.

To find $S = 1$ alternating chain or spin-Peierls systems, both presenting a singlet ground state different from the Haldane one, would be welcome, together with the corresponding theoretical characterization.

Various classical model predictions can be tested

experimentally on systems with a large value of S. The very well-known Mn^{++} systems, however, always have an antiferromagnetic intrachain coupling, giving rise to a large zero-point spin-reduction. No ferromagnetic chain systems with a large S (>3/2), a (nearly) isotropic or XY-like interaction and good one-dimensionality character ($J'/J < 0.01$) are available at present.

There is much interest in the properties of magnetic systems with XY anisotropy. For instance, the behavior of an S = 1/2 chain with $J^{xx} = J^{yy} = J$, $J^{zz} \ll J$ can be described in terms of weakly interacting fermions. On the other hand, the behavior of two-dimensional systems with an XY-like interaction can be confronted with Kosterlitz-Thouless theory. For a meaningful comparison of theory and experiment in the latter case it is necessary that the XY character of the system be ideal (e.g., resulting from hexagonal symmetry).

When the magnetic moments of a system, in antiferromagnetic interaction, draw odd cycles, it can be impossible to satisfy simultaneously all the interactions and the magnetic system becomes frustrated. While this situation has been widely studied in three-dimensional systems and is completely solved for magnetic clusters, very few examples are known in one- or two-dimensional systems. For a better understanding of the magnetic phenomena in these types of systems much effort, both theoretical and experimental, to produce simple models is required. Desirable examples would be, a linear chain having antiferromagnetic both nearest and next nearest neighbor interactions, a corner-sharing chain formed by tetrahedral units, a *trans*-corner-sharing chain formed by trigonal bipyramids, or tripled chain systems with inter-chain $J_1 \gg$ intra-chain J_2. If interactions between different frustrated chains could be controlled, a passage from one to higher dimensionality in this type of systems would be feasible.

Random exchange effects in linear chains have been theoretically and, to smaller extent, experimentally studied. So far, only systems with a random distribution of antiferromagnetic interactions have been synthesized and more sistematic effort on the preparation of this type of systems should be desirable. Random distribution of two ligands, similar but not identical, along a chain, or chains of two randomly distributed different metals with the same spin value are suggested synthetic approaches. To find a system in which the sign of the interaction varies at random would be of high interest. Such a system might consist in a zig-zag chain in which bond angles are variable.

One should note that the list of interesting magnetic low-dimensional systems given above is not, and cannot be, complete. For instance, if systems could be synthesized containing well-isolated finite chains with equal but in principle variable length, a nice opportunity would exist to test various numerical simulations of finite chain systems and to understand better finite chain effects in linear chains. In the same way many more required 1-dmm could be listed since there are quite a few experiments that need to be either made or confirmed and the list would extend beyond the finite size of the present report.

NEW DEVELOPEMENTS

In contrast to new systems required for specific experiments with precise physical properties clearly specified, there are interesting lines of synthetic activity producing many new 1-dmm. Interest on both, lines and materials, stays in that they may provide an increasing control of the chemical and crystallographic properties necessary to build magnetic lattices with predetermined properties. While such control is, by any means, not complete the enormous progress achieved in the last few years along this direction is clear. Besides, some new and in occasions unexpected magnetic properties appear in some of these new materials which, in turn, stimulate theoretical activity. From the actual situation it is possible to list the areas of current interest and to predict which could be some of the challenges in the next future. In some cases results are already appearing as exemplified in most of the posters presented at the meeting.

Although many 1-dmm are already known, the majority of them are concentrated in only a few transition metal ions, mostly divalent Mn, Fe, Co, Ni and Cu. Early transition metal ions have shown to possess very interesting representatives and much activity is detected in the synthesis of many others. Very few 1-dmm as derivatives of the second and third row are yet known.

The intense activity in recent years on the preparation and study of alternating chains has produced a large variety of different systems depending on the alternating parameter. Thus, structural alternating chains can be understood as a chain of dimers with inter-dimer magnetic interaction. Exchange constant is then the alternating parameter and examples in which intensity or sign of the interaction varies are known. Novel topological cases of interest are chains of trimers or multiple connected ladder or zig-zag chains since both intra- and inter-chain exchange pathways can be controlled to some extent.

When the magnetic moment is used as alternating parameter a bimetallic ferrimagnetic chain is obtained. Either spin value or magnitude of the g-factor can be alternated and examples of both types of ferrimagnetic chains are already known. An interesting possibility now is to connect these bimetallic alternating chains with each other to form a bimetallic layer. Depending on the exchange pathway a ferrimagnetic layer could be prepared. A very different although attractive approach is to generalize alternating bimetallic chains as *domain chains*, that is chains which can be defined by the formula $(A_nB_m)_N$, where A_n and B_m are segments of different length containing different metallic ions regularly repeated along the chain. Some simple representatives of this type of chains with short values for n,m have already been studied.

Interest on *triple chains*, i.e., linear chains clustered by groups of three, has already been emphasized, since they can present frustration phenomena. Double bimetallic chains are also known, most of them permitting a large variety of different magnetic systems depending on which transition metal ions are forming the chain. Thus, magnetic systems ranging from isolated trimers to ferrimagnetic multi-connected chains have been obtained within the same structural type. An important possibility is to cluster a higher number of chains, since many more interesting magnetic properties would appear. For example, cyclic clusters with odd number of chains would present novel frustration effects, while multi-chain systems could also be considered as a (controlled) transition from one to two dimensions.

The possibility of preparing bimetallic layers has already been mentioned. If those layers could be interconnected with each other by a third magnetic ion, the resulting system would be of great importance. The partial sustitution of the third magnetic ion by a diamagnetic analogue would permit a continuous transition from two to three dimensions in the magnetic system.

It seems worth reporting on significant progress that has been made in the area of ferromagnetic interactions in organic, molecular and polymeric materials areas that will probably see intense activity in the near future. Two polymeric molecules have been reported as organic ferromagnets, both with Curie temperature above room temperature. These are significant in that strong ferromagnetic interactions have been incorporated for the first time into a system with only p-electrons. Unfortunately, the chemistry of both of these systems is very poorly understood, the resulting reactions are not reproducible and the yields of ferromagnetic material are low (2% and 0.1%). In addition, it is not known which is the polymeric structure responsible for the observed ferromagnetism. Improvements in understanding these materials would be of significant interest. However, physical characterization, molecular engineering or potential applications cannot reasonably be anticipated until some more examples have been prepared.

NEW SYNTHETIC APPROACHES

In contrast to new systems that are *better* than current examples, there are opportunities for new magnetic systems that are *fundamentally different*. The possibilities of molecular and organic ferromagnets have already been mentioned. Other possibilities lie at the interfaces between low dimensional inorganic and organic magnetic systems or in any form of mixing transition metal ions or complexes, organic free radicals or charge-transfer molecules and lanthanide series ions. The field is just starting and it is very difficult to give precise recommendations. However, it seems that, as a general trend, organic radicals which allow large delocalization of the unpaired electron(s) would lead to interesting materials. Some 1-dmm have recently been prepared establishing the interest of the approach.

Traditionally, interest on organometallic clusters has been on the side of synthetic chemists. Very recently, however, these materials have proved to possess interesting physical, and indeed magnetic, properties. Therefore, from the point of view of 1-dmm, the synthesis of finite metallic clusters can be considered as a new synthetic approach to the search of new zero-dimensional magnetic materials which should be emphasized.

FINAL CONSIDERATIONS

The above list of conclusions, neither complete nor comprehensive, presents an impressive mosaic of some of the achievements which have already been or can be accomplished in the next future. However, when the progress in the last four or five years is evaluated, it seems that the *next future* is very near indeed and if some room is always left to the fantasy, the recent progress made in the fundamental understanding of the nature of interactions between paramagnetic centers and the increased chemical mastery to build more and more complex molecula or solid state systems, allow to think that low-dimensional chemistry and physics are leaving the era of studies of *natural* systems to enter the era of the design of frameworks with new but predictable properties.

ROUND TABLE 4. COMPETITION BETWEEN (SUPER)CONDUCTIVITY, CDW/SDW, (SPIN)PEIERLS TRANSITIONS AND MAGNETISM

H.B. Brom (Chairman/Leiden), D. Chasseau (Oxford),
C. Coulon (Bordeaux), L. Ducasse (Bordeaux),
V.J. Emery (Brookhaven), H.W. Helberg (Göttingen),
J.P. Pouget (Orsay), D.J.K. Schweitzer (Heidelberg),
M. Steiner (Berlin) and M. Weger (Jerusalem)

INTRODUCTION

The discussions between the participants of this round table were mainly focussed on open questions and future developments in their fields of interest. During the sessions the need was felt to summarize also the main achievements in the period after the last gathering in 1978.

PAST AND FUTURE DEVELOPMENTS

To cover the past and to indicate the future developments a division into theory (super)conducting systems, magnetic phases of 1D-conductors and experimental achievements was made. Because of the presence of relevant review articles in this volume, there is no need for an extensive list of references. As another source for most up to date references the ASI-meeting in Magog[1] might be useful.

Theory

Analytical theories of the one-dimensional electron gas and spin chains, which are relevant for much of the field of low-dimensional materials, were extensively developed during the 1970's. Field theory techniques and the widescale use of mappings to other quantum and classical systems led to a good understanding of the asymptotic behaviour of correlation functions and to a characterization of "phase diagrams" of a variety of models.

During the 1980's, these developments were greatly extended by the use of numerical methods - the study of finite size systems by Monte Carlo techniques and by direct diagonalization, both supplemented by finite size scaling extrapolation to the thermodynamic limit. It proved possible to study (a) the intermediate coupling regime (b) the approach to the long-distance asymptotic behaviour (c) critical amplitudes of correlation functions (d) more complicated Hamiltonians, with longer range interactions of many degrees of freedom (e) higher spin systems (f) the interplay between electron-phonon and Coulomb interactions. All of these are relevant for application to real systems and, although many problems remain to be studied, it can be said that the techniques are available to work out the properties of a wide range of systems that may become important in the future.

The same cannot yet be said about two-dimensional problems which are likely to play an increasingly important role in the field. A start has been made by Monte Carlo studies of the Hubbard model and of a restricted class of spin systems. But the increase in the number of degrees of freedom, and the greater difficulties in dealing with fermions present a real challenge to the present numerical methods. A considerable analytical and numerical attack on variety of two-dimensional problems may be expected in the coming years.

As concerns band structure calculations, they were performed for several low dimensional conductors e.g. the Bechgaard salts, their S-homologs and the BEDT-TTF-compounds. Special features like the incommensurate SDW wavevector in the Bechgaard salts required the use of the crystallographic structures at the relevant temperature and pressure. The SCF band structure obtained for β-$(BEDT\text{-}TTF)_2I_3$ shows a 2 dimensional character, although the Fermi surface still appears to be open. Furthermore there is some contribution from the anion to the density of states at the Fermi level. Because of the importance of the electron phonon coupling in the low dimensional conductors, the inclusion of this interaction in the band structure calculations is foreseen in the near future.

(Super)Conducting Systems

Several low dimensional organic crystals are presently known to have a superconducting ground state, a few, like $(TMTSF)_2ClO_4$ and $\alpha_t(BEDT\text{-}TTF)_2I_3$ are superconducting at ambient pressure, others under a few kbar, like $(TMTSF)_2PF_6$, $(TMTSF)_2AsF_6$ $(TMTSF)_2ReO_4$ and $TTF\text{-}[Ni(dmit)_2]_2$ or after a non trivial pressure-temperature history, like β-$(BEDT\text{-}TTF)_2I_3$.

Application of a pressure of 1.2 kbar for the β-phase crystals of the I_3-series of the electrochemically prepared radical salts of the donor BEDT-TTF results in a stabilization of a crystal structure which becomes superconducting even at 7.5 K. The stabilization of the structure by a special pressure temperature cycle results in a metastable superconducting state at 8 K and ambient pressure. A stable superconducting state at 8 K (bulk effect) is observed in α_t-$(BEDT\text{-}TTF)_2I_3$, obtained from α-$(BEDT\text{-}TTF)_2I_3$ crystals by tempering at about 80 °C for 3 days. After this treatment the structure seems similar to the β-phase.

For the $(TMTSF)_2ClO_4$ superconductor the superconducting and SDW ground states are competitive. A cascade of field induced SDW transitions is seen in large magnetic fields. In the Bechgaard salts it could also be shown that in certain cases the anionic potential substantially influences the electronic groundstate and might supress the metallic state either by 4 k_F electronic localization effects or by a 2 k_F structural dielectric transition involving the ordering of the non-centrosymmetrical anions. In the last category of materials the metallic/superconducting state can be restored by application of pressure.

In the near future a further characterization of the α_t-$(BEDT\text{-}TTF)_2I_3$ compounds is expected. For the synthesis of new superconductors it will be favourable to enhance the coupling between neighboring chains in order to stabilize the chains against structural changes. Another stabilization route might be found in the synthesis of chains with larger molecules as building units. Although at present there is no real understanding of the high T_c of the recently found superconducting oxides like $YBa_2Cu_3O_{7-\delta}$ the ideas developed will have a large impact on strategies for both organic and inorganic superconductors.

Magnetic Phases of 1D-Conductors

Besides the Peierls, CDW and superconducting ground states seen in some low dimensional materials, an antiferromagnetic or SDW ground state is observed in a number of organic compounds with structures closely related to the TMTSF (or Bechgaard) salts. In this series antiferromagnetic resonance is quite easy, due to the narrowness of the resonance lines, and this method together with other techniques like NMR and magnetic susceptibility on single crystals, have led to a quite complete understanding of the magnetic ground state, including the strength of the exchange and anisotropy energies. However due to the small size of the crystals neutron diffraction cannot be used and much information about the dynamics of the spins is still missing.

In $(TMTTF)_2PF_6$ a spin-Peierls ground state has been discovered and both precursor effects and the ordered state have been extensively studied by different techniques (magnetic susceptibility, ESR, NMR and X-ray diffuse scattering). The competition between spin-Peierls and AF-ground states has been discussed in connection with the transport properties of the various compounds. The present conductors do not allow the exploration of the incommensurate high field phase like in the insulator TTF-Au-BDT, which requires the synthesis of a new compound with a spin-Peierls temperature of a few K. Also the study of spin (or lattice) dynamics will depend on the chemists: deuterated single crystals of ≈ 0.5 cc are required for such a study.

Experimental Achievements

Experiments are performed under increasingly more complicated circumstances, like the structural determination by neutrons under pressure and/or at low temperature and by X-rays on single crystals under pressure. The (very weak) superstructure reflections have been seen in the CDW condensed phase by neutrons and X-ray scattering on single crystals under pressure and at low temperature. Magnetic fields up to 30 T are applied. Transport properties are measured up to pressures of 15 kbar and down to temperatures of 50 mK. Low temperature specific heat is determined on a few needles. High frequency conductance is measured to frequencies above 100 GHz etc.

Also new techniques became available, like high resolution NMR, from which we expect a considerable contribution in the coming years both in static and dynamic applications. Further future developments will benefit from the availability of polarized neutrons: the magnetic part of the scattering can be analyzed. The same holds for the increasing resolution in inelastic scattering experiments (100 neV). Both applications have to wait for the synthesis of sufficient large single crystals! The new pulsed neutron sources will give the opportunity to resolve the structure by using even powders. Finally synchrotron radiation-facilities will certainly find application in low dimensional organic research, although the question of radiation damage has to be considered.

TRENDS

For the future certain developments can be foreseen, of which a number are already mentioned in the preceding paragraphs (a) the numerical calculations for 2D-systems, (b) the refinement of band structure calculations, (c) the study of the static and dynamic properties of CDW and SDW modulated phases under pressure and high magnetic fields by neutrons (need for larger samples) (d) the search for a new "conducting" spin-Peierls system with a transition temperature of a few K, (e) the stabilization of 1D chains for the study of coherence and localization effects and as a requisite for superconductivity.

Although for structural instabilities 1D will be preferable, also new transition phenomena can be seen in 2D e.g. the Kosterlitz-Thouless transition requires a 2D system with a 2D-order parameter. The 2D-order parameter in CDW's makes 2D-CDW-compounds suitable candidates for study of this transition. Also the study of low dimensional superconductivity will require a shift to 2D-systems. Even more emphasis in this direction is expected with the coming of models for the explanation of the new high T_c superconducting oxides. For these reasons increasing importance of 2D-problems is the general trend, expected for the coming years.

REFERENCES

1. Papers at the NATO ASI on Low-Dimensional Conductors and Superconductors, Magog, Canada 1986, edited by L. Caron and D. Jérome (Plenum, New York, 1987).

ROUND TABLE ON POTENTIAL TECHNOLOGICAL APPLICATIONS

A. Barraud (Leader-France), P. Delhaes (France),
L.J. De Jongh (Netherlands), H. Gudel (Suiss),
O. Kahn (France), J.P. Launay (France),
R. Metzger (USA), J.S. Miller (USA), and
J.P. Renard (France)

ABSTRACT : This round table itemized the present (batteries, antistatic coatings, food tags) and the short future applications of low dimensional materials (5 years from now : magnets, sensors, non linear optics, electrocatalysis, superconductivity), as well as the long term (10-40 years) ones, commonly grouped under the name "molecular electronics". The advantages of low D materials were examined as well as the problems specific to each application, especially the long term ones. Finally the needs in research for the next years were examined.

INTRODUCTION

The goal of this round table was to itemize the possible application domains for low dimensional materials and the problems related to these applications. Many ideas sprang out so that the topics were divided into three :

- 1. Present applications
- 2. Near future applications (next 4 or 5 years)
- 3. Long term potentialities (next 40 years or so)

1. PRESENT APPLICATIONS

As a matter of fact, most present applications of low D materials use rather trivial properties :

conductivity for antistatic coatings (Xerox, Kodak), color changes in thermal dosimeters stuck to deep frozen food (polydiacetylene from Allied Chemicals). Batteries on the contrary make use of the wide range of stoechiometries attainable by electrochemistry in mixed valence conductors : the first organic batteries (based on TCNQ and TTF derivative working between - 40 °C and 70 °C from China and also from Japan) are appearing. Except for food tags (in this case diacetylenes exhibit a 1D polymerization kinetics remarkably well suited to the task), low dimensional materials bring little advantage compared to the competing inorganic ones. A conducting latex from BASF has not found application yet. The possibility of making transparent plastic heating windows has not been exploited either. This is presumably because most low D materials have not been characterized fully. A few of them have bad reports on their stability and their reliability, so that mass industry is reluctant towards the whole class of compounds.

2. NEAR FUTURE APPLICATIONS

Participants were optimistic for near future applications because organic chemistry has wide possibilities for adjusting the properties to those required. Battery electrodes is an example of this evolution : polyacetylene was chosen for early batteries because it was the best known material, but other more suited conducting polymers have now come out. In batteries the advantage lies in the high energy to weight ratio. Organic Xray gratings do not seem promising because of the fragility of organics to radiation. On the other hand, advantage can be taken of the low D in optical polarizers and fast shutters, whose market is increasing with laser development. Organic ferromagnets have not been proved yet and it is too early to know whether they will exhibit advantages compared to those made of conventional materials.

There is one field which all the participants considered as very promising and very well suited to low D materials : sensors. Extremely encouraging results have been obtained in several NATO countries as well as in Japan concerning high sensitivity selective sensors based on thin organic conducting films. Several years of adjustments are expected to be necessary before mass commercialization, but here the variety of conductors seems well suited to the variety of chemicals to be detected.

Electrocatalysis was thought to be also a domain in which both the low dimensionality and the redox properties of mixed valence compounds would be advantageous (redox surfaces, clusters).

Another very promising field for highly polarizable low D materials was found to be non linear optics, but this topic did not fall in the scope of this conference.

It was largely emphasized that, because of the presently poor knowledge of the properties of most of these materials, it was very difficult for us to forsee what specific property will fit the needs of the next years. This was especially true for high critical temperature (Tc) superconductors which will either know a "medium" size extension if Tc does not reach room temperature (use in liquid nitrogen), but will trigger a complete revolution in road and car technology (energy supplied from the road, battery free electric cars) and in electric power supply if Tc comes above room temperature.

3. LONG TERM APPLICATIONS

These are mainly included in the term "Molecular Electronics". Although most participants were aware of the importance of M-E in the next decades, the subject remained controversial on practical considerations such as noise, crosstalk, showing that there is presently no enough knowledge on the basic phenomena involved in active devices (enhanced tunneling, multiple quantum well resonance, tunneling in double center molecules, etc.). The situation is similar to the one in the fourties when the first non performant transistors appeared while Electronicians were still used to valves. One of the problems in designing M.E. is that of forgetting classical electronics and its specific circuit architecture, and designing new, redundant architectures in which the effect of noise or failures is of little importance.

It appears that, whatever its form, M.E. should not be designed to compete with silicon whose technology will be very much advanced by the time M.E. will start. On the contrary, M.E. should be considered as complementary of silicon in the fields poorly covered by it, especially parallel processing. Architectures such as cellular automata proposed by von Neumann in 1944 are expected to take large advantage of highly anisotropic compounds and should allow very fast parallel processing. Size does not appear to be a direct requirement but electronics affords maximum speed at the molecular level (intramolecular electron transfer commonly takes place in the picosecond range). But above all the possibility of new architectures appeared to the participants as the main advantages expected from M.E. This will turn a new page in electronic technology.

However some big problems remain, the solution of which is not in our hands to-day : no "active" systems have been made successfully yet (few attempts have been made so far), and adressing single molecules remains unsolved (even if the tunnel microscope allows us to adress one molecule, it will not solve parallel adressing).

Clearly the variety of the magnetic and conducting low D materials makes them suitable to help us solving the many problems of molecular circuitry and active systems. But the knowledge of the basic conducting and magnetic phenomena in these materials is very poor : except in a few well studied families the behavior of a new compound is hardly predictable. Even its being conducting or insulating cannot sometimes be predicted. Hence we need a prolonged multidisciplinary effort of research on these materials including :

- extensive chemical synthesis and purification.
- the development of methods of crystallization, thin (even monomolecular) film production such as the Langmuir-Blodgett method.
- the development of methods for the direct observation of the molecular arrangement (tunneling scanning microscopy for instance).
- groups of interdisciplinary laboratories providing the necessary feedback to chemists, via the study of film or crystal growth, the systematic study of magnetic and electrical properties.
- groups studying active devices and new architectures, and accepting a high percentage of failures.

So far organic low D semi-conducting materials seem as interesting for M.E. as high conductivity higher dimensional molecular materials. Anisotropic magnetic materials, if able to be operated at higher temperature in the future, will also be of great help. The present need is clearly basic research to know where the limits of these materials are. Here the amount of work to be done is very large, because of the counterpart of what is an advantage of organics, namely the wide choice of possibilities which arises from the high versatility of organic chemistry.

PARTICIPANTS

A.D. BANDRAUK : Dpt. de Chimie, Fac. Sc. Univ. de SHERBROOKE, QUEBEC JIK 2R1 (CANADA)

A. BARRAUD : D.P.C./S.C.M., C.E.N. Saclay, 91191 GYF SUR YVETTE (FRANCE)

J.C. BAYON : Dpt. Quimica Inorg., Fac. Ciencias BARCELONA (SPAIN))

C. BELLITTO : Inst. di Teoria del CNR, CP 10.00016 MONTEROTONDO (ITALY)

D. BELTRAN : Dpt. Quimica Inorg., C/Dr Moliner 50 VALENCIA (SPAIN)

A. BENELLI : Inst. di Teoria del CRN, CP 10.00016 MONTEROTONDO (ITALY)

H.B. BROM : K.O.L. Nieriwoteeg 18, 23115 B LEIDEN (THE NETHERLANDS)

R. BURRIEL : Instituto Ciencias Materiales Aragon, Fac. Ciencias, ZARAGOZA 5009 (SPAIN)

A. CANESCHI : Dept. of Chemistry, Via Maragliano 25/27, I-50144 FIRENZE (ITALY)

R.L. CARLIN : Dept. of Chemistry, Univ. of Illinois, CHICAGO, Illinois 60680 (U.S.A.)

R.J.H. CLARK : Christopher Ingold Lab., Univ. College London, LONDON WC1HOAAJ (ENGLAND)

D. CHASSEAU : Chemical Cristallography Lab., 9 Parks Road, OXFORD OX13PD (ENGLAND)

E. CORONADO : Dpt. Quimica Inorg., C/Dr Moliner 50 VALENCIA (SPAIN)

C. COULON : C.R.P.P./C.N.R.S., Domaine Univ. Bordeaux I, 33405 TALENCE (FRANCE)

J. DARRIET : Lab. de Chimie du Solide, C.N.R.S., Univ. Bordeaux I, 33405 TALENCE (FRANCE)

P. DAY : Oxford Univ., Inorg. Chem. Lab., South Parks Road, OXFORD (ENGLAND)

P. DELHAES : C.R.P.P./C.N.R.S., Domaine Univ. Bordeaux I, 33405 TALENCE (FRANCE)

M. DRILLON : D.S.M., EHICS, 1 rue Blaise Pascal, 67008 STRASBOURG (FRANCE)

L. DUCASSE : Lab. de Physico-Chimie Théorique, Univ. Bordeaux I, 33405 TALENCE (FRANCE)

V.J. EMERY : Dpt. of Physics, Brookhaven National Lab., UPTON, New York 11473 (U.S.A.)

G. FEREY : Lab. des Fluorures, Fac. des Sci., Route de Laval, 72017 LE MANS (FRANCE)

D. GATTESCHI : Dept. di Chimica, Univ. di Firenze, Via Maragliano 75/77, 50144 FIRENZE (ITALY)

R. GEORGES : Lab. de Chimie du Solide, C.N.R.S., Univ. Bordeaux I, 33405 TALENCE (FRANCE)

H. GUDEL : Inst. für Analytische und Phys. Chem., CH 3000 BERN 9 (SWITZERLAND)

D.G. HAASNOOT : Gorbens Lab., Leiden Univ., 2300 RA LEIDEN (THE NETHERLANDS)

W. HATFIELD : Dpt. of Chem., Univ. of North Carolina, CHAPEL HILL, N.C. 27514 (U.S.A.)

H.W. HELBERG : Drittes Phys. Inst. Bürgerstr. 42-44, D 3400 GOTTINGEN 1 (WEST GERMANY)

M.T. JONES : Dpt. of Chem. Univ. of Missouri, SAINT LOUIS, Missouri 63121 (U.S.A.)

L.J. DE JONGH : K.O.L., P.O. Box 9506, 2300 RA LEIDEN (THE NETHERLANDS)

O. KAHN : Lab. de Spect. des Eléments de Transition, Univ. Paris Sud, 91405 ORSAY (FRANCE)

C. KARATAS : Orta Dogu Teknik, Universitesi Inonu Bulnari, ANKARA (TURKEY)

H.J. KELLER : Anorg. Chem. Inst. der Univ., 69 HEIDELBERG (WEST GERMANY)

K. KOPINGA : Dpt. of Physica, Eindhoven University, NL 5600 MB EINDHOVEN (THE NETHERLANDS)

C.P. LANDEE : Dpt. of Phys., Clark Univ., WORCESTER, Massachusetts 01610 (U.S.A.)

J.P. LAUNAY : Lab. de Chimie des Métaux de Transition, 4 Place Jussieu, 75230 PARIS (FRANCE)

W. MASSA : Fachbereich Chem. der Univ., D 3550 MARBURG (WEST GERMANY)

R. METZGER : Dpt. of Chem., University of Alabama, TUSCALOOSA, Alabama 35487-9671 (U.S.A.)

J. MILLER : Du Pont de Nemours, Cent. Res. and Dev. Dpt., WILMINGTON, Delaware 19898 (U.S.A.)

F. PALACIO : Inst. Ciencias Materiales Aragon, Fac. de Ciencias, ZARAGOZA 5009 (SPAIN)

G. PAPAVASSILIOU : National Hellenic Res. Found, ATHENS 116/35 (GREECE)

J. PEBLER : Dpt. Chem., University Marburg, D 3550 MARBURG (WEST GERMANY)

J.P. POUGET : Lab. de Phys. des Solides, Univ. de Paris Sud, 91405 ORSAY (FRANCE)

J.P. RENARD : Inst. d'Electronique Fondamentale, Univ. de Paris Sud, 91405 ORSAY (FRANCE)

P.M. REY : CENG-DRF, Lab. de Chimie 85X, 38041 GRENOBLE CEDEX (FRANCE)

C. RILLO : Inst. de Ciencia de Materiales, Fac. de Ciencias, G.S.I.C., 5009 ZARAGOZA (SPAIN)

T. ROBERT : Serv. de Chimie Générale et Inorg., Univ. de Mons, B - 7000 MONS (BELGIUM)

T. ROJO : Dpt. Quimica Inorg., Fac. Ciencias, Univ. del Pais Vasco, 48080 BILBAO (SPAIN)

J. ROUXEL : Lab. de Chimie des Solides, Univ. de Nantes, 44072 NANTES (FRANCE)

D.J.K. SCHWEITZER : Max Planck Inst. für Med. Forschung, 6900 HEIDELBERG (WEST GERMANY)

E.J. SOLOMON : Dept. of Chem., Stanford Univ., STANFORD, California 94305 (U.S.A.)

Z. SOOS : Dpt. of Chemistry, Princeton Univ., PRINCETON, New Jersey 08540 (U.S.A.)

M. STEINER : Hahn-Meitner Inst., Berlin., D-100 BERLIN 39 (WEST BERLIN)

F. STUDER : Crismat, 15 MRA, Univ. de Caen, 14032 CAEN CEDEX (FRANCE)

J.B. TORRANCE : IBM Almaden Res. Center, 65 Henry Road, SAN JOSE, California 95120 (U.S.A.)

A. UNDERHILL : Dpt. of Chem. Univ. College of BANGOR, Gwynedd LL57 20W (GREAT BRITAIN)

M. VERDAGUER : Lab. S.E.T., Univ. Paris Sud, 91405 ORSAY (FRANCE)

G. VILLENEUVE : Lab. de Chimie du Solide, C.N.R.S., Univ. Bordeaux I, 33405 TALENCE (FRANCE)

J.U. von SCHUTZ : Phys. Inst., Univ. Stuttgart, 57 D 700 STUTTGART 80 (WEST GERMANY)

M. WEGER : Racah Inst. of Phys., Hebrew Univ., JERUSALEM (ISRAEL)

M.H. WANGBO : Dpt of Chem., North Carolina St. Univ., RALEIGH, N.C. 27695-8204 (U.S.A.)

YU PEY : Lab. de Spect. des Eléments de Transition, Univ. de Paris Sud, 91041 ORSAY (FRANCE)

INDEX

GPSR Compliance

The European Union's (EU) General Product Safety Regulation (GPSR) is a set of rules that requires consumer products to be safe and our obligations to ensure this.

If you have any concerns about our products, you can contact us on ProductSafety@springernature.com

In case Publisher is established outside the EU, the EU authorized representative is:

Springer Nature Customer Service Center GmbH
Europaplatz 3
69115 Heidelberg, Germany

Batch number: 10761599

Printed by Printforce, the Netherlands